THE WHOOPER SWAN

For Dennis and Mavis

THE
WHOOPER SWAN

MARK BRAZIL

Illustrated with line drawings
and colour plates by
DAFILA SCOTT

T & A D POYSER
London

Published 2003 by T & A D, an imprint of A & C Black Publishers Ltd.,
37 Soho Square, London W1D 3QZ

Copyright © 2003 text by Mark Brazil
Copyright © 2003 illustrations by Dafila Scott

ISBN 0-7136-6570-X

www.acblack.com

Typeset by J&L Composition, Filey, North Yorkshire

Printed and bound by Midas Printing International Ltd China on behalf of Compass Press Ltd

10 9 8 7 6 5 4 3 2 1

The Wild Swans At Coole

The trees are in their autumn beauty,
The woodland paths are dry,
Under the October twilight the water
Mirrors a still sky;
Upon the brimming water among the stones
Are nine-and-fifty Swans.

The nineteenth autumn has come upon me
Since I first made my count;
I saw, before I had well finished,
All suddenly mount
And scatter wheeling in great broken rings
Upon their clamorous wings.

I have looked upon those brilliant creatures,
And now my heart is sore.
All's changed since I, hearing at twilight,
The first time on this shore,
The bell-beat of their wings above my head,
Trod with a lighter tread.

Unwearied still, lover by lover,
They paddle in the cold
Companionable streams or climb the air;
Their hearts have not grown old;
Passion or conquest, wander where they will,
Attend upon them still.

But now they drift on the still water,
Mysterious, beautiful;
Among what rushes will they build,
By what lake's edge or pool
Delight men's eyes when I awake some day
To find they have flown away?

W. B. Yeats (1865–1939)

Contents

List of Figures

CHAPTER 6

CHAPTER 7

List of Tables

Preface

Variously known as one of the great heralds of winter in the west, and even as the angel of winter in the Far East, the Whooper Swan is a beautiful, inspiring, and fascinating bird. It is a bird that I have been fortunate to have a long association with across its range, and one that I have enjoyed following from its wintering grounds to its breeding haunts. Wide-ranging, it has, as we shall see, the most extensive and most expansive natural range of any of the swans, from as far west as Iceland in the North Atlantic, and Fennoscandia, right across the Russian subarctic, east to the Kamchatka Peninsula facing the Bering Sea and the Pacific Ocean.

A journey that begins and ends in two of the most densely populated countries on Earth, Japan and the Netherlands, takes one across almost all of the Whooper Swan's range, over the drainages of the Bikin and Amur rivers, across Yakutia, the Lena, the Yenisei, the Ob, the Urals, the Pechora, and from northwestern Russia to southern Finland, Sweden and Denmark. That same journey also traces and re-traces two of the great routes of Whooper Swan migration, those from parts east of the Urals reaching east Asia and northern Japan and those west of the Urals heading westwards via the Beloye More, the Gulf of Finland and the Baltic, to southern

Scandinavia, Poland, Germany and the Netherlands in winter. Even at speeds in excess of 900 km/h, such a journey by modern aircraft takes more than ten hours and, on occasions when breaks in the cloud cover permit, one may gaze from more than 10,000 m on one of the largest and least populated regions on Earth—Siberia, the vast expanse of tundra and taiga that comprise the contiguous domains of Whooper and Bewick's Swans.

I have been fortunate to have crisscrossed Russia repeatedly by plane at each season. During the brief days of winter, even late into spring, for hours one passes over frozen wastes, white barrens, crystal deserts, meandering rivers of ice, snow-covered taiga forest and mountains. Only in early summer do the rivers break free of ice and the whiteness of the vast landscapes turn to green, then within a few months rusts, reds and browns return. Even at the height of summer, in the end-less days of mid June, pack ice lingers close to the northern coast and by mid September the mountains are, once more, powdered white and the rivers freezing again. Into the heart of this enormous wilderness come the bugling flocks of Whooper Swans intent on reaching ancient traditional breeding grounds as soon as they thaw, to nest in remote territories amid the taiga before fleeing from winter's freeze with their young, returning again to the more heavily populated parts of Europe and Asia.

In a volume of this scope it is impossible to address equally research from all parts of this vast range and accessing materials in some languages has proven diffi-cult and exceedingly time-consuming, thus my approach is, of necessity, a biased one. Having spent many years studying and watching Whooper Swans, my initial bias is inevitably towards regions at the extremes of their range where I know them best: England, Scotland, Iceland, Sweden and Finland in the west, and South Korea and Japan in the east. There is further bias in that far more detailed research has been undertaken in Europe on this species than elsewhere in its range—so most of the literature I have accessed pertains to this small portion, amounting to less than a quarter, of its range. I have addressed rather more fully than is perhaps usual in a book of this kind the vagrancy aspect of the species' natural history. So often ignored by ornithologists (though the life-blood of birders), today's vagrants may represent the pioneers of future range extension. The Whooper Swan's population and range has been undergoing considerable change over the last century making such analysis of particular interest for the future. Finally, I have deliberately included much that is speculative and conjecture in the hope that this will stimu-late further research. I hope that readers and researchers will either forgive the extent of my bias or endeavour to help me balance it should a later edition be possible.

Mark Brazil
Ebetsu, Hokkaido, Japan, 2003

Acknowledgements

This book is the culmination of many years rather varied experience, initially as a field biologist and latterly as a natural history writer and photographer. It presents me with the pleasant dilemma as to just where should I address my thanks, as so many different people have contributed to my pleasure in studying, watching and photographing swans.

Firstly my thanks go to Dr Janet Kear, of the then Wildfowl Trust, who unwittingly convinced me at Slimbridge, England, when I was still a schoolboy, that ornithology really was 'it' and set me off on a long and fascinating track (contrary to the hopes of an ignorant careers advisor at my Warwickshire grammar school, who hoped to dissuade me of such an outlandish notion!).

Later, my early fascination with raptors was adroitly side-stepped by Dr Cliff Henty, of Stirling University, who set me on a very different route by awakening my interest in the Whooper Swan, and indirectly made it possible for me to extend my Scottish ornithological studies overseas to Iceland and then Japan.

My early studies of Whooper Swans in Scotland, Iceland and Japan were made possible by the support of my parents, by a University of Stirling student scholarship, and by significant grants from the Vincent Wildlife Trust, for all of which I am eternally grateful.

Among the many people who helped me during those and later studies, I give special thanks to Prof Arnthor Gardarsson, of the University of Iceland, who not only welcomed me to Iceland to conduct field work there during three summers, but who also provided invaluable help, advice and logistical support, in addition to engaging me in many fruitful discussions about Whoopers. Without his considerable help, my studies could never have succeeded. In the UK, the late Sir Peter Scott, his family and colleagues at the Wildfowl Trust (now the Wildfowl & Wetlands Trust), with their fascination for Bewick's Swans, provided the inspiration to pursue the Whooper Swan through four cold Scottish winters.

I have been immensely fortunate in the extent to which I have been able to travel in search of wildlife. In Argentina, Brazil, Chile, and the Falklands, in Alaska and Canada, in Australia and New Zealand, in Russia, China and Korea, and especially in Japan, the swan researchers, swan lovers, and naturalists who have helped me pursue each of the swans in turn, and other wildlife too, are too numerous to mention individually but to them all I also give thanks. In particular, my thanks go to the swan-viewing public of Japan who proved that the most exciting way of viewing and photographing Whooper Swans was not through binoculars or a telescope, or telephoto lenses at long range, but at arms length (literally!) and with a wide-angle lens.

The completion of such a book is invariably the result of numerous ornithologists diligently pursuing field studies and publishing their findings. Many friends and colleagues have kindly helped me trace those published materials in a number of languages, and have also provided me with invaluable input by reviewing and commenting on the developing manuscript. First and foremost, I am most grateful again to Prof Arnthor Gardarsson, who generously gave considerable time and invaluable advice on reading two entire versions of the manuscript.

For their help in various ways I am grateful to the following. Desmond Allen, of Tsurumi University, Yokohama, Japan, for providing helpful input into drafts of early chapters. Nyambayar Batbayar, of The Peregrine Fund, The World Birds of Prey Center, Idaho, USA, for information from Mongolia. Patrick Bergier of the Go-South Organization, for information relating to the Mediterranean region, North Africa and France. Michel Bertrand of Québec for help with records of vagrants. Dr Dennis Brazil, of Alvechurch, Worcestershire, for reading and commenting on the entire manuscript. Geoff Carey for reviewing the China section. Dr Cho, Sam-rae, of the Department of Biology, Natural Science, Kongju National University, South Korea, for information relating to the Republic of Korea. Chris Cook, of *The Japan Times* for careful proof-reading. Dr William Duckworth of the Wildlife Conservation Society, Pyongyang, North Korea, for help with literature relating to that region. Dr Josef Kren, of the St George's University School of Medicine, Wisconsin, USA, for information pertaining to the Czech Republic. Dr David Lont, of Otago University, New Zealand for considerable help with literature searches. Dr Ma Ming, of the Chinese Academy of Sciences, Urumqi, Xinjiang, China, for providing access to the Chinese literature and reviewing the section on China. Carl Mitchell, Wildlife Biologist with the US Fish and Wildlife Service, Gray's Lake National Wildlife Refuge, Idaho for providing literature on Trumpeter Swans and reviewing the North American section. Steven Mlodinow, author and ornithologist, of Everett, Washington, USA, for providing information on Whooper, Trumpeter and Whistling Swans in North America. Harry Nehls, author and ornithologist, of Portland, Oregon, USA, for providing information on Whooper Swans in Oregon. Dr Leif Nilsson, of the Department of Animal Ecology, University of Lund, Sweden, for information on that country's Whooper Swans and for input to Chapter Four. Dr Eileen Rees, of the Wildfowl and Wetlands Trust, who helped considerably with information and literature, and also reviewed various chapters. Dr Dafila Scott, who provided input on early chapters. Dr Jevgeni Shergalin, of Tallinn, Estonia, for his considerable help with recent Russian-language literature and for valuable comments on an early draft of Chapter Five. Dr Chris Spray, MBE, of Northumbrian Water for providing encouragement and reviewing an earlier draft. Dr Tadeusz Stawarczyk, of the Museum of Natural History, Wroclaw University, Poland for recent information and comments concerning status in Poland. Dr Ludwik Tomialojc, also of Wroclaw University for reviewing and commenting on the same. Dr Teresa Tomek of the Institute of

Systematics and Evolution of Animals, Polish Academy of Sciences, Kraków, Poland for help relating to North Korea. Dr Wim Vader of the Tromsø Museum, Norway, for information and help with literature relating to that region. Michael Walters of the Natural History Museum, Tring, for literature assistance. Dr Maria Wieloch of the Ornithological Station, Institute of Ecology, Polish Academy of Sciences, Gdansk, Poland, for recent information, literature and helpful comments concerning Poland. I apologise for any omissions, these are unintentional, and acknowledge that despite considerable input from friends and colleagues any inaccuracies remain my own.

Dafila Scott thanks Dr Eileen Rees for her help with reference material on which some of the illustrations are based and also thanks Keith Shackleton for inspiration and encouragement concerning the illustration of 'Leda and the Swan'.

The setbacks to the completion of the manuscript, which took several years longer than anticipated, were many and varied, and now appear amusing in their variety, ranging from international relocation and emotional upheaval, to the unwanted attentions of the unknown thief, of Kota Kinabalu, Malaysia, who stole not only an entire draft but more devastatingly also all of the correspondence relating to it. I offer my condolences on such a poor haul (given the lack of swans in the region!), but also my thanks as considerable further literature, particularly from Russia, came to light while I was reproducing the manuscript.

Finally, I would like to thank my editor at Academic Press/T. & A. D. Poyser, Dr Andrew Richford, for his continued encouragement and infinite patience, and Marianne Taylor at A & C Black for taking on the project in its late stages and bringing it to fruition. I especially thank Dr Dafila Scott who has illustrated the text so attractively with colour plates and line drawings, and Robert Gillmor for his splendid cover.

CHAPTER 1

A swan's world

INTRODUCTION

Among the world's nearly 10,000 bird species, swans number a mere seven[1] yet their ranges span over half the globe, making their success more than worthy of our attention. Swans occur across an enormous geographical area—North and South America, Europe, Asia, Australia and New Zealand, only Africa and Antarctica lack indigenous swans, but even in southern Africa Mute Swan has been introduced and Black-necked Swan has reached the sub-Antarctic Islands. Swans occur in a fascinating diversity of habitats. In the Northern Hemisphere they range from the high-latitude Arctic tundra of northern Russia to temperate lakes, rivers and marshes, and even visit the warm climes of the Mediterranean region. In the Southern Hemisphere they occur from the swampy pampas to the semi-desert plains of southern South America, to the remote Falklands and from the seasonal wetlands of arid Australia to the coastal bays and lagoons of New Zealand. One swan above all others has the largest range—the Whooper.

Whooper and Bewick's Swans range across the largest landmass of all, Eurasia and its associated islands. They occur from Iceland, Ireland, Britain and Scandinavia in the west to Chukotka, Kamchatka, China, Korea, and Japan in the east. The Mute Swan, so long domesticated and frequently introduced as an ornamental species that tracing its native range has become difficult, certainly occurs naturally across much of south and central Eurasia, and can now be found well beyond that region, in Japan and North America. I was surprised by an encounter with a party of Mute Swans in the coastal saltmarshes just south of Vancouver, British Columbia, seen from a whale-watching boat. Unlike the introduced Mute Swan, the Bewick's Swan's conspecific relative, the Whistling Swan, and the Whooper's close relative, the Trumpeter Swan, both range naturally across northern North America, from the Pacific coast of Alaska to the eastern part of Hudson Bay, Canada, in the case of the former, and from western Alaska southeast across Canada to the great plains in the latter. In a broad belt across the southern parts of South America, from the wetlands of south Brazil to rugged Tierra del Fuego, the Black-necked and Coscoroba Swans can be found. Their larger relative, the Black Swan is abundant in Australia and, after introduction over a century ago, also thrives in New Zealand. In fact, as I began writing this chapter a wintering party of Black Swans was cruising the shallows of Portobello Bay, outside my window.

In thinking of swans, I immediately imagine a flurry of powerful images: the dawn stillness and calm reflections of autumnal Scottish lochs; the lush summer greenness and stark volcanic rocks of Iceland; condensing mist rising over thermal springs amid frigid icy wastes in northern Japan (Fig. 1.1); enormous reed-fringed lagoons in Romania's Danube Delta; the vastness of wide open skies condensed and reflected in shallow lagoons in Patagonia; wind-torn peaty pools in the Falklands, and the sun-glinting shallow coastal lagoons of South Island, New Zealand. These are by no means the only haunts of swans, nor perhaps where they are most conspicuous or most regularly seen. In many parts of the western and eastern world the 'feral' Mute Swan graces town and city parks. Introduced Black Swans swim regally around the Imperial Palace moat in downtown Tokyo, but even the wildest of wild swans occasionally occurs in close proximity to people, for example, Whooper Swans can be seen at close quarters on Tjornin pond in the Icelandic capital of Reykjavik, and there are many other sites used by swans within urban habitats. In colonial Buenos Aires, Black-necked Swans breed at Costanera Sur right in the city, passed by hordes of loafers and joggers, and in over-populated Japan, man and swan are at their closest at innumerable sites in northern Honshu and Hokkaido, where clamouring hordes of Whooper Swans are fed, not merely by reserve wardens and sanctuary managers, but by the public, by hand!

Swans have a special place in the human psyche, long entrenched in legend and myth, but not all have been the subject of special scrutiny. Of the world's seven species only four, Mute, Black, Tundra and Trumpeter have, over a period of decades, been extensively studied during the breeding season. Mute Swan, because of its wide distribution and popularity around man, Bewick's Swan (a subspecies of the Tundra) because of pioneering work done by Sir Peter Scott and the then Wildfowl Trust, Whistling Swan (the other subspecies) because of its abundance, Trumpeter Swan due to its rarity in North America, and Black Swan in the Antipodes because its abundance was a threat to agriculture, have all attracted much interest and research since the mid 20th century. The Coscoroba and

Figure 1.1.   Amidst morning mist beside a hotspring.

Black-necked Swans have received least attention, not because they are any less interesting, but simply because of the relative scarcity of ornithologists within their ranges. They are still relatively little studied despite a recent increase in efforts. For political and logistical reasons it has proven rather difficult for Western scientists to study Bewick's Swan extensively on its breeding grounds[2]. The Whooper Swan too, although a northern swan, and notwithstanding it being the most wide-ranging and abundant was, until the 1970s, surprisingly poorly known. It, how-ever, has been readily accessible in the western part of its range yet, in spite of this, only in Finland had it been studied extensively on the breeding grounds until detailed studies commenced in northern Iceland in 1978, encouraged by Prof Arnthor Gardarsson, with broader research continued by him and his col-leagues into the 1980s and 1990s. In recent decades, the Whooper Swan has begun to receive the attention it deserves, with significant collaborative work in Iceland between the Icelandic Institute of Natural History and the Wildfowl & Wetlands Trust (WWT), and in Ireland with the development of the Irish Whooper Swan Study Group making three-way collaborative monitoring of the population throughout its migratory range possible. Meanwhile, in continental Europe, Bjarke Laubek has made considerable efforts to develop studies of the species. I hope that this book will encourage a further generation of naturalists and ornithologists to take an interest in this magnificent bird.

Despite its wide distribution, its conspicuousness and beauty, and its appear-ance in the folklore and mythology of many regions the Whooper Swan had, by the 1970s, when it first caught my interest, only been studied in detail in Scandinavia. Very little had then been published about the species in English, other than brief descriptive accounts, the most widely known of which were those on the Whooper Swans of Shetland by Venables & Venables (1950) and of the Lake District by Airey (1955). A paper by Boyd & Eltringham (1962) on the species' status and distribution in the British Isles was a turning point in our developing knowledge, since for the first time information on population size, distribution and composition were collected in an accessible form.

The first detailed field studies in the UK were those of Hewson (1964, 1973) and Henty (1977); the former concerned mainly with flock composition and the latter with diurnal movements. Apart from these few descriptions of the species in its wintering haunts in Britain, the most important observations had been in Finland, where three landmark publications by Haapanen *et al.* (1973a,b, 1977) presented the results of the only previous major study and provided almost all that was then known of the species on its breeding grounds. The last of these three described the summer behaviour of wild-breeding Whooper Swans from incubation to fledging and, as the only paper on those aspects at the time it formed the background to my own work on breeding biology and behaviour (Brazil 1981b, 1981c).

Even by the late 1970s, vast parts of the Whooper Swan's ecology and behaviour remained unstudied. Virtually nothing was then known, for example, about migration or moult. Very little was known of its daily or seasonal movements, its wintering ecology, its feeding ecology or flocking behaviour, and less still of the effect increased use of agricultural land was having on feeding behaviour. No data were available on the ecology of non-breeding birds and very little information was then available from the Icelandic population, which from a UK perspective was *the* most important population, as the source of 'our' wintering birds. I wanted to help redress the balance.

Life rarely runs smoothly, however. Prior to my arrival in Scotland, Whooper Swans in the Stirling area in the late 1970s had shown an interesting pattern of movement, with several small flocks each utilising several fields during a single day. Initially, I planned a detailed examination of this situation and of the factors affecting habitat selection. On my arrival, however, my preliminary observations soon revealed that, for reasons unknown, the pattern of usage had changed to a rather more regular system. Often a single flock would utilise a single field, often for several days, thus virtually eliminating the phenomenon I had planned to study!

Equally as quickly, I began to realise just how widespread our ignorance was of this magnificent bird. Wherever I turned for answers the literature merely left me with more questions. Rapidly altering my focus, I set out to provide a wider description of its wintering behaviour and ecology. The breadth of my research was increased when it became possible to make prolonged visits to the breeding grounds in Iceland in three consecutive summers in the late 1970s and early 1980s and, further, by my opportunities first to visit and then live in Japan during the 1980s, and to return there annually ever since. Increasing the breadth of study inevitably meant reducing its depth, although a general theme of interest in feeding behaviour and ecology was central. Along the way I was sidetracked into studies of geographical variation in bill patterns, of morphology, and of migration and movements.

I took a primarily descriptive and broad approach to the Whooper Swan's annual cycle, from its wintering behaviour and ecology in Britain, to its migration, and its breeding ecology and behaviour in Iceland, in order to extend, as much as possible, our understanding of the species. As with other studies in behavioural ecology, my dominant concern was behaviour as an ecological factor affecting the distribution and/or abundance of individuals (see Krebs 1972), for example habitat selection, temporal distribution of behaviours, group or flock composition and dynamics, territoriality and parental care. I was also concerned with the interactive effect between such factors and environmental factors such as habitat availability and weather. Meanwhile, much work in behavioural ecology, especially on waterfowl, was already concentrating on the fine details of social behaviour (e.g. of Bewick's Swans; see Scott 1980a,b,c,d), or feeding ecology (e.g. of geese; see Owen 1980), but such detailed studies were only relevant in the context of the wider knowledge of the species in question.

Since my own more general study of behaviour and ecology, many more detailed studies have been made of specific aspects of Whooper Swan biology. Similarly, while my own experience is limited to extreme western Europe and eastern Asia, much work has been done over the last 20 years in the vast regions in between. In this book, I have done my best to bring together my own experience with that of others, and to weave this detail into a broad account of the life history of one of my favourite birds, while also trying to bring out at least some of the excitement and pleasure gained from studying not just the Whooper Swan, but also the wild places in which it breeds and to which it migrates.

The visible grace and the mysterious movements of the swans, of the Northern Hemisphere in particular, have inspired poets, artists and storytellers down the ages, and latterly they have provided inspiration for the photographer. These aspects are described in more detail in Chapter Two. Though I have travelled widely to watch each of the world's swans, it is the Whooper that has provided me with my most deeply impressive wild memories, and endless inspiration as the subject of this book. But it cannot be viewed in isolation. Just where do each of the swans belong, how are they related, and how do they differ?

SWAN CLASSIFICATION

Given that so few species are involved, swan taxonomy has been surprisingly controversial. In form there are resemblances between certain species. Black-necked Swan of South America, Mute Swan of Eurasia and Black Swan of the Antipodes share similarities in structure, form and deportment. Differing fundamentally from these three, the 'northern swans', the Trumpeter and Whistling Swans of North America and the Whooper and Bewick's Swans of Eurasia have many features in common, as a close-knit evolutionary complex (Johnsgard 1974), that they have variously been considered by naturalists and taxonomists as four, three, two or merely one species.

Studies of parasitology, especially those of some of their external parasites[3] indicate an ancient separation between these two major groups of 'northern' and 'southern' swans (Timmerman 1963), nevertheless, the extent of more recent divergence among the 'northern swans' has remained confusing.

The 'northern swans' can be excused for suffering crises of identity, as ornithologists have repeatedly split and lumped them. Delacour & Mayr (1945), for example, considered them to be just two polytypic species, one larger (Trumpeter and Whooper) and one smaller (Bewick's and Whistling). That position was followed until as recently as the early 1970s, although in the year that Wilmore (1974) still plumped for two, Johnsgard (1974) took the most radical, and now outmoded, view, that all four forms belonged to a single quadri-typic species, while simultaneously acknowledging that the degree of tracheal convolution within the sternum differed between Whooper and Trumpeter, which argued against their conspecificity. Meanwhile, and at the opposite extreme, Whooper, Trumpeter, Whistling and Bewick's were considered to be four species (see Boyd 1972).

Given the various possible relationships between these four taxa, two theories of origin are imaginable. In the first, an ancestral Holarctic 'northern swan' may have diverged along north/south lines into two species: one occupying broad arctic-tundra habitats and the other subarctic/northern temperate ranges. In the second, an ancestral form may have become separated between east and west, i.e. Eurasia and North America, then split further into two taxa adapted ecologically, one to tundra and the other to taiga-like habitats (Johnsgard 1974, 1978). Of the two possibilities, however, the former seems the far more likely.

The two largest and most magnificent of the 'northern swans', Trumpeter and Whooper, differ markedly and have been isolated so long that their distinctiveness is undeniable (Palmer 1976), whereas the smaller Whistling and Bewick's are less readily distinguished. These are now considered not as species but as two closely related subspecies, collectively known as Tundra Swan, showing a cline in certain characteristics across their enormous range.

In what was the first modern attempt to list all of the world's birds in a comprehensive single volume for the layman, Gruson (1976) recognised just six extant swans (as had Boyd 1972), lumping Whistling with Bewick's, and Whooper with Trumpeter. Gruson's rationale was followed by Todd (1979) in his book on the wildfowl of the world, who pointed out that the 'northern swans' can hybridise, with even Trumpeter and Whistling being crossed in collections. However, that stance would lead to gross lumping as crossing between swans and geese has also been recorded in captivity.

Illustrative of the disagreement over swan taxonomy is that Walters (1980), reviewing the world's post-Pleistocene period (since the last ice age), included ten species of swans. He split not only all four northern taxa as species, but also recognised David's Swan of China (*Coscoroba davidi*), and Chatham Swan of New Zealand (*Cygnus chathamicus*) as recently extinct species. In the same year, Howard

& Moore (1991), taking a more conservative approach, listed just six extant species (as had Gruson 1976), lumping *cygnus* and *islandicus* with *buccinator* (Trumpeter), and *bewickii* and *jankowskii* with *columbianus* (Tundra).

More recently, Madge & Burn (1988), Clements (1991), Dunning (1993), Monroe & Sibley (1993) and Clements (2000) have all listed seven extant swan species, accepting the separation of *buccinator* from *cygnus*, but including *bewickii* with *columbianus*. Surprisingly Madge & Burn (1988) included all three 'northern swans' in the genus *Olor*, rather than in *Cygnus*, whereas Monroe & Sibley (1993) retained all swans, with the exception of the Coscoroba, in the genus *Cygnus*. The bizarre Coscoroba Swan stands alone, resembling what one might imagine as an identi-kit put together by someone unsure of the differences between adult shelducks, geese and swans, while its cygnets share resemblances with young tree ducks.

WHOOPER SWAN *CYGNUS CYGNUS* (LINNAEUS, 1758)

The Whooper Swan was the swan species described first by the famous codifier of taxonomy Carl von Linné in 1758 (as *Anas cygnus*) and it is therefore the type species of all the swans, though Linné considered Mute Swan to belong to the same species, distinguishing only between the wild (Whooper) or *ferus* and the tame (Mute) *mansuetus* forms (Scott & The Wildfowl Trust 1972) (Plate A). It was later reassigned to the new genus *Cygnus*, as *Cygnus islandicus*, by Brehm in 1831 (Vaurie 1965), only later being treated as *Cygnus cygnus*. Not only is it the type species of swan taxonomically but also genetic research indicates that the Whooper Swan diverged slightly earlier than either the Tundra or Trumpeter Swans (Harvey 1998).

The 140–165 cm long Whooper Swan is rather similar to its North American cousin the Trumpeter Swan *C. buccinator*, from which the Whooper differs most noticeably in its bill pattern, which is predominantly yellow, rather than black, and in its voice. Whooper Swans are large, and among the heaviest of all long-distance migrant birds. Male Whoopers average larger and heavier than females, 9.8 kg in summer in Iceland, 10.7 kg in summer in Finland, 10.2 kg in winter in Britain and 11.0 kg in winter in Denmark. The heaviest individual recorded was a male that weighed 15.5 kg in winter in Denmark. Females are somewhat lighter, averaging 8.2 kg in Iceland, 8.9 kg in Finland, 9.0 kg in Denmark and 9.2 kg in Britain (Rees *et al.* 1997a; see also Appendix 2).

Though now generally considered monotypic, early measurements suggested that Icelandic and particularly Greenland[4] Whooper Swans were, on average, smaller than European specimens, and it was even suggested that they represented a distinct smaller subspecies, *C. cygnus islandicus* (Schioler 1925). This position could not be upheld because it proved difficult to assign specimens consistently to either race (Witherby *et al.* 1940; Hørring & Salomonsen 1941; Howard & Moore 1991).

Whooper Swan is essentially a trans-Palearctic species breeding disjunctly across an enormous range of the forested boreal zone of Eurasia, from Iceland (here in tundra-like habitat), Norway, Sweden and Finland, across northern Russia to the Okhotsk Sea and Sakhalin, the Bering Sea and Kamchatka, and even on the outer Aleutians (USA), generally at 50–55°N to 64–71°N, but also south to *c.*47°N in the eastern Palearctic. Whooper Swans breed mainly in the taiga zone, but are locally extending north into forested tundra and true tundra, and south where there is suitable habitat into steppe. Its northern extremity is generally that of the limit of the taiga, although in some regions it ranges north into forested tundra and the willow scrub zone. At its southern extent it breeds to the southern limit of the taiga throughout most of its range, and, though more sporadically, at wetlands on the steppes of Central Asia. Increasingly, in the west at least, it has been colonising northern temperate regions. It breeds south to the Caspian Sea, Mongolia and northern China. In winter it ranges south across western Europe, to Ireland, Britain, Denmark and Germany, and mainly in cold winters further south. It also winters on the Black and Caspian Seas. In eastern Asia it winters in coastal eastern China, the Korean Peninsula, Japan and Kamchatka (e.g. Vaurie 1965; Dement'ev & Gladkov 1967; Brazil 1991).

A bird of northern pools and shallow lakes, where it favours reedbeds, occasionally slow-flowing rivers, and bogs in the taiga, it breeds during the northern summer in May–August. It winters along coasts, on freshwater lakes, marshes and low-lying agricultural land. In late July–early September flightless moulting birds congregate near the breeding grounds. From mid to late September, depending on the region and severity of the season, they leave their breeding grounds moving mainly southwest or southeast to reach their wintering grounds in October–November. Cold weather can force them south, well beyond their normal wintering range. They may remain on the wintering grounds until March–early May, then migrate northeast or northwest to arrive on the breeding grounds during May. Overall, the species appears to have been re-expanding its range and its population during the 1980s and 1990s.

Within its normal range from Ireland to Japan, it is likely to be confused only with the smaller Bewick's race of Tundra Swan, though Whooper has a longer, more distinguished profile. Whooper's bill usually has such a large yellow patch on the lores, extending in a point to or beyond the nostrils (Boyd 1972), that it appears as if it is yellow with a black tip, whereas that of Bewick's appears black with a yellow patch. In proportion Bewick's Swan's bill is shorter and commonly shows a pinkish line along the side of the lower mandible (Alström & Olsson 1989), which Whooper always lacks. Adult Whooper has a variable amount of yellow on its black bill, though the extent increases across the Palearctic from west to east (Brazil 1981d). It is also polymorphic in eye colour. The irides, though typically brown, like those of Trumpeter and Tundra Swans, exhibit a range of variation from various shades of brown to grey and pale blue.

Although the plumage of adults is pure white, heads and necks often become stained quite strongly orange by iron in the water of their breeding grounds or, in some regions, of the wintering grounds. This orange iron staining (found on other 'northern swans' too) fades or is lost during moult of neck and head feathers in winter (page 13; Plate 5). Cygnets are mostly dusky grey-brown in their first autumn but become steadily paler throughout their first winter and noticeably whiter towards spring (Plate 10). Over the same period, their bills turn gradually from pinkish to whitish with a yellow wash. Whereas the adult's legs and feet are black, those of cygnets are flesh-grey at first, darkening during the first winter to black.

The Whooper Swan is an extremely sociable species outside the breeding season, when it is commonly found in flocks, often of considerable size. It is highly vocal both during the breeding season when on territory and in non-breeding or wintering flocks where it gives a range of different calls for contact and in display. The voice ranges from soft to loud whooping notes, and is often described as having trumpet- or bugle-like qualities (e.g. Todd 1979). Usually double-noted, the second is higher pitched than the first, and it is these calls that afford the species its English name. When alarmed, vocalisation is generally confined to single loud harsher *whoops*, whereas for contact it uses softer, quieter notes. In flight the call is a deep, resonant *hoop-hoop*.

The Whooper numbers among the most abundant swans, along with Mute and Black each is estimated to have populations totalling *c.*500,000 by Sladen (1991), though in the case of Whooper Swan this may be more likely between 150,000 and 200,000 based on current estimates. Its breeding range is virtually contiguous with its more northerly, tundra-breeding relative, Bewick's Swan. It also breeds, or has bred less commonly, as far south as southern Sweden, Poland, Germany, Scotland, Ireland, the Aral Sea, Mongolia, and Lake Khanka. It has also bred in North America, in the outer Aleutians.

In comparison with Bewick's Swan (which is restricted to tundra), Whooper breeds in a greater range of habitats and at more varying altitudes. On migration and in winter it readily uses brackish and saltwater sites as well as fresh water and farmland habitats. As a result of its considerable weight, it needs a clear 'runway' for take-off, limiting the minimum size of pools it may occupy. It runs, on land or on water, while flapping in order to obtain sufficient speed and lift for flight.

In spring, Whooper flocks commonly arrive at staging posts near their breeding grounds in April–May. Its breeding range encompasses cool northern temperate areas and subarctic taiga, where it nests at a wide range of shallow fresh waters from steppe lakes and pools to marshes and riversides, and also locally by coastal inlets and estuaries. In Iceland, the breeding range spans the boreal lowlands (now treeless through human activity, but originally with birch forest) to the arctic–alpine tundra[5] above 300–500 m (Gardarsson *in litt.* 2001). Young of the previous year disassociate themselves from their parents or are driven off by the adults, which occupy large well-spaced territories, with one pair per waterbody, where they build a bulky nest mound close to water or on an island. The typical clutch of 3–7 eggs is incubated solely by the female, for 35–42 days (Rees *et al.* 1997a). Generally, the colder tundra zone of the Palearctic is inhabited by the smaller bodied Bewick's Swan, because its more condensed breeding cycle enables it to breed successfully there. Body size and related parameters influence the length of the breeding cycle, while latitude limits the ice-free period available. Bewick's, like Whistling Swan, which has an incubation constancy[6] of 97–99% (Cooper 1979), can condense its breeding cycle into the shorter Arctic summer because of its smaller size and because males incubate during female absences, thus shortening the incubation period.

Non-breeding Whoopers spend the summer in moulting flocks and moult earlier than breeding birds. Moult occurs during or after breeding among breeders and failed breeders. Thereafter small flocks of non-breeders, failed breeders and successful parents with their families begin to move away from the breeding range during September, gathering into larger flocks that fly in lines or V formation and arrive in their wintering quarters mainly in October–November. These gatherings may be of several hundreds or even thousands, especially at migratory staging posts.

The Whooper Swan's wintering range is almost equally as vast as its breeding range, spanning coastal lowlands of western Europe to lakes, estuaries and bays of eastern Asia and innumerable scattered localities between (Plate 3). In the west it typically winters inland at lakes and on floodplains, but also on open lowland farmland, usually in coastal regions. Only less commonly does it occur on salt or brackish water, mainly in Scandinavia. Some of the Icelandic population remain there throughout the winter, merely moving to coastal regions or spring-fed waters. In the Far East, it is typically found in winter on large bodies of freshwater, rivers or lakes, where these remain ice-free, or at coastal sites ranging from estuaries to shallow coastal lagoons, where not excluded by sea-ice (Fig. 1.2).

For British and Irish readers, it is the Icelandic population that is most significant as the vast majority of birds wintering in Britain and Ireland breed or summer

Figure 1.2.   As long as there is access to food and open water Whooper Swans can survive severe northern winters.

in Iceland. Ringing has revealed that there is interchange with those on the continental mainland, with some from Iceland wintering on the continent and some continental birds wintering in Britain, but this is relatively limited (Brazil 1983a; Black & Rees 1984, Gardarsson 1991; Rees *et al.* 1991a; Laubek 1998).

The Scandinavian and west Russian population winters on Baltic and Norwegian coasts and southern coasts of the North Sea. A smaller proportion continues further south into western and southern Europe, with numbers in southern regions apparently strongly dependent on the severity of the winter. Coastal lowlands of the Black and Caspian Seas represent another major wintering area, and smaller numbers, perhaps from the same population, winter south to southeast Europe and from Turkey east to the Aral Sea. The eastern Russian population migrates southeast to coastal east Asia, on the Chinese coast and Korean Peninsula, but particularly in Japan. Very small numbers move east to the Aleutians and Pribilovs of Alaska (Sladen 1991). Extralimital movements have brought vagrants to western mainland Alaska (where they have even bred and moulted), California and New England (LeValley & Rosenburg 1983; Sladen 1991; Mitchell 1998). Northbound migrants in Europe have overshot to Bear Island, Svalbard and Jan Mayen Island; and they have wandered as far south in winter as North Africa and the northern Indian subcontinent (Madge & Burn 1988).

The enormous breeding range makes estimating the population difficult. Though large, and currently common, Whooper Swan is nonetheless susceptible to hunting and loss of its essential wetland habitats, and historical declines have occurred. The Scandinavian population was greatly reduced by hunting, though it is now recovering. In parts of Siberia it has suffered from considerable habitat loss, and impacts on populations elsewhere mean that its range is more fragmented than it appears. The apparent Greenland population was seemingly exterminated by killing of adults and young during the flightless period.

Counts and estimates are made of certain populations, thus some figures are available. The Icelandic population, after increasing to 16,700 in January 1986 and just over 18,000 in January 1991, was thought to have stabilised at *c.*16,000 in 1995. However, the most recent coordinated count, in January 2000, revealed a record total of 20,645 (Cranswick *et al.* 2002). Other counts of wintering birds in northwest Europe suggest that there were *c.*45,000 in January 1995 (a trebling since 1974) and that the population from Fennoscandia and northwest Russia wintering in the North Sea/Baltic regions has increased further to 59,000, with *c.*17,000 in the Caspian/Black Sea regions (Salmon & Black 1986; Kirby *et al.* 1992; Cranswick *et al.* 1996; Rees *et al.* 1997a; Delany *et al.* 1999; Laubek *et al.* 1999). Data from Asia are few, though some 31,000 winter in Japan (Fig. 1.3) and 2,000–3,500 in Korea (Won 1981; EAJ 1999; Miyabayashi & Mundkur 1999; Cho *in litt.*). The numbers using large areas of the species' range are unknown, but whereas a world estimate of as many as 100,000 appeared reasonable in 1980 (Brazil 1981c; Madge & Burn 1988), the considerable increase in the Icelandic and northwest European populations and the scale of the 'guesstimates' from Russia indicate a far larger overall population, perhaps even as many as 200,000,

Figure 1.3. Kussharo-ko, a regular winter haunt in northern Japan.

though presumably not the 500,000 estimated by Sladen (1991) and various Russian authors.

Whooper Swans are generally regarded to show little regional variation, but Ma & Cai (2000) cite evidence that the karyotype of Whooper Swans in China is 2n = 78, whereas in Japan 2n = 80 has been reported, indicating that there may be cryptic differences, and Albertsen *et al.* (2002) suggest that there may be slight morphometric differences between east Asian (Japanese wintering) and European Whoopers.

TRUMPETER SWAN *CYGNUS BUCCINATOR* RICHARDSON, 1832

At 150–180 cm in length and with wings of 60.5–68.0 cm, the Nearctic Trumpeter Swan is the largest, though not the heaviest, of the extant swans (Plate A). Surprisingly, it was the last of the seven to be described. Males are usually larger, 11.9 kg, whereas females average 9.4 kg, with wingspans of 1.8–2.4 m (Bellrose 1976) and along with the similarly sized, though slightly smaller, Whooper and Mute Swans, the Trumpeter is one of the largest and heaviest flying birds.

Like the Whooper Swan, the adult Trumpeter is pure-white, however, unlike the Whooper it has an all-dark bill with a reddish streak along the cutting edge of the lower mandible. Very rarely it has a small yellow or greenish-yellow loral spot (Banko 1960). Trumpeter Swan cygnets are greyish-brown, somewhat darker on the crown and hindneck than on the underparts. They gain full-adult plumage by their second winter, though their tails and wings whiten much earlier. Cygnets have flesh-pink bills rimmed with black at the base, the tip and sides of the lower mandible. The adults' legs and feet are black, while those of the immature are greyish-flesh coloured. The iris is dark brown. From 1.8% to 13% of cygnets are leucistic, being entirely white on hatching and having yellow feet and tarsi, and an entirely pink bill (Mitchell 1994).

Though similar in general appearance to Whistling Swan, the Trumpeter is both larger and proportionately longer necked, with a longer, thicker bill, which is both large and deep. The black facial skin tapers to a broad point at the eye, thus the eye is not distinct from the bill (Lockman *et al.* 1990) while the profile is straight in an even slope from forehead to bill tip, its forehead is flatter than the Whistling Swan's, giving it a more imperious profile. The length from the eye to the nostril equals the distance from the nostril to the bill tip (Boyd 1972). The feathering on the adult's forehead extends in a broad wedge (whereas it is straighter or slightly rounded in Whistling Swan), although there is some variation in profiles where the forehead feathers meet the upper bill, with some having an inverted triangle or V, a U or a squared shape (Mitchell *in litt.* 2001). When relaxed, a Trumpeter tends to drape its neck so that the kink appears to rise from the forepart of the back, and the breast protrudes forwards. In contrast, Whistling Swan, like Bewick's Swan, carries its neck more erect from the base.

Quite readily distinguished by voice, Trumpeter utters a deep single or double-bugling honk *ko-hoh*, quite different from Whooper Swan or the higher pitched, more barking call of Whistling Swan.

The former breeding range is believed to have been extensive, stretching from central Alaska east through central Yukon and southwest Northwest Territories, north Saskatchewan and Manitoba to Hudson Bay. It also bred in Ontario, Québec and east to Nova Scotia, New Brunswick and Newfoundland. To the south it bred in the Carolinas, west Tennessee, northwest Mississippi, east Arkansas and Missouri, and in the west to south Nebraska, Wyoming, north Utah, south Idaho, northeast Oregon and possibly California. It formerly wintered from southern Alaska along the Pacific coast to southern California and possibly east through Arizona, New Mexico to the lower Rio Grande Valley and the Gulf coast of Texas. Before being decimated, Rocky Mountain Trumpeter Swans very likely wintered in the Great Salt Lake marshes of north Utah. On the Atlantic coast it wintered from the southern ice limit to Florida (Bellrose 1976; Mitchell 1994; Mitchell *in litt.*).

Once widespread and abundant across northern North America below *c.*800 m, Trumpeters are greatly reduced in numbers, having been widely hunted by settlers for food, and slaughtered commercially for their meat, feathers and their quills (1600s–1800s). Some measure of the slaughter can be seen from the fact that a single company marketed over 108,000 skins in 1820–1880 (although this figure probably included some Whistling Swans). Numbers had already fallen drastically by the 1870s, and by 1935 a mere 69 were thought to survive in the only known population of the contiguous states (Todd 1979; Mitchell 1994). For a while there was concern that it might have become extinct, but a tiny remnant population was then discovered in Yellowstone National Park (Montana, Idaho and Wyoming).

Translocation of some of the survivors to additional sites enabled a slow recovery, and thereafter a previously unknown Alaskan population was discovered. Though increasing once more as a result of strict protection, Trumpeter Swans are still primarily restricted to Alaska, north-western Canada and other north-western states, where they occur on freshwater and brackish lakes, ponds and marshes. Most of those in the northern USA are derived from reintroduction programmes. The increase in numbers since the 1930s has been spectacular, and is lauded as a conservation success, being the result of intentional translocations, natural increase given suitable habitat and protection, and general wetland conservation (which restored many once-degraded or lost wetlands now used by swans, and other waterfowl). The Trumpeter is gradually re-expanding its range, both naturally and by reintroduction, into some of its former breeding areas.

Its current natural breeding range includes much of central and southern Alaska south of the Brooks Range and east of the Yukon-Kuskokwim Delta, locally south from south Yukon and Northwest Territories east to south Saskatchewan, south to north-central Nebraska, western South Dakota, northwest Wyoming, east Idaho, south Montana, central Nevada and south-central Oregon. Its natural wintering range now includes south Alaska, west and south-central British Columbia, west Washington, south to west Oregon, east Nevada, west Utah, south Montana, east Idaho, northwest Wyoming and southwest Dakota (Mitchell 1994). In addition, it has been reintroduced and become re-established locally in north, northeast and southeast Michigan, east and northwest Minnesota, north and central Wisconsin, south Ontario and east Saskatchewan (Mitchell 1994; see Fig. 1.4).

Trumpeter Swan population estimates indicate that it increased steadily during the 1970s and 1980s. By 1995 numbers had reached *c.*16,300 along the Pacific coast, 2,500 in the Rockies, and *c.*900 in the interior (Caithamer 1995), with many interior birds presumed to be from reintroduction attempts. These increases have continued, with the total population increasing by *c.*3,900 (20%) since 1995 and by 20,000 (>500%) since 1968. Regionally, numbers increased between 1995 and 2001 from 16,300 to 17,551 (8%) along the Pacific Coast, from *c.*2,500 to 3,666 (46%) in the Rockies, and from *c.*900 to 2,430 (150%) in the interior (Caithamer 1995, 2001). In Alaska alone the population increased from 1,924 in 1968 to 13,934 in 2000 (Conant *et al.* 2002).

Some Trumpeters are non-migratory and these may commence territory defence as early as February, with copulation and nest building occurring from March (Banko 1960; Mitchell *in litt.* 2001). Others, however, particularly those from Alaska, are migratory and do not commence breeding until late April or May. Trumpeter Swans occupy large territories and, like other 'northern swans', build large nest mounds near water or on small islands. The clutch of 5–8 eggs is usually incubated by the female alone for 33–40 days, in a nest often built atop a Muskrat den. The cygnets grow rather rapidly, fledging in *c.*100 days (Walters 1980;

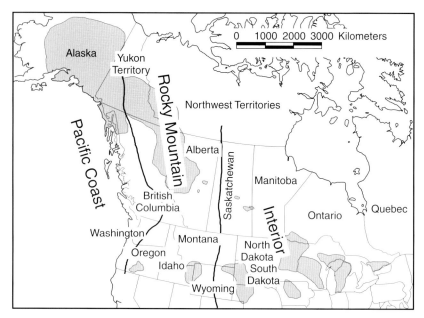

Figure 1.4.   Current breeding range of the Trumpeter Swan shown as shaded areas (after Hawkings et al. *2002).*

Mitchell 1994). They usually forage in shallow water and feed on common vegetation available depending on the season in such habitats, including eelgrass, sedges,
grasses, pondweeds, water buttercup, water lily (*Nuphar* spp.), bulrushes and
arrowleaf (*Sagittarria* spp.).

The species' scarcity has perhaps precluded vagrancy much beyond its normal range;
however, it has strayed as far south as Mexico, in Chihuahua and Tamaulipas (Drewien
& Benning 1997), and even to Kanchlan Bay, Russia, and northern Honshu, Japan,
where one wintered twice in the early 1990s (OSJ 2000; Syroechkovski 2002).

TUNDRA SWAN *CYGNUS COLUMBIANUS* (ORD, 1815)

Tundra Swan is the specific name given to both the Nearctic and Palearctic
populations of Whistling and Bewick's Swans by those modern taxonomists who
currently regard them as conspecific.

WHISTLING SWAN *CYGNUS COLUMBIANUS COLUMBIANUS*

The inappropriately named Whistling Swan (it does not whistle!) of the New
World was the last but one to be described, by Ord in 1815 (Plate A). It has been

renamed the Tundra Swan in North America because that population consists of the nominate subspecies (e.g. National Geographic 1999). Here, however, I have retained the older name because it is helpful in easily distinguishing the Old World Bewick's Swan *C. c. bewickii* and New World Whistling Swan *C. c. columbianus*.

At 120–150 cm, Whistling Swan is slightly larger than Bewick's Swan, although measurements do overlap. Nevertheless, it is still a rather small delicate 'northern swan' with a more slender bill, a far more rounded head than either Trumpeter or Whooper Swans, and a slightly concave bill profile.

The only reliable field characteristic distinguishing Whistling from Bewick's is the former's black beak and essentially black facial skin. This extends back in a point to the eye and across the forehead. At the base of the bill, close to the eye, there is often a small teardrop spot of yellow which is variable in size, but always very small in proportion to the black. The yellow is also significantly smaller than even the smallest yellow patch on Bewick's; it may even be absent or appear so at a distance. Whistling Swans have, on average, just 3% of the bill yellow, whereas Bewick's have 31% (Boyd 1972; Evans & Sladen 1980). In comparison with its larger and confusingly similar Nearctic relative, the Trumpeter Swan, Whistling Swan has a more slender bill, a slighter forehead, a concave profile, the eye is distinct from the base of the bill, it has a proportionately shorter neck, and the distance from the eye to the nostril is longer than that from the nostril to the bill tip (Boyd 1972; Lockman *et al.* 1990).

The voice includes a high-pitched whooping, and a range of honking, clanging calls recalling those of Canada Goose or even American Wigeon. Its yelping flight call is a musical, trisyllabic *wow-wow-wow* and it is equally vocal on the water where, in winter, it is highly sociable and excitable, engaging enthusiastically in noisy greeting ceremonies as individuals join and leave flocks.

Whistling Swans breed across the Arctic, from coastal Alaska east across northern Canada to Baffin Island, usually below 300 m. It arrives on its coastal tundra and river valley breeding grounds in the latter half of May. It nests close to water on the banks of ponds or lakes, sometimes on small islands, laying a clutch of usually 3–7 eggs incubated by the female alone for 30–40 days. Fledging is rapid, taking only 60–75 days (Walters 1980; Ma & Cai 2000). Rapid breeding and growth appear possible in part because of the relatively high proportion of animal matter (insect larvae, fairy shrimps and molluscs) in the diet.

After the breeders have moulted they start southward migration during October. Like other 'northern swans', migratory flocks form wavering lines or V formations, maintaining contact between flock members quite noisily. They follow specific traditional migration routes and use several stopovers, arriving on their wintering grounds in November–December and remaining there usually into at least March.

In winter, Whistling Swan favours coastal brackish and freshwater marshes where it feeds mainly in shallow water, although it has adopted a relatively new habit of foraging over farmland, especially on arable crops and winter cereals. The Whistler's quicker feeding action helps distinguish it from Trumpeter Swan.

Whistling Swans winter in two main regions. Those breeding in western Alaska follow the Pacific coast to California and also occur at several inland wetlands, east to Idaho and Utah, but are concentrated mainly in the valleys of central California. Birds breeding in much of the rest of the range migrate through the Canadian Northwest Territories south to the west of the Great Lakes, then east across northeastern USA to winter at Atlantic coastal marshes from Maryland to North Carolina, with some travelling up to 3,250 km (Todd 1979). Although the sizes of the catchment areas for wintering birds vary enormously between east and west, the wintering numbers are almost equally divided, with perhaps slightly larger numbers on the Atlantic coast.

In the west, Whistlers winter primarily from southwest British Columbia south to central California along the coast, and inland in northern and central California, southernmost Oregon, northern Utah and northern Nevada. They are rare as far south as southern Arizona and southern California. In the east, they winter primarily from southeast Pennsylvania and northern Maryland to North Carolina.

Whistling Swan is not confined to North America. In common with a number of other waterfowl and shorebirds strongly identified with the Nearctic, its range actually extends west across the Bering Strait into northeast Asia. In Asia, it has been expanding west and south in recent decades, and now occurs along *c.*500 km of the north Chukotka coast, with 600–1,000 estimated to be summering there (Syroechkovski 2002).

Vagrants have wandered in winter to north and central Florida (40–50 records all 16 November–12 March; Stevenson & Anderson 1994), Texas (where it is rare throughout; Texas Ornithological Society 1995) and northern Mexico where it is irregular and rare (December–February) south to Baja California Sur, Durango and Tamaulipas, but with most records from Chihuahua (Howell & Webb 1995; Drewien & Benning 1997). It has also straggled in winter to Bermuda, Cuba, Puerto Rico, St Thomas (Virgin Islands) and Antigua (AOU 1998; Raffaele *et al.* 1998). Whistling Swan is accidental in Newfoundland and east across the Atlantic to Ireland, Britain and Sweden, and west across the Pacific to Hawaii (Midway and Maui) (Madge & Burn 1988). It is an annual, albeit rare, winter visitor to Japan where hybrids between Whistling and Bewick's Swans have also been reported (Mikami 1989; Brazil 1991).

Whistling Swan is likely to have always been more abundant than Trumpeter. An estimate made in the early 1970s put its population at *c.*146,000. As with Bewick's Swan, however, total numbers are highly dependent on recent breeding success. A 1996 estimate of the North American population indicated that it had increased but was now relatively stable, with as many as 166,000; 86,300 in the east and 80,400 in the west (U.S. Fish and Wildlife Service and Canadian Wildlife Service 1996). The eastern population subsequently increased further and now exceeds 100,000 (Serie *et al.* 2002). Alaska is the single most significant breeding area being home to *c.*60% of breeders, with the greatest density in western coastal areas, and the majority migrating to California to winter. This sizeable population is well protected, but supports limited hunting and harvesting by indigenous peoples. A highly recommended review is that by Limpert & Earnst (1994).

BEWICK'S SWAN *CYGNUS COLUMBIANUS BEWICKII* (YARRELL, 1830)

At just 115–140 cm, Bewick's Swan is the smallest of the 'northern swans' (Plate A). It is smaller, relatively shorter necked and shorter bodied than Whooper Swan, with which it often occurs in winter, and has a more distinctively rounded head, shorter bill with a slightly concave profile giving it a cuter, retroussé look, and smaller, rounded or square-ended yellow bill patch not usually reaching the nostrils (Boyd 1972). Their superficial similarity in the pre-binocular era led to confusion, and the smaller Palearctic 'northern swan' species was only first separated from Whooper Swan in 1830. It was named in memory and honour of the consummate wood engraver and famous illustrator of birds, Thomas Bewick (1753–1828).

As with all 'northern swans', adult Bewick's are pure white, while cygnets are grey-brown. Although most of the cygnets' duskier feathers are lost by their first spring migration and first summer, they retain some greyish feathers about the head and neck, making it possible to age them well into their second winter. Young are difficult to separate from young Whoopers, except on structure. In flight, however, Bewick's have quicker wingbeats and are more agile on take-off and landing than Whoopers.

The bill pattern is typically black with a yellow area at the base of the bill. This does not, however, extend as far along the bill sides as it does in Whooper. Although showing less yellow overall than Whooper, the bill pattern is so variable that individual identification by trained observers is possible, and such methods have been used in long-term studies of the species in Britain. Unlike adults, cygnets have pink bills, only the tip and cutting edge are black. The bill becomes steadily duskier in spring as adult-like bill pattern develops. The legs and feet of adults are black, those of cygnets flesh-grey. The iris of both adults and young is normally dark brown, but can be pale brown or occasionally blue.

Bewick's Swan is even more vociferous than Whooper. It gives a range of honking and yelping calls, which are both higher pitched and delivered more rapidly than those of Whooper, although there is some overlap. The calls are usually softer, faster, more yelping and less bugling than in Whooper, and have been described as *hoo-hoo-hoo*.

It is a very wide-ranging subspecies, breeding across almost the entire Palearctic or Old World Arctic, from the Kanin Peninsula to Chukotka. Its breeding range lies north of the Whooper's in a narrower band, as it is restricted to nesting locally in low-lying areas of open grass or swamp with scattered pools, lakes and rivers in the tundra. Territorial, it arrives on its breeding grounds between mid May and early June, and nests on dry hummocks, either in open coastal tundra or beside lakes, rivers and estuaries, often in loose, well-scattered colonies. The clutch, often 2–5 eggs (range 1–6) is incubated mainly by the female for 29–30 days (Rees *et al.* 1997c).

Like Whistling Swan, Bewick's consumes considerable animal matter as well as vegetable matter in the breeding season, although it is almost exclusively vegetarian in winter. The shortness of the ice-free summer at the high latitudes where it nests dictates that the breeding season is condensed. It remains on the breeding grounds only until September or early October, by which time the cygnets must be fledged and capable of migrating south or they will be caught by the early-autumn freeze. Family parties join others en route, forming large flocks which remain together over the winter.

Bewick's Swan is a long-distance migrant. The western population, breeding from the Taimyr Peninsula west, migrates southwest through the White and Baltic Seas to the coastal lowlands of northwest Europe. There they inhabit low-lying wet or flooded grasslands, and nearby crop fields, particularly those of winter cereals. They also occur on lakes, reservoirs, coastal bays and estuaries, perhaps more so on migration than in winter, feeding by grazing, dabbling, dipping or upending, for terrestrial and aquatic vegetation. Particularly numerous in Denmark, the Netherlands, England and Ireland, smaller numbers occur in Belgium and France, even south to the Camargue (Rees *et al.* 1997c). Some, of uncertain origin, winter along the southern Caspian Sea in Iran and in the former southern Soviet Union. Eastern birds, from the Taimyr Peninsula east, winter in similar habitats in eastern Asia, but are more often found on inland and coastal waters than crop fields (though they do visit rice fields), with large numbers in the Korean Peninsula and, especially, Japan, and scattered south on the Chinese coast to Guangdong Province, and occasionally Taiwan.

They typically arrive on the wintering grounds from mid October after a relatively rapid journey, stopping briefly en route (more time is spent at stopover sites in spring). Stopover sites may be occupied for days or weeks until they are forced to continue their migration by falling temperatures. Thus, the most distant wintering sites may only be used in midwinter, with birds leaving as early as mid February and returning again via various prolonged stopovers along their northbound route. Very sociable and vocal, winter flocks are typically rather noisy as members maintain an almost constant babbling and indulge in demonstrative greeting and aggressive displays as families or groups join or depart the main flock. Families not only remain together during their first winter, as in Whooper Swan, but may re-unite during second and even subsequent winters.

Migrants occasionally reach Mongolia and inland northern China, and vagrants have been recorded from as far from their usual breeding and wintering areas as Iceland, Bear Island, Svalbard, most of southern Europe, Algeria, Libya, Israel, Iraq, Pakistan, northwest India, Nepal, the Ogasawara Islands, Alaska, British Columbia and Saskatchewan in Canada, and Oregon and California in the western USA (Brazil 1980; Madge & Burn 1988; OSJ 2000; McKelvey *in litt.*).

The population size varies according to annual breeding success. Western populations, which are counted annually on their European wintering grounds, have increased from 16,000–17,000 birds in the mid 1980s (Dirksen & Beekman 1991) to 29,227 in 1995 (Beekman 1997). Eastern populations are less well known, but must considerably exceed 35,000 individuals, as 26,684 wintered in Japan in January 1999, a decrease from the peak of 31,198 in 1996 (EAJ 1999; Albertsen & Kanazawa 2002). Some ornithologists have, in the past, considered the eastern birds as a separate race, Jankowskii's Swan *C. c. jankowskii* Alpheraky, 1904, which allegedly has a larger bill, higher at the base and broader near the tip. There is considerable overlap in measurements, apparent intermediates between Bewick's and Jankowskii's Swans exist, and as they cannot be clearly distinguished on morphology or range, there is no justification for recognising *jankowskii*. This taxon is not, on the whole, accepted (e.g. Cramp & Simmons 1977; Rees *et al.* 1997c; Clements 2000); although the OSJ (2000) retain it, they do not provide data for its separation. Several *columbianus/bewickii* hybrids, with intermediate amounts of yellow on the bill have been reported in the wild, and mixed pairings between Bewick's and Whistling Swans are known from Japan and Russia (Evans & Sladen 1980; Mikami 1989; Murase 1990, 1991; Syroechkovski 2002). Furthermore, it has been suggested that there is a cline in bill patterns (with decreasing yellow from west to east) and there may be regular gene flow between *C. c. bewickii* and *C. c. columbianus* across the Bering Strait (Scott 1981) even further weakening any argument for *jankowskii* being distinct.

MUTE SWAN *CYGNUS OLOR* (GMELIN, 1789)

The Mute Swan is one of the world's heaviest flying birds, reaching weights in excess of 15 kg and ranking alongside Trumpeter Swan as the largest of the waterfowl (Todd 1979) (Plate B). Its specific identity was not recognised until 1789, although Linnaeus appears to have been aware of its existence three decades earlier (albeit mistakenly considering it a domesticated form of Whooper). Though originally a breeding bird of steppe lakes in Central Asia, the inappropriately named Mute Swan (it is not silent) is perhaps now the most familiar of all swans worldwide. It occurs on lowland waters, ponds, lakes, reservoirs, streams and rivers with plentiful vegetation, and also on brackish or saltwater in sheltered coastal bays, estuaries or lagoons, particularly in winter (Madge & Burn 1988). It has been widely introduced outside its native range, from Canada to South Africa and Japan, and is a tame and confiding component of city park lake avifaunas, even in the heart of some of the busiest cities such as London and Tokyo. It feeds primarily by dipping or upending, using its long neck to access deeply submerged vegetation, though it also dabbles and, in some regions, quite commonly grazes on land. Its favoured foods include the various parts of aquatic vegetation (roots, stems, leaves, seeds) particularly of eelgrass, pondweed, milfoil and algae.

Now, as a feral breeding bird of more or less domesticated origin, semi-wild or wild, it is generally resident across much of temperate Europe from southern Scandinavia to Austria and east through Russia, China and Mongolia, though it is absent from much of the far north and far south. It nests on banks or islets in swamps, laying 4–7 eggs (rarely up to 12), which are incubated by the female for

35–38 days (Walters 1980). Fledging is prolonged, at 120–150 days (Todd 1979), and thus it is unable to breed in subarctic boreal or tundra regions occupied by Whooper and Bewick's Swans. Eurasian Mute Swans share with Black-necked Swans the common habit of carrying their cygnets on their back between the wings, behaviour not observed in Whooper Swan.

Mute Swan has a scattered distribution across central regions of Asia to northern China and northern Japan, with some birds of domesticated or feral origin, though it is no longer possible to distinguish these. Those that breed in the north of the range, in southern Scandinavia in the west, and in Hokkaido, Japan, in the east, are migratory, with Scandinavian birds wintering in Baltic coastal regions, and north Japanese birds wintering in central Honshu. Most of the remainder are resident. Mute Swans winter along the shores of the Black and Caspian Seas, on Turkish coasts, in lowland China, and Honshu, Japan (Madge & Burn 1988).

It has been introduced by man more widely than all of the other swans together and now occurs in western Europe, eastern North America, South Africa, Australia and New Zealand. Vagrants have occurred to the south of the normal range, in the Azores, Algeria, Egypt, Israel, Jordan, Iraq, Afghanistan, Pakistan, northern India and southern Japan (Madge & Burn 1988).

Mute Swan is readily distinguished from the other white swans of the Northern Hemisphere by its bright reddish-orange bill, which is black at the base and has a black knob, and because of its habit of swimming with its wing feathers raised in a strong arch, like sails, and its neck curving gracefully in an S. Adults are, like 'northern swans', entirely white. Young are dark greyish-brown and have grey-pink bills with a black-base, although the albino form, known as 'Polish', is all white, even as a cygnet, and has pinkish-grey legs and feet unlike the adults which have black, and normal juveniles, which have grey legs and feet.

By no means silent; Mute Swan is only mute in comparison with the other northern species, which have haunting, whooping and bugling calls. It has its own repertoire of hisses, snorts and grunts. Furthermore, whereas the wingbeats of other swans are silent or merely make a quiet whooshing audible only at close range, like the wings of other large heavy birds, the beating wings in Mute Swan make a highly audible, distinctive, throbbing *waou, waou, waou*. And, whereas other swans, tend to be highly vocal in flight, the Mute Swan's most noticeable noises are the thrumming sounds of its reverberating pinions, and a few grunting notes. While the four 'northern swans' have rather short tails (in flight their feet reach just to the end of the tail), Mute Swan has a longer, pointed tail that extends well beyond its feet.

In its native range it is shy and unapproachable, but it has long been domesticated (by the Greeks and Romans, and later in Britain) as a bird for the table, and becomes extremely tame, readily feeding from the hand, especially in winter. Holloway (1996) considered it likely that the Mute Swan was an indigenous British species, breeding wild in and around East Anglia until hunted to near extinction by the 10th century; it was saved from extinction there by its being bred in a semi-captive state from at least 966.

Throughout most of its range it is strongly and aggressively territorial while breeding. Males, on the water, raise their inner wing feathers in an arch, and hiss aggressively with their beaks open, or may rest the neck on the shoulders and swim powerfully and jerkily to intimidate and drive away intruders. On land, with the neck feathers raised and beating their wings vigorously, territorially aggressive males are most alarming and easily capable of driving dogs and even people away from their nests. Breeding begins in April when huge nest mounds consisting of vegetation are built near water, often among reeds or other tall growth. At some localities, e.g. Abbotsbury Swannery in Dorset, on the south coast of England, and in parts of Denmark, large numbers breed in colonies. Large numbers also gather in post-breeding moult concentrations.

European populations are considered to be increasing, though locally there have been serious declines through poisoning caused by ingesting spent lead gun-shot or by swallowing lead weights lost or discarded by anglers. Asian populations are far less well known, being widely scattered and almost relict in distribution. Some of these populations, like other waterfowl of the region, are clearly threatened to an unknown extent by hunting and habitat destruction (Madge & Burn 1988). The world total may be around 500,000, with *c.*180,000 in western and central Europe, *c.*300,000 in Russia and adjacent states and 8,000 in North America (Ma & Cai 2000).

BLACK SWAN *CYGNUS ATRATUS* (LATHAM, 1790)

The Black Swan is like a photographic negative image of the Mute (Plate B). It has a relatively restricted range, but is highly successful and abundant. It was once symbolic of extreme rarity when, in fable form, tales of it reached 17th-century Europe. When swimming it appears entirely sooty black, but it has in fact bright white primary feathers, which are concealed at rest; however, they flash in bold contrast when stretching its wings or in flight. In general, it is similar to the Mute Swan though it is less heavy and has an even longer neck, which is more slender just below the head, and has a more steeply peaked forecrown. Its wings are not merely highly distinctive because of the white flight feathers contrasting with black wing-coverts, but are made further remarkable by the broad crinkly edges to its black greater coverts, giving it the appearance of wearing a windblown cloak. Juveniles are greyer than adults and considerably darker, and less brown, than cygnets of other species, and in flight they show dusky tips to their flight feathers.

The bill and bare skin between the bill and eye varies from scarlet to a deep waxy wine red or crimson, and, in adults, there is a white subterminal band and a pink nail. Cygnets have a dark grey bill with an entirely grey tip. The iris is white or reddish in adults but brown in young, while the legs and feet are dark grey in young birds, becoming black in adults.

Black Swan is a sociable species and very much more vocal than Mute Swan. It uses a range of soft conversational contact calls, and a somewhat musical high-pitched bugling note, although this does not carry far except in calm conditions. It is, nevertheless, less vocal than the 'northern swans'.

Prolific and abundant, it occurs as a native across the western part of Western Australia, over southeastern Australia and Tasmania, and widely throughout New Zealand. Its large population is difficult to estimate, but given that *c.*50,000 occurred on the Coorong, South Australia, in 1957, it can be, locally at least, extremely abundant, with perhaps 300,000 in Australia and 100,000 in New Zealand (Sladen 1991). Though essentially protected by law throughout Australia its large local concentrations do cause problems in agricultural areas and, in response to considerable crop damage, a short hunting season has been introduced in Victoria and Tasmania.

The species was introduced into New Zealand in 1864, but wild individuals may have crossed the Tasman Sea and colonised naturally. It is now widespread and common on both main islands and has even reached the remote Chathams, where it presumably occupies the same ecological niche once occupied by the long-extinct swan of New Zealand. The New Zealand population of Black Swan has been estimated at *c.*60,000 (Madge & Burn 1988), though this is considerably lower than in the recent past. So common was it in some regions that eggs were collected commercially and certain undesired colonies were deliberately disturbed. Prior to the devastating storm 'Wahine' in 1968, the population of South Island's coastal Lake Ellesmere alone exceeded 100,000 birds (Todd 1979).

This highly gregarious bird, of both fresh and brackish water, shallow (<1 m) lakes, bays and lagoons, feeds commonly on pondweed, wild celery and Elodea. It prefers to nest beside large, relatively shallow lakes. Nests are built near water within screening vegetation or on small islands. The least territorial of all the swans, it regularly nests in colonies, so dense that nests are built as close together as a neck-reach apart, and at densities quite unknown among other swans. The clutch of 4–8 eggs is incubated by both sexes for 34–45 days, and notably, not only do males participate in incubation, but may even do so longer than their mates. Cygnet growth is prolonged, with fledging taking up to 180 days, during which time the cygnets may gather in a crèche (Walters 1980; Williams 1981; Ma & Cai 2000). Outside the breeding season, and away from nesting areas, it may occur on flooded agricultural land, coastal lagoons, estuaries and sheltered coastal bays. It feeds on aquatic vegetation mainly by dipping (submerging its head and neck) and dabbling, though it sometimes upends in deeper water, and will also graze on waterside pastures.

The timing of breeding varies locally and, when conditions permit, it may nest at any time of year. Typical breeding seasons are, however, February–May in northeast Queensland, June–August in Western Australia, and June–December in New Zealand.

Classifying the status of the Australian population is extraordinarily difficult as it consists of a confusing mixture of resident, dispersing and even nomadic birds. Following the breeding 'season', it gathers into large moulting flocks. Considerable post-breeding dispersal also occurs, as a result of which the species has occurred throughout Australia and most of New Zealand. In Australia, such extreme and long-range dispersive movements may be connected with the search for new breeding areas following the continent's local and highly irregular rains.

BLACK-NECKED SWAN *CYGNUS MELANOCORYPHUS* (MOLINA, 1782)

Like a diminutive Mute Swan in form, the South American Black-necked Swan measures just 102–124 cm (Plate B). Although it is the second smallest of the swans, it is South America's largest waterfowl. First described by Molina in the same year that he described the Coscoroba; they were the first swans to be described to science after Linnaeus' description of the Whooper Swan.

This white-bodied bird has entirely white wings, and a strikingly contrasting velvety black neck and head, with a fine white line curving back from above the eye. The large red caruncle over the bill base also serves to lend it a superficial resemblance to Mute Swan. Young are similar to the adults, but are brownish-grey where the adults are black. Birds attain adult plumage by the end of their first year, though they do not develop a full frontal knob until 3–4 years old. The bill and caruncle are contrasting in adults. The former is blue-grey, with a pinkish-white nail, while the bare facial skin and large frontal knob are bright red. Juveniles have grey, dull reddish-based bills and lack the frontal knob. Whereas the skin of both the legs and feet of other swans are black, those of the Black-necked are pink in adults and grey in cygnets. Young have a pale brown iris, becoming dark brown in adults.

Black-necked Swan occurs on shallow freshwater lakes, brackish marshes and saltwater coastal lagoons, estuaries and sheltered bays. It appears equally at home in remote wetlands as in densely inhabited areas. It feeds mainly on aquatic vegetation in shallow water by dabbling and dipping, and is rarely seen on land, where it moves awkwardly. Unlike the only other South American swan, the long-legged Coscoroba, Black-necked has relatively short legs and a long body, with the legs positioned well to the rear, making walking awkward, and take-off and landing clumsier than other swans, although swimming is powerful.

It ranges widely in southern South America, breeding from Cape Horn and the Falklands north to central Chile, extending north into Paraguay and extreme southern Brazil; it is particularly common in Argentina, Chile and Uruguay.

The breeding season varies regionally and with latitude, thus in central Chile and Argentina it breeds in July–August, while farther south, e.g. in the Falklands, it nests from early August to mid September. Aggressive and strongly territorial when breeding (although several nests have been reported in close proximity in parts of the range), males chase almost any intruders from the vicinity of the nest. They do not, however, use the arched-wing threat posture typical of both Mute and Black Swans. They are only weakly vocal and the main call is a wheezy whistle that fails to carry.

Nest mounds are usually constructed in dense vegetation near water, on small islands or even partly floating. However, where aquatic vegetation is sparse, I have found nests as far as 100 m or more from water in the Falklands. The clutch of 4–7 eggs is incubated by the female alone for 34–36 days (Walters 1980), the male defending the territory. Black-necked Swans carry their cygnets on their backs more frequently than other swans, and may continue to do so even when quite large. This form of brooding is a habit presumably well suited to their extremely aquatic lifestyle.

Black-necked Swan is most gregarious when not breeding. Large concentrations gather to moult and these may reach 5,000–6,000 birds. While populations in warmer temperate and subtropical areas are essentially resident, many southern mainland birds disperse north from cooler temperate regions in March–April for the winter, and may reach northern Paraguay. Vagrants have reached the Juan Fernández Islands off Chile and even the South Shetlands just north of the Antarctic Peninsula!

The overall population of this little-studied species is poorly known, but seasonal variation in food supply apparently causes significant fluctuations. There may be around 100,000 individuals, with about half in Argentina, up to 20,000 each in Chile and Uruguay, and 2,000–3,000 in Brazil (Schlatter *et al.* 1991). Though widespread and reasonably common, local drainage of lowland wetlands for agriculture has doubtless adversely affected its population in some regions. In contrast, in Chile at least, it is now reported to be increasing following a decline in numbers resulting from persecution.

Coscoroba Swan *Coscoroba coscoroba* (Molina, 1782)

The smallest and most unusual of the seven swans, the Coscoroba is sometimes considered to be related to whistling ducks (particularly because of its clearly patterned downy young), but has until very recently been generally accepted as a true swan, albeit with distinctly goose-like affinities and a duck-like bill (Scott & The Wildfowl Trust 1972; Wilmore 1974; Todd 1979; Madge & Burn 1988) (Plate B). Recent work by Zimmer *et al.* (1994), however, pointed to it being just as likely to have separated prior to the divergence of geese and swans, as to there being a sister group relationship between *Coscoroba* and *Cygnus*. Further work, by Harvey (1998, 1999) who sequenced the mitochondrial cytochrome *b* gene, helped further clarify the Coscoroba's position as a relative neither of swans *Cygnus* spp. or geese *Anser*/*Branta* spp., but of the Cape Barren Goose *Cereopsis novaehollandiae* of Australia, thus we may soon become familiar with regarding the Coscoroba as a non-swan, despite its name, thereby reducing the world's complement of swans to just six.

Coscoroba Swan takes its strange name from its equally bizarre, and un-swan-like, voice. It emits a rather loud toy-trumpet-like *cos-cor-oo*, of which the first syllable is longer and higher pitched.

At just 90–115 cm, and with a wing length of 42.7–48.0 cm, Coscoroba Swan is even smaller than a Bewick's, and it feeds like a large duck by dabbling in shallow waters or grazing on land. All white, with black-tipped wings, it has a rounded goose-like head but a duck-like bright waxy red bill with a paler nail. Unlike other swans, the entire bill base is feathered and lacks any form of basal knob or caruncle. As a consequence, the face has a quite different, softer, even more comical expression than any other swan. The legs and feet are bright rosy pink, and because they are relatively longer than in other swans, it is the only one capable of taking off without a run. Also unlike other swans, the eye is not typically dark but ranges from yellowish to reddish-orange. The sexes are essentially similar, although males are typically larger and heavier (males average 4.6 kg and females 3.8 kg). Cygnets are whitish, but have patchy greyish-brown feathering on the crown, back and wings, rather like shelducks or Cape Barren Geese, and have blue-grey, rather than red or pink, legs and bill. They become whiter during their first winter, but do not gain adult plumage until the second autumn.

Coscoroba Swan is a partly sociable species that occurs in small flocks rarely exceeding 100. It occurs at wetlands throughout southern South America, from Cape Horn north to *c*.45°S, particularly in Chile and Argentina, though its range also reaches north to Uruguay, Paraguay and southern Brazil, especially during post-breeding dispersal.

It breeds alone or in small colonies at wetlands that are widely scattered, isolated and threatened. The breeding season varies regionally and with latitude, October–December in Chile, but June–November in Argentina. They build bulky nests, typically near shallow water or on small islands. The 4–9 eggs are laid in a ground or floating nest and incubated by the female for *c*.35 days (Walters 1980). Though it has a wide range, the total population probably does not exceed a few thousand individuals. Its poorly understood migrations are further confused by regional variation in the breeding season, followed by dispersal and moult gatherings. This is the most unusual and least known of the extant swans.

Although almost universally considered a monotypic genus, Walters (1980) included the supposedly extinct species, David's Swan of China, within the same genus, as *Coscoroba davidii* (though see below).

EXTINCT SWAN RELATIVES

While seven species (eight forms) are currently extant, other recognisable swans have existed in the not-distant past, and other species are also attested to (though their identification and taxonomic distinctness are problematical). Two, *Cygnus equitum* and *C. falconeri* (the latter about one-third larger than a Mute Swan) are considered probably to have arisen from the same stock as Whooper Swan (Northcote 1982, 1988) with which they must have co-existed. They are both thought to have become extinct during or after the Pleistocene (Wilmore 1974).

C. equitum was abundant on Malta from the Ipswichian Interglacial (*c*.125,000 before present). It was a broad-bodied dwarf swan with some goose-like features, closest to Whooper and Bewick's Swans. Northcote (1988) considered it not only to be flightless, but that it walked well and unlike other swans may have habitually taken off and alighted on land. It may have used both fresh and brackish water, feeding on high-calorie food plants in largely terrestrial habitats. Morphologically distinct from both Whooper Swan and the giant, flightless, extinct *C. falconeri*, it would have co-existed with both. The Giant Maltese Swan *C. falconeri* was, according to Northcote (1982), a giant land-feeding swan, resembling a large Whooper weighing *c*.16 kg. Despite having a *c*.3 m wingspan it was well adapted for walking, but it seemingly had little or no flight capability.

Tracing other extinct species has proven difficult and I have come across few references and these only fortuitously, thus no doubt other swans have been described that I have not traced. Of those to which I found reference: *C. bilinicus* occurred during the Tertiary (66.4–1.6 million years ago) and has been found in the Czech Republic (Mlikovsky & Svec 1989), *C. herranthalsi* in the Pliocene (between 2–13 million years ago) and known only from a toe bone found in Belgium, while *C. hibbardi* (a relative of Whistling Swan) and *C. paleoregonus* (a relative of the Mute Swan) occurred in the Americas, and *C. lacustric* and *C. nanus* were both relatives of Black Swan from Australia's Pleistocene (1.6 million years to 10,000 bp) (Wilmore 1974; Price 1994). The most recently described species (from a bone fragment) is *C. verae* of the early Pliocene, found in Bulgaria in 1995 (Boev 2000). More recently lost was the Chatham Islands Swan, or 'Poua', *C. sumnerensis* Forbes, later reclassified as *C. chathamicus* by Oliver (1955), which was slightly larger than the otherwise similar Black Swan, and is presumed to have become extinct between 1590 and 1690 through hunting (Williams 1964; Scarlett 1972; Wilmore 1974; Walters 1980).

Truly lost, and even more recently, is the extraordinary David's Swan of China, which was described by Swinhoe (1870) from a single specimen as an all-white swan smaller than a Bewick's, with feathering between the bill and eye. The bill was vermilion with a black nail, while the legs and feet were orange-yellow. So poorly known is it that even its identity is questionable, the specimen having been lost, and some authors consider it probably a hybrid goose/swan (Jabouille 1936; Germiny 1937; Boettischer 1943; Grote 1943). Others accept its identity, with Walters (1980) even placing it with Coscoroba Swan as *Coscoroba davidii* (Swinhoe), although Scott & The Wildfowl Trust (1972) and Evans & Lebret (1973) considered it probably a genetic anomaly such as a leucistic juvenile Bewick's Swan, as the original description resembles that of young leucistic Trumpeter and Bewick's Swans, and of 'Polish' Mute Swans.

Thankfully, the inclusion last century of Trumpeter Swan among the list of extinct swans was narrowly averted, and now all surviving species seem secure, although local populations may be threatened.

Urdarbrunnr—Spring of Urd

CHAPTER 2

Swan culture

'The Silver Swan, who living had no note,
When death approached unlocked her silent throat,
Leaning her breast against a reedy shore,
Thus sung her first and last, and sung no more,
Farewell all joys, O death come close to mine eyes,
More geese than swans now live, more fools than wise.'

Orlando Gibbons, *Madrigals & Motets*, 1612.

NAMING AND KNOWING

Few sights or sounds are as stirring as that of a flock of Whooper Swans clamouring in flight as they plane down, their wings angled, heads raised and feet spread wide as brakes, onto their feeding or roosting grounds. In a subarctic winter setting, crisp with a brilliant blue sky, where deeply snow-crusted dormant volcanoes ring ice-covered lakes and steam curls skywards from thermal springs into the frigid air, little else can match the beauty of the Angels of Winter or be pure enough to grace such sacred waters.

The striking beauty of the archetypal swans has lodged itself so powerfully in the human imagination that their massed annual leave-takings and arrivals have served only to heighten their mystery. As peoples and their oral traditions travelled across Eurasia, ancient tales of swans, swan maidens and swan knights were carried and intermingled, some spreading in mythical form far beyond the range of the birds themselves. These creatures of myth and legend somehow became linked with the cult of the sun but now it is impossible to unravel the origins or meanings of many of the ancient tales (Evans & Dawnay 1972).

Even without the evidence of Stone Age art, we could presume that the size, colour, sounds and seasonal movements of the wild swans would have attracted the attentions of early man. It is not surprising then that they feature so powerfully in folklore as creatures of mystery and magic across their Palearctic range. Ancient and widespread beliefs have it that human souls assume the form of a bird after death (the species varies regionally), and wild swans were strongly associated with this final transformation. In particular, bird-maiden mythologies, which are of very ancient lineage throughout northern Eurasia, gave rise to the swan-maiden motif, among others, as an explanation of the bird ancestry of humans (Armstrong 1958), a belief enshrined in ritual among the Siberian Buryats (see below).

The Swan-Maiden theme, among the most widespread of all legends describes, in its commonest form, the arrival of a flock of swans at a lake. Shedding their feathers, they take human form and bathe in the guise of attractive maidens. A man spying on them steals one of their garments and when the others flee one remains and becomes his wife. Years later, she finds the hidden garment and, re-acquiring feathered form, flies away. Variants on the tale abound, the oldest coming from India, though it is unlikely that they originated there, being themselves fragments of older tales, but common to all is the inherent implication that the incident relates to a patriarchal society in which beauty is revered, making a man's dream of the bird of happiness come true (Armstrong 1958). Despite the tortuous and long-lost routes, both geographical and psychological in the dissemination of this and the related Magic Flight tale, the Swan-Maiden tale nevertheless probably evolved in the far north where swans were common and nakedness rare.

Not only has the graceful swan achieved a very special place in human hearts as a creature of silvern beauty, yet somehow, in some bygone age, it was also raised to the status of a powerful symbolic figure representing, in apparent contradiction, virginity, sexuality and even death. Certain owls and several crows have also been associated with death. Their attributes, blackness in the case of crows or their association with darkness in the case of owls, were perceived by simple minds as clear links with the fearful realm of death. To associate graceful white swans with death, however, requires a perverse combination of literary learning, superstition and biological ignorance.

The Whooper Swan was a very special bird to the ancient Norse storytellers, who revered it as the progenitor of all swans, and the only creature pure and distinguished enough to grace the sacred waters of the Spring of Urd (Brazil 1990) (see Chapter Title Page), which flowed below the sanctuary of the gods, the great sacred Ash Tree, the three-rooted Yggdrasil. One of its roots extended into the underworld, Niflheim, another into the land of the giants, Jötunheim, and the third to the home of the gods, Asgard, and supported the sky. The three Norns or Fates, the maidens who wove the fate of men, Urd (goddess of the past), Verdandi (goddess of the present) and Skuld (goddess of the future) (Hutchinson 1996), are associated with the Whooper Swan, because it was their duty to water Yggdrasil, around which two pure swans swam. The spring of Urd was considered so holy that everything that came into contact with its waters became 'as white as the film that lies within an eggshell'. Thus, the original swans were pure white, and from them, all other swans were reputed to have originated (Evans & Dawnay 1972; Wilmore 1974). So, the Whooper Swan originated in the very heart of purity and became the progenitor of all swans, a mythical feature later echoed in Linnaean taxonomy.

In the great saga-weaving societies of Scandinavia, Whooper Swans were also directly associated with the gods, particularly with Freyr, the god of sunshine, rain, and of fruitfulness or fertility. The white cumulus clouds that formed Freyr's chariot were believed to be swans. The Whooper was also linked with the Valkyries, who chose the slain from battlefields to be carried to the great hall of Valhalla to be received by the great Norse god Odin (Wilmore 1974). Perhaps unsurprisingly therefore, the early Icelanders (according to Todd 1979) also revered Whooper Swans as possessing supernatural powers and as undertaking an annual journey (their migration) to their ancestral homeland in Valhalla[1].

It comes as no surprise that Finland, where the Whooper Swan has long been a widespread breeding bird, also has swan legends or that a region as vast as Siberia has ancient tales that must also relate to the species, but it is far less obvious why India should also possess tales, with both Brahma and Buddha linked to swans in legend, although little distinction was seemingly made in some regions of Asia between swans and geese. Not merely a migrant in the biological sense, moving seasonally between regions, the Whooper Swan has also travelled metaphorically, via various ancient traditions, down long-lost routes of cultural transmission. En route, not only has its identity perhaps been intermingled with other swans (conceivably even other birds), but its symbolic significance has shifted and varied too. While many symbolic meanings have overlapped or intertwined, others have differed dramatically, such as its surprising and incongruous link with death. In Ireland, swans were believed to embody the spirits of the dead and were not to be disturbed (Armstrong 1958). A sensible reason for their reverence and protection, but on what was it based?

Yet to jump straight to such serious symbolism is to bypass much of modern interest. Whoopers, and the other white swans, have had an impact on man through traditional culture. Swans, largely Whoopers, were known in the far north as symbols of summer, in the south they arrived with the impending winter, and between their haunting voices would have been heard, and their graceful white forms seen, only on migration. Today, although reverence has waned and beliefs in their symbolism faded, they nevertheless exercise a much more popular appeal, being a focus for the general public, birdwatchers and photographers alike, across their range. At sites where Whoopers gather in winter, from the UK to Japan, people are attracted by their grace and the air of mystery that still surrounds them. But whereas in the west those visitors are likely to come to watch them from afar with binoculars, in the Far East they are much more likely to bring food (for it is the thrill of hand-feeding that is the main draw) or expensive cameras hoping to create the perfect image of this perfect bird.

Swans have played a significant role in the mythology, literature and art of widely differing cultures spanning many millennia. Archaeological remains and ancient drawings of swan-like birds date to the Stone Age and perhaps beyond. Cro-Magnon rock carvings in northern Russia from *c.*18,000 BC (Wilmore 1974) very likely relate to the Whooper Swans those early people encountered on their boreal breeding grounds. Beliefs in the association between the soul, the swan and death were already established and left their echoes down the ages. Thus, Whooper Swans were already Whooper Swans before modern humans existed!

Much later, tribal Tungus and Yakuts, and shamans from Russia, revered the swan as a totem; Siberian tribes, according to Wilmore (1974), believed that swans harboured human souls, making hunting them taboo, in a traditional belief remarkably akin to that mentioned earlier in Ireland. The Siberian Khorinskie Buryats apparently considered the eagle to be the father and the swan the mother of their people, and they greet swans as they arrive in spring with gifts of tea and milk.

The Buryat bird ancestry myth is but another variant on the swan-maiden theme, though at a considerable geographical extreme. There, a hunter watched as three swans arrived at a lake to swim, leaving their feathered cloaks on the shore. Taking one of the cloaks he married the woman to which it belonged. Many years later when he allowed her to try on her cloak once more she flew out of the smoke-hole of their tent[2] crying that each spring and autumn when the swans fly north or return south her husband must honour them with ceremonies, which were retained in the giving of prayers and libations to the first swans to arrive each year. The conjunction of ritual or ceremony with myth typically indicates a more archaic survival than either one without the other. Even if not actually the case among the Buryats, where the garment-theft motif may have been incorporated later, it is among the likes of these Siberian peoples, with their animistic bird–animal–human links, that the original swan-maiden legend may have originated (Armstrong 1958).

Whooper Swans were similarly considered sacred by the Lake Baikal and other eastern Siberian tribes, many of who also believed that they were descended from wild swans. The belief that swans embody the souls of humans, often of women, is widespread among Slavic peoples, as was the belief that it was a sin to shoot one and that misfortune would follow (Boreiko & Grishchenko 1999). Such beliefs from within the Whooper Swan's range could conceivably have provided the origins of that well-travelled and widespread myth of the swan maiden.

By the time of the Iron Age (around 500 BC) the taboo against killing Whoopers had clearly lessened or perhaps it had never spread much beyond the range of the Buryats and the Irish. Whooper Swan bones have been identified in middens of that time, although images and symbols of them were also appearing in stone, metal and legend, bearing witness to their continued reverence as sacred and super-natural creatures (Wilmore 1974) while more pragmatically being welcomed as a source of food.

Iron Age records of Whooper Swan remains in Britain range from the dour stone brochs of Orkney to the wooden lake-village of Glastonbury. Whoopers were still plentiful in southern England during the Roman era (AD 43–400) as evidenced by bones found at Colchester. During the subsequent Dark Ages, the only proof of the species' continued presence in the British Isles comes from Ireland. Midden remains reveal that it was still being eaten, and continued to be so during the Viking invasions. Its significance as a valued food source should not be under-estimated; the meat is rich, dark, well flavoured and one bird is sufficient to feed several families. The Mute Swan meanwhile was being domesticated, also for its meat, and it is that species that presumably forms the basis of the Swan-Knight myths (see Wilmore 1974), but while the latter became favoured as food and pros-pered in semi-domestication, it seems that the wild Whoopers retreated north pos-sibly in response to climatic cooling (Fagan 2000; Gribbin & Gribbin 2001). Perhaps during historical periods, it was only ever a winter visitor to the once vast wetlands of England. Much earlier of course, as glacial ice retreated from northwest Europe in the steadily warming millennia following the last great glacial period that ended about 10,000 years ago, swathes of boreal forest patch-worked with wetlands would have spread across much of western Europe, including Britain, at much lower latitudes than today, placing those regions well within the species' range.

Swans are perceived as having innumerable characteristics, almost invariably pos-itive or beneficial. Among Slavic peoples the close association between male and female swans was the embodiment of matrimonial fidelity and beauty[3]. In some regions, they are considered able to foresee the future, in others, they are portents. Their appearance over the sea betokens quiet and favourable travel for sailors, and in many areas they are honoured as the protective patrons of mariners. Elsewhere, their arrival is believed to foretell rain or cold weather (Boreiko & Grishchenko 1999), and certainly their arrival and departure from many parts of their wintering range seems not only weather dependent but also prophetic of changing conditions.

I have found only one account of swans representing a powerful negative force and its rarity makes it worth recounting. In this poem from the Tartars, the hero Kartaga grappled with the anti-heroine Swan-woman year after year. Despite his strength, Kartaga was unable to vanquish the evil Swan-woman, because her soul was not in her body. Only with the aid of two magical horses that traced her soul to a golden casket contained in a black chest at the foot of a copper rock was he finally able to defeat her (Frazer 1922).

The intermingling of relatively modern Christian traditions with more ancient pagan rituals is widely prevalent; one only has to consider the various symbolic features of paganism that form elements of both Christmas and Easter. This pagan/Christian melding even incorporates swans. In parts of northern Russia, the birthday of Mary, the virgin mother of Christ, was celebrated on 21 September with the release of a pair of swans into the wild (Boreiko & Grishchenko 1999).

Throughout much of Scandinavia, across Siberia and in northern Japan, the overlap between the ranges of earlier tale-weaving peoples and swans provides sufficient justification for believing that even general references to swans may be taken as relating quite specifically to the Whooper. Elsewhere, such tales may have related to white swans in general. For all the confusion that exists over the precise origins of literary, musical or artistic references to swans, the viewer, once exposed to the inspiring beauty of the Whooper Swan, is unlikely to be too concerned about just how precisely the cultural image relates to the species being observed. Knowledge of such literary and musical material serves only to heighten the personal experience of encountering the bird in a wild setting. But first what of that name?

The Whooper Swan bears a noble name of considerable antiquity. This is the definitive swan, the first to be given a Latin binomial by Linnaeus (1758), who originally named it *Anas cygnus*. Today, both its generic (*Cygnus*) and specific (*cygnus*) names are derived from the Greek for swan, kuknos 'the honker', its generic name having been reassigned, from *Anas* to *Cygnus* in 1836, by Bechstein (del Hoyo *et al.* 1994). In its borrowed form, in Latin, its name took two forms, either *cycnus* or *cygnus*, though in modern ornithology we use only the latter. Thus the scientific name now in use traces its roots to classical Greek and ultimately the words used for swans in the language of the ancients whose descendent languages travelled west to Europe from regions far to the east. Modern names for Whoopers in the Romance languages (French, Italian, Spanish and Portuguese) are closely derived from Latin as will be seen below, while even in English the Latin link is evident in the origin and continued use of the word 'cygnet' for a young swan. The scientific name has not, however, been immutable, for not only was it changed from *Anas cygnus* to *Cygnus cygnus*, but for a while it was, rather poetically, also *Cygnus musicus* (Hudson 1926).

The English word 'swan' is also of interest as it dates back, still in the same form, to both Middle and Old English in which it meant 'the sounder'. 'Swan' is derived from the root 'swen-', 'to resound', and is the base of the Latin for a sound, 'sonus', and of an array of related English words such as consonant, sonorous and sonnet (Potter & Sergeant 1973).

Many place names in England begin with the form Swan, e.g. Swanmore, Swanbourne, Swanley and Swanage, but such names, attractive as they are, are potentially ambiguous and may not refer to the bird at all. The Old English word for a farmhand or herdsman, generally a swineherd, was *swän*, and this epithet has also provided us with English place names. Yet in the old Indo-Aryan language Sanskrit, the archaic precursor of Indo-European languages dating from *c.*1800 BC, the word for swan is *svanati*, and perhaps from this distant source we have derived the Germanic *svanur* and the modern English swan. Place names beginning with swan are all the fewer than they might be because another word for swan also competes for eponymous commemoration. In Old English, an alternative name for the bird was *ielfetu*, with *elpt* its equivalent in Old Norse, which formed, presumably, the origin of *álft*, which is used in modern Icelandic. This alternative name for swan can be found in such places as Elterwater, the name of a Lancastrian lake derived from Old Norse 'Elptarvatn' meaning swan lake (Airey 1955) and thus while Swanmore in Huntingdonshire and Swanbourne in Berkshire illustrate the use of one name, the other is more frequent, as in Elveden, in Suffolk (Potter & Sergeant 1973). This alternative tracing via Old Norse does, I feel, relate specifically to the Whooper Swan. As Kear (1990) remarked, such names are commoner in the north and east of England where Whoopers would, and still do, occur, and thus in Yorkshire we have Aldmire and in Durham Elvet Hall, although Gardarsson (*in litt.*) has raised the pertinent consideration that this might be coincidental and those areas could correspond with old Norse colonies, to which familiar names may have been carried.

Toponymic celebration of the significance of swans is by no means confined to Iceland and Britain and probably occurs in each of the Old World regions where the 'northern swans' occur. In Russia, for example, where 'Lebed' is the term for swan, the memory of swans lives in place names such as Lebedin City and Lebedivka, and the rivers Lebedinka, Lebedishka and Lebedinets in the Carpathians (Boreiko & Grishchenko 1999).

It was the Whooper Swan that past occupants of Scandinavia would have been most familiar with, so for them *elpt*, as in the modern Icelandic *álft*[4], meant not just any swan, but specifically the Whooper. Similarly in Britain, the earliest swan reference appears to be to migratory Whoopers, *ylfete* in the Anglo-Saxon poem *The Seafarer*, of *c.*700 AD, concerning the Bass Rock in the Firth of Forth, Scotland (Stanford 2000; Stanford *in litt.*). In Iceland, place names bear witness to the long-established presence of these birds, as at Alftavatn and Alftafjördur (Gardarsson & Skarphedinsson 1984), although the swans long preceded human settlers there. Doubtless the same is true of place names in many other Indo-European languages. Could one trace an ancient range for the species based on its legacy in place names?

Old Celtic, the language of one of the large groups of human migrants who reached southwest Europe, gives us the word *elaio* for the Whooper Swan meaning 'the rising one' (perhaps based on observations of the 'wild' swan's nervousness prompting it to take flight quickly when disturbed). From *elaio*, which is allied in characteristics with both *ielfetu* and *elpt*, come the Gaelic *eala*, and the Welsh *alarch*. It was from *elaio* that a Celtic source derived the poetical Latin word *olor*, which is now, somewhat confusingly, best known as the specific appellation of the Mute Swan *Cygnus olor* (Potter & Sergeant 1973), which is just another way of showing that such swans are related.

As already mentioned, modern Icelandic has retained the use of the Old Norse *elpt* for the Whooper Swan in the form *álft*, but uses the alternative Icelandic word for swan *svanur* for other species. *Svanur,* apparently equally old in both Indo-European languages as a whole and in Icelandic, refers to sound. In Icelandic, it is associated with more affected and written language (it was used for example in medieval tales and 19th-century and early-20th-century literature, and even sounds foreign to some ears. When Finnur Gudmundsson stabilized official Icelandic bird names in the 1960s, he retained *álft*, but used *svanur* for the genus and as the suffix in *Dvergsvanur* (= dwarf swan, for Bewick's Swan) and Hnúðsvanur (= knob swan, for Mute Swan) (Gardarsson *in litt.*), neither of which breed in Iceland, indeed the former was first recorded there only in 1978 (Brazil 1980).

Thus, from a few ancient sources, we now have a host of modern names for just one species—the Whooper Swan. Sandberg (1992) gathered names from 15 of Europe's languages. Among some of them linguistic links and affinities are clear, thus in the Germanic languages: *Sangsvane* (Norwegian and Danish), *Sångsvan* (Swedish) and *Singschwan* (German); while in the Romance tongues it is the *Cygne chanseur* or *Cygne sauvage* in French, *Cigno selvatico* in Italian, *Cisne cantor* in Spanish and *Cisne bravo* in Portuguese. In English it is currently the Whooper Swan, but it was previously also know as the Wild Swan, and nearby, in the Netherlands, it is still the *Wilde Zwaan*.

Languages descendent from the Slavonic branch of the northwest European group have a very distinctive form for the Whooper Swan, but one that is also derived from the same source by the early translocation of the letter l between Slavic and western Indo-European languages. Thus, in modern Polish we have *Labedz krzykliwy*, in Czech *Labut zpevna* or *Labut spevava*, in Serbo-Croatian *Labud*, in Bulgarian *Lebed* and in Russian *Lebed klikun*. Of their origins and derivations, however, I have failed to find explanation. Moreover, in Romanian (a modern Romance language) it has retained the name *Lebada*, perhaps from a time before that country's conquest by Rome. Among Ural-Altai languages, in Hungarian it is *Enekes hattyú*, in Finnish *Laulujoutsen* and related Estonian *Laululuik*, while in Lithuanian it is *Gulbé giesmininke* and Latvian *Ziemelu gulbis,* names that diverge more radically from other modern forms. In Manx it is *Olla chiaullee*. In the Celtic tongues it is *Eala* in both Irish and Gaelic, *Alarch* in Welsh, *Alar'ch* in Breton, and *Elerch* in Cornish (Potter & Sergeant 1973; Lipsberg 1983; Kear 1990; Leibak *et al.* 1994; Zhalakevichius 1995; Gorman 1996; Hagemeijer & Blair 1997). In Asia it is *Gangar Khun* in Mongolian, *Da Tian'E* (meaning Great Swan or Greater Heaven Goose) in Chinese (in Mandarin at least), *Huang zui Tian'E* (Yellow-billed Swan or Yellow-billed Heaven Goose) in Taiwan, *Keungoni* in Korean, and in Japanese it is *O-hakucho* (the great white bird) (Brazil 1991; Bold & Fomin 1996; MacKinnon & Phillipps 2000; Lee *et al.* 2000). In the Chinese, and perhaps in the Japanese name too, implicit in the reference to large size is an association with heaven, thus in these cultures the name can equally be considered to mean 'heavenly bird' (Ma & Cai 2000).

The sonorous flight-call of the Whooper Swan is one of the most far carrying of any bird. Its bugling notes serve during both day and night, foul weather and fair, to keep families and flocks together throughout their long migrations. They are also highly vocal on their wintering grounds, a fact that cannot have been ignored by early observers. Many of the vernacular names used for Whoopers are simply descriptive of the fact that it is beautifully and hauntingly vocal or of its size and demeanour. In English, Whooper Swan is clearly onomatopoetic, derived from the similarity of its loud, deep *whoop whoop* or *whoop-a, ahng-ha* call to the Middle English 'whope' or 'whowpe', the whooping cry given by falconers and huntsmen at a kill. Thus it generally became the Whooper Swan—the swan that whoops or shouts (Potter & Sergeant 1973).

Previously known as the Hooper (e.g. Coward 1920) and even the whistling swan, wild swan, whopper, elerch and elk(e), the name Whooper appears to have been in use by 1566, although the alternatives were evident until the early 1800s (OED 1971). Despite the early usage of the name Whooper Swan, two centuries later the name issue became more confused when, in 1768, Pennant chose the Dutch-like Wild Swan as his standard name. Just under two decades later, in 1785, he altered this to Whistling Swan, a name which Coward (1920) believed, reasonably I feel, was derived from the sound of its wings and not from its call. Thereafter, until Yarrell chose Hooper in 1843, all three names, Hooper (and its variant Whooper), Wild Swan and Whistling Swan, were widespread (Martin 1993). Since Yarrell, however, the change to Whooper in English appears to have stuck, though when reviewing the historical record one must wonder for how long!

BIRD OF MYTH, LEGEND AND SYMBOLISM

Swans have provided inspiration for countless generations of artists, musicians and poets, even choreographers, and little wonder for archaeologists have discovered murals of swans dating to before the Stone Age (Bradshaw 1986). Among the more famous tales from Greek mythology are those concerning swans, the most popular being that of 'Leda and the Swan', which relates how the nymph Leda, on the eve of her wedding to King Tyndareus of Sparta, was visited by the great god Zeus disguised as a swan (Fig. 2.1). Seduced by Zeus, Leda bore just two eggs, but each contained twins! One pair was said to be the godchildren of divine Zeus, the Dioscuri, Castor, and Pollux, and the other the human children of mortal Tyndareus, one of whom was Clytemnestra, the other the famed Helen of Troy. Zeus' miraculous seduction of Leda in swan form was immortalised in the heavens by the creation of the largest symbol to male ardour known—the constellation Cygnus.

Mythical imagery of the archetypal swan has permeated the minds not merely of musicians and poets. The fanciful, serene and dramatic images created affect all who are exposed to them, and, repeated in variation upon variation, paintings of Leda and Zeus in swan-form hang in the world's top galleries, the most famous representation being the sexually aggressive sculpture of Michelangelo (1475–1564) depicting the most bizarre of all hybridisations. The poet W. B. Yeats (1865–1939) describes that human–swan mating that Michelangelo painted, in his own poem entitled *Leda and the Swan*.

> 'A sudden blow, the great wings beating still
> Above the staggering girl, her thighs caressed
> By the dark webs, her nape caught in his bill,
> He holds her helpless, breast upon his breast.'

Leda and the Swan

Not only in galleries, but also in concert halls and theatres are swans commemorated. In performances of Wagner's *Lohengrin* and *Parsifal*, the swan-knight will live on forever, while Tchaikovsky has perhaps done more than anyone in recent centuries to immortalise the swan through his world famous ballet *Swan Lake*.

The physical form of swans has been associated not merely with beauty, inspiration, and virginal purity and sexuality, but, as mentioned before, also with death. This surprisingly unattractive and unfortunate symbolic link between beauty and mortality probably arose among the early myths of Neolithic Europe (Bradshaw 1986). These myths apparently featured a divine hierarchy dominated by a powerful mother goddess, wherein the king was merely her temporary consort, being sacrificed at the year-end to be replaced by another. The passing of the solar year was measured then using a turning wheel, the wheel of Nemesis. Nemesis, the daughter of Nyx or night, was the personification of the fate that metes out happiness to the just and punishment to the unjust, relentlessly pursuing the sinner. At the midway point, when the wheel of fate had turned halfway and the solar year was half gone, the hapless king was acknowledged as due 'to meet his Nemesis', that is to die. How the gods were cheated! For in practice, the king, not satisfied with his allotted temporary role, avoided divine fate by having a boy serve in his place for that day and thus be sacrificed in his stead! The soul of the ritually sacrificed king was carried away by migratory swans, when they flew north each summer to their unknown breeding grounds. For their pious role they were considered sacred. To the imaginative mind, the haunting calls of the wild swans, as they disappeared north, were cries of mourning for the Mother Goddess' consort (Evans & Dawnay 1972). For them to have been so visibly migratory and so vocal, there is no doubt they were 'northern swans', giving me some justification in laying claim to these early myths as relating to Whoopers.

This early myth, linking sacred swans to the death of the Mother Goddess's consort was embellished in later eras, when it was generally believed that the soul of the dead king was conveyed to paradise by Apollo, the Greek sun god. Apollo was the son of Zeus and Leto/Leda, and a twin with Artemis. Was this then the origin of the intricate link between the swan and the sun or merely an example of that myth? Apollo travelled in a swan-drawn chariot to and from the land of the Hyperborians, the paradise that was believed to exist 'at the back of the north wind', where men lived for a thousand years. There, in everlasting spring, the Hyperborians dwelt under Apollo's tutelage; he was like a father to them and he retreated there each year when the sun withdrew in winter. Swans were not merely Apollo's aerial steeds, but are also credited with having been present at his birth, and thus shared his gift of prophecy. And, coincidentally, in Slavic mythical traditions swans formed the escort for the sun god (Evans & Dawnay 1972; Boreiko & Grishchenko 1999).

Entwinements of symbolism begin to become as tortuous as the stems of a climbing wisteria, with the beauty of swans honoured in their association with gods, death and the sun at one and the same time. There are further links too, for Apollo was also god of music, and thus his association with swans that sang as they disappeared north each year is made even more powerful. This link with the migratory habits of these wild birds is further strengthened by the belief that Apollo's visits to the Hyperborians were never for less than three months, and thus akin to the breeding period of the swans. Did such ancient myths among the Greeks really relate to migratory wild swans? Swans are not common in Greece today, but perhaps in the past they were. Could such myths date back much further, closer to the end of the last ice age in Europe when the ranges of northern species must have extended much further south?

The mournful sounds of the wild swans as they disappeared carrying off the dead king's soul led inevitably to their association with death and perhaps also helped to engender the peculiar belief that they only sing just before they themselves die. Thus the classical Latin proverb: 'The swan sings before death'.

Yet for the most part it was believed that swans were silent or poorly voiced, and perhaps as much for that reason as any other, the swan reared in semi-domestic form and which became most familiar, became known in English as Mute. Only at the end of life, at the moment of direst stress, were swans able somehow to force a sound. There are many references to the swan in this form, either as a bird of poor voice, or as a moribund songster:

So Virgil (70–19 BC), in the *Aeneid* (XI 458) described:

'Dant sonitum rauci per stagna loquacia cycni'

('The hoarse-voiced swans are calling where the pools are noisy'; Potter & Sargent 1973). Much later, the English poet Samuel Taylor Coleridge (1772–1834) was to write, with no great originality, in his *Epigram on a Volunteer Singer*, 'Swans sing before they die' (perhaps taken directly from the Latin tag) though his conclusion was perhaps more amusing than most:

'Swans sing before they die—'twere no bad thing
Did certain persons die before they sing.'

Another theory regarding the supposed silence of the swans has been expounded; where a drab, unadorned nightingale's beauty lies in its voice, the brilliant peacock's beauty lies in its tail, while its call is a horrible scream. Similarly, the swan may not look *and* sound beautiful, thus its first song must be its last (Bradshaw 1986).

There are many references to swans singing their death songs, and this myth was so powerful among the ancient Greeks that Plato (428–348 BC) thought it worthy of being put into the mouth of Socrates (*c.*470–399 BC) on the day of his death (*Phaedo* 85). The legend necessarily enshrines both fancy and fact, and may have begun with the inspiring sight of these migratory birds gleaming high overhead in the sunlight. In this way the migratory Whooper Swan became, for the ancients of Europe, the bird of Phoebus Apollo. In its fall to earth, as its life ebbed away, the connection between the swan and the Lord of the Sky, the Inventor of the Lyre, was revealed by its utterance of sweet music as it died. Prosaically, it is the Whooper that is believed to have given rise to the 'swan-song' legend, the music is the imaginative equivalent of the death-rattle, occasioned by the last prolonged exhalation through its long trachea producing a series of musical notes (Potter & Sargent 1973; Martin 1993). Martin (1993) considered, reasonably, that it is the Whoopers extra loop of the trachea within the sternum that makes its voice deeper, more powerful and farther carrying than either Bewick's or Mute Swan. Yet, whereas in general the concept of the swan-song contains the sadness of ending, in Aesops' fables (6th century BC), the dying swan was considered to be rejoicing at finally being delivered from the tribulations of life,

Plate A.   Bewick's Swan (top left), Whistling Swan (top right), Whooper Swan (bottom left), and Trumpeter Swan (bottom right) (juveniles behind adults).

Plate B.   Coscoroba Swan (top left), Black-necked Swan (top right), Mute Swan (bottom left), and Black Swan (bottom right) (juveniles behind adults).

Plates

Plate 1.   The longer neck of the male is recognisable when pairs are together.

Plate 2.   Poised, but alert.

Plate 3.   Freshwater lakes are a major wintering habitat.

Plate 4.   Wintering in a caldera lake, Japan.

Plate 5.   An orange-stained aggressive male, his mate behind.

Plate 6.   The author and a strongly defensive nesting pair, Iceland.

Plate 7.   A watchful resting Whooper Swan reveals its typically blackish-brown iris and yellow eye-ring.

Plate 8.   A small proportion of birds have pale blue and grey eyes.

Plate 9.   Unusually extensive black on this bird's bill renders it individually recognisable.

Plate 10.   A first winter bird with traces of pink on its yellowing bill.

Plate 11.   Coming in to land.

Plate 12.   Spread feet serve as airbrakes.

Plate 13.   Social interactions are vociferous affairs though rarely violent.

Plate 14.   A full-scale grappling fight.

the hunter's snares and guns, and from hunger, and thus its song was of release from travail.

Human endeavour in the literary arts has been associated with swans since at least Pythagoras' day (582–500 BC) when it was believed that, on death, skilled poets became swans. So Virgil became known as the 'Mantuan Swan'; and, much later, William Shakespeare 'The Swan of Avon'. Poets, bards, storytellers, all have served to spread tales of swans far beyond the ranges even of the swans themselves, and it is fitting then that they themselves should be dubbed swans. Tales have crossed innumerable cultural and international boundaries changing in the telling, perhaps even shifting between species. Mythologies of the Mediterranean should perhaps be associated with the more southerly Mute Swan (although did those tales really arise there? And if they did, did they do so in a bygone age with a different climate and very different natural history?). Those myths of Iceland, Scandinavia and much of northern Russia, however, were doubtless inspired by the grace of the most wide-ranging swan of all—the Whooper.

The ancient poetic expression of many an accurate observation shames the abilities of most modern writer/biologists. Thus Homer (8th or 9th century BC) wrote in his *Iliad*:

> 'Not less their number than the embodied cranes,
> Or milk-white swans in Asius' watery plains.
> That, o'er the windings of Cayster's springs,
> Stretch their long necks, and clap their rustling wings,
> Now tower aloft, and course in airy rounds,
> Now light with noise; with noise the field resounds.'
> (Translated by Alexander Pope).

Aristotle (384–322 BC), the founder of natural science, in his *Historia Animalium*, also revealed the perspicacious quality of observation, albeit mixed with a little fanciful imagination, in noting the social and migratory behaviour of swans. In much the same way that modern biologists pose theories based on the knowledge available to them, so Aristotle's thoughts, though they may seem quaint and ignorant to us, did at the time, represent the dizzy heights at the forefront of swan biology.

'Swans are web-footed, and live near pools and marshes; they find their food with ease, are good-tempered, are fond of their young, and live to a green old age. If the eagle attacks them they will repel the attack and get the better of their assailant, but they are never the first to attack. They are musical, and sing chiefly at the approach of death; at this time they fly out to sea, and men, when sailing past the coast of Libya, have fallen in with many of them out at sea singing in mournful strains, and have actually seen some of them dying.'
(Translated by D'Arcy Wentworth Thompson)

No doubt some great depth of understanding lies behind simple proverbs. However, the meaning of the 16th-century proverb, 'All his geese are swans', remained opaque to me until I learned that in modern Icelandic there is the saying 'Everyone thinks his own bird is beautiful—although it is both dirty and skinny', referring to the common human self-deception in which we may represent our own belongings or attributes as better than they are.

Much later, poets seem to have relied less upon direct observation of the bird and more on pre-existing mythology and on their own literary imaginings, thus Alfred, Lord Tennyson (1809–1892), English poet laureate, penned the line: 'The wild swan's death-hymn took the soul. . .' and in *Tithonus, I*, he wrote:

> 'The woods decay, the woods decay and fall,
> The vapours weep their burthen to the ground,
> Man comes and tills the field and lies beneath,
> And after many a summer dies the swan.'

The English poet James Elroy Flecker (1884–1915), in The Old Ships, seemed at least to have admired the swan's grace on the water in person, though I feel that he was most likely referring to the raised-winged Mute Swan:

> 'I have seen old ships sail like swans asleep'

Ben Johnson (1572–1637), another English poet laureate, in his *To the Memory of Shakespeare* wrote of the 'Sweet Swan of Avon!' His work, and that of various other poets, is ambiguous. There seems today an obvious link with the Mute Swan, long established on the River Avon which runs through Stratford, yet the tradition of referring to a bard as a swan is much older, and perhaps reached British shores from much further afield. Other poets more likely referred to the wild swans of the north, to Whooper and Bewick's, but which? They were only distinguished by ornithologists in the early 1800s (Yarrell 1830), and how soon did poets become aware of such a discovery?

William Wordsworth (1770–1850), also an English poet laureate, referred to swans in his *Yarrow Unvisited*.

> 'The swans on still St. Mary's Lake
> Float double, swan and shadow.'

But were they Whoopers? The swan was so powerful an image that many of the great English poets used its symbolism in their works. Many, no doubt, referred merely to the symbolic, indeterminate swan using it as a device; others were inspired by the familiar feral Mute Swan, but perhaps not all.

Other versifiers, William Butler Yeats in particular, referred to the haunting vision of the wild Whooper Swans returning each winter to a favourite place of the poet, and gave a sense of one of the eternal seasonal patterns of nature in his *The*

Wild Swans at Coole (see Yeats 1994 and the poetic dedication of this book on page 5).

More symbolically inclined poets knew their birds esoterically, thus in the now unfamiliar Beth-Luis-Nion order of familiar vowels (A, O, U, E, I) wherein the order is represented by both a numerical and a seasonal sequence, in Bardic lore, a bird stood for each of the seasonal references. Thus E, the day of the autumn equinox, was represented by Ela (the Whistling = Whooper Swan, and was close to the Gaelic name) and by Erc (rufous-red), because at this season, the migrant Whoopers were preparing to leave their northern breeding grounds, and because rufous-red is the colour of autumnal foliage and the swan's iron-stained neck feathers (Graves 1961). Coincidentally, there is an astronomical link too as a number of the stars in the region of the constellation Cygnus are orange or red.

Swan tales are innumerable and, not surprisingly, abundant among those peoples familiar with the birds. In Finland, for example, a country that had several different tribal cultures, the Whooper was viewed differently in different regions: in western Finland it was considered a gamebird, while in the east it was sacred. Despite their sacred status, in a pattern repeated over many parts of the world for numerous species, the spread of firearms led to their steady extermination, until by the 1950s just a few pairs remained in northern Finland. Popular endeavour was its saviour through another shift in technology and of attitudes. Yrjö Kokko, a vet in Lapland, dreamed of shooting a nesting Whooper Swan, but with a camera, not a gun. His book, *Laulujoutsen* published in 1950, documenting the life of the Whooper Swan was the culmination of his experiences, and made modern Finns aware of the bird. This made its protection possible, and the population has increased a hundredfold in only 40 years. From being revered as sacred to a target of hunters and once more protected, the fortunes of the Whooper have swung back and forth. Once associated with swimming on the 'River of Death', it was brought back from the 'death' of near extinction in Finland by people. It has now taken on a new symbolism, no longer considered sacred, it is however still associated with cleanliness, purity and naturalness, and is commonly used in Finland in that modern form of mythology—advertising. In Nordic countries a swan label indicates to consumers that a particular product is environmentally sound (Finland.org 2000).

The enormous subject of swan tales, mythology and symbolism is so well addressed by Armstrong (1958), Graves (1961), Evans & Dawnay (1972), Wilmore (1974), Boreiko & Grishchenko (1999) and others that I have only lightly touched on it here. Those interested in the subject would do well to refer to those authors, who have distilled many of these tales, which, to a large extent, are similar, linked or even repetitive. In some cases they may refer to swans other than Whooper, although the Russian, Scandinavian and Irish tales in particular, are more than likely to have referred to Whoopers specifically.

Though the details vary, there are, no doubt, as many forms of the swan-maiden tale as countries to which they have travelled and centuries through which they

The Swan Maiden story crosses cultural boundaries

have survived. The central theme is relatively constant, concerning the metamorphosis of God-like innocence to human frailty, from supernatural swan[5] to human form. The transformed heroine, half mortal, half supernatural, is safe in her human guise and becomes the willing, gifted wife of the human who finds or saves her as long as she alone remains in possession of her secret or of her hidden feathered skin. The twist lies in her fate, for she is immediately transformed into a swan again when her secret is exposed or some particular taboo is broken, and she must then depart her husband forever.

Swans figure powerfully in Irish legend, and there in the 'swan-abounding-land', it is the Whooper that is abundant and no doubt progenitor of taboos and beliefs that made it sinful to injure or kill a swan. Interesting parallels exist between Irish and Nordic myths and those of Greece, even though the swans to which they relate are presumably different, Whooper in the former case, and Mute in the latter. If, however, the myths date back several millennia to a time *before* the Celts moved west then they may relate to the same birds after all. Thus, in both Irish and Greek literature the swan is a metaphor for maiden because of the swan's inherent purity, and in Ireland swans were believed to harbour the souls of dying virgins (Wilmore 1974).

The swan is linked not merely to virginal purity, due to its whiteness, but also to sexuality, because of its physical shape, and as a female symbol through its V-formation flights. Swans were linked with gods, being sacred to Venus (Aphrodite, goddess of beauty), drawing her chariot, and being associated particularly with Leto or Leda, mother of Apollo, and victim of the infamous seduction by Zeus, and also sacred to Apollo drawing his chariot across the sky.

The swan was also an erotic symbol of opposites. In its whiteness and purity it represented the nakedness and chastity of womanhood. In its rounded body it represented the receptive feminine form, while at the same time, in the phallic length of its enormous neck[6] it also very potently encompasses the erotic image of the productive male, emphatically encompassed in the image of Zeus (Evans & Dawnay 1972; Scott & The Wildfowl Trust 1972).

Swan myths and legends of Greece, probably themselves originating in the north where other swans abounded, also spread throughout Europe, with golden crowns and chains, and human/swan transformations linking tales such as those of the Children of Lir with those of the swan knight, and the innumerable variants of the swan-maiden theme, perhaps the best known of which is the Hans Christian Andersen version. The tale of Lir's Children[7] is of interest from various perspectives. Like so many tales it carries echoes of others, evidence of its long transmission both geographically and temporally, in this case of the tale of Zeus and Leda, in as much as Lir's wife, Eve[8], also bore two sets of twins, a daughter (Finola) and son (Aed), and two sons (Fiacra and Conn). Unfortunately Eve died, and Lir married her sister Eva. His first marriage was fated with sadness and his second with jealousy, for Eva resented the love universally held for her sister's children, so she cursed them to be swans for 900 years, 300 on the Waters of Moyle, 300 on Lake Derravaragh and 300 on Inish Glora on the western sea, until a Christian bell was heard bringing the light of new faith to the land or a Prince of the North married a Princess of the South. Jealous Eva finally repented but could not repeal her spell and thus release her sister's children from their transformation. She was, however, able to ease their lot, by giving them the powers of human speech and making sweet music that cast a spell on all who heard it. The enchantment finally ran its course and having found sanctuary with a Christian monk, and linked with silver chains, they were restored, ancient and decrepit, living only long enough to be baptised. Seemingly an ancient myth had gained approval by the church (Armstrong 1958).

In Celtic mythology and in the folklore of many other regions, a binding chain motif often appears in association with a swan indicating that it is a creature transformed, a fairy or a god in disguise, often, like Zeus, in amorous pursuit. The joining or harnessing of swans with gold or silver chains is so widespread a motif that perhaps it has its origins in some belief or ritual. A Tibetan story tells of a more pragmatic use for such a chain, the catching of a swan maiden by a hunter (Armstrong 1958). Did the tellers of such tales simply pass on a compelling oral tradition or were they also aware of the inspiring creatures? Were the tellers of the Children of Lir, in which King Lir's children are turned into swans and condemned to swim between the Irish and Scottish coasts for 900 years, aware of the annual winter movements of Whoopers between these two regions? The originator certainly gave us an imaginative transformation, not merely from humans into swans, but from the silent swans of previous myths to full-voiced birds capable of 'making sweet plaintive music' to soothe the minds and hearts of those willing to listen.

Though untransformed, in Indian folklore wild swans were already voiced. In fact they were able to speak and so performed the service of carrying messages between Princess Damayanti and her suitor, Rajah Nala. Swans as messengers are not new in folklore, but it is interesting to speculate how such a tale should exist in India, where swans are very rare vagrants.

In the West there is a romantic attachment to swans, with dozens of myths and folktales surrounding them. Perhaps the best known is Hans Christian Andersen's Eleven Wild Swans. Many involve humans turning into swans or swan maidens taking human form, but in Japan similar tales are applied to the Japanese Crane (Brazil 1990). Swans appear infrequently in classical Japanese art and literature — it is the crane that holds sway — thus the situation at Kominato, in northernmost Honshu, is unusual. There *O-hakucho*, the great white bird, is revered as a messenger of Raiden, god of thunder and lightning. It was believed that harming birds would earn divine punishment; so sick swans were nursed to health at the local shrine, where hunting was prohibited as early as 1896. In 1920, the area was designated a sanctuary and in 1931 became a natural monument (Brazil 1991). Today thousands of people visit Kominato to feed the wintering swans from eastern Russia.

Of the three species of swans in China, it is the Whooper that is particularly famous, because of its elegance, size and pure-white plumage. It is considered symbolic of angelic beauty, nobility, purity, good luck, and representative of honesty, loyalty, faithfulness and courage (Li 1996; Ma & Cai 2000). Documented allegorical references to virtuous swans date back more than 2,300 years in both Europe and China, where Liu Bong (*c.*202 BC), the first emperor of the Han dynasty, even wrote a 'Song of the Swans' (Ma & Cai 2000), but ancient Chinese authors were perspicacious observers of migrating swans too, thus Guan Tze (d. 645 BC) observed that 'Wild swans fly to the north in Spring, to the south in the Autumn and they never miss their timing' (Ma & Cai 2000). We should not be surprised therefore at a very much more ancient association between people and swans. Furthermore, because the outline of the northeastern province of Heilongjiang is considered to resemble a standing Whooper Swan, it has (since 1985) been designated the provincial emblem, confirming that despite the passing of millennia the universal appeal of wild swans is retained and that the appeal is trans-cultural.

In Japan, swans are not so anciently recorded, though they do appear as early as 712 AD in the *Kojiki* and in 720 AD in the *Nihonshoki,* among the earliest Japanese literature. Heavenly swans flying overhead were credited in the *Nihonshoki* as having helped the mute son of Emperor Suinin to speak and so began the tradition of the imperial court rearing swans. To this day, swans grace the moat surrounding the Imperial Palace in Tokyo. From the indigenous Ainu in the north of Japan (where Whoopers are common on migration and in winter) to the Okinawans in the south (where any swans are accidental; Brazil 1991) there are stories revering swans. Their qualities were seen as including courage and strength, and people offered prayers to them in the hope of emulating their virtues. Remnants of that reverence are retained at a handful of 'Shiratori' (literally white bird) shrines in Tokyo, Nagoya, and Miyagi Prefecture where altars honour swans as protective deities. Birds depicted at 'White Bird' shrines in China are clearly Whoopers (Ma & Cai 2000), and one cannot help wondering whether the reverence arose independently in Japan or whether it was carried there, as has been the case with so many elements of supposed Japanese culture. As in Iceland, Britain and Russia, in Japan place names and people's names reflect an ancient link with these birds; Shiratori is not an uncommon surname for example, just as Lebed is in Russian.

Not all swan 'mythologies' are ancient, and not all depend on the time-honoured oral tradition. In Japan, where the Whooper is known as the 'Angel of Winter', it has achieved a tremendous following through different media, and it has a degree of popular appeal unknown elsewhere in its range. Its living beauty is captured on what must be hundreds of km of 35 mm film each winter (Fig. 2.1), and is nowhere more effectively embodied than by the photographs of Teiji Saga, now in his mid 80s, the doyen of Japanese swan photographers, who for decades has travelled throughout Japan in search of the ultimate image (Brazil 1984c).

At an increasing number of localities, visitors gather to feed and photograph clamouring hordes of Whooper Swans that have learned that man at close range brings no harm. This was not, however, always the case. Hunting of swans was once as widespread in Japan as that of geese and other waterfowl (Brazil 1991) and during and immediately after the Second World War, when protein was in seriously short supply, waterfowl were viewed as a vital resource (Austin 1949).

Figure 2.1. Ice, snow and dawn mist are vital ingredients when photographing The 'Angels of Winter.'

In January 1950 a few timid Whooper Swans took refuge at a small reservoir in Niigata Prefecture. Then obscure, Hyoko is now internationally renowned. Against a backdrop of snow-covered Japanese Alps, the swans found a resting place and an ally. Juzaburo Yoshikawa was smitten by the magic of the swan flock. Not only did he succeed in persuading the mayor and local hunters into declaring obscure Hyoko a swan sanctuary, he also began feeding them and protecting them from dogs and disturbance. By 1954, the swans at the lake were declared a natural monument and, thanks to the dedication of one man, they increased to over 1,000 birds.

By the time of my first visit in 1980, Juzaburo's role as 'swan keeper' had been passed to his son Shigeo (himself now retired), who dressed in the same familiar attire as his father so as not to disturb the established tradition of feeding. Protection continued and had already spawned other newer traditions—those of visitors feeding and photographing the swans. Film and television have made the Whooper Swan a familiar image to the Japanese people, and their arrival each autumn is announced in the media, like the first cuckoo in spring in England.

Another tradition began in Hokkaido as recently as 1988. In that year Mr Yoshinao Yoshida was inspired to entice swans to visit Onuma, a large shallow lagoon between the twin capes of northernmost Hokkaido by calling them. His passion for swans, his use of broadcast calls and the subsequent provision of food, has transformed Onuma into a significant stopover point for many thousands of Bewick's and lesser numbers of Whooper Swans migrating between Japan and Sakhalin, which is just visible hazily from Hokkaido on the northern horizon.

Swan legends are interlinked throughout Eurasia, and some elements at least are ancient. Frequently depicted in Bronze and Iron Age art, it seems that the earliest beliefs were modified by successive waves of cultural transmission and invasion. There is good reason to assume that the swan-maiden myth travelled part way with the megalith builders of Europe, but that its origins can be dated to at least the early Bronze Age (Armstrong 1958).

When the plain metal iron, rather than the DVD, represented the height of modern technology, there were already ancient links between supernatural swans, gold, fairy peoples, and the warding strength of iron against the older generation of magicians and craftsmen, with links even to Vulcan, and that great smith Wayland, and to the swan maiden. These fascinating mythical tales are so convoluted, the elements so intertwined between cultures, that they stretch the imagination almost as much as the biological aspects of a species the range of which spans the Old World, that visits the New, and migrates between temperate and subarctic zones (Fig. 2.2).

Figure 2.2.   Cultural transmission in stone, begun in the Stone Age continues today: Odaito, Japan.

CHAPTER 3

The Whooper Swan, a closer look

'The Whooper is a ten times handsomer bird than a tame swan in the eyes of an ornithologist, but is not really so graceful—its neck is shorter, and its scapulars not so plume-like. Instead of sailing about with its long neck curved in the shape of the letter S, bent back almost to the fluffed-up scapulars, the Whooper seemed intent on feeding with its head and neck under water.'

W. H. Hudson (1926) quoting Henry Seebohm.

DISTRIBUTION

The Whooper's natural range ranks as the largest among the seven swans. Measured on the grand scale, its extensive breeding range spans almost the entire boreal or taiga zone of the Old World, from the Atlantic to the Pacific. Only seldom found beyond the northern treeline, where the taiga gives way to the tundra, its broad, belt-like range barely overlaps the southern limit of the narrow-ribbon-like breeding range of Bewick's Swan. Most Whooper pairs from Scandinavia to eastern Siberia breed at wetlands within regions naturally forested with conifers, with smaller numbers in the birch-forest zone. Typically they occupy breeding pools and lakes amid forest, as well as wetter bogs, marshes and fens, as is the case across Norway, Sweden, Finland and much of Russia. In Iceland, where there has been much work on the breeding biology since the late 1970s (Brazil 1981b,c; Rees *et al.* 1991a; Einarsson 1996), the habitat is somewhat atypical as Whooper Swans occur there in a virtually treeless[1] boreal environment and, where they breed in the highlands, they do so in an arctic–alpine tundra-like environment, but one dominated by an oceanic climate. Though atypical in this sense (differing markedly from continental boreal and tundra regions), the Icelandic breeding population is the focus of considerable attention because most Icelandic birds winter in the British Isles[2].

In the British Isles small numbers summer in Ireland, Scotland and England and have been doing so with greater regularity during the last decades of the 20th century, particularly in Scotland. Breeding was once regular in the Orkneys, where it was exterminated, but in recent decades isolated breeding attempts have been mainly in central and western Scotland. An increasing number of records of summering or even breeding outside the normal range are available from other countries over the same timescale, with breeding having occurred recently in Germany, Poland, Lithuania and Latvia.

The wintering range is even more extensive. Whoopers migrate to more southerly regions, fanning out to locate suitable wetlands in regions of relatively mild winters. In the west they occur in Iceland and southern Scandinavia, the British Isles and across northwestern Europe from the Netherlands to Denmark, Germany and Poland, and occasionally, in cold winters, south through France to Iberia and Italy, and even to Mediterranean North Africa. In its central range it winters south to the Black and Caspian Seas, while in the east it winters in eastern China, on the Korean Peninsula and in the northern half of the Japanese archipelago. The species' range and population size are described and discussed more fully in Chapters Four and Five.

DESCRIPTION

PLUMAGE AND POSTURE

The Whooper shares with all three of the other 'northern swans' (Trumpeter, Whistling and Bewick's) entirely snow-white plumage, with no apparent plumage differences between males and females or variation between seasons (Fig. 3.1; Plate 1). Some acquire, to varying extents, a rusty-orange tinge to the head, neck, chest and belly. In some this iron staining can be deep and strong, in others merely a hint of colour.

Newly hatched cygnets are covered with a pale grey down, somewhat darker on the head, with the cheeks, throat, foreneck, upper neck and lower back paler, and

Figure 3.1.   Poise and grace are among the attributes of the pure white Whooper Swan.

two white spots (disappearing with age) between the scapulars reported in Russia (Dement'ev & Gladkov 1967). Underparts are whitish. The bill base is feathered. Young commonly seen on their wintering grounds are greyer in their cygnet plumage than Mute Swan, being ash grey or grey-brown throughout the autumn (Fig. 3.2; Plates A & 10). They become steadily paler throughout their first winter and spring and typically by April they are nearly all white, and by the following autumn indistinguishable from adults.

This large-bodied waterbird is graceful on the water, with the front of the body slightly lower than the rear, its neck held rather stiffly erect, its head pointing forward and the bill horizontally. On water it holds its long, pointed wings close to its sides, the primaries mostly obscuring the tail, which is short and rounded. Its exposed back is convex in profile (Plate 1).

Figure 3.2.   First winter birds are ash grey with a white-based bill.

In addition to obvious differences in bill shape and pattern, with Whooper lacking the knob of Mute Swan, Whooper Swan also differs from Mute in general carriage. It does not arch its wings in a fan over its back, as does Mute when aggressive. Instead, it partially spreads its wings, holding them out from the body and moving them vigorously at the carpal. When aggressive it lowers its head for attack or in a ground-staring display and may even submerge its head at such times when on water (Witherby *et al.* 1940). Nor does its neck form the graceful S-shaped curves of Mute Swan; instead it carries the neck in a stiffer, straighter manner (Witherby *et al.* 1940) or kinked back over the body and then straight (Cramp & Simmons 1977), as does Trumpeter Swan, and where the Mute might be described as regal, the overall demeanour of a swimming Whooper is imperious (Fig. 3.3).

On water and the wing the Whooper Swan is an embodiment of grace. Though the transition between the two is somewhat laborious it is less so than the take-off from land. Take off from water requires a rapid swim, rising to a laboured, pattering take-off run (during which 'it flogs the water'; Coward 1920), as it beats its wings, which frequently strike the water, while its feet thrust against the water until it has acquired sufficient speed for flight, rising with the neck outstretched and the wings beating powerfully, though slowly. The Whooper's powerful legs are just sufficiently long to enable it to walk strongly on solid ground with a high stepping action, albeit with a marked sideways sway, though nowhere near the pronounced awkward waddle of Mute Swan. When walking, it holds its neck erect or with the lower neck arched back and the upper neck straight. However, its legs are not sufficiently long to afford the bird an easy take-off from land, which is achieved with a short run, the head and neck thrust forward and the

Figure 3.3.   The profile and imperious demeanour of the Whooper Swan are distinctive.

wings beating down as far as possible, to achieve lift (Fig. 3.4). This need for a clear runway for take-off limits places where they land. Once airborne the Whooper flies powerfully with regular wingbeats. Although flocks rise en masse, they soon settle into a pattern and if their movement is more than a few hundred metres or if they are on migration, families and flocks form wavering lines at varying altitudes from barely treetop height to elevations at which they can easily pass over mountain ranges.

At close range, in flight, the Whooper's long straight-quilled wings produce an audible swishing or whistling, much like the wings of any other large bird, but considerably quieter than the 'singing metallic throb' of the arched quills of a Mute Swan, which may carry a considerable distance (Witherby *et al.* 1940; Cramp & Simmons 1977). Flight may be sustained over very long distances, though the Whooper Swan is not particularly agile in the air. Rather it appears sluggish, turning in wide arcs, climbing slowly, and even descending far less rapidly for example than do geese; finally using its feet as airbrakes (Fig. 3.5, Plates 11 & 12).

In flight the long neck is fully extended straight forwards, like other swans, geese and cranes, not hunched as in herons and egrets, and the head bobs gently up and down in time with the wingbeats. The large black webbed feet, though held tucked beneath the relatively short tail, are normally visible in flight. In especially cold weather the feet are more carefully tucked among the body's contour feathering for extra insulation and may be almost completely hidden. Family parties and flocks typically take flight together in quick succession following ritualised pre-flight calling and head bobbing. In low-altitude flight, which may also be over short distances, Whooper parties often appear disorganised, but at higher altitudes,

Figure 3.4.   Grace in flight is achieved only after a labouring run.

Figure 3.5.   Whooper Swans commonly circle their chosen landing site descending slowly.

indicative of longer flights, they typically assume long lines or V formations. The bird temporarily in the lead is generally assumed to generate slipstream, from which the others benefit. When watched carefully from directly below, head- or tail-on, younger individuals can be distinguished within flocks, especially in autumn, by their noticeably shorter wingspan.

Where the Mute Swan moves in a very ungainly fashion on land, the Whooper walks better, faster, in an almost goose-like fashion. The neck which appears longer and more slender than that of the sympatric Mute, may be carried stiffly erect on land, although the lower part is also often bent back some distance with the rest vertical, and it can also be gracefully curved, though less arched than the Mute (Witherby *et al.* 1940; Cramp & Simmons 1977).

Whooper Swan is readily identified among other waterfowl in its Palearctic range merely by its size and pure-white plumage. Where it occurs with other swans it can be distinguished on form alone. Mute Swan looks humpbacked, an appearance emphasised because it carries its wings rather loosely across the back, and holds its neck curved. In comparison, Whooper's carriage is generally more erect, it typically carries its head high on a straighter neck, especially when alert, though it is frequently held curved when feeding or in aggression (Soothill & Whitehead 1978; Martin 1993), and its apparently more streamlined body is lower, lying closer to the water's surface. As Coward (1920) aptly described it: 'From behind the wings show like two smooth cushions; the short pointed tail is carried horizontally.' The characteristic horizontal carriage of the tail is a useful and reliable means of separating Whooper from Mute, as the difference between the Whooper's horizontal tail and the Mute's tilted upwards at *c.*45° is discernible even over considerable distance (Kennedy *et al.* 1954). It is readily distinguished from the smaller Bewick's by the latter's smaller size, shorter neck, more rounded head and, at closer range, by its bill pattern.

On land and in its behaviour, the sociable Whooper Swan is the most goose-like of the swans. Bewick's is a smaller, more petite version of the Whooper, with a more rounded head and a shorter bill, which has a concave profile, in comparison with Whooper's longer straighter bill, which may measure in excess of 10 cm (Ogilvie & Pearson 1994). To the experienced eye, males and females can also be separated, but only by minor characteristics. Males, for example, tend to have longer, thinner necks than females, but this generally requires side-by-side comparison (Plate 1).

The Whooper's legs are short, stout, and they and the large, broadly webbed feet are black. The bill too is partially black, particularly the nail, the tip, the ridge and the sides, but the most distinctive feature is the elongated triangle of deep egg-yolk yellow, broadest at the base and tapering to a point on the sides towards the tip. The bill appears hard and inflexible, but in the hand this is belied by the soft skin covering the hard bony upper mandible and spanning the lower jaw, which sometimes bulges prominently if food is trapped beneath the somewhat spiny tongue.

Though Whoopers are generally pure white some northern breeding grounds, such as parts of Iceland, are rich in iron compounds that gradually stain the head, neck and belly a colour ranging from pale orange to rich rust (Fig. 3.6; cf Plates 2 & 5). The head, because of its frequent and repeated immersion and exposure, is typically darkest. Frere (1846) was perhaps the earliest to comment ornithologically on the deep, rusty-red staining on the common (Bewick's) swan, though he thought it resulted from feeding among weeds. However, in the much older poetical

Figure 3.6.   Orange or rusty staining of the head and neck is a common feature on the breeding grounds.

tradition, staining on the neck of Whooper Swans has long been known and was traditionally associated with the autumn equinox (Graves 1961). Very much less poetically, Höhn (1955) demonstrated that for Snow and Ross's Geese, and the 'northern' swans, the orange-red stain was the result of deposition of iron oxide on the feathers. Such staining occurs widely among waterfowl; it can be seen for example on Northern Pintail, but is obviously more conspicuous on white-plumaged birds feeding in wetlands rich in such minerals. The head, neck and chest feathers are moulted during late autumn and early winter, so that by mid-winter many Whooper Swans that were stained on the breeding grounds are pure white again, although in east Asia, particularly in parts of Honshu, Japan, some wintering sites must be iron-rich as some swan herds develop a deeper stain during the winter. Something similar has been noted from Ireland. Kennedy *et al.* (1954) reported that, on first arriving there, Whoopers show no or minimal staining, yet by December it has become noticeable and by March the red may be so intensive that 'many necks are virtually black and even the under-tail coverts become discoloured'.

The Whooper's all-white plumage is undistinguished in that it lacks adornment of any kind, although rare exceptions do occur. In Japan, Kakizawa (1977) observed a female with unusually elongated crown feathers that formed a noticeable crest.

The Whooper Swan is gregarious even in the breeding season when non-breeders may gather into sizeable moulting flocks. At other times it is particularly sociable and although commonly found in family parties, these typically associate forming larger flocks often of a 100 or more, and even 1,000 or more at certain favoured staging or wintering localities. Martin (1993) described Whooper Swans (with the exception of young migrants, which can be naive) as generally wary of man. While this may be true of some areas, especially in the west, elsewhere it becomes extremely approachable, though never tame, through the regular provision of food (Brazil 1991). Though gregarious, it is inevitably sometimes aggressive towards other flock members, particularly when feeding birds intrude too closely on a family. It is usually less aggressive to other species than is Mute Swan, though at sites where food is provided and different swans compete, Whoopers tend to displace Bewick's and even Mutes at times (Rees *et al.* 1997a), and is known to kill a range of other waterfowl (see Chapter Six). Martin (1993) noted it killing birds as large as crows in defence of its young.

The Whooper Swan's scientific, English, and most of its northwest European names are derived from its vocalisations, which range from soft, subdued whooping contact notes that can be heard all night from roosting flocks, to prolonged loud bugling displays between pairs, particularly on the communal wintering grounds. Flock members are also particularly vociferous when preparing for flight, as a kind of pre-flight signal (Brazil 1981c; Black 1988), when pairs or families join a flock, and calls are also commonly given while airborne. As Ogilvie & Pearson (1994) put it, given their far-carrying calls, Whooper Swans have little need for any further contact sounds, although in fact their wingbeats are audible, and this too may serve a purpose.

Though flight is vital for their long-term survival, flightless injured Whooper Swans have summered in many locations from England to Japan. In the British Isles pairs have bred of which at least one member may have been injured and unable to fly, and therefore, unable to migrate. This flightlessness is of course exceptional. Nevertheless, all Whoopers experience an annual period of enforced flightlessness. As in other large wildfowl, swans have evolved a solution to the necessity of wing moult that involves shedding the main flight feathers largely simultaneously. Until all the new feathers have re-grown most of their length, they are flightless, though by condensing the process, they are able to reduce the length of that flightless period, and hence a period of major risk, to a minimum. In Whooper Swan, this period may occupy 5–6 weeks during which time they are particularly unwilling to leave bodies of water. For the smaller Bewick's Swan, living where the tundra summer is itself short and where the arrival of autumn/winter is abrupt and potentially life-threatening, this period may last no more than four weeks, while the Mute Swan, at home in the milder climate of the temperate zone with a prolonged summer fading gently into autumn, holds the record among all the wildfowl having, at up to seven weeks, the longest flightless period (Ogilvie & Pearson 1994). It is this annual flightless period that was first exploited by local swan hunters who caught them for food and feathers, and now by swan biologists, as it makes it possible to catch them for banding during the summer.

SOFT PARTS

The bill is longer than the head; it forms a gently sloping profile from the crown to the bill tip in an almost straight line. It is higher than it is wide at the base and flares slightly to form a somewhat rounded tip, and is strong and well adapted for grasping and tearing vegetable matter. There is a single vertical row of lamellae along the upper mandible edge, and both external and internal rows in the lower mandible. The nail is large. The lores and bill are devoid of any feathering or pro-tuberances (Witherby *et al.* 1940). The bill base is bright egg-yolk yellow or yellow-orange, except in young birds and occasionally adults when scratches and damaged areas appear whiter. In some individuals the entire yellow area appears a pale watery yellow, perhaps in individuals that are sick.

Bare yellow skin extends from the eye around which it forms a narrow ring almost horizontally across the forehead, but angling forwards and across the lores to the base of the lower mandible (Plate 7). Normally, the yellow extends along the culmen beyond the rear edge of the nostril, and perhaps further on the sides of the bill, beyond the nostril towards the tip. The bill pattern is variable, but far less so than in Bewick's (Brazil 1981d), though it may still occasionally be used to identify individuals (Einarsson 1996) (Plate 9). There is also apparently a cline in bill patterns, with those in the east overall yellower than those in the west, such that birds in Japan have the yellow at the base of the bill typically unmarked with black, whereas in the west (Britain and Iceland) there is often a narrow band of black skin at the feather line (Fig. 3.7); those in Fennoscandia have intermediate

Figure 3.7.   The black-based bill pattern is common in the west, but scarce in the east.

markings. Their bill markings differ most markedly from their nearest relative, Trumpeter Swan, where the breeding ranges are closest, across the Bering Sea (Brazil 1981d; Ohtonen 1988). The tip of the bill, the strong nail, and the edges of both the upper and lower mandibles are black, and the pinkish line along the edge of the lower mandible in Trumpeter and Tundra Swans is lacking.

The strong legs and large webbed feet of adult Whoopers are black, while those of newly hatched cygnets are greyish becoming steadily darker. By their first autumn migration the legs and feet are mostly black, but some arrive on the wintering grounds with noticeably grey feet. The bill is initially short and rounded; the tip black, the mid portion a deep dull red, with down covering the base. With age the bill elongates, the red fades to pale pink, eventually white, and the down is lost from the base of the bill. During the first winter, the non-black portion of the bill steadily changes from white to yellow from the feather line forwards, and by spring it is virtually impossible to distinguish them from adults. A rare case of immature-type bill patterns has been observed in late autumn (on 25 November 1993, at Kilcolman Wildfowl Refuge, Co. Cork, Ireland), when two out of four birds had white bill bases, one with faint pink tinge, indicating that they developed far more slowly than normal, perhaps as a result of poor health or nutrition. If the accompanying pair was their parents this would also indicate a rare case of Whooper Swans associating with their parents during their second winter.

For a species so uniform in plumage, it is perhaps not surprising that, despite initial appearances, the colours of the soft parts are variable. Although iris colour was widely described as only brown or blackish-brown (e.g. Witherby *et al.* 1940; Dement'ev & Gladkov 1967), and this certainly applies to the overwhelming majority, a small proportion has eyes ranging through many shades of brown and blue even to pale blue and grey (Plates 7 & 8).

Though leg and foot coloration of all 'northern swans' is typically described as black (e.g. Cramp & Simmons 1977; Madge & Burn 1988), there is, even there, some variation in intensity, for example at Martin Mere, UK, Rees *et al.* (1991b) observed three Bewick's Swans with orange or yellow legs, and among Whooper Swans in Iceland birds with mottled brownish and pale feet have been seen. It is in their bill patterns, however, that the greatest variation among Whoopers is seen (Brazil 1981d; Rees *et al.* 1991b).

At first sight, the most striking contrast between the bill patterns of the Whooper Swan and its old-world relative, Bewick's Swan, is that where the Bewick's has an essentially black bill with a reduced area of yellow towards the base, the Whooper has a yellow bill with a black tip. This is of course simplifying the difference, and some differences are far subtler. Alström and Olsson (1989) (best known for their keen evaluation of the identification criteria of Asian passerines) noted a previously unnoticed or undescribed feature, a small, but particularly significant difference between the pattern on the sides of the lower mandible of Whooper and Bewick's swans. In the Whooper, the base of the lower mandible shows a small yellow triangular mark. In Bewick's, the sides of the lower mandible at the base of the bill (i.e. the gape) are orangey or pink, and contrast clearly with the black of the upper mandible. However, they vary considerably; some individuals have extensive pale pink in the gape, others have little, while some have dark pink, and others no pink at all.

When first comparing the bill patterns of Whooper and Bewick's in Britain, being aware of the considerable work that had been put into individual recognition of the latter, I was mainly cognisant of the striking contrast in the variability between the two. I was initially of the opinion that Bewick's Swans are individually recognisable but Whoopers are not (Brazil 1981d). On closer examination I was wrong. Certainly there is not the same degree of variability as in Bewick's, but variability there is. Its usefulness is, however, a matter of degree. To the casual bird-watcher and average ornithologist, the variations in the bill patterns of Bewick's Swans are noticeable. They fall into several broad categories, known as 'darky', 'pennyface' and 'yellowneb', but the individual details are too complex to be memorised or catalogued. To the highly trained eye of a swan researcher, particularly those of the artists and researchers, the late Sir Peter Scott, his daughter Dr Dafila Scott, Dr Eileen Rees, and other staff of the swan research programme at the Wildfowl & Wetlands Trust (WWT), those patterns revealed individuality, family linkages and family trees, and facilitated enormous amounts of detailed research (Evans 1977; Scott 1978; Rees 1981).

Although studies by the WWT in Britain have shown that the bill patterns of Bewick's Swans show great individual variation, and extensive records of these exist, distinguishing between them requires a highly trained eye and extensive experience. Furthermore, the bills of Whooper Swans are much less variable, despite their larger size, and far less easily distinguished than Bewick's (Rees *et al.* 1990b; Fig. 3.8). Consequently, individual recognition has been even harder to achieve, and as a result there are relatively few individual bill pattern records. Yet by the same account, though I realised that individual recognition of Whooper Swans was beyond me, I could easily discern differences in overall Whooper Swan bill patterns when looking at many birds from widely different geographic regions.

GEOGRAPHICAL VARIATION

Obviously, in a uniformly white bird there can be no regional variation in plumage, variation, if any, must be limited to aspects of morphology, anatomy or soft-part coloration. The Whooper Swan, it transpires, does in fact differ across its immensely broad range.

In the case of the Tundra Swan, sightings of intergrades between Bewick's and Whistling swans, out-of-range occurrences of both subspecies (Evans & Sladen 1980) and apparent hybrids sighted in winter in Japan, are all indicative of some interchange between even these two clearly recognisable populations. Scott (1981) suggested that there is a cline in the extent of yellow on the bills of Bewick's Swans

Figure 3.8.   Variation in bill pattern among yellownebs *in Whooper Swans (above) and Bewick's Swans (below).*

across the Russian Arctic, with those in the Far East, wintering in Japan, being significantly blacker, while those in the far west wintering in Britain have significantly more yellow. This cline is quite probably continuous, with similar variation across the Canadian Arctic in Whistling Swan (Scott 1981) and indicative of how closely related the two taxa are. In contrast, on first observing flocks of Whooper Swans in Japan, members of the easternmost population of the species, I was immediately struck by the fact that they seemed to show considerably more yellow on their bills than their western counterparts. By collecting quantitative data on the bill patterns of Whoopers during field work in Iceland, Scotland, and Japan, it was possible to compare eastern and western birds.

Initial examination showed that Whooper bill patterns did not fit usefully into Sir Peter Scott's classification for Bewick's Swans. Among Whoopers I found that the dark forms, where the culmen is black from the tip along the centre line to the feathering, was rare; 'pennyfaces' with a yellow spot on an otherwise black culmen, were uncommon; and that 'yellownebs', where the black of the tip is separated from the base by yellow, constituted the bulk. For Whooper Swans, the scarcity of birds with the first two bill-pattern categories made it useful to devise a further category by dividing 'yellownebs' into two types, black-based 'yellownebs' (continuous black pigmentation along the feather line from eye to eye, and sometimes extending some distance on the culmen), and yellow-based 'yellownebs' (bill base yellow or there are

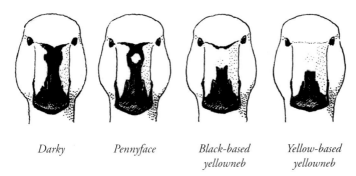

| Darky | Pennyface | Black-based yellowneb | Yellow-based yellowneb |

Figure 3.9.   Classification of Whooper Swan bill patterns.

only a few spots of black along the feather line). Using these four categories made classification and comparison easy (Brazil 1981d; Fig. 3.9).

It transpired that among Whoopers from the Icelandic population wintering in the British Isles, less than 1% were 'darkies', only 3% were 'pennyfaces', while 74% were black-based 'yellownebs' and 22% were yellow-based 'yellownebs'. By contrast, among those from the eastern Russian population wintering in Japan, there were no 'darkies' at all, fewer than 1% 'pennyfaces', just 4% were black-based 'yellownebs' and an astonishing 96% were yellow-based 'yellownebs' (see Table 3.1). Thus Whoopers at the western extreme of their range have considerably more black on their bills than those in the east. Of particular interest was that this highly significant difference varies in the opposite direction from that in the Tundra Swan (Brazil 1981d; Scott 1981).

Table 3.1.   Regional variation in Whooper Swan bill patterns (after Brazil 1981d and Ohtonen 1988).

Whooper Swan populations and bill patterns	UK/Irish/Icelandic	Finnish/northwest Russian	Japan/east Russian
'Darky' (dark-billed form)	<1%	0%	0%
'Pennyface'	3%	2%	<1%
Black-based 'yellowneb'	74%	38%	4%
Yellow-based 'yellowneb'	22%	60%	96%

Scott (1981) considered that the Bewick's Swan's range of bill patterns, with a higher proportion of dark-billed forms in east Asia compared with western Siberia, was 'consistent with the likelihood of significant gene flow across the Bering Strait between Bewick's and Whistling Swans'. It is now known that a considerable number of Whistling Swans occur in northeast Russia (Syroechkovski 2002). After all, until about 12,000 years ago the Bering Land Bridge connected northeast Asia to northwest North America, and the current Bering Strait is no barrier to migratory waterfowl. For example, the entire population of Snow Geese that breeds on Wrangel Island, off northeast Russia, can be considered Asian because they breed in Asia, yet almost all of them winter in the western USA. While the darkest-billed Bewick's Swans of Asia are geographically closest to the yellowest-billed Whistling Swans of North America, in direct contrast, the yellowest-billed Whoopers are in east Asia, leaving them maximally different in bill coloration from Trumpeters (which does not vary across its Nearctic range) where their present ranges are closest. The combination of their relative geographical proximity with their maximal difference in bill patterns perhaps indicates that these taxa diverged much earlier or that they have been more consistently isolated (Brazil 1981d). It seems that the most recent gene flow between these two is more likely to have operated across the North Atlantic, from Europe to North America and vice versa, rather than across the Bering Strait (Ohtonen 1988), although the only known records of vagrant Trumpeter Swans outside North America were in Yakutia and Japan.

In recent decades, an increasing number Whooper Swans have been found in North America, though as yet I have found no evidence of them hybridising with Trumpeters, although a mixed pairing between Whooper and Tundra Swans has been widely reported.

If the proportions of the various bill patterns at the easternmost and westernmost extremes of the Whooper Swan's range were really indicative of a cline, one would expect that over intervening regions the proportions would reflect differing degrees of similarity or difference. Only one attempt has, to my knowledge, been made to examine this, in Finland (Ohtonen 1988), a region not so greatly removed geographically from the Icelandic/British Isles population I had studied, yet one that is nevertheless largely, though not wholly, isolated from the latter population. There, the large number of Whoopers appearing during migration represents a Finnish and northwest European/Russian breeding population. Migrants through Finland winter mainly in Sweden, Denmark and Germany.

Ohtonen (1988) studied the large numbers of Whooper Swans (1,000–1,900) that pause around Hailuoto Island in the Gulf of Bothnia near Oulu, western Finland, on autumn migration. He found that yellow-based 'yellownebs' were the most abundant comprising 60%, while the proportion of black-based 'yellownebs' was 38% and 'pennyfaces' just 2%. Ohtonen (1988) observed no pure dark-billed forms (see Table 3.1).

Although in geographical terms the Icelandic/British Isles and Finnish/west Russian populations are not widely separated, Ohtonen (1988) found highly significant differences between them with respect to the frequency of yellow- and black-based 'yellownebs'. Less surprisingly, because of their great geographical separation, he also found that difference between the Finnish/Russian and far eastern populations was highly significant. Ohtonen's (1988) results from Finland clearly support the premise that geographical variation in Whooper populations is such that the proportion of yellow in the bill pattern increases from west to east (Fig. 3.10). Czapulak & Plata (1998), who studied bill pattern variation in winter 1994–1995 among

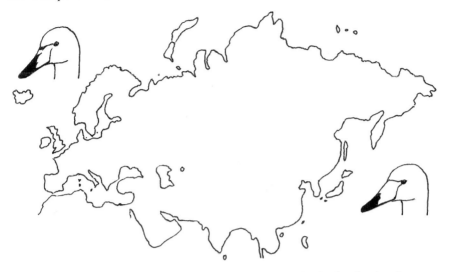

Figure 3.10. Black-based yellownebs are most common in UK and Iceland, Yellow-based yellownebs are most common in Japan.

Whooper Swans wintering in Poland found that 62–68% were black-based 'yellow-nebs' and that the relative proportions of bill patterns were similar to those of birds wintering in Germany and Denmark (although this contradicts Ohtonen's findings), but different from those on migration through Finland. They concluded that any clinal variation is not particularly obvious with the proviso, however, that Poland is still very far west and that the breeding grounds of Polish wintering birds are still uncertain. It would be of great interest to discover whether birds in central Siberia also differ from those further west and east, and it remains to be seen whether there is a similar variation in any other physical parameters—clearly an area ripe for research. It may yet be that the recognisable populations of this wide-ranging species, as defined by their migration routes, may be recognisable in the field.

HABITAT AND HABITS

In defining the habitat of Whooper Swan, it is important to bear in mind the enormous geographical range of the species, and the broad scope of habitats available to it. What may be a common habitat choice in say Iceland, Ireland or Britain, may be less commonly selected in Scandinavia, and may not even exist in another part, such as China, the Koreas or Japan, let alone be preferred. Thus generalisations should be taken as exactly that, with room for considerable local and regional variation. Overall however, they are most likely to be encountered on fresh water or at least in wetland habitats from deep to shallow water, deep lakes or rivers to shallow lagoons, marshes and flooded meadows (Fig. 3.11). While in most regions they are most likely to occur on fresh water, in other regions, even within the same countries, they occur on brackish or saltwater (in some cases at brackish or saline inland sites, as on the Central Asian steppes). Japan is a good example of this, in east Hokkaido it is particularly common in coastal areas, while in central Hokkaido it is most often encountered at lakes or rivers, usually where these are kept ice-free during winter by geothermal activity (Plate 4). In some parts of its range, e.g. central and other parts of Scotland, where Henty (1977), and I studied it intensively (Brazil 1981a,c), and in various parts of northwest Europe, e.g. Sweden, Denmark and the Netherlands, it quite commonly grazes on arable farmland, and in marshes, in a rather goose-like fashion. Yet in other areas, such as Japan and Korea, field feeding is uncommon, except occasionally on rice paddies, and more rarely on pastures (themselves scarce in the Far East).

Whoopers are generally much wilder than the Mute Swan and, as a result (in the west at least), often difficult to approach. They tend to frequent areas away from regular human activity during the breeding season. In Europe it is commoner during winter on more remote waterbodies and agricultural land than at similar sites close to heavily populated areas, whereas in the Far East, particularly Japan, there are numerous sites where Whoopers have little fear of man and come readily to be fed.

During the breeding season it prefers boreal forest or taiga selecting freshwater lakes, small pools, slow-moving rivers, marshes and swampy sites to nest, where crucial criteria are that the waterbody should be sufficiently long to provide room

Figure 3.11.   Winter woodlands; A rare and only temporary habitat choice for these Whoopers in Hokkaido.

for take-off, and that it should remain ice-free for the entire breeding cycle, from nesting to fledging.

The very large and bulky nest is usually located near water, certainly within sight of it, and often only a matter of a metre or so away, on an islet or headland jutting into a lake, on a river bank or on a slightly raised area close to marshland pools. The nest is of moss, grass, sedge or whatever aquatic plants are readily available. Mud has occasionally been reported in the fabric of the nest (Soothill & Whitehead 1978), though I have not observed this myself. Nests are frequently, though not always, crushed flat by the end of the breeding season as a result of the cygnets returning to it to roost in their first days after hatching. However, the same site may be reconstructed and re-used annually (see Chapter 8).

The clutch, frequently of 4–6 large creamy white eggs, averaging 11.3 × 7.3 cm, is laid in April–June depending on region, latitude and altitude. Laying occurs later in colder, higher or more northerly regions. Parental duties are shared, with the female undertaking incubation for *c.*35 days while the male guards the territory until hatching, often remaining vigilant close to the nest, especially during the female's brief absences for comfort or foraging (Brazil 1981c). Both parents then take close care of the cygnets, which are able to fly after 8–9 weeks. The length of this process, from egg laying to fledging, which takes *c.*130 days, is critical in determining the northern limit of the species' breeding range. Whereas in southern Finland, all summers are ice-free for more than 140 days, making it a suitable area for breeding, in the north of the country only 50% of summers are sufficiently long for the bird to complete its breeding cycle (Owen 1977) making it a marginal habitat.

Breeding pairs are strongly territorial, defending an area sufficiently large to supply the feeding needs of a growing family. This may be as large as an entire lake, in upland areas, or merely a section of a larger, more prolific lowland marshy area. In some areas, appropriate breeding sites do not provide sufficient food and there comes a time when the parents must move their family or see them starve.

Following breeding, family parties coalesce, and join non- or failed breeders in flocks ranging from a few individuals more than a family unit, to hundreds or even thousands. On migration, these flocks adopt V, echelon or line formations, and commonly fly relatively low, often following river valleys when traversing hilly or mountainous districts. They commonly arrive after flights across water at rather low altitudes and, in Hokkaido, northern Japan, I have observed them climbing steadily from staging sites at sea level along the Okhotsk Sea to 600–800 m in order to cross volcanic calderas and avoid mountains inland; this style of migratory flight is presumably typical. Their migrating formations pose a dramatic and stirring sight.

The Whooper Swan is one of the highest flying of all birds, and has been recorded at heights above 8,000 m over the British Isles, travelling at ground speeds of up to 139 kph and enduring temperatures lower than −48°C (Stewart 1978; Elkins 1979; Martin 1993). Perhaps their thick down is as much for altitudinal, as seasonal, protection! Their almost simultaneous arrival at widely separated wintering sites in the British Isles may be indicative of regular high-speed flight (they may cross from Iceland to Ireland in as little as seven hours), although there seems to be no evidence that high-altitude flight is common. On the contrary, recent research confirms that low-altitude flight, even over the sea, is normal (Pennycuick *et al.* 1996, 1999).

During winter, Whoopers prefer shallow coastal areas, bays and inlets, inland lakes, reservoirs, and even small pools and rivers. In areas where fresh water is regularly frozen for prolonged periods, they are attracted to almost any isolated ice-free areas of standing water or to those rivers that remain ice-free. In Hokkaido, for example, many small flocks winter on such ice-free sections of fast-flowing rivers where the surrounding landscape is entirely frozen.

On the water, Whoopers feed in a variety of ways, either dabbling at the surface for floating or emergent vegetation, or dipping the head and neck beneath the surface for submerged vegetation down to depths of 40+ cm, while the body remains horizontal on the surface. Swans have especially long necks, a specific adaptation enabling them to feed in deep water. In such situations it is quite common to observe Whoopers not merely dipping, but also upending, tipping forward until its balance shifts it into a vertical position tail pointing skywards while reaching into the water with its neck stretched straight to reach deeply submerged vegetation. Whoopers are capable of upending for 10–20 seconds (Ogilvie & Pearson 1994) but though feeding bouts may be lengthy and intensive, upending requires effort and balance, and is energy intensive, forcing them to pause periodically between bouts. Young cygnets in particular have difficulty with this behaviour; they are able to sustain it only for very short periods and often tumble over completely. Diving as a feeding behaviour is normally associated with smaller ducks, e.g. scaups *Aythya* spp., goldeneyes *Bucephala* spp. and sawbills *Mergus* spp. While I have never seen Whoopers dive for food, they are capable of submerging, as I have observed them dive in Iceland to escape aggressive interactions involving other swans, in Hokkaido to avoid attack by White-tailed Eagles, and also to escape humans attempting to catch them during the flightless moult period.

Foraging occupies the most significant proportion of the Whooper Swan's daily regime, but is interspersed with regular bouts of other behaviours that require varying amounts of time, including brief comfort movements and occasionally intense social interactions. Preening and stretching, though infrequent, are conspicuous, especially among wintering flocks, as the proper arrangement of their feathers is necessary to maintain insulation (and waterproofing). And vital this insulation is, for even where birds engage regularly in appropriate preening behaviour they are still occasionally at risk from hypothermia. Whoopers have been known to freeze to death overnight, sometimes in large numbers, as at Odaito in eastern Hokkaido, Japan, in the 1960s (Abe 1968).

As with other waterfowl, preening is not just an important aspect of self-maintenance, but also a crucial aspect of waterproofing. The long reach of the Whooper Swan's flexible neck enables it to preen almost its entire body from either a standing or a floating position. The chin and lower mandible are used to spread waterproofing secretions from its uropygial gland at the base of the tail (Fig. 3.12).

Figure 3.12. Preening, essential maintenance behaviour.

The only areas it cannot easily reach are the head and upper neck, but these areas are scratched and preened using the toes (Fig. 3.13). It is impractical to preen and waterproof its belly while on water, without much rolling, and birds regularly use secluded locations at the water's edge to complete the preening process. Waterproofing typically involves taking oil first onto the chin, then vigorously rubbing the chin over the feather tracts to be treated.

Other comfort movements are similar to those described for Trumpeter Swan (Banko 1960; de Vos 1964; Mitchell 1994), and include 'body shake', 'wing shake', 'wing flap', 'tail wag', 'head shake', 'head flick', 'head roll', 'foot shake', also 'wing and leg stretch', 'wing stretch', 'gaping or yawning', 'scratching' and 'foot pecking'. These may be performed while standing, sitting or floating. Bathing varies in intensity. It may involve washing just the foreparts by dipping or thrusting the head, neck and breast into the water so that it also runs across the back. More intensive bathing involves beating the partially open wings in the water and rolling sideways, while kicking and splashing with one leg. In its most extreme form, bathing consists of rolling over while wing thrashing. Bathing is followed by shaking, from the tail to the head and often involves wing flapping.

Figure 3.13.   Scratch-preening.

Stretching, another brief but regularly observed behaviour, is more difficult to categorise in terms of its immediate value, but presumably serves to relax and ready muscles and joints for movement or flight. The wings are extended sideways and downward, one at a time, and may be either nearly or fully opened. Stretching one wing is often followed by the other, and it is not unusual for a wing stretch to be combined with a leg stretch while balancing on the other foot. In such situations, the leg and wing on the same side are stretched together (Fig. 3.14). Both wings may also be stretched together, in which case they are raised above the back and opened either partially or fully, while the bird stretches its neck forwards. Such stretching may involve a single or multiple wing flap, which can be quite audible. This posture is maintained only briefly before the bird relaxes to a normal stance. Wing stretching and flapping occur commonly prior to taking flight, and thus may be observed most frequently as birds prepare to move from their foraging grounds to their roosting sites. To some extent, stretching may also be performed in flight, with birds momentarily altering the natural set of the neck or wings.

Swans in general often raise one leg and stretch it backwards, even while swimming. Instead of returning it to the water and continuing to use it after stretching, Whoopers sometimes rest it across the tail or within their flank feathers to keep it dry and warm. They sometimes even swim for a while with a foot raised and the web spread. This may be to dry it, or, as Ogilvie & Pearson (1994) have suggested, it may be a way for an overheated bird to lose surplus heat. However, subjectively at least, this odd behaviour seems no more common at sites where swans swim regularly

Figure 3.14.   Combined wing- and leg-stretch.

and at length in geothermally warm waters, where overheating might be expected to occur more commonly, than in cold waters. Contrary to Ogilvie & Pearson's (1994) suggestion, perhaps where winter temperatures and humidity are extremely low but rates of insolation are high, as in northeast Asia, the black-webbed foot may serve to absorb heat, thus the behaviour may serve for positive thermoregulation.

When resting or sleeping, which Whooper Swans do on land, water or on ice (Fig. 3.15), they turn and curl their neck over the back, draping the neck along the back between the wings and face to the rear, usually tucking the bill up to the eyes beneath the scapulars. In this way they are, particularly in winter, able to make themselves more compact, reducing the surface area exposed to potentially subzero temperatures (overnight air temperatures regularly fall to –20°C and lower in some parts of the wintering range), yet still able to open their eyes without having to move anything except their eyelids (Fig. 3.16; Plate 7). On seeing something untoward, however, they are quick to raise the head and neck to the

Figure 3.15.   Whooper Swans will roost far out on frozen lakes.

Figure 3.16.   The waters of this coastal bay have frozen around these roosting birds during the night.

typical vigilant posture. They may sleep sitting or standing, on one or both legs, or floating. In warm sunshine, even in winter, they appear to occasionally nap, allowing the neck to loll back so that the head and bill still face forwards but rest on the lower curve of the neck, while closing the eyes. They generally choose protected locations where they may sleep without risk of disturbance or predation. Wolves, feral dogs and Red Foxes appear to be the most significant non-human mammalian threats during winter, with only the latter two now common in regions where swans winter. But surely an adult Whooper Swan would be capable of avoiding any potential predator, unless it were already weak or injured, or caught by surprise.

SOCIAL BEHAVIOUR

The family remains the basic social unit from fledging through autumn migration, the winter and even the return migration in spring until the pair is ready to breed again. Thus, there is a prolonged period during which cygnets may acquire knowledge from their parents. The fate of cygnets that lose both parents is generally unknown, although there are rare reports of adoption. In winter flocks, most social interaction is between members of pairs, family members, and pairs and families, though competition over food or position sometimes leads to vociferous interactions between individuals (Plate 13). While posturing and bluff are fairly common among birds feeding close together, and while it is common for one to lunge at or even peck another, more demonstrative aggression, such as running at, swimming at, or flying at, is less common, and physical aggression within a feeding flock is relatively rare. When it does occur, it can be prolonged and severe, with the aggressor gripping its victim by the wing, tail or body with its bill and pursuing and even thrashing it with its wings for minutes on end, or being towed along by the fleeing victim. Sometimes, when birds are evenly matched these interactions develop into impressive grappling fights (Plate 14). Ogilvie & Pearson (1994) considered fights to be commoner among breeders than in feeding flocks. However, this is not true over most of the range as breeding pairs are commonly isolated, thus precluding direct interactions (whereas in feeding flocks such behaviour can, and does, occur at any time). Only where several pairs nest in close proximity, which occurs at certain upland lakes, and certain lowland marshes in Iceland, Russia and China, is this true. Perhaps, as among Bewick's Swans in Russia, there may be a short intense period of fighting during territory establishment in spring. On the wintering grounds, close meetings are frequent, even extremely so, and interactions are essentially face-offs between rivals for food. Individuals, pairs or families face each other, their necks stretched forwards, often with their wings spread or half spread, wing flapping, head pumping and calling noisily. A rival may be driven off by combining this calling display with the physical threat of moving quickly in its direction. If birds feel evenly matched then they may fully spread their wings, point their heads downwards so that their bills almost touch the ground or are underwater (Ogilvie & Pearson 1994). The victor or victors in such interactions may then indulge in a full-blown triumph ceremony.

Such displays, especially between families, are exciting, noisy affairs. Their bugling calls ring out and carry far across the water or over frozen lakes, and in dense wintering flocks their excitement may trigger interactions between other birds. Calling seems infectious, often spreading quickly through flocks until the cold winter air is filled with the loud stirring sounds.

If neither party is willing to concede, then the encounter may escalate to a very brief physical attack though, far more often and for no apparent reason, interactions end abruptly and within moments the previously aggressive birds are preening or feeding again and the volume of calling within the flock has subsided. Where flocks are relatively cohesive, remaining together over days, weeks or months, the same pairs or families may meet and interact frequently, establishing status or priority between them. In contrast to these encounters between families and pairs or individuals, according to Ogilvie & Pearson (1994), 'Encounters between pairs and single birds are comparatively rare, because both are fairly low down on the pecking order'. Although this may be true in flocks where natural food is relatively dispersed, at sites where food is supplied, and often extremely concentrated, there are (contrary to Ogilvie & Pearson 1994) numerous aggressive interactions between pairs and singles, though these are generally very brief and take the form of simple displacement, easily overlooked mild threat or careful avoidance.

Immediately following aggressive interactions, especially with other pairs or families, victorious pairs quickly turn their attention to each other in a highly expressive and vocal triumph display (described more fully in Chapter Seven). This ceremony is much more vigorous and dramatic than that of the Mute Swan, and is far more noticeable, because unlike the latter's display, it is accompanied by loud bugling calls. Members of the pair may begin side by side, but often end facing each other with their wings partially or fully spread, bent at the carpals and the neck and head stretched forwards like a bowsprit, at *c.*45°. During this ceremony the pair alternate their calls in a superb antiphonal duet. In the calm, freezing winter air of an east Hokkaido dawn, at the caldera-filling Lake Kussharo, my favourite site for swan-watching, calls can be heard over distances over the ice in excess of 1 km. The antiphonal calls of the displaying pair gradually shift until they synchronise, making the blaring crescendo a stirring and powerful blending (Hall-Craggs 1974).

This calling pattern is extremely sophisticated, and reveals a remarkable degree of auditory awareness and control. During the antiphonal phase the duration of each call is gradually lengthened from *c.*1/10 second to 1/2 second. Not only do the birds make this sequential change, but they do so in response to each other, with one taking the lead and the other responding almost instantaneously. It is presumed that the male Whooper takes the lead, though, as is the case in other duetting species, such as Blakiston's Fish Owl (Brazil & Yamamoto 1989) reversal of calling order doubtless occasionally occurs among Whoopers. There are, according to Ogilvie & Pearson (1994), conflicting reports as to whether the male always produces the lower note, although being larger, perhaps it might.

Whooper Swan courtship resembles that of Mute Swan in that birds swim breast to breast, turning their heads from side to side, head-flagging, with their neck feathers fluffed out, though the displays of Whooper and Bewick's Swans, prior to copulation, are less elaborate than those of Mute, and are limited to just a few seconds of mutual head-dipping before the male mounts (Ogilvie & Pearson 1994). However, as Soothill & Whitehead (1978) noted, and as I (Brazil 1981c) observed, the Whooper's courtship display is often a very elaborate and dramatic affair that may begin relatively modestly, with the male moving his head up and down, but which escalates to include bow-spritting, and vociferous calling, concluding with vigorous full- or half-wing flapping. The pair, approaching each other, may even rise in the water beating their wings, with heads and necks extended forwards, even prior to independent, then synchronous, head-dipping. Immediately afterwards, the male spreads his wings; the female usually, although not always, follows suit, and the two call as they rise in the water, turn quickly and then settle down to preening.

The gathering of Whooper Swans at traditional wintering sites has pre-disposed them to be provisioned with food at this season. While not tame, they are at least more familiar with the people that feed them (Fig. 3.17). At some sites this involves familiarity with a specific individual or the clothing that the provisioning person wears (e.g. at Hyoko). In the west, at WWT centres in Britain for example, provisioning tends to be very controlled, by particular people at certain times, and the public are not involved. In the east, however, particularly in Japan, Whooper and Bewick's Swans are viewed as sightseeing targets. Whoopers are particularly favoured and though at some sites most food is supplied by a designated swan-keeper, at these and elsewhere the birds are also fed by visitors, who may arrive by the coach load, at the water's edge. At such places (and one could easily list a hundred),

Figure 3.17. A dedicated swan-keeper in Hokkaido.

Whoopers come to be fed by hand. The breeding grounds are a different matter. During summer, Whooper Swans generally prefer remote sites away from human traffic. While some pairs may stand their ground during research visits to the nest, this represents nest defence not tameness (Plate 6). Other breeding pairs will leave the proximity of the nest, the female temporarily abandoning the eggs, as soon as they become aware of one's approach, even at a ranges of 500 m or more. This natural propensity to shyness and to avoid people during the nesting season has led to them being symbolic of unspoiled nature (Haftorn 1971), and though the description comes from Norway it matches my own experiences in Iceland and the reports of swan biologists elsewhere in the breeding range. Yet this particular habit seems to be changing, in north Norway at least. There, over the last 25 years, Whoopers seem to have been losing their fear of man and sometimes now nest in proximity to well-travelled roads and villages (Vader *in litt.*). Perhaps their shyness resulted from past hunting but now, with increasing numbers in many parts of their range, we may see more instances of birds safely associating with people.

HYBRIDISATION

Hybridisation is more common among waterfowl than other avian groups. It is not unusual among captive waterfowl, even between only distantly related species, and is even reasonably common in the wild. Hybrids among swans are less common, but a number have occurred, perhaps the most bizarre being the 'chess-board swan', a hybrid between a Black Swan and a Mute Swan that appeared in Germany in 1982 (Randler 2001). Whoopers are known to have hybridised with Black, Mute, Trumpeter and Whistling Swans, and even Greylag and Canada Geese, in captive collections (Johnsgard 1960; Scott & The Wildfowl Trust 1972) and with Mute Swans in semi-wild or wild conditions. Similarly, captive Trumpeters have interbred with Mute, Whooper, Whistling and Bewick's Swans, and Canada Geese, with the hybrid Trumpeter × Tundra young being fertile (Mitchell 1994). Hybridisation among wild waterfowl is far less common, though regularly reported in dabbling duck of the genus *Anas*, and diving ducks of the genus *Aythya*. Hybridisation among larger species in the wild is very rare.

In general, the natural breeding ranges and ecological requirements of the Palearctic swans, Whooper, Bewick's and Mute, do not overlap. Furthermore, their habitat selection, behaviour, migration and energetic needs make it unlikely that they would mix except on their wintering grounds. However, with the introduction and spread of the feral Mute Swan and the release or escape of captive Whoopers outside their normal ranges, opportunities for meeting during the breeding season have arisen repeatedly in recent decades. Mixed pairs have been reported, to my knowledge, between Whooper and Mute Swans in a semi-wild situation in Europe, and between Whooper and Tundra Swans in North America, and in some cases hybrid offspring have resulted. Nothing is known, however, about their fate or fertility (Fig. 3.18).

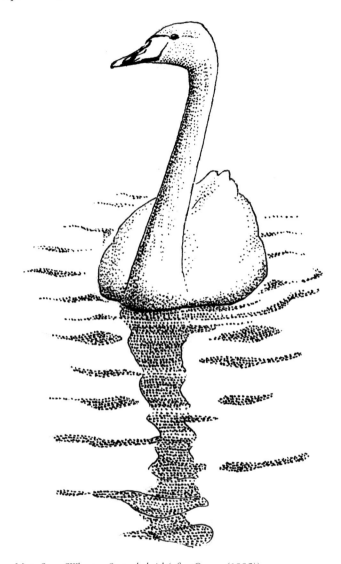

Fig 3.18.   Mute Swan/Whooper Swan hybrid (after Owens (1995)).

Ingremeau & Ingremeau (1992) wondered if the mixed pairing they observed of a Whooper Swan with a Mute Swan at Cherine Nature Reserve, France, had produced the first known hybrids. Perhaps they were the first for France, but the first record of which I am aware involved mixed pairing between a male Whooper and a female Mute Swan producing hybrid young on Lough Corrib, Co. Galway, Ireland, 20 years earlier, in 1972; breeding may have occurred in previous years too. Then, in 1976, a Whooper was found mated with a Mute Swan on Lake Inishmore,

Co. Galway (Hutchinson 1979), and a similar mixed pair was in Galway also in 1976, and mixed pairs have also been seen in the Scottish Hebrides (Spray 1980). Hybridisation between Whoopers and Mutes has also been reported elsewhere in the UK, in Sweden and Poland (Mathiasson 1992; Gibbons *et al.* 1993; Martin 1993; Sikora 1995; Wharfedale Naturalists 1999).

CHAPTER 4

Range, habitat and population–Europe

INTRODUCTION

With a species such as Whooper Swan, with such an enormous Palearctic range, breeding from Iceland to Kamchatka, and wintering from Iceland and Ireland, south to Biscay, the Adriatic, the Aegean and Black and Caspian Sea coasts, and as far as east Asian coasts, it is clearly impossible to adequately address all aspects of its range, habitat and population in all areas. Attention has focused, therefore, on a review of the available and most pertinent literature and a treatment of the better known regions. Investigation is ongoing and there are many gaps in existing knowledge to be filled.

If any country can lay claim to the bulk of the species' range it is Russia, and thankfully much Russian literature has become available in English as a result of the sterling work of Jevgeni Shergalin (see Brazil & Shergalin 2002, in press). China, too, represents a large swathe of the Whooper's range, but there it is far less well known, and rather than a plethora of publications available information depends almost entirely on Ma & Cai (2000). Furthermore, the depth of available information varies enormously between regions, and treatment here is, of necessity, variable. I hope that the reader will bear with these anomalies. Perhaps the more obvious ones will prove a spur to researchers to fill the gaps. My discussion of these topics, proceeds, as far as possible, from west to east and north to south within each region.

Earnst (1991) has pointed out that in the 1970s–80s most swans have increased in numbers, perhaps also in range, as a result of effective habitat protection, hunting restrictions, the shift to feeding on agricultural crops, and increased availability of ice-free waters in industrial areas. Earnst (1991) also considered that while Mute, Tundra and Trumpeter Swan populations were definitely increasing, some local populations of Whooper and Bewick's were increasing and others decreasing, the net result of which was unclear. Unfortunately, because of the patchiness of the information available it is still impossible to be clearer, but as a species Whooper Swan is certainly common and is certainly not threatened, except locally, and breeding has been reported for the first time from several countries in the last two decades.

Owen *et al.* (1986) considered there to be four largely discrete migratory populations: *the Icelandic population* wintering primarily in the British Isles, *the west Siberian or west Palearctic population* (in which I assume they included the Fennoscandian, which winters in continental Europe, particularly Germany, Denmark and the Netherlands), *the central Siberian population*, which winters in the Black Sea, and *the east Siberian or east Palearctic population*, which winters on eastern coasts of Asia, particularly in Japan and South Korea, though *not* in coastal southeast Asia, despite Martin's (1993) assertion. Because of habitat loss it is also very much scarcer in coastal China than Martin (1993) would have us believe. These four populations are not completely isolated, as an apparent cline in their bill patterns and some mixing on the wintering grounds seems to indicate (Brazil 1981d; Ohtonen 1988). It is estimated that at least 200 Finnish-breeding birds now visit southern Britain each winter, while as many as 600 Icelandic-breeding Whoopers may winter in Europe (Gardarsson 1991; Laubek *et al.* 1999).

RANGE

The Whooper Swan's summering and wintering grounds combined make its range the largest among swans. The northern extent of its breeding range is limited by the length of the summer ice-free period and the length of the Whooper's own breeding cycle, which is partly related to its large body size. The factors delimiting the southern extent of its breeding range are unclear and difficult to determine. Though its range during the last 100 years has encompassed Iceland, northern Norway, northern Sweden, northern and central Finland, it has been extending its range south in those countries and even begun to breed in Britain, Germany, Poland and various others in recent decades. Could competition with Mute Swans be a limiting factor? Could climate or habitat availability be an influence? Perhaps as global warming influences the limits of the boreal zone we will see another shift in its distribution. In Asia, Whoopers used to breed as far south as the Aral Sea in Kazakhstan (Kovshar 1996) and it certainly breeds in Mongolia, and in very small numbers at Lake Khanka, but otherwise in the east the limits to its range are perhaps more influenced now by human interference than natural factors.

In line with Earnst's (1991) supposition that range extensions may have accompanied increases in swan numbers since the 1970s, Whooper Swan does seem to have been undergoing a minor range extension in Europe. New summering and breeding records have occurred recently beyond its normal breeding range from Ireland to Latvia (Baumanis 1975; Bobrowicz *et al.* 1986; Illison 1990; Soodla 1990, 1992; Bonisch 1992; Deutschmann & Haupt 1992, 1994; Lewartowski 1992; Murphy 1992, 1993; Sikora 1994; Deutschmann 1997). Furthermore, at the easternmost outpost of its range, extension has taken it into the Nearctic, where it has become a regular winter visitor to the Bering Sea islands over the last 50 years and wild birds bred on Attu[1], Alaska, in 1996 for the first time (Tobish 1996).

HABITAT

BREEDING SEASON HABITAT PREFERENCES

The Whooper Swan is a long-distance migrant, trading its largely northern temperate wintering grounds for boreal breeding haunts. Following arrival, typically in April–May, at resting or staging areas from shallow coastal lagoons to particular sections of rivers or large inland waters (their crucial aspect being that they are already ice-free), breeding pairs leave their young of the previous year and move into the taiga or forest-tundra. Older birds quickly re-occupy traditional nesting sites, while younger birds seek vacant ones. These sites must provide a sufficient sense of remoteness and security, and possess enough submerged aquatic plants, algae and emergent vegetation to feed both members of the pair and their brood in the months ahead. In some areas, e.g. Scandinavia, where food is deficient but nesting habitat suitable, pairs may move young broods from the nesting grounds to suitable feeding areas within a few days or weeks of hatching. Even so, these must still be within just a few kilometres for this to be possible, as the move is energy consuming and dangerous for the cygnets (Haapanen *et al.* 1973a,b, 1977).

In extreme northeast Russia, in the Anadyr-Penzhina lowlands, Whooper Swans establish territories at comparatively small shallow-depression lakes, often in raised hilly tundra. Such depressions are relatively warm and productive, and these wetlands produce an abundance of sedges (*Carex* spp.), and horsetails (*Equisetum* spp.). Further south, in the Yana-Taui depression and the Loyma basin, they breed at remote taiga lakes and are extremely secretive (Kondratiev 1991). In Iceland, some pairs are secretive and shy, while others are bold and demonstrative, perhaps depending on their age and status (Brazil 1981c).

Non-breeders often form very large moulting flocks and moult early. In some parts of Iceland, and perhaps elsewhere, these flocks gather on freshwater lakes, brackish coastal lagoons and shallow estuaries or sea bays. In other parts of their range, even in other areas of Iceland, and in eastern Russia, non-breeders also form small groups on lakes or river channels (Kondratiev 1991; Rees *et al.* 1991a).

WINTER HABITAT USE

Whooper Swans range extraordinarily widely in winter. Thus, while generalisations are dangerous, their winter range can be defined in two very broad categories, climatically severe and climatically mild. In the former, according to Ravkin (1991), Whoopers from the 'West Siberian Plain' winter on large and small lakes, unfrozen rivers and even polynias (unfrozen patches in ice-bound rivers). In east Asia they gather at major river mouths (e.g. the Nakdong in South Korea), on the sea, at major bays and coastal lagoons (e.g. Odaito and Furen-ko in east Hokkaido) provided these remain ice-free, at inland waters including volcanic crater lakes and rivers where these are kept ice-free by geothermal activity (Lake Kurilskoye in southern Kamchatka and Kussharo-ko in east Hokkaido), and at large ice-free lakes, marshes and rivers (Brazil 1991) (Fig. 4.1). I have watched them, in dawn temperatures below –25°C, gathering at narrow strips of water kept ice-free by geothermal activity along the shore of frozen Kussharo-ko, but as most other sites freeze in northern Japan they are forced to leave.

Figure 4.1.   Geothermal activity makes wintering at some northern sites possible.

Further south, in areas with less severe or even mild winter climates, their habitats range from most non-rocky coastal sites, including sandy shores, bays, estuaries and river mouths, to most inland waters: pools, lakes, rivers, and marshes. In many parts of northwestern Europe, they even occur regularly on farmland.

In central Scotland I found birds using areas of flat farmland with a mixture of arable and ley fields. They preferred to feed on waste grain in stubble fields from late October until late January/early February, when they switched to grass. A similar preference for stubble in autumn and grass in spring has been observed elsewhere in Scotland, and in Denmark (Brazil 1981a,c; Laubek 1995c), where they feed on winter wheat, oilseed rape, stubble, pasture, aquatic vegetation, and waste potatoes; however, food selection is seasonally variable and follows a particular pattern. Laubek (1995c) found that soon after their arrival, in October, almost 80% foraged at wetland sites, particularly shallow marshes, and fjords. Submerged vegetation such as *Zostera*, *Ruppia*, *Chara* and *Zannichellia palustris* were significant foods. In November–December, the numbers feeding in aquatic habitats declined to 30% and thereafter to 15%. From early November they shifted to arable land where winter wheat *Triticum aestivum*, winter barley *Hordeum vulgare* and winter rye *Secale cereale* were all important, and then from late November they began foraging on winter oilseed rape *Brassica napus* with c.20% of birds taking the latter in December–January and c.10% thereafter. In a year when severe frost and snow in February devastated the oilseed rape, grass and winter cereals, the Whoopers moved to flooded stubble and grass fields (70%) and potatoes (10%). In Ireland too, Colhoun *et al.* (1996) found a pronounced seasonal shift in habitat/food preferences from October to April, initially at aquatic sites, shifting to stubble/ potatoes, to winter cereals, to grasslands, then back to winter cereals, suggesting a rather widespread pattern, but one as likely to be dependent on similarities in agricultural activities and crop availability as on real preferences by Whooper Swans.

Field feeding by Whoopers and other 'northern swans' is a relatively recent phenomenon. Traditionally, Whoopers spent the winter on open freshwater, on marshes or shallow tidal areas where eelgrass *Zostera* was a particularly important food source. In Britain, they apparently only switched from primarily coastal habitats to feed on agricultural land in the 1930s and 1940s, in response to a combination of the dramatic crash in eelgrass stocks and hard winters. During that period they began to visit fields specifically to feed on potatoes, probably in response to changes in agricultural practices (Kear 1963; Pilcher & Kear 1966; Owen & Kear 1972). The first report of grazing appears to be that of Anderson (1944), despite Floyd's (1946) assertion that Whoopers were never seen any distance from water, although he had watched up to 25 Bewick's Swans grazing on stubble fields after the harvest. Anderson (1944) saw Whoopers in Scotland feeding on grass fields up to 5 km from the nearest loch, after flooding had subsided. Use of stubble and grass fields may have developed out of flights to potato fields, or from the use of flooded stubble fields or pastures, or as a response to hard weather (see Chapter Six). A further shift in habitat use has been noted in Ireland, where Wells *et al.* (1996) reported a general impression that Whoopers there have been feeding less at traditional sites on inland waters, and more on land, particularly improved grassland, and a similar move has been taking place in southern Sweden (Nilsson 1997).

A similar shift in foraging behaviour may be widespread as areas of natural wetlands, coastal or inland, have declined, and as cultivation has steadily increased throughout much of the wintering range of the Icelandic and West Palearctic populations of Whooper Swan. In Japan field feeding is relatively uncommon, but where it occurs it is mainly on rice stubble (where these are close to large lakes suitable for roosting; Brazil 1987). Such feeding is very rare on other types of farmland. That it does occur, at least occasionally, is exemplified by my sighting in early spring 1984 of a large flock on snow-free meadows in central east Hokkaido, far from the nearest wintering areas (Brazil 1991). But as an example of how few Whoopers had adopted the field-feeding habit in Japan, only 217 of 7,225 birds were recorded on fields, at just two of 81 study sites (WBSJ Research Division 1982).

Whoopers wintering in Britain use a range of different habitats, which are only rarely affected by snow or ice cover. Many still occur on permanent inland freshwater bodies, mostly large lakes, except in October when 55% occur on improved pasture, and during the late winter when they also use improved pastures (Rees *et al.* 1997b) though there may be extreme local variation. During the January 1995 census, habitat details were collected for 12,033 Whoopers. Of these, 38% were on improved pasture, 32% permanent inland water, 17% flooded pasture and 7% arable land (Cranswick 1996), and thus 62% were on farmland. Improved grasslands are especially attractive to Whoopers and other swans, with root crops, such as potatoes and sugar beets, being preferred. In autumn, cereal stubble fields are particularly favoured as these hold substantial quantities of spilt grain after the harvest. Many then move to root crops during midwinter, before switching to freshly shooting cereals in late winter (Brazil 1981a,c; Colhoun *et al.* 1996; Rees *et al.* 1997a, 1997b).

POPULATION

Icelandic population overview

For Whooper Swan watchers in Ireland and the UK it is the Icelandic breeding population that dominates thoughts. In Iceland it is a widespread breeder at most wetlands in both lowlands and uplands, from sea level to *c.*700 m (Gardarsson & Skarphedinsson 1984). Though the Icelandic is arguably the best-known population now, it was poorly known as recently as the 1970s and unknown to British researchers at that time, and estimates of its numbers on the wintering grounds were wildly erroneous.

Maxima include estimates of up to 2,500 breeding pairs and 14,000 birds in autumn. However, the estimate for the Iceland/Ireland/UK population is generally assumed to be the same as the wintering population, which was 16,000 in January 1995 (Cranswick *et al.* 1996; Scott & Rose 1996). Census figures depend on coverage, which varies regionally, but 15,842 were recorded in the international census in January 1995, with 7,799 (59%) in Britain and Northern Ireland, 7,072 (45%) in the Republic of Ireland and 971 (6%) overwintering in Iceland (Cranswick *et al.* 1996). Counts at a more restricted range of sites in January 1995 and 1996 indicated a decline between years attributable to both reduced coverage and to summer 1995 having been a poor breeding season (cygnets represented 15.6% of flocks in 1995, but 17.9% in 1994) (Cranswick *et al.* 1996; Delany *et al.* 1999).

Of this population, the Ouse Washes in eastern England, and Loughs Neagh/Beg in Northern Ireland each hold more than 5% of the total population in midwinter making them the most important wintering sites. At the time of the international census, six other sites also held more than 1% and, more recently, as many as 14 sites throughout Ireland exceeded the 1% threshold (Delany *et al.* 1999). Clearly then, Whoopers move to some extent between years, favouring different sites annually, thus even sites with few or no Whooper Swans in a given year may prove internationally significant at other times.

The Icelandic breeding population appears to have been largely stable between the mid 1970s and mid 1980s, followed by a steady increase in the late 1980s with the population seemingly stabilising at a higher level or perhaps beginning to decline, until a considerable further increase was revealed by the January 2000 census. International censuses in Iceland, Ireland and Britain, found an increase from 16,760 in January 1986 to 18,035 in January 1991, declining to 15,842 in January 1995 (Salmon & Black 1986; Kirby *et al.* 1992; Cranswick *et al.* 1996), but rising by over 30% to 20,645 in January 2000 (Cranswick *et al.* 2002). However, these censuses may be of more birds than just those from Iceland, as unknown numbers of continental birds winter in Britain. This remains a pitfall when attempting to reconcile population estimates made in winter with numbers breeding in Iceland.

THE ICELANDIC BREEDING POPULATION

Population counts were rare in the past, and in the early 1970s the Icelandic summering population was thought to be 5,000–7,000 birds with no apparent increase over the previous 25 years, but only a small proportion breeding (Gardarsson 1975). Earlier in the century there may have been an increase or at least a shift in moulting and wintering distribution in response to the widespread eelgrass crashes in the 1930s–50s (Gardarsson *in litt.*). Better coverage and considerable improvements in counts in the early 1980s yielded over 9,000 birds and led to the total population being then estimated as possibly 11,000 (Gardarsson & Skarphedinsson 1984; Owen *et al.* 1986; see Table 4.1), though this did not reflect a population increase so much as improved surveying.

When Gardarsson & Skarphedinsson (1984) made the first attempts to survey the Icelandic population, they discovered almost twice as many birds as were previously known! They surveyed after the breeding season, during the short period in late September and early October before the majority leave Iceland, and maximum numbers are concentrated at relatively few conspicuous and well-known moulting and staging areas (see Fig. 4.2). There, prior to the emigration of most of the population to the British Isles, they are easy to count, though undoubtedly some would have been missed as the earliest arrivals in the UK and Ireland appear during this period. Gardarsson & Skarphedinsson's (1984) results were astounding: 12,557 in early October 1982, 13,686 in early October 1983, and 14,247 in early October 1984, more than double the previous estimate of 5,000–7,000. They considered that their very much larger figures were the result of better coverage, and that they did not necessarily indicate a population increase (see Table 4.2), yet comparisons revealed a statistically significant increase in the Icelandic population wintering in the UK and Ireland over the previous ten-year (more than double) and 20-year (more than treble) periods, an increase that was apparently accelerating. Coincidentally, marked increases in other populations were also documented. For example, the continental west European wintering population doubled in the 20 years after 1974 (Rose 1995), while the Japanese wintering population more than doubled in 1982–92 (WBSJ 1992a), though this could conceivably have been concentration as a result of habitat loss elsewhere in east Asia. Whether the Icelandic

Table 4.1.   The Icelandic population (after Gardarsson & Skarphedinsson 1984).

Region	Total seen	Estimated number missed	% young in the regions
South	1,646	100–200	23.0
Southwest	1,934	50–100	18.9
Northwest	234	100–200	4.1
North	2,227	200–300	25.0
Myvatn	716	0	17.3
Northeast	296	400–600	–
Southeast	2,004	0	7.9
Total Iceland	9,057	850–1,400	18.2
British Isles		200–500	
Total missed		1,050–1,900	
Corrected total		10,107–10,957	

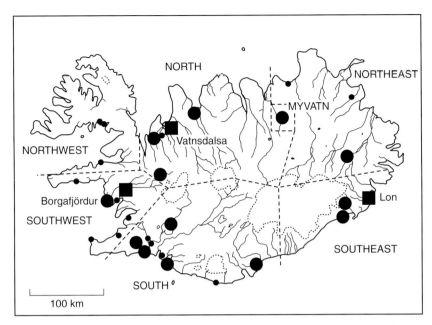

Figure 4.2. The distribution of main staging sites of Whooper Swans in Iceland in October 1982. Small circles indicate 11–100, large circles 101–1,000 and squares over 1,000 birds (after Gardarsson & Skarphedinsson 1984).

Table 4.2. Recent estimates of the Icelandic summering population.

Year	Total counted	Total estimated	Reference
1950s–1970s		5,000–7,000	Gardarsson 1975
October 1982	12,557		Gardarsson & Skarphedinsson 1984
October 1983	13,686		ibid.
October 1984	14,247		ibid.

population has increased or not, it is clearly an extremely successful one with approximately 1,500–2,500 breeding pairs (Rees *et al.* 1997a) at a density of up to 14 birds per 100 km^2, i.e. much higher than Finland where there are just 0.9 per 100 km^2 (Rüger *et al.* 1986; Haapanen 1987, 1991) although interpreting such densities is notoriously difficult given that large areas within its range are actually unsuitable habitat.

In the light of previous estimates from Britain (Brazil & Kirk 1979 had found a minimum of about 7,000), these were unexpectedly large numbers, prompting the Wildfowl Trust to coordinate a complete midwinter census in 1985–86 with the aim of covering all of Ireland, Britain and Iceland, to revise estimates of the various wintering populations, and provide a baseline figure against which to compare future trends.

Gardarsson & Skarphedinsson's (1984) observations in Iceland were extremely exciting. In the southeast they confirmed the significance of the brackish lagoons of Lón and Alftafjördur as staging sites, as had been suspected by Brazil & Gardarsson when, in following up Boswall (1975) who had presumed the 535 Whoopers he encountered at Lón to be an annual moulting flock, they observed 3,300 there on 19 May 1979, during an especially late spring. Spring totals in the southeast are now known sometimes to exceed 4,000, most on Lón, where there were 7,000 in April 1998 (Gardarsson *in litt.*), though they typically occur earlier (e.g. 2,600 on 6 April 1980). Furthermore, the two are also important moulting grounds with *c.*1,200 on Lón and 200–400 on Alftafjördur. They also play a role during the autumn prior to emigration. Observations in autumn 1982 showed that the main exodus was on 14–23 October, coinciding well with peak arrivals in the British Isles. Of particular interest was Gardarsson & Skarphedinsson's (1984) observation that families appeared to depart after adults without young, as indicated by the rising percentage of cygnets among totals recorded immediately prior to emigration (see Table 4.3). It seems likely that preferential migration occurs among the various status classes in this species, a subject that will be discussed in Chapter Ten.

Ringing of Whoopers in Iceland, Denmark and the British Isles confirmed what was long suspected: that the bulk of the Icelandic population winters in Ireland and the UK, with a smaller, but significant, proportion staying in Iceland. Some Icelandic Whoopers also cross the North Sea to winter in Norway, Denmark and the Netherlands (Gardarsson 1991; Kirby *et al.* 1992).

While the Asian population has been extending into the northwestern New World, in its westernmost range it formerly also had a toehold in the New World, in Greenland. Hørring & Salomonsen (1941) refer to a number of records of Whoopers there in May–October and, most interestingly, in February–March, indicative perhaps of overwintering, but possibly of midwinter movements/early-spring migration (which may begin in mid March in Iceland). They considered that Whoopers may have bred on the southwest coast of Greenland last century but, as elsewhere, the length of the summer ice-free period would have limited any extension north. Whoopers almost certainly bred in Greenland as recently as 1929, as a cygnet was shot at Kûngmiut in the southeast in September that year. The Greenland birds doubtless reached their breeding grounds via Iceland.

Table 4.3.   The changing proportion of cygnets at an autumn staging area (after Gardarsson & Skarphedinsson 1984).

1982	12 October	14 October	23 October	1 November
Total no. of Whooper Swans	1,060	1,178	416	243
Proportion of cygnets	3.8%	4.3%	18.5%	13.6%

WHOOPER SWANS WINTERING IN ICELAND

Following their revelations concerning the Icelandic breeding population, Gardarsson & Skarphedinsson (1985) continued by elaborating on the species' status in Iceland in winter. Though it had long been assumed that the population might be only partially migratory, with some small numbers, hundreds or perhaps even up to 1,500, wintering on south and west coasts and at certain springs (Owen 1977; Owen *et al.* 1986), this subject had unsurprisingly received little attention. Gardarsson & Skarphedinsson (1985) demonstrated that Whoopers regularly winter in distinct habitat types, on spring-fed open freshwater in the northeast and south, on shallow coastal waters where ice seldom forms in the southwest, and brackish lagoons in the southeast when these are ice-free, and more recently also man-made habitats both urban and agricultural. Many favoured sites represent safe havens that are always ice-free. Numbers and distribution are dependent on favourable conditions, but up to 1,300 remaining in some winters (Kirby *et al.* 1992; Cranswick *et al.* 2002). If mild conditions permit, swans disperse further afield, to shallow fresh water, marshes and even cultivated land, much as one might expect, further south.

There appears to have been a marked increase in the numbers wintering from the mid 1970s to the mid 1980s. Between 1976 and 1979, Gardarsson & Skarphedinsson (1985) estimated that there were unlikely to have been many more than 500 wintering birds, whereas that number had increased to 700–800 by 1982–83 and 1,300 by 1984–85. Numbers wintering inland in the south and northeast were thought to have been stable, but counts on west and southwest coasts doubled in 1976–1985. Such variation was considered dependent on weather conditions. In January 1995, the wintering population was recounted and a total of 971 estimated, representing 6.2% of the Icelandic population (Cranswick 1996). Similarly in January 2000, 6% wintered in Iceland (Cranswick *et al.* 2002; Fig. 4.3). Some changes had taken place historically, but these had led to noted decreases, e.g. those moulting in Gilsfjordur declined from 'very large numbers' in the 1930s to just a few dozen in recent years, because of a decrease in eelgrass resulting in a local food shortage. Elsewhere, positive changes have been well documented. At Lón, for example, wintering numbers were very low until 1984–85 when they rose suddenly to over 500. However, wintering there and at other southeast lagoons is only possible if they remain ice-free, which is only in some winters (Gardarsson *in litt.* 1996). Such short-term increases are likely in response to recent mild winters. Overall, however, Gardarsson & Skarphedinsson (1985) believed that the former wintering population would have been on a similar scale to that currently. It will be fascinating to observe in coming decades whether numbers wintering there (and in other areas traditionally affected by severe winter weather) continue to increase as winters become progressively milder. It remains true, of course, that with Iceland's northerly position, and the winter climate there being variable and at times extreme, Whoopers, like other waterfowl, may initially opt to winter in Iceland but subsequently move south in response to severe conditions (Elkins 1983).

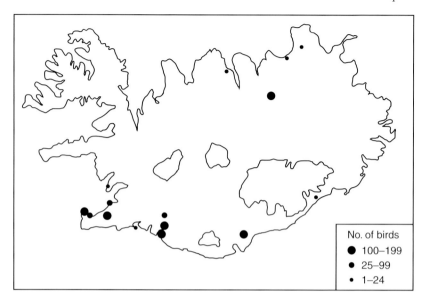

Figure 4.3.   Distribution of Whooper Swans wintering in Iceland, January 2000 (after Cranswick et al. *2002).*

THE UNITED KINGDOM'S BREEDING POPULATION

Whoopers have a long and traceable history in the British Isles, dating from the Cromerian Interglacial period of the Middle Pleistocene. In the past 15,000 years, as glaciation first became less severe and then as the ice sheets retreated north, much of the region would have consisted of tundra and boreal-like habitats. Perhaps, in the not-distant past, Whooper Swan was a common breeder at latitudes considerably further south than currently, including much of Britain. According to Martin (1993), it occurred during both warmer and colder periods, and it is assumed to have bred widely during suitable climatic periods, such as the early part of the 'little Ice Age' (1550–1850) when it may still have been a fairly common breeder (Fagan 2000). Gibbons *et al.* (1996), in analysing the population trends of breeding birds in the UK since 1800, considered that in 1800–49 the Whooper Swan was a non-breeder, that it bred during the period 1850–99, but was then a non-breeder in 1900–69, returning as a breeder in 1970–95.

Whoopers currently occur mainly in October–April and chiefly north of a line from the Wirral to the Humber, though in recent decades up to 1,700 have regularly wintered on the Cambridgeshire/Norfolk border at the Ouse Washes (Batten *et al.* 1990; Andrews & Carter 1993; Bowler *et al.* 1992, 1993, 1994, 1995; Rees *in litt.*). In addition, it was formerly a regular, if unappreciated, breeding bird. 'When they again retire northward to the arctic regions to breed, a few, indeed, drop short, and perform that office by the way, for they are known to breed in some of the Hebrides, the Orkney, Shetland, and other solitary isles; but these are hardly worth notice.' So wrote an unimpressed Bewick (1797–1804). It continued as a well-established breeding bird in the Orkneys at least, until as late as 1760–80 (Gibbons *et al.* 1993) where it was extirpated by disturbance and egg theft (Fea 1775; Neill 1806; Reynolds 1982). Subsequently, and for 150 years, it bred only very rarely in Britain (almost invariably in Scotland; see Gibbons *et al.* 1993). It seems quite likely, however, that the British penchant for secrecy concerning the location and frequency of rare breeding birds may well have hidden a significant number of other records. It was, for example, considered by Batten *et al.* (1990) that breeding in recent years involved only 1–2 pairs, some of which were considered feral, but birds have been recorded summering almost annually, and this pattern continues (Ogilvie & the Rare Breeding Birds Panel [hereafter RBBP], 1995, 1996a,b, 1998a,b, 1999a,b, 2000). Some of these no doubt involved wounded birds, which Buckland *et al.* (1990) noted occasionally summer at coastal sites, and on the River Don. After many decades of merely sporadic attempts, the Whooper seems now to have re-established itself as a rare breeding bird. A very small proportion of wintering birds may, therefore, be resident, although the vast majority of the UK's wintering Whoopers are immigrants from Iceland.

As with the wintering population, which is concentrated in Scotland and Ireland, the summering population is also distributed primarily in the north and west, with only a few in England. Despite the numbers of birds now involved, some at apparently suitable nesting sites, the proportion attempting to breed is very low (Thom 1986). In the Western Isles, summering non-breeders may number 20 birds, but pairs are rare. Low productivity appears typical of all Whooper Swan breeding populations (only 30–40% of the Icelandic population is thought to breed annually) and is not a consequence of a lack of suitable breeding locations. Away from the Western Isles, summering birds are scarce and scattered. Some probably remain because they or their mates are unable to migrate through physical injury or the effects of sub-lethal poisoning (Thom 1986; Owen *et al.* 1986). In one case a breeding pair included an escaped pinioned female (Owen *et al.* 1986). The historical record, combined with the fact that Whoopers now regularly occur in southern Sweden, where they breed at lower latitudes than in Scotland, indicate, however, that neither summering nor occasional breeding should be regarded as abnormal (Gibbons *et al.* 1993).

Recolonisation of northern parts of the UK appears distinctly possible through a combination of factors, including the current trend towards cooler springs and summers, the tendency of some northern species to breed in Scotland, the species' range extension in Fennoscandia since the late 1950s, and that it is now protected in both the UK and continental Europe (under Schedule 1 of the Wildlife and Countryside Act 1981; Annex 1 of the European Union's Birds Directive; and Appendix II of the Berne Convention; Batten *et al.* 1990). The direct cause for it having thus far proved unable to re-establish itself as a regular breeder, may be its greater susceptibility to disturbance than the Mute Swan for example. Disturbance is a difficult issue and hard to prove, and of course begs the questions as to whether breeding studies involving nest visits are causing the reduction in output. If disturbance *per se* is not significant, then in the UK a further factor may be limiting its capacity to spread. Its conspicuousness makes its nests easily found by egg collectors; at least one Scottish nest was robbed during the 1988–1991 period (Gibbons *et al.* 1993).

Batten *et al.* (1990) considered that although single birds or pairs have summered in Scotland in most years between 1900 and 1980 eggs or young had been produced only *c.*10 times. Gibbons *et al.* (1993), however, provided a much fuller picture, revealing that an injured pair bred in Sutherland annually in 1910–18, but was shot by men returning from the war. Another pair bred in the West Highlands annually in 1912–21 producing four eggs in May 1921 (Gordon 1922). A Perthshire pair was less fortunate, having bred for three consecutive years (years unknown) they were killed by the poisonous outflow from a lead mine. The next record is perhaps the most surprising. Witherby *et al.* (1940) refer to a pair that bred in Norfolk in 1928, although it is unknown if wild birds were involved. A pair then bred successfully on Benbecula in the Western Isles in 1947 (Thom 1986), but between 1948 and 1977 no breeding records were reported (Gibbons *et al.* 1993), though Batten *et al.* (1990) refer to breeding in Inverness-shire in 1963. Apart from this slight contradiction, there is agreement that there was a lull in breeding records until 1978 when a pair reared three cygnets on Tiree (Sharrock 1980), and two further cygnets in 1979. The same pair was still present in 1980, but apparently without breeding. Also in 1979, a feral pair (believed to be a wild male and an escaped, pinioned female), reared a cygnet at Loch Lomond, Dunbartonshire, and may also have bred in 1978. The female is known to have originated from a captive population at Wards, Gartocharn, in the 1960s–70s. Two young were reared again at Loch Lomond in 1980 and in 1982, when two feral pairs nested (J. Mitchell *in litt.*; Thom 1986). Since then birds have bred there regularly. It is increasingly difficult, since the late 1970s, to distinguish wild birds from feral. Breeders occupied no fewer than three sites in 1980 (Owen *et al.* 1986), and it seems that since 1980 1–2 pairs (some feral) have bred or attempted to breed almost annually.

Of the four breeding pairs reported in 1988–91 for the new atlas (Gibbons *et al.* 1993), two pairs were probably wild, one feral, and one involved a mixed pairing with a Mute Swan. More records were reported to the RBBP, however, and thus it is clear that at least five pairs bred in Britain in 1990 alone (and that not all previous records may have been reported to the atlas). The most recent figures indicate that slightly larger numbers may now be involved (Spencer & RBBP 1993; Ogilvie & RBBP 1994) or that these may have gone unreported previously. Whooper Swan is currently considered to be a rare and erratic breeder, perhaps 2–5 pairs breed annually (Gibbons *et al.* 1993; Ogilvie & RBBP 1995, 1996a,b, 1998a,b, 1999a,b, 2000) but are generally a well-kept secret. For their security exact localities are not given. As an example, let us examine the 1990s.

According to Spencer & RBBP (1993), 1990 was a remarkable year, both for the total number summering and the large number of breeding pairs, many of which appeared to be genuinely wild, in notable contrast to many of those during the 1980s that were clearly 'pricked' birds unable to migrate. Whoopers were observed at 15 localities, only one in England, in the southeast, where four adults were present on 27 May, and a pair with a cygnet on 31 May. The remaining 14 were in Scotland. One 'tame' individual summered in the south. In 'Mid Scotland', two localities were occupied, one had two feral pairs, each hatching three young, while at the other a pair raised four cygnets. In 'North and West Scotland', Whoopers were present at 11 sites. At the first, four adults and a first-year were observed. At sites two, seven and eight, lone birds summered while, at the third, a single injured bird remained and, at the fourth, four summered. At the fifth, one was seen on a nest that was subsequently deserted. At site six a pair summered, one of which was injured. At the ninth two adults summered while at site ten a pair was present until at least 12 May, and at the final locality two pairs were observed. One pair produced five eggs (though sadly none hatched), and 13 other singles all summered. In all, 5–32 pairs were thought to have bred, at least three of them feral (see Table 4.4).

The following year, 1991, Ogilvie & RBBP (1994) reported fewer birds but at more localities. Whoopers summered at 18 localities, the majority in Scotland. At one locality in southeast England, there were up to three feral individuals all summer; at one in north England a single feral pair bred, raising three young; and at one in south Scotland, a single, presumably injured bird remained all summer. In mid Scotland, there were three localities: at the first, two feral pairs raised four

Table 4.4. Whooper Swans in the UK during summers 1990–98.

Year	Summering localities	Pairs	Pairs bred	Cygnets
1990	15	5–32	5+	11
1991	18	4–21	4+	7
1992	11	7+	7	1
1993	20	4+	4	1
1994	14	5+	5	11
1995	9	1	1	0
1996	5	4	4	6
1997	9+	9	4	3
1998	3	2+	2	1+

young; at the second, a single pair bred but the eggs were stolen in June and one of the adults died in August; at the third locality, a single remained all summer. The remaining 12 localities were all in north and west Scotland. At the first, three adults were seen and display was observed in June–August, but they did not breed; at sites two and three, four adults summered at each; at sites 4–6 two adults summered at each (though Ogilvie & RBBP 1994 make no judgement as to whether these were pairs); at sites 7–12, single birds summered. In all, 4–21 pairs were considered to be breeding in the UK in 1991, of which four at least were presumed feral (see Table 4.4). However, as Ogilvie & RBBP (1994) pointed out, only one breeding pair consisted of apparently wild individuals. They considered that the potential for nesting by feral individuals, particularly in England, is increasing, a factor of some concern given the recent history of other escaped wildfowl and their impacts.

In 1992, Whoopers were found at 11 localities. Six pairs bred, of which four were feral. In southeast England, a feral pair nested but failed, and two tame adults remained on 8 June–1 July, and a pair bred at another locality. In southern Scotland, a feral pair raised one young (the male of a second pair died during the winter) and at a second locality a pair held territory where they had bred in 1990–91, but did not nest. In north and west Scotland Whooper Swans were present at five localities. At one site, a pair may have bred but conclusive proof was lacking; at a second, a lone adult built a nest; at the three others singles were present during the summer. Thus, there was indication of a further decline in the number of summering birds in northern Scotland, with only one nesting attempt, and feral pairs were continuing to breed at a handful of localities (Ogilvie & RBBP 1995; 1996a).

In 1993, the species summered at 20 localities and four pairs bred, although only one was considered wild. Just one cygnet was raised. In southeast England, a pair bred and fledged one young at a site where they had bred in 1992 (additional to Ogilvie & RBBP 1995), and seven other adults summered. At another site a single remained most of the year. In southern Scotland, they occurred at three localities: at one a pair bred but their nest was flooded, at a second an injured pair and a single (also injured) summered, and an injured individual summered at the third. In mid Scotland, there were also three localities. At one site a pair bred but their brood died during bad weather, at the second a pair (probably feral) were seen displaying on 24 May but did not breed, and at the third a single was on a suitable breeding loch on 13 June. In north and west Scotland, there were 12 localities. At one site a wild pair nested and probably laid eggs, but failed due to flooding following heavy rain. At a second, four adults (probably two pairs) were present for at least a week in early June. At nine other sites singles summered, some or all of which were probably injured. At the twelfth a single was seen on 10 June. The increase in the number of summering birds compared with 1992 was thought to be due to better recording (Ogilvie & RBBP 1996a).

In 1994, summering occurred at 14 localities and involved five breeding pairs (two deemed to be wild) raising nine young. In southeast England, they were found at two localities. A released pair fledged five young at one site, and a released pair (that did not attempt to breed) and three singles were at the 1992–93 breeding site. In southern Scotland, a pair (probably released) fledged three young. At a second site another pair, also possibly released, built a nest but did not lay. At a third site a single (probably injured) summered, and at a fourth site an adult seen on 26 June may also have been injured. In north and west Scotland they were present at eight localities. At one a pair fledged three cygnets, at another a pair laid four eggs, which did not hatch, at a third a pair laid but their nest was robbed, at a fourth a pair summered and may have attempted to breed, while at four others singles summered. The successful 1994 season (the best since the species was included in the RBBP's report in 1978) appeared to involve a number of genuinely wild pairs in north and west Scotland, perhaps indicative of natural colonisation. Monitoring this is, however, made difficult by apparently uncoordinated artificial introductions (Ogilvie & RBBP 1996b).

In contrast to the successes of 1994, 1995 saw a return to the usual scattering of injured birds, with summering at just nine localities. Only one pair bred, and that failed. In southeast England, single released birds were at two localities. In southern Scotland, a pair that had probably been released laid eggs but none hatched, and at a second location one probably injured bird summered. In north and west Scotland, a single built a nest and sat for a while, and at four other sites singles summered (Ogilvie & RBBP 1998a).

In 1996, there was even less activity, with Whoopers present at just five localities, involving two successful breeding pairs and two others that failed. None was reported from England. In 'Mid Scotland' two escaped pairs bred at one locality in Dunbartonshire but both nests were flooded. In north and west Scotland a pair bred and fledged three young at one locality, at a second, a pair bred and had raised three half-grown young by July, at the same site where a pair had failed the previous year. At a third site a pair held a territory, and at a fourth site a pair stayed the summer (Ogilvie & RBBP 1998b).

In 1997, a pair of feral Whoopers raised three young in Luton, Bedfordshire; a free-flying pair was at Windsor Great Park, Berkshire, in May–July; a feral pair was at Marsden Quarry, Northumberland, and at other sites nearby in June–August, several escaped/released birds were present at localities in England, and seven were at the regular locality in Dunbartonshire, though none apparently bred (Ogilvie & RBBP 1999a). In north and west Scotland apparently wild pairs nested and laid eggs at three localities but none succeeded in hatching. In Northern Ireland, three pairs were present at one locality (where they had also been present in 1996) but were not thought to have bred (Ogilvie & RBBP 1999b).

In 1998, the most recent year for which information is available, two apparently wild pairs bred in Scotland, and the only report from England was of a juvenile (origin unknown) in Berkshire from 25 June (Ogilvie & RBBP 2000). In the Shetlands, Whoopers bred in three consecutive years, 1994–96 (Nightingale & Allsopp 1997), and a small number (1–4 birds in 1975–80) regularly summer in Orkney, these often being injured or immatures.

One might expect that with global warming the ranges of some breeding birds will retreat north and that we might lose the Whooper Swan as a British breeder (Moss 1998), but despite climatic patterns being generally disturbed, there has been an increase in overall Whooper Swan populations in the west, which seems to be leading to increasing numbers of birds breeding in regions and countries south of their previous range.

THE UNITED KINGDOM'S WINTERING POPULATION

Over 6,500 Whooper Swans currently winter in Scotland, England and Wales (representing *c.*33% of the Icelandic population) and a further *c.*3,700 in Northern Ireland (Cranswick *et al.* 2002). The vast majority originate in Iceland, although an unknown but probably small and annually variable number reach southeastern sites from continental Europe (Owen *et al.* 1986; Salmon & Black 1986; Batten *et al.* 1990; Kirby *et al.* 1992; Cranswick 1996).

The vanguard of autumn migrants arrives in small groups as early as September, although the main influx occurs in early October to mid November. They generally remain until their spring northbound departure, which, for most birds, is in March–April (Kirby *et al.* 1992), although they may move between wintering localities several times.

Whoopers winter widely throughout the UK north of a line from the Humber to the Wirral, but most commonly in the northern and western isles of Scotland, mainland Scotland, Northern Ireland, and less commonly, though still widely in northern England. Small numbers reach Wales, particularly Anglesey, and southern England. However, the flocks in Norfolk at the Ouse Washes (particularly Welney), at Wolferton, and the Norfolk Broads (Hickling/Horsey) are the southernmost regular concentrations (Kemp 1991), although Martin (1993) mentioned the Kent marshes and Hampshire Avon as holding above-average numbers for southern Britain. The increasing importance of more southerly sites, such as the Ouse Washes and Martin Mere, is attributed (Martin 1993) to the particular attractiveness of the Wildfowl & Wetlands Trust centres, as a result of the provision of supplementary winter food, though their attractiveness may also be as safe roost sites (Dafila Scott *in litt.*; Fig. 4.4). WWT Martin Mere and WWT Welney now attract the largest flocks in England. Typically, however, in the south they are unusual winter visitors.

Whooper Swan reaches the southern limit of its usual wintering range in Britain in mid Wales, in contrast to Bewick's Swan which famously extends further south, e.g. to WWT Slimbridge, Gloucestershire, but with a far more restricted range in the north (Lack 1986). Nevertheless, as a rare visitor Whooper Swan is also known from the southern counties of England and even as far southwest as the Isles of Scilly. The difference in winter distribution between the two, with the smaller bodied Bewick's wintering further south than the larger Whooper, is even more marked where the winter climate is severe, as in Japan, where the significance of relative

DKS

Figure 4.4. A European winter refuge.

body size, low ambient temperature and thus heat energy loss, are crucial factors affecting their distribution.

The somewhat isolated group of Whoopers in East Anglia was earlier thought to originate from the Scandinavian or Siberian population (e.g. Cadbury 1975), and the arrival of several oiled Bewick's and Whoopers on the Ouse Washes following an oiling incident in the Netherlands lent support to this (e.g. Ogilvie 1979). Following the marking of just 46 Whoopers with neck-collars in northeast Iceland in 1980, however, four (of 22 re-sighted widely across Britain) reached the Ouse Washes, confirming that some, and perhaps many, East Anglian Whoopers were of Icelandic origin (Brazil 1983a; Owen *et al.* 1986). More recently, the arrival of Finnish-ringed birds in southeast England (Beekman *et al.* 1999) confirms un-equivocally that birds from the northwest European population do also reach England, but the degree of interchange is considered low in relation to the population sizes (Delany *et al.* 1999).

According to Owen *et al.* (1986) 'The vast majority of British Whooper Swans have always been in Scotland. . .'. Thom (1986) described it there as a 'Widely dis-tributed winter and passage visitor, often in small parties but occasionally in flocks of several hundreds'. Scotland does hold important numbers, though perhaps less significant than these authors would have considered. At that time, the British and Irish wintering population at its November peak was estimated to be a mere 7,000, thus inflating the significance of those in Scotland, which amounted to 62% of that figure in 1979 (Brazil & Kirk 1979; Thom 1986).

The first Whoopers arrive early in Scotland, sometimes even in early September, although most arrive in late September–November (Dymond 1981; Thom 1986), with surprisingly little variation from Shetland to Norfolk. The supposed peak in numbers in November was formerly linked to the idea that Scottish winterers were swollen by birds en route to Ireland (e.g. Owen *et al.* 1986). Given the larger numbers recorded in recent censuses in Scotland in January, and that large numbers also make first landfall in northwest Ireland and as early as anywhere in the UK, the in-transit theory is no longer convincing. The latest midwinter census in January 2000 showed a reversal of the older pattern with 20% of the total wintering in England and only 13% in Scotland (Cranswick *et al.* 2002).

Though one might expect northern sites to receive the earliest arrivals and south-ernmost the latest, it appears that even sites as far apart as northern Scotland and eastern England (a journey of up to 800 km) may receive their first visitors within just a few days. In the UK's northernmost wintering areas, the Shetlands and Orkneys, the main autumnal influx occurs in late September–early October, although numbers continue to increase during November, with maxima at Loch Spiggie, the main site in Shetland, ranging from 100, following a very poor breed-ing season, to 300 (Dymond 1981; Okill 1987a,b, 1988, 1990). In Orkney mean numbers reach in excess of 500 (Reynolds 1982). It is tempting to relate this increase with the decline in Shetland that occurs from late November. Numbers in Orkney decline markedly between December and January with many emigrating, presumably south. In the early year the trend is not clear, but during March numbers increase again, presumably as birds move north from the mainland. The main exodus north is during the last weeks of April and in early May (Reynolds 1982), and is over a period very similar to that at most other localities.

Rees *et al.* (1997b), analysing the distribution of Whoopers across Britain, found that in October, 36% were in Shetland and Orkney, 25% in northeastern Scotland and 29% in Northern Ireland. Although 35–45% of those wintering in the UK then remained at Scottish sites in midwinter, they found a significant decrease in numbers at more northerly sites and a corresponding increase in southern Scotland, northwestern England and east-central England. In the UK, during April, as numbers further south decreased, high proportions occur in southeastern (25%) and northeastern Scotland (62%) as birds visit staging areas on their return to Iceland.

Buckland *et al.* (1990), in noting that at the Loch of Strathbeg, Scotland, adults arrive first, followed 3–4 weeks later by family parties, confirmed Gardarsson & Skarphedinsson's (1984) observations that families leave Iceland some time after adults without young, and lent support to the assumption that preferential migration occurs among the various breeding status classes (see Chapter Ten).

Flocks are deceptive; at first sight they may seem a cohesive unit, but they do not necessarily remain together or move as a unit over longer timeframes. Also, consid-erable interchange in membership may occur, with some moving extensively within Britain, as shown by neck-collar and colour-ringing studies, although in some areas individuals may indeed show high rates of return to a favoured area, such as

Caerlaverock or the Ouse Washes[2] (Brazil 1983a; Black & Rees 1984). The structure and movement of flocks seems to vary considerably, not only between regions, but also within a region over different periods. A case in point is the difference in the Stirling area (Scotland) before and after 1977.

Henty (1977) found that flocks utilising the area east of Stirling split up considerably more often than was found after 1977 by Brazil (1981a,c). Henty (1977) observed several small flocks in the area simultaneously which moved regularly between fields and apparently sometimes used separate roost sites. After 1977, however, the pattern of daily movement changed. It became rare to find Whoopers other than in a single feeding and roosting flock, rather than in several, and, although flocks changed in size and even disappeared for varying periods, any flock present tended to use a single field for several days as well as a single roost site (Brazil 1981c). Selection of feeding fields by flocks also changed. Whereas in 1974–76 selection occurred several times in one day, if the flocks moved; after 1977, selection occurred only once per day, as the birds left their roost site, unless the flock was disturbed. Whoopers leaving a riverine or lacustrine roost site must climb above trees and powerlines before they are able to fly safely out to fields to forage. From this height (*c.*30 m) they are able to view a considerable area of the fields available, as well as where earlier birds have already landed. The appearance of fields varies greatly from the air, thus both geese and swans may be capable of recognising good feeding fields while in flight. Abandonment of previously favoured fields may be due either to disturbance or declining food. If birds were disturbed, they always moved to another field providing the same food type.

Severe weather affects the continental and national distribution, with cold-weather movements accounting for extreme wintering totals (e.g. Bowler *et al.* 1993), vagrancy, and far larger wintering numbers in southern countries. Severe weather also affects local and regional distribution and disrupts the pattern of daily movements. During the late 1970s, the pattern of field use and movement of swans in the Stirling area changed markedly. In contrast to Henty (1977), Brazil (1981a,c) found the area to be used by medium-sized flocks on their way south or by flocks resident for short periods when conditions elsewhere had caused them to move. Use of the area in late winter and spring by smaller flocks seemed erratic, and swans were sometimes absent for prolonged periods. Perhaps field feeding was a relatively new phenomenon in the area at that time, and perhaps this was part of ongoing change as the birds developed their use of habitats in the area (Henty 1977; Brazil 1981a,c).

Cold weather may also cause temporary enlargement of flocks as birds concentrate at fewer locations, thus Cranswick & Pollitt (1994) noted that 856 at Welney during a cold snap in February 1993 was the largest ever recorded in England, while elsewhere peaks reached 883 at Loughs Neagh/Beg, 1166 at Lough Foyle, 389 at Lough Eye/Cromarty Firth and 612 at Upper Lough Erne and 840 on the Ouse Washes.

Thom (1986) noted that fewer Whooper Swans occur in Scotland generally in midwinter (not just Orkney and other northern sites; Reynolds 1982), this being apparently linked to cold-weather movements to Ireland. In March, numbers increase again as the return to the breeding grounds begins, with most having departed by late April, although occasionally some linger into May. Buckland *et al.* (1990) provided further evidence for this pattern with figures from Loch of Strathbeg, where the largest autumn count was 822 in October 1967, but where in midwinter typically only 50–100 remain, with numbers peaking at around 100 in March (302 on 15 February 1981 was exceptional).

Patterns of distribution and numbers in Scotland changed markedly during the 20th century, although the causes were unclear. Bannerman & Lodge's (1957) statement that numbers vary 'from year to year according to weather conditions' was based more on supposition than fact, as the true status and distribution in Britain was not then known. Their description of the species, as 'a very regular hard weather visitor, not only to Scotland, but also to many English counties', may, however, be taken at face value, and that they considered the Outer Hebrides to be their 'chief wintering place' seems to have been generally accepted during the first half of the 20th century. This was to change markedly, however, as the Hebrides declined considerably in significance and wintering became much commoner on the Scottish mainland, particularly in central Scotland. By the 1960s, however, this region too had declined in significance, whereas numbers increased markedly in the Moray Firth area, and particularly in England, where many more sites were occupied by the 1980s, and where by the turn of the century numbers exceeded those in Scotland (Boyd & Eltringham 1962; Brazil & Kirk 1979; Owen *et al.* 1986; Cranswick *et al.* 2002).

Boyd & Eltringham (1962) considered that the Scottish population increased considerably in 1948–52, but then remained fairly static in 1952–1960. The 1979 survey (Brazil & Kirk 1979) indicated a further increase and another change in distribution, with much larger numbers in the Northern Isles and Tweed Valley, and fewer in central Scotland. Thom (1986) considered that the explanation for any apparent increase lay in variations in census coverage. Although many Whoopers use freshwater habitats on the Scottish mainland, the increase in numbers there since the time of Bannerman & Lodge may be partly related to the development and increase in the use of fields (Brazil 1981a,c). Thom (1986) also noted that the largest recent concentrations were at the Cromarty Firth. Numbers there exceeded 500 annually in 1977–81 and reached a record 1,300+ in November 1979, presumably because of an exceptionally abundant supply of the submerged herb wigeon grass, a rich source of food. A repeat high of 1,115 at Loch Eye/Cromarty Firth in November 1990 was attributed to the same cause (Martin 1993).

In Scotland, the large number of remote lochs, rivers and marshes, coupled with the prevalence of field feeding, provide ideal conditions for Whooper Swans to winter in a widely dispersed form. Thus regular counts of the better known wildfowl concentrations inevitably under-record the species. Two attempts to rectify these shortcomings, namely the Wildfowl Trust and British Trust for Ornithology enquiry of 1960–61 and the Brazil & Kirk survey of 1979, also failed to achieve full coverage. The latter, however, indicated that birds wintering in Scotland in

1979 amounted to 62% of the (then known) British and Irish total (Thom 1986). Though under-recording was seen as a pitfall, the considerable extent of under-recording, especially in Ireland, was not imagined, and it is clear now that Scottish wintering birds in fact represent a very much smaller proportion of those wintering in the British Isles than previously thought.

THE SIZE OF THE WINTERING POPULATION IN THE UK AND IRELAND

Estimates of the numbers of Whooper Swans wintering in Britain and Ireland have increased considerably as coverage has improved since the 1960s. Boyd (according to Harrison 1973) had considered there to be 2,200–3,000 in England, Scotland, and Wales in 1952–60. In perhaps the earliest attempt to survey the British wintering population, Hewson (1964) recorded 2,200 in November 1960 and 3,100 in November 1961, and it was suggested that the population probably exceeded 4,000 (Boyd & Eltringham 1962), a total higher than in previous decades, although this was based on sparse information. By the early 1970s the population in Scotland was still estimated to be *c*.2,000 with 500 in England and Wales. In 1977, Owen (1977) considered that numbers had probably not greatly changed since the 1962 census (i.e. still *c*.4,000). In November 1979, however, Brazil & Kirk (1979) recorded at least 4,751; 4,145 in Scotland, and 606 in England and Wales, and 6765 including Ireland, although in the same year Ogilvie (1979) estimated the population at just 3,500. Brazil & Kirk (1979) considered that their total was not unusual but that it certainly missed many birds, and therefore suggested that numbers were larger than previously estimated. The difference between Brazil & Kirk's (1979) minimum of 6,765 and the estimated 9,000–11,000 population illustrated the poor coverage in Ireland and complete lack of coverage in Iceland (Owen *et al*. 1986). Comparable figures were obtained in 1990–91, when maxima in Britain and Northern Ireland alone were 8,908 in November 1990, 7,686 in December 1990 and 8709 in January 1991 (Rees *et al.* 1997a).

Although perhaps the most comprehensive by that date, our 1979 survey was frustratingly incomplete, particularly in remoter parts of Scotland and much of Ireland (Brazil & Kirk 1979), and while the figures may have been useful, no attempt was made to estimate the numbers missed. Yet the numbers for Scotland, England and Wales (4,751) alone were remarkably close to the figure of 5,000 quoted a decade later by Batten *et al.* (1990), suggesting that perhaps coverage was not as poor in 1979 as had been thought, except in Ireland. In Ireland, in the same year, Hutchinson (1979) estimated the wintering population to be 4,000–6,000, considerably more than previously thought.

Unfortunately, because national wildfowl counts miss so many Whoopers, those annual figures also do not reflect the full total, and Owen *et al.* (1986) noted that national trends could not be considered reliable on an annual basis. A 23-year run of annual counts did however reveal a statistically significant upward trend, as predicted given the three estimates then available. Given the inherent weaknesses in the national surveys previously attempted and that Whoopers are extremely mobile and often frequent sites not covered by formal wildfowl counts, this similarity may have been purely coincidental. The annual wildfowl counts, however, did show a dip during the mid 1960s (perhaps related to the series of very cold years in Iceland between 1965 and 1970), followed by an increase in 1970–79 at a mean rate of 6% per year, which Owen *et al.* (1986) attributed to a decreased mortality rate. Why mortality should have declined over that particular decade is unclear. Owen *et al.*'s (1986) rationale was that the species has a consistently low reproductive output ranging from 6% to 26%, averaging just 18% in 1948–61, with no indication of an improvement. Indeed, the 5% of the 1979 total represented by cygnets (Brazil & Kirk 1979) was the lowest ever recorded, but in a year infamous for being uniquely cold in Iceland (the coldest in a century of observations). For a population to sustain itself with such a low output its annual mortality must of necessity also be low, and to allow for the recorded increase in numbers, mortality may have decreased (Owen *et al.* 1986). That assumes of course that past population estimates were meaningful, which may be far from the truth.

With an increasing understanding of the true Icelandic population, estimated at *c.*14,000 by 1984–85 (Gardarsson 1991) it became clear that the numbers previously considered to be wintering in the UK and Ireland were still likely to be greatly underestimated. Given this, the WWT coordinated the first complete census, which included Iceland, Ireland, and Britain on 11–12 January 1986 (Salmon & Black 1986). Coverage was considerably greater than in all previous attempts, again making it difficult to compare the results. Unsurprisingly, given the then recent Icelandic estimate, the January 1986 total of over 16,700 was the largest count ever made, and a credit to the organisers and the comprehensive coverage they achieved. To what extent this should be seen as an increase in numbers, given the status of previous surveys, is uncertain. Perhaps, instead, the 1986 survey should be viewed as the first reliable and repeatable baseline survey. Ireland, in particular, had not previously been so thoroughly surveyed, although Salmon & Black (1986), still perhaps thinking from the same perspective as Thom (1986), that Scotland held the highest numbers, felt it necessary to qualify the high count in Ireland as being partially explicable by birds having left Scotland during cold weather early in January, a tendency originally suggested by Black & Rees (1984).

A follow-up international census was made in January 1991, when an even higher total of 18,035, thought to represent almost all of the Icelandic breeding population, was counted. As some flocks were definitely missed and some areas under-observed, it is likely that the true figure was closer to 19,000 (Kirby *et al.* 1992; see Table 4.5). Ireland held the vast majority with 3,484 (19.3%) in Northern Ireland and 8,490 (47.1%) in the Irish Republic. Britain held 5,225 (29%) (barely more than the 1979 total), and 831 (4.6%) were in Iceland (Kirby *et al.* 1992).

In 1991 the British and Irish Republic totals were very similar to those in January 1986. However, numbers in Iceland were 36% lower and in Northern

Table 4.5. *Recent winter estimates of the Icelandic/Irish/British population.*

Year	Total Counts	Ireland	Scotland	England and Wales	Iceland	Estimated total	Reference
November 1960						2,200	Owen et al. 1986
November 1961						3,500	Atkinson-Willes 1964
			2,000	2,000		4,000–6,000	Boyd & Eltringham 1962
early 1970s		2,000			500	2,500	Ogilvie 1972
1979						3,500	Ogilvie 1979
November 1979	6,765	2,014	4,145	606			Brazil & Kirk 1979; Owen et al. 1986
		4,000–6,000			1,000–1,500	5,000–6,000	Atkinson-Willes 1981 Hutchinson 1979
October 1985						5,000+	Batten et al. 1990
January 1986	16,742	10,306	5,136		1,300	17,000	Salmon & Black 1986
January 1991	18,035	11,974	5,225		831	19,000	Kirby et al. 1992
January 1995	13,947	8,942	5,005		797	16,000	Cranswick 1996
January 2000	20,645	3,716 and 9,084	2,684 and 4,129		1,238		Cranswick et al. 2002

Ireland 47% higher than in 1986. Whereas in 1986 excuses were still being made for the unexpectedly high figures in Ireland (10,320; Salmon & Black 1986), generally assumed to be the effect of a cold-weather influx from Scotland, in 1991 even higher numbers (11,974; Kirby *et al.* 1992) occurred in Ireland indicating that large numbers there were not unusual (see Tables 4.5, and 4.6).

Hewson (1964), who studied flock sizes in the 1960s, showed that most Whooper flocks held fewer than ten and only 5% of 1,252 flocks contained more than 50 birds. Kennedy *et al.* (1954) also found it usual to see parties of fewer than a dozen, and during the Irish survey of January 1986, Merne & Murphy (1986) found that the majority of the 361 sites held fewer than 50 each. The January 1991 census reconfirmed that Whooper Swans in Britain and Ireland tend to occur at widely scattered localities in rather small flocks. Although in some areas, especially on migration, large flocks may occur, in Britain any wintering concentration of over 100–200 is major. During the 1991 survey, Whoopers were recorded at a total of 514 sites, but 42% of these held fewer than ten, while 84% held fewer than 60 (Kirby *et al.* 1992) (Fig. 4.5). There is also a significant association between geographic location and flock size, with the largest flocks in Scotland (especially the northeast), Northern Ireland and northeast England, with smaller flocks occurring in more southerly parts of Ireland and England (Rees *et al.* 1997a).

In January 1995 the third international census achieved excellent coverage in Britain and Ireland, and covered many of the most important Icelandic sites too. The overall total of 15,702 counted or estimated[3] represented a reduction of 13% against the 1991 census. Cranswick (1996) considered that the reduction was partly due to double-counting in 1991, but also partly due to a genuine decrease. On the basis that an unknown, but probably small number, of Scandinavian and Russian Whoopers winter in Britain and that as many as 400 Icelandic birds winter on the European continent, Cranswick (1996) recommended that a figure of 16,000 be considered as the Icelandic population in 1995.

By January 2000, the population had increased to its highest level known— 20,645, a 30% increase over the 1995 total, with 44% in the Republic of Ireland, 18% in Northern Ireland, 13% in Scotland, 20% in England and 6% in Iceland. Increased numbers of flocks and larger flock sizes were noted from Ireland, while

Table 4.6. Numbers of Whooper Swans counted (estimated 1995) in Britain, Ireland, and Iceland (after Salmon and Black 1986; Kirby et al. 1992; Cranswick 1996; Cranswick et al. 2002).

	1986	1991	1995	2000
Britain	5,136	5,225	5,016	
Northern Ireland	2,363	3,484	2,783	3,716
Republic of Ireland	7,943	8,490	6,932	9,084
Iceland	1,300	831	971	1,238
Total	16,742	18,035	15,702	20,645

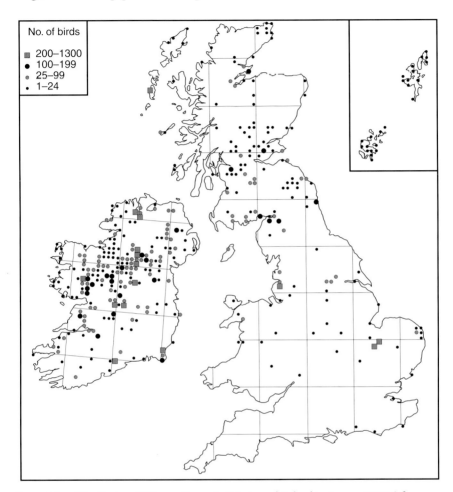

Figure 4.5. Distribution of Whooper Swans in Britain and Ireland in January 2000 (after Cranswick 2002).

in England birds were more concentrated and primarily at two sites, WWT Martin Mere and WWT Welney (Cranswick *et al.* 2002).

THE PROPORTION OF CYGNETS IN THE WINTERING POPULATION

Breeding success varies enormously annually. Over the period 1948–84 the proportion of cygnets among those wintering in Britain and Ireland was 5–26% with a mean of 19.6%. Whereas the 1979 survey found that cygnets contributed just 5%, a low proportion following an extremely poor breeding season, the 1986 survey found 22.9%, a high proportion, clearly following a very successful breeding season (Boyd & Eltringham 1962; Brazil & Kirk 1979; Gardarsson & Skarphedinsson 1984; Salmon & Black 1986), but little is known of the factors affecting annual variation in breeding success. Of particular interest, however, is that even after a good breeding season there is marked regional variation in the proportion of cygnets among the wintering population, with some regions having considerably smaller proportions than others, indicating some differing degree of selection by families and non-families (see Table 4.7).

In contrast to the 1986 survey, that in 1991 clearly followed a poor breeding season, and the overall proportion of cygnets in the wintering population was, at just 9.8%, less than half that in 1985. The proportion in Britain and Northern Ireland was somewhat higher than the mean, at *c*.14%, but that in the Republic of Ireland was a remarkably low 5.9%, almost the lowest on record. A comparison with the 22.9% in Britain and 23.1% in Ireland during the 1986 census provides a stark reminder of the great variation in productivity that Whooper Swan experiences annually. In winter 1992–93, following a reasonable breeding season, Cranswick & Pollitt (1994) found that even WWT reserves, when analysed alone, attracted wintering flocks with a wide range of proportions of juveniles (9–14%), indicating that despite their provision of food these reserves do not necessarily attract high proportions of families, as might be expected. Such annual variation can be particularly impressive at a single site. For example, Black & Rees (1984) found that at Caerlaverock, Scotland, the annual variation was from 7.7% to 31.0%, though at 17.5% the overall mean was very similar to Boyd & Eltringham's (1962) estimate of breeding success in the total British population in 1948–61.

Table 4.7.   Regional variations in the proportion of young Whooper Swans at wintering sites in Iceland, Ireland and Britain in January 1986 (after Salmon & Black 1986; Kirby et al. *1992).*

Region	Number 1986	% of cygnets	Number 1991	% of cygnets
NE Scotland	1,028	25.8		
NW Scotland	537	24.1		
EC Scotland	429	23.0		
Clyde Basin	527	17.3		
Solway Basin	881	17.2		
NW England/N Wales	480	17.1		
SE Scotland/NE England	655	26.9		
S England/S Wales	588	23.8		
Iceland	1,300		831	18.6
Northern Ireland	2,363	16.7	3,484	14.0
Republic of Ireland	7,943	23.1	8,490	5.9
Britain	5,136	22.9	5,225	14.0
Total	16,742		18,035	

Kirby *et al.* (1992) speculated that the low proportion of cygnets in the Irish Republic in 1991 might have been due to differences in habitat quality selected by family parties, which Scott (1980c) and Black (1988) have shown are usually dominant in winter flocks. Alternatively, Kirby *et al.* (1992) considered that it may have resulted from yearlings and single birds undertaking exploratory dispersal to sites further from the breeding range, although that supposition was not supported by the 1986 census which showed that the proportion of cygnets in flocks was higher in the Irish Republic than elsewhere in the British Isles.

In Iceland, in 1991, the proportion of cygnets was a relatively high 18.6%, supporting Gardarsson & Skarphedinsson's (1985) view that birds overwintering there include a higher proportion of family parties than migrants (Kirby *et al.* 1992). Further supporting Gardarsson & Skarphedinsson's (1985) view is that not only was there considerable regional variation in the proportion of young in flocks within Britain, Ireland and Iceland during the 1991 census, but there was also considerable variation in family size. In Iceland, the majority of adult Whoopers were accompanied by three cygnets (36%) as opposed to two (37.9–40.9%) in Ireland, and just one (41.1%) in Britain (Kirby *et al.* 1992). Of the 522 families recorded by the 1991 census, 36.2% were of two young and 86.5% had 1–3 young, indicating the relative rarity of large broods, especially in a poor breeding season (see Table 4.8). However, in contrast to the earlier situation, in January 2000 the proportion of young was slightly lower in Iceland than in Britain (Cranswick *et al.* 2002).

During the January 1995 census, of the 10,156 for which ages were assessed 17.9% were cygnets, indicating that the 1994 breeding season had been reasonably successful. Regional variation in the proportion of young was again observed, with southwest Iceland and northwest Scotland having the highest values and Northern Ireland the lowest (Cranswick 1996). Overall brood size was 2.32. During the January 2000 census there was a high degree of consistency between countries and

Table 4.8.   Proportion of young, and frequency of different brood sizes of Whooper Swans in Ireland, Britain, and Iceland (January–March 1991) (after Kirby et al. 1992).

Region				Brood size				
	Total	1	2	3	4	5	6	7
Britain	n = 1,810	25	13	11	11	1	0	0
	% = 14.5	41.1	21.3	18.0	18.0	1.6	0.0	0.0
Northern	n = 2,097	32	39	20	12	0	0	0
Ireland	% = 13.5	31.0	37.9	19.4	11.7	0.0	0.0	0.0
Republic of Ireland	n = 6,092	74	126	69	35	4	0	1
	% = 5.9	24.0	40.9	22.4	11.4	1.3	0.0	0.0
Iceland	n = 806	14	11	18	7	0	0	0
	% = 18.6	28.0	22.0	36.0	14.0	0.0	0.0	0.0
Total	10,805	145	189	118	65	5	0	0
	9.8	27.7	36.2	22.6	12.5	1.0	0.0	0.0

regions in terms of proportion of young in the population, which was 16.8%, reflecting a good breeding season. Overall brood size was 2.3. The proportion of young in Iceland in winter was slightly lower than in the British Isles (Cranswick *et al.* 2002).

THE SIGNIFICANCE OF CERTAIN SITES AND THEIR PATTERN OF USE

Ireland and the United Kingdom hold significant numbers of Whooper Swans both nationally and internationally. Sites of international importance are defined as those holding at least 1% of the total population throughout the species' range over a five-year period. The critical level for the Icelandic-breeding population was set in the early 1990s at 170 (Kirby *et al.* 1992), although given the current estimate of 20,000 this should now be increased to 200. Sites with 60 or more qualify as sites of national importance in Britain.

In Britain in 1991, flocks containing more than 60, and therefore nationally significant, were mainly in Scotland: in Orkney, in the Western Isles, along the coastal plain of eastern Scotland, on the Clyde and Solway firths, and at a smaller number of sites across England. The largest flocks in Northern Ireland were concentrated around Loughs Foyle, Neagh, Beg and Upper Lough Erne, while in the Republic of Ireland the largest were in the interior of Co. Mayo, the Shannon Valley and in Cork and Wexford (Kirby *et al.* 1992).

On the basis of counts made during the 1986–87 to 1990–91 period, several sites also qualified as internationally significant: Lough Foyle (mean 1,157 birds), Loughs Neagh and Beg (1,152), Upper Lough Erne (739), Loch of Harray (669), the Ouse Washes (594) Loch Eye/Cromarty Firth (588), Martin Mere/Ribble Estuary (538), Solway Firth (246), Loch of Strathbeg (245) and Lock Leven (209) all qualified in Northern Ireland and in Britain. In the Irish Republic, Lough Swilly (750), Wilkinston (367), Blackwater Callows (245), Glen Lough (236), Middle River Shannon (211), Lough Oughter (207), Little Brosna River (201), Blindwell Turlough (186), Stabannon (185) and Rahsane Turlough (179), all qualified (Kirby *et al.* 1992).

Andrews & Carter (1993) in their conservation and monitoring review of 1990–91 found that ten sites held a five-year mean of >200 (with peak numbers in 1990–91): Lough Foyle (988 in November 1990), Lough Neagh/Beg (1,110 in December 1990), Upper Lough Erne (896 in January 1991), Loch of Harray (927 in January 1991), Ouse Washes (578 in March 1991), Loch Eye/Cromarty Firth (1,115 in November 1990), Martin Mere/Ribble Estuary (538 in November 1990), Solway Estuary (196 in March 1991), Loch of Strathbeg (129 in January 1991) and Loch Leven (180 in December 1990). Another two held more than 170 in 1990–91 (though not scoring on the five-year average): these were the Montrose Basin (501 in February 1991) and Loch of Skene (314 in December 1990).

Wintering sites can be considered of significance, not merely because of the numbers they attract, but also because some may attract the same birds annually.

Caerlaverock serves, again, as a good example. There, Black & Rees (1984) found that the return rate among ringed birds was very high (both males and females), with 78% returning for at least a second winter, and 40% returning for four consecutive winters after first being ringed in 1979–80. Pairs and families were found to be most likely, and single adults and yearlings least likely, to return. The high proportion of ringed cygnets that returned (78.6% of 42) was surprisingly high compared with juvenile Bewick's Swans, which showed less attachment to their first wintering site.

Cranswick (1996) noted that in January 1995, the overall distribution was similar to that in 1991; however the proportion wintering in England had risen slightly. Furthermore, eight sites held more than 1% of the population (Swilly/Foyle, Ireland/Northern Ireland; Shannon Callows, Glen Lough and Lough Atedaun, Ireland; Loughs Neagh and Beg and Upper Loch Erne, Northern Ireland; Ouse Washes and Martin Mere, England), while a number of Scottish sites important in the early 1990s had, for unknown reasons, declined dramatically in significance.

THE PATTERN OF USE AT WILDFOWL AND WETLAND TRUST CENTRES

It is not practicable to attempt to review the pattern of usage of all the sites that are of international or national significance, even in a country such as the UK, let alone throughout the wintering range of this species; however, three of the Wildfowl & Wetlands Trust's reserves hold flocks that are highly significant on an international scale, and they provide suitable examples. They are: WWT Caerlaverock, beside the Solway Estuary in southwest Scotland, WWT Martin Mere, near the Ribble Estuary in Lancashire, England, and WWT Welney, on the Ouse Washes in Norfolk, England (Fig. 4.6).

The WWT has, since 1970, maintained records of Whooper Swans at their centres and the published material covering 1990–2000 gives an indication of activity in the UK (though it should be borne in mind that this applies only to the relatively small proportion of swans actually at WWT sites (*c*.12% of those wintering in the British Isles in 1994–95; Rees & Bowler 1997), which are to some extent atypical as they are provisioned with food. The majority wintering elsewhere may behave differently.

Interestingly, the Caerlaverock population only developed in the 1970s, even though there was already a small population nearby at Islesteps, 10–15 in the 1950s and 60–80 by the late 1960s. The Caerlaverock flock did not grow by recruitment and relocation from Islesteps. In fact, numbers at both sites increased simultaneously. From the mid 1970s, the maximum number at Caerlaverock increased annually so that by the mid 1980s Caerlaverock had outstripped Islesteps, and already qualified as a site of international importance (Black & Rees 1984; Rees & Bowler 1997).

The wintering populations of both Bewick's and Whooper Swans at Martin Mere in Lancashire began to develop only in the late 1970s, rising steadily from

Figure 4.6.   The locations of three internationally significant winter flocks in the UK.

fewer than 50 Whoopers in 1980 to almost 500 by 1990. The development of strong and persistent flocks at both Caerlaverock and Martin Mere may have resulted from the increase in the Icelandic population in the 1960s and 1970s, whereas the continued increase in numbers at Martin Mere since 1990, to more than 1,500, despite censuses indicating a stabilisation of the Icelandic population, are likely due to the attractiveness of the reserve drawing birds from other sites (Rees & Bowler 1997).

The Ouse Washes population as we know it now, like that at Caerlaverock, is a relatively recent phenomenon. Whereas the Caerlaverock flock was on the increase during the mid 1970s, that at the Ouse Washes began to build-up slightly later. Though remains of Whooper Swans in post-glacial peat deposits indicate that it occurred frequently in the Cambridgeshire fens *c.*2,000 years ago (Cadbury 1975), at some point after that it underwent a serious decline, perhaps as a result of fenland drainage. In the early 1960s, Cadbury (1975) found none there, and in 1970–75, the average January maximum was a mere 31. Even at that level it was already the southernmost regular wintering flock in Britain, and usage was increasingly significantly.

At the Welney centre, it was the provision of grain for Bewick's Swans already wintering on the reserve that initially encouraged Whooper Swans. Numbers were low at first, and only grew slowly during the 1960s and 1970s. Whoopers first exceeded 100 birds in 1979–80 when a peak of 106 was attained. During the 1980s numbers increased rapidly, completely altering the species' status in the county (Table 4.9). The more recent rapid rise coincided with the further provision of food, in this instance the increased tipping of waste potatoes. In contrast to the majority of Bewick's Swans, which typically fly out to feed on arable land regard-less of the availability of food at Welney, Whoopers there are strongly attracted to waste potatoes, being reluctant to leave the reserve while adequate supplies are available (Kemp 1991). Whooper Swans had clearly discovered a good combination of fen feeding, a safe roost site, and the additional bonus of intermittent supplies of potatoes.

Beginning in 1990, the pattern of usage at these three important sites was as follows (Fig. 4.7). In autumn 1990, despite some early arrivals in September, Whooper Swans were, on the whole, later than expected, with the main influx not occurring at Caerlaverock, Martin Mere or Welney until November. The first birds reached Caerlaverock on 21 September (a record early date there), but also reached Martin Mere and Welney, considerably further south, just three days later on 24 September (a record early date at Welney). At Caerlaverock numbers built up slowly, remained at 170–180 throughout January–February, then surprisingly peaked very late with a maximum of 192 on 14–15 March. They soon declined, and by 31 March 1991 only 24 were present. Thus the peak presumably involved birds on their northbound migration from sites further south. At Martin Mere, numbers reached a new site record of 448 on 30 November, rose slightly to 469 on 6 December and to another record, the season maximum, of 473 on 16 December 1990. Welney too attained another record in 1990–91 season with a maximum of 561 on 11 January 1991 (Rees *et al.* 1991b). That winter the proportion of juve-niles was very even and confirmed that the 1990 breeding season had been a good one. Juveniles represented 17.6% (63 of 357) of the Martin Mere population on 18 December, 16.9% (26 of 154) at Caerlaverock on 5 November, and 16.4% (of 373) at Welney on 12 January.

In autumn 1991, arrivals followed a more typical pattern, with small numbers arriving at Caerlaverock, Martin Mere and Welney in late September, but with the main influx during October and early November. Spring migration was most

Table 4.9. Whooper Swans on the Ouse Washes, UK (after Kemp 1991).

	1979–80	1980–81	1981–82	1982–83	1983–84	1984–85	1985–86	1986–87	1987–88	1988–89	1989–90	1990–91
Adults	106	130	161	223	248	240	320	520	582	603	686	600
Juveniles (%) in total flock							20	8	18	10	11	16

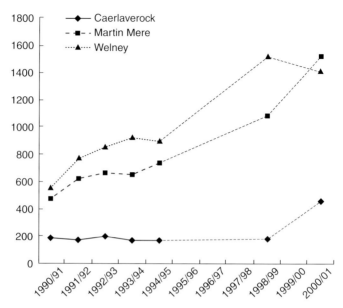

Figure 4.7.   Peak Whooper Swan numbers at WWT Caerlaverock, Martin Mere and Welney, 1990–2001.

noticeable during the second half of March 1992. Again in autumn 1991, first arrivals were closely grouped, even at sites far apart. The first eight adults arrived at Caerlaverock early on 27 September, and coincidentally the first two reached Martin Mere the same day, while the first to arrive at Welney, a pair, did so next day. Numbers rose very slowly at Martin Mere until a "swanfall" on 22 October, after which 300 were seen from 5 November, rising to 479 on 15 November and 631 on 22 November (a new Lancashire record). Meanwhile, at Welney, after the main influx in early November, by 11 November 477 had gathered, but more continued to arrive throughout November and by early December 500 were present, rising to 606 by the year-end and 716 on 8 January 1992 (724 were counted over the entire Ouse Washes that day)[4]. January saw a further increase, but February the peak with 778 on 26 February representing the highest count of the winter, and a new record for England and Wales. Numbers only dropped to 624 by 16 March with the main exodus taking place in the following two weeks to leave just *c*.60 by 30 March (Bowler *et al.* 1992). Numbers at Caerlaverock were slightly lower than in the previous season, peaking at 171 on 15 December 1991 (Bowler *et al.* 1992).

Arrivals and departures from Welney follow a pattern rather typical of the country as a whole, even though it is one of the southernmost regular wintering sites in Britain. Kemp (1991) found that the earliest Whoopers typically arrive before the end of September, that numbers increase during October but the main influx is typically in November, with peak numbers in December or January. In contrast to the smaller, and more northerly breeding, Bewick's Swans, Whoopers are slower to depart in spring. They are seemingly in no hurry, even when conditions are mild. Bewick's are quick to leave during mild winters with relatively few remaining on the Ouse Washes even in late February. In contrast, Whoopers, perhaps because they migrate a shorter distance to their breeding areas, linger into mid or late March as a matter of course.

The number of cygnets reported during the winter indicated that 1991 had been another good breeding year, although the spread was uneven; 14.6% at Caerlaverock on 15 December were juveniles (25 of 171 aged); 22.6% (121 of 535) were juvenile at Welney; and a high of 25% at Martin Mere were cygnets (Bowler *et al.* 1992).

The following winter, 1992–93, continued the trend with yet more Whoopers wintering at WWT centres. Caerlaverock received its first four on 23 September 1992, and subsequently numbers there reached 200 on 17 November, of at least 294 in the area. A mass migration occurred during late October and early November with a large influx resulting in high counts at all three WWT reserves (Bowler *et al.* 1993). On 25 December 1992, Welney set a new record, when 830 were counted. After a slight fall, numbers rose again. Then, when Welney experienced a cold snap on 28 February 1993, not only was the winter's peak reached, but yet another new reserve record, a phenomenal 856 were counted, making this the largest-ever concentration in England. Martin Mere similarly broke its own previous site record with 666 on 23 December 1992, also during a period of cold weather (Bowler *et al.* 1993). The main exodus from Welney occurred during March as usual, with clear evening skies encouraging departure. By 29 March just 80 remained (Bowler *et al.* 1993). Summer 1992 had been less productive than those of 1990 or 1991 yet, with estimates of cygnets ranging from 9.1% (of 173) at Caerlaverock on 21 December, and 13–14% at Welney and Martin Mere, it was by no means a disaster and, despite this reduced output, the species fared better than did Bewick's which was estimated to have just 3–5% cygnets (Bowler *et al.* 1993). Furthermore, for a second winter, the proportion of cygnets recorded at Caerlaverock was lower than at the southerly sites, a trend that I have also found in Japan (Brazil 1983c).

Certain parts of Britain have seen considerable increases in numbers in recent years, and not merely at WWT sites (see above). In Norfolk Whooper Swan has traditionally been regarded as a scarce passage migrant and winter visitor, becoming more widespread during severe weather. That county, however, saw a dramatic increase in numbers during the 1980s (Kemp 1991). Always scarce along the north Norfolk coast with, in the past, rarely more than 5–10 individuals involved, Kemp (1991), however, found up to 90 on the Norfolk Broads, at Hickling/Horsey, up to 686 on the Ouse Washes and erratically up to 40 at Wolferton, and over 700 wintering in the county as a whole. These numbers have since been exceeded (especially at Welney, with 856 in 1993; Bowler *et al.* 1993).

The first arrivals of the 1993–94 winter were eight at Martin Mere on 26 September (Bowler *et al.* 1994), and numbers rose to further record-breaking totals at WWT reserves during that season. Welney exceeded its record again when 924 gathered there on 6 February 1994, the record for the entire Ouse Washes was also broken with 986 on 22 February, 723 of which were at Welney. Meanwhile, Martin Mere just missed its previous record (666) when 650 gathered there on 18 February (Bowler *et al.* 1994). The percentage of young was 11–14% in the flocks that winter (Bowler *et al.* 1994).

The first Whoopers of the 1993–1994 winter arrived at WWT Caerlaverock on 28 September. A mere ten had gathered by 7 October, but a large influx followed, with 71 on 14 October and at least 129 on 24 October. Numbers continued to increase, to 143 on 15 November, 157 by 6 December and 163 on 26 December, including 24 (14.7%) cygnets. Cold weather in mid February attracted more to the reserve, with maxima of 162 on 15 February 1994 and 175 on 18 February. On 26 March, 120 were still present, but numbers declined rapidly thereafter (Bowler *et al.* 1994).

The earliest migrants arrived at WWT Martin Mere on 26 September 1993, but they did not stay, two of them being found at Welney next day. The next to arrive were on 7 October. Thereafter numbers built up quickly to 137 on 15 October, 428 on 22 October and 483 by the end of the month. More arrived steadily during November, reaching 624 on 22 November (Bowler *et al.* 1994). Numbers rose slightly in the New Year, reaching 630 on 23 January 1994, 651 on 3 February and 655 on 18 February. Of the 589 whose ages were determined on 11 January, 104 (17.7%) were cygnets (Bowler *et al.* 1994).

The first arrivals of the 1993–1994 winter at WWT Welney were a pair on 27 September that had been at Martin Mere on 26th. The next few arrivals (four) were on 9 October, followed thereafter by more almost every day, with 126 present on 17 October and 320 by late October, slowly rising to 460 on 14 November and 560 on 11 December. The proportion of juveniles was rather low with 47 (13.4%) cygnets among 350 on 2 December. Numbers continued to rise in the New Year, to 775 on 30 January 1994 and a new record for Welney and for the entire Ouse Washes of 924 on 6 February. Numbers continued to increase in the area and although numbers at Welney fell to 723, a new record of 986 was set on the Ouse Washes on 22 February, the largest concentration ever in England and Wales. Numbers remained high well into March, with 800 still present on 20 March Bowler *et al.* 1994).

The first Whoopers of the 1994–1995 winter arrived at WWT Caerlaverock on 28 September, but further arrivals were slow, presumably because of unfavourable winds. There were 12 by 4 October but only three next day, numbers reached double figures again on 23 October (14). Conditions changed and birds arrived suddenly on 26 October with 37 present by the end of the day and 45 by dawn the next. By late on 28 October 100 were present including 27 cygnets. The largest roost count was 176 on 26 November. About 100 remained for most of December, whereas in January 1995 they fluctuated at 70–150, with new birds still arriving. On 4 January, of 150 present 32 (20.9%) were cygnets—confirmation of an excellent breeding season in Iceland in 1994. Numbers remained high into late March, but fell quickly from 142 on 24 March to 117 on 31 March, 76 on 3 April, 26 on 9 April, 16 on 16 April and just six on 23 April (Bowler *et al.* 1995).

The first four Whoopers arrived at WWT Martin Mere also on 28 September, but numbers built up slower than usual with only 32 on 21 October and 345 by 31 October. Numbers climbed quickly during November, with 533 present at dawn on 5th, 651 by 24th, and 738 by 13 December setting a new reserve and county record. On the same day a further 39 were found nearby pushing the new Lancashire record to 777. The good breeding season was reflected here too with 125 (17.3%) cygnets among 723 on 22 January 1995. Numbers began to decline in late February, from 734 on 21 February to 666 on 1 March, 514 on 9 March, and falling rapidly to 390 on 11th, 106 on 23rd and only 26 by 30 March. By 9 April one adult and two cygnets, all presumed injured or ill, were all that remained (Bowler *et al.* 1995).

Predominantly southwest airflow and heavy cloud cover delayed departure from Iceland, such that the first to reach WWT Welney did so only on 4 October 1994 (two) with few further arrivals until later in the month, when numbers suddenly rose to 446 by 23 October. Continuing mild conditions did not hasten new arrivals, by 20 November only 492 were present. A further influx in late November pushed numbers up to 853 by 4 December. On the same day a further 289 were on the RSPB Ouse Washes reserve, the total of 1,142 shattering the previous record of 986. The peak count at WWT Welney was 905 on 21 January 1995, and more than 1,000 were still present on the washes as a whole on 19 February. The proportion of cygnets was estimated to be 17% and the largest brood was of six. Numbers fell during late February–March with 386 on 26 March. Favourable conditions for migration in late March led to a rapid drop in numbers to 163 on 1 April, 53 on 4 April and only 12 by 23 April (Bowler *et al.* 1995).

Publication of the annual use of Caerlaverock, Martin Mere and Welney in the journal *Wildfowl* ceased after the 1994–95 winter making it more difficult to track, but the first Whoopers of the 1998–99 winter arrived at both Caerlaverock and Martin Mere on 5 October 1998, about a week later than previous years. Numbers at Caerlaverock rose to 188, while at Martin Mere a new reserve record of 1,085 was set. Welney also set and broke a new reserve record of 984 by peaking at 1,526 on 6 December 1998, when a further 1,055 were in the Ouse Washes area (Anon 1999). For the most recent winter for which I have information, that of 2000–01, both Martin Mere and the Ouse Washes set new records, 1,530 at the former and 1,797 at the latter (1,421 at WWT Welney), while Caerlaverock and nearby Islesteps also attracted 466 (Eileen Rees *in litt.*).

IRELAND

The extraordinary significance of Ireland as the winter recipient of the bulk of Iceland's breeding Whoopers was long overlooked by English ornithologists, though no doubt not by Irish ones! Early in the last century, Coward (1920) considered the Whooper rarer in Ireland 'than the smaller species', while of the Bewick's he wrote 'most numerous in Scotland and the north and west of Ireland'. In retrospect, either an extraordinarily symmetrical shift has occurred or lingering confusion over their identification led to the transposition of information that coincidentally matches remarkably the current distribution of the Whooper! According to Kennedy *et al.* (1954) and Hutchinson (1979), however, the status of the Whooper did change considerably after about 1900. Whereas previously Bewick's had been more numerous, after 1900, the reverse became true. In Co. Galway, Mayo and Cavan, Whoopers suddenly became widespread in the early 1940s and by the 1970s it was common throughout the west and midlands of Ireland, with fewer in Munster and the east coast counties (Ruttledge 1974; Hutchinson 1979). As in Scotland and England, in Ireland they appear commonly in October, sometimes in September and remain until March–April.

Ogilvie (1972) had estimated that the westernmost wintering population of Whoopers, essentially derived from Iceland, was of *c.*2,500 in Britain, 2,000 in Ireland and 500 in Iceland. Though now clearly wildly out, based on knowledge at the time it seemed reasonable. With hindsight, it is most surprising that he was almost closer with his winter figure for Iceland than any other. Gardarsson (1975) estimated that during the 1970s Icelandic Whoopers numbered 5,000–7,000 and that the population had changed little over the previous 20 years. Hutchinson (1979) was the first to hint at Ireland being more significant than previous authors, such as Ogilvie (1972), had indicated, with his estimate of 4,000–6,000 birds. In some locations large flocks of more than a 1,000 have been observed (e.g. Lough Swilly). Keen to clarify the situation, Brazil & Kirk (1979) attempted in November 1979 to census the population in Britain and Ireland but failed to achieve the degree of coverage they had hoped for, recording only 2,014 in Ireland (coincidentally almost identical to Ogilvie's estimate, but with the knowledge that our coverage in Ireland was very poor and hence that our figure was a considerable underestimate).

Not until 1986 was anything like full coverage achieved, and then, even during the January 1986 survey, the surprising total of 10,320 was considered to be abnormally high due to hard-weather movements from northern Britain (Merne & Murphy 1986) rather than a true reflection of the significance of Ireland. The

January 1991 census of both the Republic of Ireland and Northern Ireland, however, finally confirmed Ireland's great importance as a wintering area. This census was linked to an international census of Icelandic breeders. The total number of Whoopers was revealed then to be 18,030, of which a staggering 66.4% (11,974) were in Ireland, making Ireland more than twice as important for the Icelandic population than Britain, where just 29% (5,225) were found, and Iceland where 4.6% (831) wintered. Not only did these figures prove Ireland's importance but they also represented a significant increase over the total of 16,756 recorded during the January 1986 census (Merne & Murphy 1986).

Censuses of the breeding population in Iceland during the 1980s began to establish that numbers there prior to emigration in autumn were significantly larger than previously thought: 12,557 in early October 1982, 13,686 in early October 1984 (of which 3,303 were young) and 14,247 in early October 1985 (3,092 young), compared with 5,000–7,000 from the 1950s to the 1970s (Gardarsson 1975; Gardarsson & Skarphedinsson 1984; Merne & Murphy 1986)[5], thus forcing a complete re-think of the size of the entire flyway. Where were the birds going, how many were going there, and how many remained in Iceland to winter?

In light of the Icelandic findings, it was essential to establish answers to some of the inevitable questions. In response the Wildfowl Trust organised a midwinter census for 1985/86 covering Britain, Ireland and Iceland. On 11–12 January the census produced 2,377 in Northern Ireland and 7,943 in the Irish Republic (Merne & Murphy 1986). At the same time Salmon & Black (1986) recorded 5,136 in Britain and 1,300 in Iceland, giving a total of 16,756 and the largest number ever recorded.

Where Hutchinson (1979) had surmised, Merne & Murphy (1986) and Hutchinson (1989) were able, with 10,320 recorded, to point with some certainty at the clear significance of Ireland as the most important wintering area for Icelandic Whoopers. The island held 62% of the population, a figure that virtually doubled Hutchinson's (1979) previous best estimate of 4,000–6,000.

Given that the general opinion during the 1960s and 1970s was that we were dealing with a relatively small population, these winter figures, had they come without the surveys in Iceland, would have been quite unimaginable. Owen *et al.* (1986) considered that the recorded increase was probably due to better coverage in 1986 than ever before, combined with a real increase in the Icelandic population as a result of reduced mortality. Any such comparisons with past figures are of course fraught with uncertainty, as no comparable surveys had achieved such coverage. To what extent the Icelandic population had been under-recorded and for how long will now never be known. Merne & Murphy (1986), no doubt as astonished as anyone at the scale of numbers in Ireland, attempted to explain the unprecedented figure by suggesting that freezing conditions over much of Britain and Ireland in early January may have caused some to move from Scotland, thus abnormally swelling numbers. Again, with hindsight, that attempted explanation was unnecessary.

The Irish survey revealed a great deal more than just a large total. There were in all 361 sites and the majority of these held fewer than 50 birds. Only 44 sites held higher numbers and just 17 held 100 or more (up to 876; Merne & Murphy 1986). Whereas overall numbers were considerably larger than previously known, the habit of tending to occur in smaller flocks appears not to have changed. For example, Kennedy *et al.* (1954) had also found it usual to see parties of fewer than a dozen, while flocks of 60–80 were uncommon and the largest flock was of *c.*140 in Co. Galway in November 1944.

It became clear too that 1985 had been a very good breeding season indeed, with a mean brood size of 2.41 (N=230) in Ireland during the winter (Table 4.10). Large numbers were aged and brood sizes recorded during the national census, and the percentage of cygnets for the whole of Ireland was found to be 23.1 (compared with 22.4% in Britain), a proportion very close to the maximum known, 26%, for the Icelandic population between 1948 and 1984 (mean 19.6%, range 5–26%). Interestingly again a north–south difference was evident, this time, in the Irish figures, with only 16.6% cygnets being recorded in Northern Ireland as opposed to 25.1% in the Republic (Merne & Murphy 1986).

Variation in annual breeding success can be almost as great as in various species of geese, and variations between localities and between countries equally as marked. When Sheppard (1981) compared the figures obtained from northwest Ireland with the only other figures available for the same period, from Sweden, he noted that the two populations were 'clearly not at all similar in their breeding performances', and concluded that they must come from separate breeding areas, those in Sweden presumably from Russia and those in Ireland from Iceland (see Table 4.11).

A comparison between Sheppard's (1981) and Nilsson's (1979) results (see Table 4.11), is particularly useful because it reveals just how extreme the annual variation is in a given region. It also highlights the differences between regions. The low proportion of young in northwest Ireland in winter 1979–80 coincided with the low proportions (5%) recorded that same winter by Brazil & Kirk (1979) for Britain.

As Sheppard (1981) pointed out, however, there are pitfalls when trying to examine such data. Hewson (1964), for example, had discovered a distinct discrepancy between the proportion of cygnets at Lough Beg in Northern Ireland and Loch Park in Scotland, with the former being much lower, even though both flocks were presumed to originate in Iceland and hence could be expected to have similar ratios. It transpired that the difference was due to the tendency for family parties to be less attracted to large flocks (such as at Lough Beg), and more attracted to small (Loch Park). Sheppard (1981) too, noted a similar tendency, finding several large

Table 4.10.   *Brood sizes in Ireland in 1986 (after Merne & Murphy 1986).*

Brood size	1	2	3	4	5	6	7
Frequency	63	76	46	29	13	2	1

Table 4.11.   *Proportion of young Whooper Swans (%) in wintering populations in Ireland and Sweden (after Sheppard 1981 and Nilsson 1979).*

1970–81	Northwest Ireland	Sweden
1970–71	24.6	7.9
1971–72	–	16.3
1972–73	14.1	9.3
1973–74	12.5	19.6
1974–75	14.7	14.6
1975–76	9.3	23.7
1976–77	–	16.4
1977–78	17.4	7.0
1978–79	17.8	–
1979–80	7.6	–
1980–81	20.1	–

flocks and many small, scattered groups which generally consisted of family parties and thus the addition of just 1–2 families could significantly increase the proportion of cygnets locally. Furthermore, there is considerable movement in some areas during the course of a winter (e.g. Henty 1977; Brazil 1981a,c) and thus the proportion of young at any particular site may vary (Boyd & Eltringham 1962).

Another aspect that changed noticeably during the 20th century in Ireland was the length of time that the birds remain. According to Ruttledge (1966), up to about 1900, it was exceptional for the first to arrive before December, whereas by the 1950s they were typically arriving in mid October, even as early as 4 September (Kennedy *et al.* 1954), a pattern that continues today. Most departed in late March, with some remaining into early April, though both Kennedy *et al.* (1954) and Ruttledge (1966) also noted an increasing tendency for summering, with both adults and immatures reported in May–June and even July–August.

Intensive study of leg-banded birds throughout Ireland has revealed that although arrival may be as early as late September, the main arrival (64%) is during October from the second week onwards, with most having left by mid April (McElwaine *et al.* 1995). The first landfall is in the northwest, typically at sites such as Loughs Foyle and Swilly (Sheppard 1981), which are also important prior to spring departure. In autumn, the early arrivals tend to remain in this area for less than two weeks before spreading widely throughout Ireland, occupying numerous wetlands and reaching the south last. As many as 28% actually proceed to England, Scotland and Wales, whereas later arrivals, particularly those in December–February (68%), remain in the Foyle/Swilly area or use it again on their return journey. Whereas arrival is prolonged, departure is swift with the majority leaving in the first week of April (McElwaine *et al.* 1995).

Despite the wide winter range, few data are available that reveal trends over time at individual sites. It is in fact from a locality in the extreme south, Kilcolman Wildfowl Refuge, Co. Cork, that the most comprehensive records come. There, daily records have been kept for 20 years (O'Halloran *et al.* 1993). Interestingly, Whooper Swan was scarce in southern Ireland, particularly in Co. Cork, until about the mid 1940s, thereafter it became more widespread and has increased, a pattern also evident in Fennoscandia, suggesting perhaps a commonality of cause—the relaxation of persecution, recovery from the eelgrass crash, and perhaps changes in agriculture. Whooper Swans first appeared at Kilcolman in the late 1940s and numbers have increased since, to *c.*20 during the early 1970s and 120 or more in the late 1980s, consistent with the overall Irish trend, according to O'Halloran *et al.* (1993).

The rate of increase has been estimated at close to 6% per year by Owen *et al.* (1986), basing their assumption on the very low mortality in the Icelandic population and thus the statistically significant increase in Britain since the mid 1960s, and presumably no doubt that in Ireland (O'Halloran *et al.* 1993). Given the limited coverage of previous counts in the UK, the increase in numbers recorded may not, however, reflect the degree of actual increase.

Prior to 1900, it was exceptional for Whooper Swans to arrive in Ireland before December. Now, however, as has been found in the UK (e.g. Kemp 1991; Bowler *et al.* 1992, 1993), the first main wave appears in October and sometimes even September. In the south the tendency is for them to appear somewhat later, although even at Kilcolman the first arrivals have been as early as 1 October, spanning a period of 46 days, with first birds also arriving as late as 15 November. Not only have numbers increased at Kilcolman, but their stay has also lengthened significantly so that they now spend a mean 174 days there each winter, longer than that reported in other studies (O'Halloran *et al.* 1993). Of particular interest, but still unexplained, is the significant relationship between their arrival and departure, such that, in winters when birds arrive early there is a tendency also to depart early; conversely if their arrival is late then so is their departure. Regardless of the date of departure, however, numbers at Kilcolman have peaked in spring in all but two (1980 and 1988) of the 20 years (O'Halloran *et al.* 1993), which, given its location in the south, is difficult to explain.

Black & Rees (1984) reported cold-weather movements to Ireland from Scotland in the 1979–80 season, at a time when Scotland was still believed to be the main wintering area and Ireland the recipient of the overflow, but changes in numbers in Ireland, both within and between seasons have yet to be conclusively attributed to cold-weather movements (O'Halloran *et al.* 1993). It is more than likely, given the known focus of wintering in Ireland, where an estimated 62% of the Icelandic breeding population occurs in winter, that any local arrivals from Scotland would be swamped in the normal movements of swans between sites within Ireland.

So far, we have seen how Ireland has risen greatly in significance as the major recipient of Icelandic breeding Whoopers during winter; however, Whoopers have also occurred in Ireland during the summer. Franklin (1947) noted three in Co. Sligo in late July 1947; one was reported from Co. Mayo in June 1951 (Jackson 1952), and Kennedy *et al.* (1954) and Ruttledge (1966) reported adults and immatures in

May–August. They have summered regularly in the west in recent years. In 1992 for example, summering birds included one at Larne Lough, Co. Antrim on 12 May, another at Lough Neagh, Co. Down, on 12 May, one at Tory Island, Co. Donegal, on 24 June, one at Quoile Pondage, Co. Down, on 5 July–3 September, three at Lough Beg, Co. Londonderry, on 18 July, and one 'summered' at Castles Range, Co. Roscommon (Smiddy & O'Sullivan 1993). There were several records of Whoopers paired with Mutes, then in 1992, breeding was finally confirmed, when a pair of Whoopers bred in 1991, 1992 and 1993 in Co. Donegal but reared only one young, another pair present at the same site did not breed, and a second pair reared five young in Sligo (Murphy 1992, 1993; Gibbons *et al.* 1993; Smiddy & O'Sullivan 1993; Anon 1995a). The pair in Donegal has apparently bred regularly since.

In January 1995, when the number of Whoopers wintering in Ireland was re-counted, a total of 2,783 was reported from Northern Ireland, and an estimated 6,932 from the Republic, giving a total of 9,715 or 61.8% of the Icelandic population (Cranswick 1996). In January 2000, Northern Ireland held 18% of the Icelandic population or 3,716, while the Republic held 44% or 9,084, a cumulative 62% of the Icelandic population (Cranswick *et al.* 2002). Although a considerable amount of movement has been recorded within Ireland, there is some evidence that birds in the northern half are relatively independent of those in the south and west (Wells *et al.* 1996). More than 13,000 re-sightings of banded birds moving around Ireland are currently being analysed (McElwaine *in litt.*) and once available that information will doubtless provide fascinating insights.

The arrival and departure of large numbers of Whooper Swans into and out of Ireland is unequivocal. However, the route they travel and whether they commonly move to and fro between other parts of the British Isles has been the subject of much speculation. Hewson (1964) believed there to be a southwest movement from Scotland to Northern Ireland, which then returned to Iceland directly. Ogilvie (1972) considered that considerable interchange occurred between Scotland and Ireland, whereas Cramp & Simmons (1977) indicated that some migrated directly to Ireland and others moved later in response to severe weather in Scotland, with movements occurring in either direction throughout the winter. Black & Rees (1984) found that many unringed families stayed only short periods at Caerlaverock and attributed their disappearance to onward movement to Ireland. That such movement does occur between the countries was confirmed by Merne (1986) who observed birds on the Wicklow coast that had been colour-marked on the Solway Firth, *c.*280 km to the northeast.

Further direct confirmation of movements into Ireland from Iceland and from Scotland to Ireland was obtained by banding birds with neck-collars in Iceland. These studies showed that some migrated to Ireland directly, while others visited Scotland first, one even making first landfall in Shetland, then moving to Ireland. They also showed that birds moved within Ireland (Brazil 1983a; Gardarsson 1991). After initial arrival in October–December, birds generally stayed for some months in one general area (Gardarsson 1991). In December–January, however, as many as half left their early-winter sites in favour of sites further south or south-west, with some moving from Scotland and the north of Ireland to the south of Ireland (McElwaine *et al.* 1995).

Banding on the breeding grounds has demonstrated that while there is mixing within the overall population, subpopulations experience a degree of isolation. Thus, among birds banded at Skagafjordur, northwest Iceland, 46.5% were subsequently observed in Ireland, a significantly higher proportion than from all of the other banding sites in Iceland combined. Conversely, only 27.8% of those from Thingeyjarsysla and 21.1% from Jokuldalsheidi (both in northeast Iceland) were subsequently seen in Ireland. On average, fewer than 10% of birds banded in Scotland or England were later recorded in Ireland, and none of the birds banded at Welney made it to Ireland (McElwaine *et al.* 1995).

The intensive field work of Graham McElwaine and Jim Wells in Northern Ireland, based on birds banded in Iceland and Britain, has confirmed or revealed many aspects of the movements into and out of Ireland and Britain. Most notably they have shown that far from Scotland being the main staging area en route to Ireland, the reverse is true, with large numbers arriving in Ireland then moving to Britain. The significance of the Lough Foyle/ Swilly area as a staging post has been clearly revealed (McElwaine *et al.* 1995). Their work has shown that many arriving in Northern Ireland during October move to Caerlaverock or Martin Mere, some making the journey in a single day, and some moving even to Welney. While movements to Britain were common among early arrivals, few late-winter movements were noted (McElwaine *et al.* 1995) suggesting that departure is more direct. In addition, birds move through Scotland to Ireland on a broad front during early winter, whereas during the latter part there is an exodus across the Irish Sea from WWT reserves (McElwaine *et al.* 1995). Thus Ireland's role is a complex one. It is the direct recipient of birds from Iceland, the recipient of many migrating through Scotland, a staging area for birds moving to Britain, and it also receives late-winter emigrants from Britain. To a certain extent the predominant wind direction in October–November affects movements between Ireland and Britain, with more moving to Britain after having been diverted to Ireland during an autumn dominated by easterly winds, and more moving to Ireland after being diverted to Britain following one of westerly winds (McElwaine *et al.* 1995). That fewer banded birds are reported further south in Ireland may be partly a result of observer bias, and partly in relation to the locations where birds have been ringed in Iceland. Few, if any, have been ringed away from the north and, just as birds from western Iceland tend to winter in Ireland while those from eastern Iceland tend to winter in Scotland and England (Rees *et al.* 1990a; Gardarsson 1991; McElwaine *et al.* 1995), perhaps those from central and southern Iceland have a slightly different wintering area. The midwinter period (spanning the New Year) is dominated by southerly movements from study areas in Northern Ireland to sites further south, while the latter part of the winter is not surprisingly dominated by movements north, particularly to the main staging areas of Loughs Foyle, Swilly, Neagh and Beg.

It seems that the Whooper Swan was scarce in southern Ireland until the mid 1940s, since when it has been increasing. Kilcolman was one of the first sites where they were first recorded regularly. In general, however, the 300+ birds (341 at nine sites in January 1991) tended to occur in widely distributed small flocks, as in Britain (Boyd & Eltringham 1962; Hewson 1964), although the largest flock was of 186 on the River Blackwater, on the Cork/Waterford border (Ridgway & Hutchinson 1990; O'Halloran *et al.* 1993; Smiddy & O'Halloran 1991; O'Donoghue & O'Halloran 1994). The increase in numbers wintering at Kilcolman (6% per year) is consistent with the increase in the international population (O'Halloran *et al.* 1993).

The Whooper Swan's status as a long-standing irregular visitor to Ireland apparently began to change during the 1930s. Prior to that only Lough Swilly was a regular haunt, whereas by the 1970s all corners of the island were being visited regularly. By the mid and late 1980s, it was regular in large numbers, but with many more in the north. Numbers have increased most markedly southeast of a line from Limerick to Dundalk, in Louth, Kildare, Tipperary and Waterford. It now seems that Whoopers may be taking over sites that were previously strongholds of Bewick's (Sheppard 1993), but why that species should have declined, remains unexplained especially given that they share wintering areas elsewhere. Until the 1980s, Whooper numbers peaked at Loughs Foyle and Swilly in autumn then decreased (Sheppard 1981), whereas they now remain there throughout the winter in numbers exceeding 1,000 (Sheppard 1993; McElwaine *et al.* 1995). By the mid 1980s, 14 sites supported internationally significant numbers and at each of the eight where Whoopers had also occurred in the early 1970s, numbers had more than doubled with, for example, those at Lough Foyle increasing from 232 to 1,349 (Sheppard 1993). Conversely, Strangford Loch, which had been famous in the 1970s for its large numbers (411), had declined in significance with numbers dropping to 118 in the 1980s and to just 60 in the early 1990s (Sheppard 1993).

Sheppard (1993) saw the Irish increase in the context of the overall increase in Iceland. Certainly the published estimates and counts—2,000 in Ireland of a total of 5,000 (Ogilvie 1972); an Icelandic population of 10,000–11,000 (Gardarsson & Skarphedinsson 1984); a British and Irish wintering population of 16,700, with 10,320 in Ireland (Merne & Murphy 1986); 9,715 in Ireland of an estimated 15,702 in 1995, and 12,800 of 20,645 in January 2000 (Cranswick *et al.* 2002)— suggest a continued rise, although just how significant this is, is uncertain. Certainly the steady increase in survey numbers helps explain the Irish increase. If, however, the increase is merely the result of better coverage in Britain and Iceland, what is the true situation in Ireland? Was the Whooper Swan really so uncommon there in the past or was it undercounted there too? Whatever the history of the population, Irish wintering numbers currently represent *c.*62% of the Icelandic total (Sheppard 1993).

Although occurring predominantly in Britain in the north and west, and common in Ireland, the Whooper Swan only occurs in Wales in small numbers and mainly in Anglesey, Caernarfonshire, Merioneth and Montgomeryshire. During the 19th century, it was only regularly recorded in the northernmost counties, with then, as now, considerable annual variation in numbers. A marked influx in North Wales in winter 1890–91 was notable. Since then, it has generally been regarded as only a rare visitor to southern counties. Since the 1950s–1960s, however, numbers have increased and it is now found in most counties, although numbers rarely exceed 60 at any locality. The largest flocks occur most regularly in Anglesey and Caernarfonshire, where more than 50 are regular (Lovegrove *et al.* 1994), although the 1991 international survey failed to find more than 45 in the whole country (Kirby *et al.* 1992).

FENNOSCANDIAN/NORTHWEST RUSSIAN BREEDING POPULATION OVERVIEW

The Fennoscandian and northwest Russian population migrates primarily to the Baltic, other parts of northwest and central Europe, and there mixes with small numbers from the Icelandic population. An estimated 4,800–6,300 pairs breed in Europe (excluding Iceland; Scott & Rose 1996), while the wintering population was previously estimated at 25,000 (Rüger *et al.* 1986; Monval & Pirot 1989; Rose & Scott 1994). This was revised to 40,000 by Scott & Rose (1996). However, a more recent coordinated count found 52,000 and estimated 59,000 (Laubek *et al.* 1999). The great difference between estimates of breeding and wintering populations relates to the high proportion of non-breeders in swan populations. Denmark, holding more than 20,000 Whoopers, is the most significant country for this population, although other areas become more significant during cold weather. Prolonged severe weather in Europe led to the Netherlands holding 3,300 birds—*c.*50% higher than normal winter numbers. Germany is also significant, with more than 14,000, and 7,500 remain in Sweden in winter (Scott & Rose 1996; Delany *et al.* 1999). Annual variation as a result of cold-weather movements is considerable and, in particular, the standard International Waterfowl Count sites miss as many as one-third in some countries. Birds occasionally wander much further south, to France and even Spain, with small numbers in northwest Spain in the late 1980s including one ringed in northern Iceland (Rees *et al.* 1997a).

Continental European birds have not been recovered in Ireland, and although as many as 22 Finnish neck-banded birds reached southeast England in the cold winter of 1995–96 (Scott & Rose 1996) it seems unlikely that the degree of interchange between these two populations is as many as the 3,000 wintering in Britain and Ireland suggested earlier by Gardarsson (1991).

In Denmark, almost two-thirds winter in just three areas: Limfjorden, Mariager Fjord and the coasts and fjords of South Zealand and Storstrom. There are other major concentrations along the Baltic coast of Germany in the Mecklenburg–Vorpommern region and along the middle reaches of the River Elbe (Delany *et al.* 1999).

There was a substantial increase in the Fennoscandian/northwest Russian breeding population in 1974–93. Based on reports from both the breeding and wintering grounds, it appears that the population doubled over this time period. Peak counts in the 1990s have greatly exceeded estimates from the mid 1980s, with as many as 4,400 wintering in Norway, 6,000–9,000 in Germany and 2,000–3,000 in Poland. However, as elsewhere in the species' range, exact numbers and trends are difficult to assess because many small flocks away from regularly censused sites are missed (Delany *et al.* 1999). Nevertheless, changes and regional differences are likely to reflect a combination of cold-weather movements and variation in breeding success between years and areas.

FENNOSCANDIA AND CONTINENTAL EUROPE

In the context of the species' enormous range, the relatively discrete Icelandic population has received considerably greater attention than all of the remainder. However, over the last two decades the extent of research and hence literature available for other regions has increased dramatically, although the degree of detail is variable. In Europe, for example, the amount of research varies greatly between countries and this does not relate directly to whether they possess large breeding or wintering populations. If the Icelandic/Irish/British population is the best known in the West, then perhaps second is the Scandinavian. The Finnish population, in particular, though it may not match the density of most of the species' range, which essentially lies to the east, does at least provide a window on perhaps a more 'typical' population (compared to Iceland) in that the habitats occupied there are more comparable to those across Russia.

NORWAY

The indigenous Whooper population of Norway was, as elsewhere in Scandinavia, greatly reduced by human activity during the first half of the 20th century. By 1950 only very small numbers bred in the northernmost county of Finnmark (Myrberget 1981). The breeding population has, however, steadily increased since the 1950s with some colonisation of coastal islands in Troms county in the early 1960s and, between the Arctic Circle and *c.*70°N, coastal breeding pairs increased during the 1970s in both Nordland and Troms (Myrberget 1981). By the 1970s they had even begun breeding in southern Norway (Godo 1986), and in 1984 they bred in central Norway on the Fosen Peninsula, where they may also have raised young in 1983 (Haldaas 1985). The increase in numbers and range extension continued such that by 1994, Whoopers were found at a total of 74 sites during the breeding season, with 46 of these in Finnmark. Most of the rest were close to the northwest coast, although three sites in the west and one in the south, east of Oslofjord, confirmed the spread southwards (Gjershaug *et al.* 1994). Rees *et al.* (1997a) put the number of breeding pairs at 100–400 in 1970–90, though given Gjershaug *et al.*'s (1994) study, this appears to be rather overestimated, there likely being fewer than 100 breeding pairs.

This increase was also accompanied by an increase in wintering birds. There seems to be some contradiction, however, in the exact numbers involved. Ohtonen (1992) estimated the wintering population to be about 1,700, even though 30 years earlier Lund (1963) had estimated there to be as many as 2,500. In late-February 1988, when the Norwegian Ornithological Society conducted a national swan census, they found 3,918 Whoopers (Hauge 1990). Furthermore, that total was considered to be an underestimate because January–February temperatures were high over much of Norway leaving most freshwater bodies ice-free, enabling swans to disperse more widely than usual and consequently making counts less accurate. As a result, the wintering population in 1987–88 was estimated at 4,200–4,400, a considerable increase over the 1960s (Lund 1963; Hauge 1990). Nygård *et al.*'s (1988) survey of wintering waterfowl found the population to be slightly higher at 4,700 and they indicated that the annual range was therefore 4,500–4,700 or 18.8% of the total northwest European population (then estimated at 25,000). Wintering Whoopers proved to be very widespread, with birds present in all counties except Finnmark and with the highest number, 778, in central Norway (More og Romsdal), followed by 551 in the southwest (Rogaland). Nevertheless, the highest densities were found in two main regions, along the northern part of the west coast, and highest numbers in the Olsofjord–Telemark area of the south (Nygård *et al.* 1988). According to Hauge (1990), in the north, they prefer shallow, sheltered marine localities where they feed mainly on eelgrass, whereas in western, southern and eastern Norway they clearly prefer freshwater localities. Nygård *et al.* (1988) noted that 'most birds stay in shallow tidal areas, but will locally form quite large concentrations in open stretches, as in the Vorma River, 40 km NE of Oslo'.

The origins of the Norwegian wintering population have not been clarified, however Nygård *et al.* (1988) pointed to the fact that two neck-collared in Iceland were seen in Rogaland in November–December 1984, with one in Troms, indicating that they may come from the west. Nevertheless, the Norwegian population may also include birds from further east too.

SWEDEN

Along with most of northern Europe, Sweden was largely covered by ice during glaciation events until 15,000–10,000 years BP. Retreating glaciers, leaving behind developing marshy ponds and lakes, must have eventually generated enormous areas of potential wildfowl nesting habitat. As the glaciers retreated ever further north, and as summers became steadily longer and milder, the Whooper Swan was able to move into this subarctic-like landscape. That Whoopers were successful colonists is evidenced by their occurrence throughout Sweden since the last glaciation (Mathiasson 1991). The first documented breeding record, proven by an eggshell find near Falsterbo has been dated to *c.*14,000 years BP (Mathiasson 1991). The Whooper has thus had a very long documented history in Sweden, as it has no doubt elsewhere in Scandinavia, but whereas it was widespread in prehistoric times, relatively recently, its breeding distribution became restricted to the north of the country, its northern limit set by the ice-free summer period (Haapanen *et al.* 1973a).

Whooper Swan was a widespread breeder in northern Sweden during the 19th century, favouring areas east of the high mountains and apparently avoided coastal areas, though its southern extent is unknown. During the second half of the 19th century heavy hunting depleted the population until by the 1920s it had reached just 20 pairs all in the remote and vast mire of northernmost Lapland (Nilsson *et al.* 1998). In 1927 legal protection was afforded and the population began a long, slow recovery, extending its range steadily in inner Lapland. By 1950, it had spread more widely in northern Sweden, but its distribution then is best described as 'very patchy'. From 1950, however, the situation began to change rapidly, with the rate of dispersal increasing and breeding occurring in Sweden's southernmost province of Scania in the 1960s. Thereafter it spread to include all of the southern provinces (Nilsson *et al.* 1998).

In 1972–75, Haapanen & Nilsson (1979) made aerial surveys of all of northern Fennoscandia including Sweden south to 65°N. Within that northern third of the country they found 310 pairs, 120 with nests. When the population was estimated in the 1980s, for the first time since the 1920s, there were considered to be 500 breeding pairs in the country (Arvidsson 1987). Information from both the breeding season (Haapanen & Nilsson 1979) and the wintering grounds (Nilsson 1997) indicated that the population increase was continuing, and Nilsson *et al.* (1998) therefore undertook a repeat of the earlier aerial surveys in May–June 1997, but extended coverage south to 64°N. The final population estimates of the 1997 survey were based on extrapolation from actual counts. A total of 2,775 pairs, 1,760 with nests, was estimated for northern Sweden. Of particular interest was that 51% had nests, 30% were paired but had no nest, and 19% occurred singly or in groups. As one would expect, there were also marked regional differences, based on habitat types. In the mountains there were just 55 pairs (25 with nests) whereas 1,100 pairs (760 with nests) were estimated in the northernmost mires and woodlands. Not only had the population increased markedly between the early 1970s and 1997, but the proportion nesting was up from 32% to 64% (Nilsson *et al.* 1998).

Haapanen & Nilsson (1979) also found 320 pairs in Finland and 130 in Norway. Nevertheless, not all pairs were breeding; only 38% of the pairs in northern Sweden had nests (see above). They estimated that throughout northern Fennoscandia only *c.*30% was breeding. Densities were low, 0.30–0.59 pairs/100 km^2, depending on region, and the birds favoured either small lakes rich in emergent vegetation or vast mires.

The Swedish Whooper population is, like those in both Finland and Norway, increasing, having enlarged its range considerably since the 1960s. With the recent increase showing a typical exponential growth curve, it has colonised (or perhaps re-colonised) both central and southern Sweden, and its population is likely to continue growing as suitable nesting areas remain unoccupied. By 1985, *c.*300 breeding sites had been reported; the number of breeding pairs was *c.*500 with about half breeding in middle and southern Sweden, and several hundred non-breeding pairs were estimated. The increase in Sweden's northernmost two provinces has been considerable, from 310 pairs with 120 nests in 1979 (Haapanen & Nilsson 1979) to 2,800 pairs with *c.*1,800 nests in 1997 (Nilsson 1997). With its continuing spread and increase, the more aggressive Whooper may soon begin to displace feral Mutes from many inland localities, making it likely that the Mute Swan population will become essentially coastal (Arvidsson 1987; Mathiasson 1991). Whereas in 1972–75 Whooper Swans nesting in northern Sweden did so at lakes and in large mires in northernmost Norrbotten province, in the 1997 survey they were found throughout Norrbotten and Västerbotten provinces, although still commoner in the north and from the mountain range to the coast. Although almost all were outside the mountainous area, at altitudes of just 200–500 m, several were found nesting in rich alpine valleys (Nilsson *et al.* 1998). Since about 1945 Whoopers have colonised Skåne (Scania), Sweden's southernmost province, and by the early 1980s the population was *c.*30 pairs (Holmgren & Karlsson 1982). Interestingly, according to Arvidsson (1987), the first Whoopers to colonise southern Sweden may have originated from zoological gardens, though how much of the current southern Swedish population might derive from feral stock is unclear.

For those Whooper Swans breeding in the Western Palearctic, western Sweden forms the northeastern extremity of their regular wintering range, with the west coast representing the most important wintering quarters in the country. Mathiasson (1991) estimated that 2,500 winter in Sweden, although Ohtonen (1992) placed the total higher, at least 3,500 or more than 16% of the continental European wintering population. Sweden's wintering population may include immigrants from elsewhere, while Swedish-breeding Whoopers may winter outside the country. Curry-Lindahl (1964), for example, considered that Swedish Whoopers probably wintered along the Atlantic coast of Norway. Thousands more pause in southern Sweden during both autumn and spring migration, most of non-Swedish origin, en route to their Finnish/northwest Russian breeding haunts.

Nilsson (1975) noted considerable annual variation in numbers of waterfowl wintering in Sweden in relation to the severity of winter weather. Whooper numbers, for example, declined after the hard winter of 1969–70. Nilsson (1979) also found marked annual variation in the percentage of first-winters wintering in south Sweden in 1960–78. This variation proved to be significantly correlated with winter temperatures, such that lower temperatures on the wintering grounds led to reduced breeding success the following summer. Whooper productivity was also significantly correlated with May temperatures in Russian breeding areas.

Estimates of the population wintering in Sweden have varied with, for example, *c.*3,000 in January 1987 during one of the coldest winters for many years, but 4,000 in January 1988 and 1989 (two exceptionally mild winters) or some 16% of the 25,000-strong northwest European population (Nilsson 1991). Nilsson (1994) found that the population, in southern Sweden at least, was divided between those wintering coastally and those inland. Numbers of coastal birds declined significantly from a peak of 1,100 in 1969 to fewer than 100 by 1993, by which time they were only wintering erratically and in small numbers. Conversely, over the same timeframe, numbers wintering inland increased significantly, with the regional index for Scania trebling from 60 in the mid 1960s to 180 in the mid 1980s, though there was a subsequent decline to under 100 (regional index) in the early 1990s (Nilsson 1991).

In January 1995, a nationwide survey of wintering Whoopers was made as part of an international census. That survey revealed that the species was widely distributed over the south of the country, mostly in Scania and along the west coast, with a considerable increase over the 4,000 in 1988–89. A total of 7,439 was counted in 1995, 2,240 of them in Scania. Taking into account areas not covered, the estimated total was 8,000–9,000 (Nilsson 1991, 1997). A midwinter count in 2000 also produced an estimate of 8,000 (Nilsson 2002). In particular, the 1995 survey revealed a considerable shift in habitat use, with more than 5,400 inland that winter, and that there had been a considerable increase in the use of agricultural land, with as many as 30% field feeding. Despite that, the majority (*c.*70%) were still on waterbodies, many on quite small streams. Furthermore, of the more than 7,000 counted in 1995, 30% were on coastal waters, 40% on fresh water and 30% on land. This general pattern varied regionally, with the majority (63%) of those in Scania feeding on arable fields and grassland, but only 35% in the southeast and as little as 12–13% in other regions, a pattern apparently related to regional snow cover (Nilsson 1997).

In Scania, fields of winter rape were used by more than one-third of birds, and winter wheat was also more popular than grassland. Elsewhere, winter wheat and grasslands were both popular, though nationally most Whoopers were on waterbodies (Nilsson 1997). In November–December as many as 30% were reported on sugar beet fields. Then, with the ploughing of those fields by late December the swans were forced to turn their attentions elsewhere. In January–February 70% were concentrated on rape fields while the rest were on winter wheat (Nilsson 1997). Unfortunately, because previous surveys concentrated on open-water sites, it is not clear when field feeding started among those wintering in southern Sweden nor at what rate it increased.

The near tenfold increase in the numbers breeding in northern Sweden over the 25-year period from the early 1970s is much greater than the increase of the wintering population in southern Sweden (from *c.*2,000 in the early 1970s to 8,000 in 1995; Nilsson 1997) and also greater than for the overall wintering population in northwest Europe (Rose 1995). Within Sweden, the population has greatly increased in number and considerably extended its range, with birds now breeding commonly in areas where 20 years earlier they were unknown, for example in Hälsingland province in central Sweden, and increasing their numbers in most regions including the south (Nilsson *et al.* 1998).

It seems that the Whooper Swan in Sweden is now close to its historical population range and status. The remnant population in the early 1920s was occupying a range close to the northernmost limit of the species—where half of the summers are too short for cygnets to fledge, i.e. shorter than 140 days (Haapanen *et al.* 1973b), presumably surviving there only because they were less hunted. Since the relaxation of hunting pressure in 1927, it has steadily recolonised assisted by improved feeding conditions on the wintering grounds, which with the introduction of new crops and farming methods has led to increased food availability. This has doubtless helped boost the population increase by enabling overwintering birds to be in better condition in spring (Nilsson 1997; Nilsson *et al.* 1998). Nilsson *et al.* (1998) noting that the preferred nesting habitat is of small lakes with lush vegetation or peatlands and mesotrophic areas, and that mires of these types are very abundant in northern Sweden and Finland, considered that habitat availability would not yet limit the population increase.

FINLAND

Just as in Sweden, the Whooper was once common in subarctic Finland, but the first half of the 20th century saw the breeding population throughout Fennoscandia decline rapidly, primarily because of the shooting of breeding birds. In Finland, it was almost extirpated by the mid-20th century, however, with more enlightened attitudes, reduced hunting and an interest in swan conservation numbers have increased such that by the late 1980s the breeding population was estimated to be 800–900 pairs, and the breeding range had re-extended to include the entire country (Koskimies 1989; Ohtonen 1992). Ohtonen (1992) considered that Whooper densities were likely to be higher in Russia and that density likely increased eastwards, partially because those regions had suffered less from hunting. Sadly, this may no longer be the case.

The decrease and increase in the Finnish Whooper population during the 20th century were related to the distinct differences between habitats in the north and south of the country, and to relative breeding success in the two. They also demonstrate the species' sensitivity to disturbance, and its potential for recovery. Southern regions are very much more productive in terms of food biomass and consequently nesting success is high. Conversely, in the north, lakes are less productive, and the nesting success is lower. As southern breeding birds were first to bear the brunt of persecution, and as the population retreated to the poorer northern peatlands, so numbers were lost very rapidly as the most productive subpopulation was eradicated. With the change in attitudes from the 1950s, the reverse trend has been equally rapid with the population extending south and southwest from the less-productive wilderness of northeast Lapland to the highly productive south. Whoopers currently breed relatively sparsely in central and northern Finland, most numerously near the Russian border, with scattered pairs in the interior of southern Finland. Because the most productive nesting ponds in the south have now been resettled, the population has increased rapidly (Koskimies 1989; Ohtonen 1992).

By the late 1940s, as a result of persecution only *c.*20 pairs are thought to have survived. That number increased slowly to 150 by 1975 when density was highest in the northern boreal zone and decreased both to the north and south. Despite

Martin's (1993) indication that there were only 80 breeding pairs among *c*.500 birds in Finland, there was in fact a considerably larger total. The population had increased very rapidly, reaching 700 pairs by the late 1980s, half of them breeding in Lapland, and the other half in the warmer south, there being a strong positive correlation between the amount of suitable habitat and numbers of Whoopers in the northern boreal zone (Haapanen 1991). The combination of a longer ice-free period and higher summer temperatures in the south is thought to contribute to the higher productivity of Whooper Swans there.

The total Finnish population, including breeding pairs and non-breeding individuals, was thought by Koskimies (1989) to have reached *c*.3,500 by the late 1980s, although Haapanen (1991) considered that the spring population had reached 3,000 by the mid 1980s, and *c*.5,000 birds by the late 1980s.

Sufficient detail is known concerning Whooper Swan habitat requirements and population levels in Finland for estimates to be made of the proportion of any region still available for further colonisation. It appears, when estimates of available breeding territories are compared with known populations, that breeding areas in Lapland are nearly saturated. Haapanen (1991) found 430 breeding or non-breeding pairs in a possible maximum of 450 territories. By contrast, in southern Finland, Whoopers still occur at relatively low densities given the number of potential territories. As their status is not related to any shortage of habitat, the population there may still be expected to grow.

The rapid growth has been accompanied by swift colonisation of unoccupied areas through dispersal. Whooper Swans have dispersed an overall distance of 580 km from the eastern border to the south coast over 23 years, at a mean rate of 25 km/year. The picture is somewhat more complex than it appears initially, however, because the overall population consists of local subpopulations with large non-populated areas in between. While the overall population has dispersed rapidly, the dispersal rate of local subpopulations has averaged only 2.1 km/year (Haapanen 1991).

Looking at Finland in its entirety, Haapanen (1991) showed that the population increased by 11.0% (±0.2%) per year from 1950 until 1977, though there were clear regional differences. In the north, in Lapland, the increase was relatively slow at just 7.8% per year. In the south, however, growth after 1950 occurred at a remarkable 14.2% (±0.7%) per year. This difference between the north and the south was shown by Haapanen *et al.* (1973a) to be a consequence of the more severe climate and consequential lower habitat quality in the north, both factors contributing significantly to limiting swan productivity. The constant rate of increase not only suggests that density-dependent factors have not limited growth, but also places the Finnish population in the unique position of being among the very few bird populations to have shown such a constant rate of increase for so long (Haapanen 1991).

Whoopers breeding in Finland, Norway and Sweden migrate away from their breeding areas. Even those that winter on coasts of Norway and southern Sweden are usually some distance from their breeding grounds. Others move even further south, many reaching Denmark and Germany. Although the majority of Icelandic Whoopers also migrate long distances, some are able to winter in their breeding areas (Gardarsson *in litt.* 1996), perhaps a unique situation for the species. Making the situation in Finland more complex, and hence more interesting, is that many birds migrate through the country, primarily to and from breeding grounds in northwestern Russia, where Ohtonen (1992) considered the population to be stable. During the period 1973–87 peak numbers staging during autumn migration at Hailuoto, in the northern Gulf of Bothnia, were 1,000–1,900 without any evidence of an increase, although in 1984, the best year, the total was 4,000–4,500 (Merila & Ohtonen 1987). Whether by coincidence or because swans are the focus for other species willing to take advantage of them is unclear, but from late September most waterfowl at Hailuoto feed along the same shores as Whoopers, and by late October this has reached over 90% (Merila & Ohtonen 1987).

A banding project undertaken in Finland by Bjarke Laubek and colleagues netted 563 birds in 1987–97. Over this period 2,993 observations of 399 individuals were made, again confirming the great value of neck-collars in studies of the species. In addition to 292 individuals re-sighted within Finland, many records have been obtained elsewhere, including two in Belgium, 184 in Denmark, 40 in Germany, 35 in Holland, 15 in Norway, one in Russia, 107 in Sweden and 40 in the UK (Laubek 1998).

The dramatic increase in the Finnish breeding population over the last 50 years, from about 15 to 1,500 pairs, has involved it in re-colonising large areas of its former range, which had in the meantime been occupied by various other waterfowl. The effects on these waterfowl communities belonging to the same foraging guild (particularly members of the genus *Anas* in the same dabbling guild) have been investigated by Poysa & Sorjonen (2000). They found that Whooper Swan shares ecomorphological similarities with several species, in decreasing order of similarity: Northern Pintail, Mallard, Northern Shoveler, Garganey, Eurasian Wigeon and Common Teal. Despite similarities in ecomorphology, Poysa & Sorjonen (2000) found no evidence of any adverse impact from recolonisation by Whooper Swan on population densities of dabbling duck. Their results suggest that in re-expanding their range, particularly into areas inhabited by boreal waterfowl communities, Whooper Swans have occupied vacant niches.

WINTERING IN CONTINENTAL EUROPE

Considerable numbers of Whooper Swans winter in Scandinavia and continental Europe with the most significant numbers in Denmark (>31% of the continental European population) and Germany (>35%). Numbers have increased since counting began in 1967. Atkinson-Willes (1981) put the northwest European population at *c.*20,000 individuals, with another 25,000+ wintering around the Black Sea and east into Turkestan. In 1980, less was known of wintering numbers in Scandinavia than now, and more than half of the population breeding in

Scandinavia and the western former Soviet Union was considered to winter in Denmark, adjacent southern Sweden, Schleswig-Holstein and Mecklenburg, on the shallow bays of the western Baltic and fjords of northern Jutland (Atkinson-Willes 1981). Some, however, reach much further south, on occasions, though perhaps not annually, even the Mediterranean countries, with records, for example, from Lombardy in Italy (Bianchi *et al.* 1969). By the mid 1980s, Rüger *et al.* (1986) estimated that there were 25,000 in Europe, a figure also used by Monval & Pirot (1989) although Owen *et al.* (1986) considered that the Scandinavian and western Siberian breeding population, which winters in Denmark, the Netherlands and central Europe, amounted to many fewer, just 14,000–15,000. It is unclear why there is such a large discrepancy between their figures. The increase in wintering numbers (20,000 to 25,000) was put down to the increase in the breeding population in Finland, Sweden, and Norway by Ohtonen (1992) (see Table 4.12).

Despite the considerable attention that the Icelandic/Irish/UK population has received since the 1960s, it was not until 1995–96 that an attempt was made to completely census the northwest European wintering population. Previously, Monval & Pirot (1989) had estimated the mainland European population, derived from Scandinavia and Russia, to be 25,000. Their estimate was based on International Waterfowl Counts, which monitor numbers at selected sites and do not attempt to census the entire population, thus underestimating by up to one-third (based on national counts in Denmark in 1992–94; Laubek 1995a,b). Laubek (1995b), summing national censuses from Norway, Sweden, Denmark, Finland, Estonia, Latvia, Lithuania, Poland, Germany, Holland, the Czech and Slovak republics, Austria, Switzerland and Belgium) put the estimated total for northwest Europe at 33,600–39,300, although this was based on counts in 1988–93. Then in January 1995, when a coordinated census of continental northwest Europe was undertaken, a total 52,000 was counted, and as coverage was incomplete the

Table 4.12.   *Five-year means (1979–1983) and estimated populations of Whooper Swans wintering in different parts of continental Europe (after Rüger et al. 1986, and Ohtonen 1992).*

Wintering area	Numbers	% of total
Finland, Estonia	500	2.3
Sweden	3,500	16.2
Norway	1,700	7.9
Poland	300	1.4
GDR	5,400	25.0
FRG	2,300	10.6
Denmark	6,700	31.0
Netherlands	1,200	5.6
Total	21,600	100.0
Estimated population	25,000	

population was estimated at 59,000 (Beekman *et al.* 1999; almost twice the previous estimate) with the majority in Denmark (37%) and adjacent Germany (26%), and smaller numbers in Sweden (14%), Norway (13%) and Poland (6%; Laubek *et al.* 1999). The population has increased greatly in number and is currently estimated at 60,000, while surveys of the Fennoscandian breeding range show an expansion and higher densities (Beekman 1998).

DENMARK

South of the Baltic Sea, Whooper Swan is essentially a winterer and each year large numbers migrate through, or congregate at, sites in several northwest European countries, particularly Denmark, the Netherlands and Germany. The first arrive in late September–early October, with the average date in northern Jutland being 28 September in 1975–93. Coordinated counts of Danish sites in October–April in the 1991–92 and 1992–93 winters have shown that although Whoopers begin to arrive during October and numbers build during November, their mass arrival is rather later, in late December, usually during cold weather. January sees by far the greatest totals, with fewer in February and March. Midwinter totals of more than 15,000 in these years were a 50% increase over those in 1965–73, when Joensen (1974) found that numbers ranged from as few as 5,700 in mild winters to 10,800 in severe ones. This dramatic increase in wintering numbers over the last 25+ years corresponds in part at least to increases in Swedish and Finnish breeding populations (Laubek 1995c).

Though occurring widely in Denmark, sites holding the largest numbers (in excess of 500) are mainly in the north and southeast, in three areas of approximately equal importance, northern Jutland, northeast Jutland and southern Zealand and Møn. Their stay in Denmark is, however, rather short with birds leaving as early as February–March, although the last do not usually leave until late April, mean 28 April in northern Jutland in 1975–93 (Laubek 1995a, 1996). Of particular interest in relation to later sections on timing of migration and breeding, is that Laubek (1995a) found most late birds to be yearlings, apparently migrating very much later than pairs and other adults.

A comparison of brood sizes between Whooper and Bewick's Swans in Denmark in 1991–93, showed that although both most frequently have 2–3 cygnets, the former more consistently produces 4–5, and occasionally broods of 6–7, while Bewick's produced few broods of 4–5 and none larger during Laubek's (1995a) study. The percentage of juveniles among Whoopers present in Denmark varied considerably throughout the winter, from just 3.6% in October 1991 to 30% in March 1992. There was, however, considerable variation between years and regions; birds wintering in Zealand, for example, regularly have a higher proportion of young than those in northern Jutland, indicating that they originate in different breeding areas. Ringing has shown that some of those in northern Jutland are from eastern Iceland and northern Finland, while birds from Russia visit Zealand (Laubek 1995a).

GERMANY

Whooper Swan is a common, even abundant winter visitor, particularly to coastal regions of the north, where as many as 15,000 may be found, particularly in Schleswig-Holstein and Mecklenburg. Degen *et al.* (1996) found 2,729 in January 1995 in the Niedersachsen region, with major concentrations in the lowlands of the Elbe, Ems, Weser and Aller rivers. In that winter the proportion of juveniles was 12.7–12.8%, which they considered indicative of low breeding success in the previous breeding season. Of the estimated 59,000 in northwest Europe, 26% (15,340) occur in Germany, making it second only to Denmark (37%; 21,830) as a wintering destination for this population. In northern Germany, Whoopers are essentially terrestrial, mostly feeding on arable land, whereas in the south they are more typically found in aquatic habitats (Laubek *et al.* 1999).

In contrast, its status in summer was that of a rarity, and despite several records (e.g. Dathe 1984) there had been no cases of real oversummering until two adults remained at Dammer Moor reserve in northeast Germany, from 23 June–12 August 1991. In 1992, four adults remained from 24 April until 11 October. Then in 1993, not only were they present again from 31 March to 18 October, but the two pairs were observed displaying and in one instance of territorial behaviour drove off a Mute Swan (Deutschmann & Haupt 1994).

In line with the population expansion in northwest Europe, particularly in Finland and southern Sweden, it has begun breeding beyond its previous range, e.g. in Poland and northern Germany. According to Profus (1999) and contradicting Deutschmann & Haupt (1994) the Whooper Swan had in fact bred in northern Germany as early as 1982 and between then and 1989 eight broods were reared; however these may have involved escapes or injured birds. Breeding began in a region close to Poland, where birds had already nested. Since 1994 it has seemingly become established as a regular wild breeder in Germany (Profus 1999). In 1994 the Dammer Moor pair produced five eggs, one of which hatched, but the cygnet disappeared soon after. The following year they produced one cygnet from nine eggs, and again in 1996 they also produced one young from nine eggs, while four pairs bred in 1997 (Deutschmann 1997). Successful breeding seems to have occurred only in the 7–8th years, and unsuccessful breeding in the sixth year after summering began, indicating that a combination of both sexual maturity and is required for breeding success.

The Netherlands is a well-known wintering area for large numbers of Bewick's Swans. Whoopers also occur but in considerably smaller numbers, at far fewer sites, and numbers are very much dependent on winter conditions in the Baltic. In severe winters, as many as 3,500 may occur whereas only 1,500–2,000 may appear in mild winters. Of the five key sites, Flevoland and the Ijssel River in River District are the most important (Koffijberg 1998). As elsewhere in Europe, they have taken to feeding on agricultural land and are commonly observed on grassland and arable land throughout the winter, especially on oilseed rape.

Lying beyond the main range, Belgium appears to have attracted few wintering Whoopers or not many documenters of swans, although some banded in Finland have reached Belgium in winter. Breeding occurred in several years, 1972–81, but all those involved were considered to be escapees (Rees *et al.* 1997a).

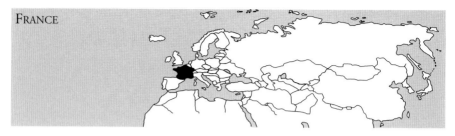

FRANCE

Although not referred to in other census reports (e.g. Rüger *et al.* 1986), Whooper Swan does winter regularly, albeit in very small numbers. Fewer than 100, typically 10–20, occur in normal winters. Sixty to 80 are reported in the northeast during cold winters, particularly in Lorraine, Alsace and Champagne between early December and late February, with some on the north coast, in coastal Brittany and elsewhere during very harsh weather (Steimer 1979; Francois 1981; Duquet 1992; Bergier *in litt.*). An influx of *c.*120 into northeastern France in January–February 1997 was exceptional (Anon 1997b).

WINTERING IN MEDITERRANEAN EUROPE

The southern limits of the wintering range in Europe are rather difficult to assess accurately, as in some countries the distinction between uncommon or rare winterer and accidental is difficult to draw. In general, however, it seems that Whooper Swan is an accidental visitor to the entire Mediterranean region with the exception of Greece (see Chapter Five). In Greece, it is a scarce and local winter visitor usually only to the Evros Delta, where as many as 1,500 have been reported (e.g. on 2 February 1969), although it apparently occurs more widely during cold winters (Bauer & Müller 1969; Handrinos & Akriotis 1997).

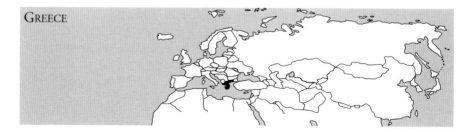

GREECE

Handrinos (1996) described the species as a regular but localised and rare winter visitor. Furthermore, following cold weather, as in 1985, 1993 and 1994, it also occurs on some islands and at sites on the mainland where they do not normally occur. Annual surveys have found Whooper Swans at only 33 of the 617 Greek sites counted during International Waterfowl Counts (IWC) in 1968–94, with birds at as many as seven sites in any given winter. In some winters (1974, 1982–84, 1986 and 1990) none was found (Handrinos 1996). Midwinter numbers (1963–94) have reached as many as 400 (in 1969). Counts of 219 in 1963, 297 in 1964 and 400 in 1969 coincided with harsh winters. But severe winters do not always lead to high numbers. Very few were recorded in the harsh winters of 1973 (65), 1974 (0), 1989 (4), 1990 (0) and 1993 (61) for example (Handrinos 1996). Whoopers are usually reported at the same sites as Mute Swans, but in small flocks of just 10–50. Outside IWC sites, large numbers (*c.*100) were apparently counted in 1963, 1964, 1967, 1968, 1978 and 1994 (Handrinos 1996). During the 21 years of the IWC in Greece covered by Handrinos (1996) no pattern of geographical distribution was apparent, although in general birds are found in Thrace and eastern or central Macedonia. The Evros Delta, on the Greek/Turkish border (40°48′N 26°00′E) is of particular significance because it regularly holds 90–100% of all those wintering in Greece (Handrinos 1996).

THE EASTERN BALTIC REGION

The eastern Baltic, comprising Estonia, Latvia, Lithuania, Poland and Belarus, is a regular winter haunt of variable numbers of both Mute and Whooper Swans, though conditions are only really favourable in mild winters, because areas of shallow water are usually ice-bound for too long. In Latvia, and especially Lithuania, swans occur mainly on inland waters. It is the coastal bays and shallows of Estonia, however, which form the main wintering sites in the eastern Baltic (Kuresoo 1991). Maxima of Whooper Swans exceeding 1,000 have been recorded in Estonia (January 1961) though numbers are minimal in hard winters, with much smaller numbers in Latvia and Lithuania. There were only a few tens, for example, in those countries in January 1988. Birds wintering in this area were considered by Kuresoo (1991) to be primarily from Finland and Karelia.

In parallel with its range expansion in Scandinavia, the Whooper has also begun to colonise the eastern Baltic countries. Breeding though still scarce is on the increase, and whereas Kuresoo (1991) knew of just three successful breeding records in Estonia since 1980, small populations have become established there, in Lithuania and Poland since.

Interestingly, in the light of Arvidsson's (1987) observations in Sweden and those of Deutschmann & Haupt's (1994) in Germany that suggest that Whooper Swans there are the more aggressive species, displacing Mute Swans, in Estonia, Kuresoo (1991) considered the reverse to be true in winter. There, Mute Swans are the more aggressive and, due to their increasing numbers, are displacing Whoopers from several wintering haunts in the west Estonian archipelago, leaving them now concentrated mainly in north Estonia.

As in Finland, the Whooper Swan is a winter visitor, passage migrant and breeder (Paakspuu in Leibak *et al.* 1994), though it has only recently become established as a breeder. The first spring migrants from further south return as early as late February and early March, with the last having passed through by late April. Large-scale migration typically coincides with the breaking up of ice cover, peaking usually in mid April (Fig. 4.8). Migration is on a relatively narrow front along the west coast, with only 10–15% of all migrants appearing inland. A total of *c.*10,000 pass through in spring, with up to 3,000 in the Väike Väin Strait in the early 1990s, up to 6,500 in Matsalu Bay in the 1980s and up to 2,000 in Haapsalu in the early 1990s. Paldiski is also an important staging area. Flock size seems to vary considerably with birds usually occurring in small groups of up to 50 but sometimes in flocks of up to 1,000 (Paakspuu in Leibak *et al.* 1994). In autumn, migrants appear from late September or early October to early December. It seems that fewer are reported at this season, although flocks may be larger (up to 200, exceptionally

Figure 4.8.   Whooper Swans staging in Estonia in April.

2,000), and they do not stay long. The western coast of Lake Peipsi, where as many as 1,000 have been reported (in 1993), is an important staging area (Paakspuu in Leibak *et al.* 1994).

In addition, Whooper Swan also winters in Estonia, but numbers vary depending on the extent of ice cover—the more ice, the fewer birds. According to Kuresoo (1988) numbers were greatest in winter 1960–61, when there were over 1,000 in west Estonia. During the 1980s, numbers ranged from 73–386, while in the early 1990s up to 850 were recorded. When individual flock sizes were studied, in January 1983, they were found to be small, averaging just 32 (Kuresoo 1988). Whereas the main wintering area was formerly in western Estonia, it is now the Gulf of Finland in the north, seemingly where smaller numbers of Mute Swans concentrate, indicative perhaps of interspecific competition (Kuresoo 1988; Paakspuu in Leibak *et al.* 1994).

Whooper Swans have also been known to summer in the country. In 1979 breeding was first attempted, then in 1988, breeding was confirmed near Lehtma, Lä, and in the following year a pair nested near Veisjärv, Vl. Subsequently, summering pairs or single adults have been recorded at various locations. The third, fourth and fifth breeding records were in Mihkli Parnu District, in 1991, at Aljarva, Muhu Island, in 1992, and Kuressaare, Saaremaa Island, also in 1992 (Anon 1994; Luiguijõe *et al.* 2002). In each case they were at small dystrophic lakes or pools in raised bogs (43% of breeding attempts), eutrophic inland lakes (31%) and coastal lakes and bays (26%). In 1992, however, single pairs also bred on coastal bays (near Aljava and Kuressaare, Sa). The breeding population in 1994 was considered to be *c.*5 pairs (Paakspuu in Leibak *et al.* 1994) making it one of the smallest national breeding populations in the western range, but it has subsequently increased to 30–40 pairs (Luigujõe *et al.* 2002). Summering seems to be on the increase, with at least 20 in Matsalu Bay, Laane District, on 3 August 1995 (Anon 1996a).

LATVIA

Given its relative abundance in Estonia and Lithuania, the numbers of Whoopers wintering in Latvia is surprisingly small, ranging from as few as 11 in 1994 to 57 in 1991, with more generally occurring in mild winters, at fish farms (Lipsberg 1983; Kalmins 1997). Although the majority of flocks on migration include only several tens, the largest was of *c.*200, on 11 April 1972 (Lipsberg 1983). It seems surprising that so few should be reported and that higher numbers do not occur on migration, as they are relatively common during spring/autumn in adjacent countries; perhaps they are currently overlooked. Despite sparse winter information, a small number of Whoopers have been reliably recorded as breeding in western Latvia since at least 1973, when a nest and eggs were found at Kazdang (a number of questionable reports of breeding relate to the 19th century at Lubana Lake in eastern Latvia, and at Tashu Lake in western Latvia in 1944; Lipsberg 1983). By 1987, six breeding pairs, two non-breeding pairs and two individuals were found in Kuldiga, Saldy and Liepaja Districts of western Latvia, and one pair may have bred on Lubans Reservoir in eastern Latvia. Breeding numbers have gradually increased since to 25–30 pairs in 1992 and there is every indication of a subsequent increase and spread to the east (Lipsberg 1983, 1988; Celmins *et al.* 1993; Anon 1996b; Kalmins 1997; Anon 2000).

LITHUANIA

In Lithuania, it is a common migrant, a scarce and irregular winterer, and a rare breeding bird. Key sites regularly support internationally significant concentrations on migration (Svazas *et al.* 1997). Common on migration (2,000–5,500 birds), large numbers appear along the western (Baltic) coast and smaller flocks inland (Svazas *et al.* 1997). Thus, a very significant proportion of the rapidly increasing northwest European population passes through Lithuania on migration in spring/autumn. Large staging concentrations are exclusively known from seasonally flooded coastal meadows and pastures, with birds apparently timing their arrival to coincide with peak spring floods of the Nemunas Delta. Smaller groups also visit coastal and inland fish ponds that are rich in suitable vegetation.

Whereas in autumn only small flocks (up to 200) are recorded at coastal sites, in spring the situation is very different. The pattern of spring migration has changed considerably since the 1960s (Svazas *et al.* 1997). Flooded meadows around the Nemunas Delta attracted large flocks of up to 4,000 as recently as the 1960s, however since land reclamation the attraction of the remaining seasonally flooded meadows declined, with only small flocks pausing there in the 1960s–1980s. Since 1990, however, large numbers have again been recorded in coastal areas, where they now seem to favour floodplains and, in particular, flooded meadows. In February 1990, 1,500 were in flooded meadows near Rusné. Large flocks (400–900) used floodplains around the Nemunas Delta in the springs of 1992, 1993 and 1995 (Zhalakevichius 1995; Svazas *et al.* 1997). In mid April 1994, the highest number in recent times was recorded: 2,000 at Meiziu Lake, Mazcikiai District (Anon 1997a), however this record was broken in March 1997 when 5,200 were found near the Nemunas Delta (Svazas *et al.* 1997).

In winter, Whooper Swan is irregular, and is mainly found at coastal sites where up to 50 were recorded annually (Zhalakevichius 1995). Winter numbers are usually very small (1–80), with the largest flock, of 76, recorded on Kursiu Lagoon in January 1995 (Svazas *et al.* 1997).

The Whooper is a rare breeding bird in Lithuania with just 20–40 estimated to summer or breed at shallow reed-fringed lakes and ponds annually (Zhalakevichius 1995; Svazas *et al.* 1997). Breeding was first recorded in 1965 in the Nemunas Delta, and again in 1967, with birds breeding irregularly there since. A nest was found at Kurshiu lagoon in 1968, and it has bred at large lakes in southwest Lithuania and certain inland fish ponds. Since 1989 they have bred regularly at Kivyliai pond, and more recently also at Voké fishpond and several other sites.

It is particularly interesting that Whooper Swans have increased as migrants, wintering and breeding birds, with the greatest increase during the migration period in 1987–97. A number of Lithuanian sites regularly hold more than 1% of the northwest European population and several often far more. In spring 1997, for example, floodplains near Silute held *c.*10% of the population (Svazas *et al.* 1997). Svazas *et al.* (1997) related this marked increase to the doubling of the northwest European wintering population since 1974, and to the importance of spring flooding in the Nemunas Delta region. Flooded meadows and grasslands are of particular importance to the large flocks of Whoopers migrating through the area. They considered the frequent flooding of the 1990s (1990, 1992–95 and 1997) also to have influenced the increase in swans using these sites.

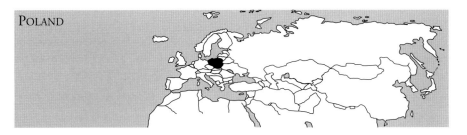

POLAND

The Whooper Swan is both a migrant and also a fairly numerous winter visitor to the Baltic coast and sometimes inland. It has become a regular summering and even breeding species, albeit in very small numbers, since 1981. In winter and on migration, it prefers to forage in shallow waters or wet coastal meadows, only occasionally using artificial feeding sites, and moving off when preferred areas become frozen and snow-covered (Tomialojc 1976, 1990; Meissner 1993). On migration, it arrives on the coast from October, exceptionally as early as August. Most depart in April, but exceptionally it has been seen in May and even as late as

23 June. The main migration periods are from early October until late December and from late February to late April (Tomialojc 1976, 1990; Wieloch *in litt.* 2001). It typically occurs in flocks of up to 30 and less often up to 70, but larger numbers have been recorded on rare occasions. About 800 were on Lake Lebsko on 13 March 1926 (Tomialojc 1976, 1990). Whoopers are less commonly encountered away from the coast, and inland records typically involve singles or small flocks. However, at different times, it has been reported widely throughout. A flock of 40 inland in March 1917 near Szprotawa was considered exceptionally large.

Despite recent advances, the wintering population is still rather poorly known, though 838 were counted on 18 December 1988 (Wieloch 1990; Wieloch *in litt.* 2001). If similar numbers occur annually, that seems a very small total for such a large country. In mid April 1990, only 67 were found at 256 waterbodies and rivers (Wieloch *in litt.* 2001). Meissner (1993) who studied those wintering on the Gulf of Gdansk, found that no more than 200 occurred there at one time during autumn/early winter. They arrive in early November and peak during December, with a very few remaining during January and February. Numbers increased again in March, when up to 60–90 occurred, then declined quickly during April indicating that they are primarily a migrant there, with about half as many in spring as in autumn; however, these figures may mask a very much more rapid turnover in spring. More recently, Wieloch (*in litt.*) found that winter numbers in January 1995 were 3,143 (estimated total 3,700) and 2,587 (estimated total 3,050) in January 2000. As in other countries, the percentage of young is enormously variable, ranging from just 6% after the 1985 breeding season to 16% after the 1984 breeding season, and even in the same regions there is inter-year variation, e.g. Wieloch (*in litt.*) found that at the main wintering grounds in northwest Poland in 1995 there was 11.6% young, while in 2000 there was 7.6%.

The Slonsk reserve, at the confluence of the Warta and Odra rivers (52°34′N 14°43′E) near the border with Germany, is the most important wintering site in Poland with birds arriving in early November and remaining until late March (with the exception of periods when the reservoir freezes completely). Whereas numbers ranged from 140–350 in 1977–81, in recent years they have reached nearly 2,000 (Bartoszewicz 1999), perhaps involving birds from Gdansk Bay, where numbers may have fallen. During winter the species is now regular in northern and western Poland, especially in the Bay of Gdansk where there are regularly up to 300–400 and exceptionally, as in 1988, as many as 660. In Western Pomerania, along the Lower Odra River, and at Szczecin Bay, there are typically up to 800 and sometimes larger local concentrations of up to 1,000 (in 1995). The middle course of the Odra River holds *c.*300, and in many other parts of the country, depending on conditions, there may be dozens wintering (T. Stawarczyk *in litt.*).

If poorly known in winter, it was completely unknown as a breeder until just over 25 years ago. The sudden appearance of breeding Whoopers well outside their normal breeding range has been noted in Finland, and apparently in both Latvia and Poland (Haapanen 1991; Hagemeijer & Blair 1997). The first Polish breeding record was an isolated case in the northeast, in Biebrza Marshes, where a pair nested in 1973 (Czapulak 1991; Dyrcz *et al.* 1991). A site in the western Barycz Valley was occupied from 1981, and a nest built in 1983 (Czapulak 1991; Czapulak & Witkowski 1996). Also in 1983, a breeding pair was found in Silesia in southwest Poland, and since then seven breeding sites have been found with up to five pairs (1986) breeding annually. Breeding has also occurred in central Pomerania and, in 1997 and 1998, in southern Poland. Numbers have now reached 12–13 pairs (Tomialojc & Stawarczyk in press). Breeding in Silesia followed the growth there of a small wintering population, from seven in 1983 to 146 by 1989, with the proportion of cygnets varying between 3.8% and 23.1% (see Table 4.13). Profus' (1999) survey of seven nests, with clutches of 4–8, averaging 6.1 eggs, and of 25 broods averaging 4.2 cygnets per successful pair (2.2 if those that failed are included), suggests a relatively high success rate in this small population. While the origin of birds breeding in Poland is unknown, Tomialojc (*in litt.* 2001) has speculated that a captive-bred female may have escaped and been joined by a wild wintering male, and thus initiated breeding.

Interestingly, as has previously been shown for Ireland and the UK, the percentage of young Whoopers varies considerably not just annually, but also regionally within the same year, for in Slonsk reserve (*c.*60 km north of northern Silesia) the proportion of cygnets reached a maximum of only 5%. Those wintering in the reserve were, however, thought to be mostly non-breeders (Beszterda *et al.* 1983), though some small family parties also occur. Czapulak (1991) thought it possible that Silesia had higher numbers of cygnets because they were reared in the region. Subsequently, however, the age structure of those using Slonsk reserve has changed, with generally higher but widely varying proportions of cygnets, from as few as zero to 43.7% (Bartoszewicz 1999). Such wide variance is to be expected in small flocks unrepresentative of the population at large, nevertheless even when only large flocks are considered the proportion still varied considerably, from 7.1% among 838 in February 1995 to 21.9% among 223 in March 1996. We may find in future that estimates made of the proportion of young in a population where migratory flocks may be coming and going, are simply meaningless.

Table 4.13. The proportion of cygnets in the Silesian wintering population (after Czapulak 1991).

Year	Percentage
1984	3.8
1985	15.6
1986	21.5
1987	23.1
1988	15.6
1989	13.7

Not only has the species increased as a breeder, but it has also become a more regular migrant throughout the whole country in recent years. It is most numerous in the north and west, but still only a sporadic visitor in small numbers over the rest of country. The largest concentrations are in the Bay of Gdansk (up to 400 in autumn), Vistula Lagoon (up to 500 in spring) and the particularly important Slonsk reserve (see above). In many other places Whoopers occur in flocks of dozens or sometimes more than 100 (T. Stawarczyk *in litt.*).

The Czech Republic has few lowland areas and does not immediately spring to mind as being within the range of the Whooper Swan. Nevertheless, it does occur there, albeit somewhat rarely. Most records are in December, February and March, although it has been recorded in October–May (Kren 2000). Whoopers have been found throughout much of the country, and in 1985–2000 were found almost annually at large ponds, reservoirs and rivers, but with no more than 25 in any winter. Typically it is recorded alone or with Mute Swans but has also been seen in small flocks. Since 1960, numbers appearing on migration or in winter have increased by 15–20% (Kren 2000), and these presumably reach the country by way of either Germany or Poland.

Information from Hungary, Romania and Bulgaria has been surprisingly scarce, but according to Gorman (1996) it is a rare but annual winter visitor to Hungary, and the Danube Delta of Romania appears to be an attractive potential wintering site. Until the mid 19th century, scattered pairs bred in the then extensive marshes along the Tisza River in Hungary, it is now, however, only a winter visitor in October–April, and most regularly in January–February. Singles, families and more rarely small flocks visit larger unfrozen wetlands each winter, particularly in the Hortobagy, Kis-Balaton, Lake Balaton and Szeged areas, with slight invasions occurring in some years (presumably originating from the Black Sea wintering population) (Gorman 1996).

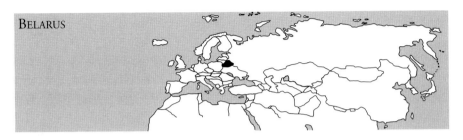

In Belarus, Whooper Swan was considered a very rare or accidental visitor on migration with no more than *c*.10 records over the previous 50 years (Nikiforov *et al.* 1997). However, the late 1990s witnessed a spate of records, with two adults and their young on a lake in Shabany, near Minsk, from early December 1997 until early January 1988, and an adult there again on 2 December 1998 (Yurko 2001). Zhuravlev & Pareiko (1999) observed several during 1998 in a flooded area near Pogonnoe and Dron'ki (Gomel Region). They were first seen in late April (23rd), when a pair was found east of Pogonnoe. Two more pairs were located on 24–27 April just northeast of Dronka (*c*.8 km from the previous sighting). On 16 and 23 May, three pairs were at the same site, and one pair was still there on 3 and 6 June. High vegetation, however, may have concealed the others. Three adults were seen again on 2 October 1998 near Pogonnoe. All of the 1998 sightings were within 7–8 km, and almost all were of pairs. Summering probably occurred, but neither nest nor cygnets were seen, so breeding, if it took place, failed. Nevertheless, the record is further evidence of continued range expansion.

An annual winter visitor in reasonable numbers, the species is clearly affected by cold weather. The only available information is from the Black Sea, particularly the northern coastal region between Odessa and the Crimea, especially the Black Sea Biosphere Reserve, with 650–1,660 swans wintering in 1984–88, their numbers fluctuating with the weather (Grinchenko *et al.* 1988; Ardamatskaya & Korzyukov 1991). Sudden falls in temperature in autumn 1987 led to the departure of considerable numbers in a southerly direction; such fluctuations in the weather may be responsible for records in Greece and Turkey. Then, in December 1995–February 1996, 1,810 wintered along the northwest Black Sea (Anon 1996b). Most recently, Rudenko *et al.* (2000) considered that both Mute and Whooper Swan wintering numbers (8,000–12,000) were stable. Figures published from the region are difficult to reconcile with estimates (published in Europe) of there being 17,000 in Ukraine (e.g. Rees *et al.* 1997a).

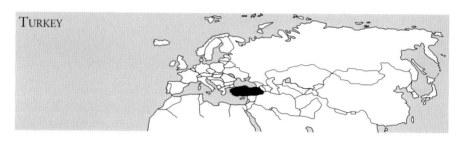

TURKEY

Whooper Swan is an uncommon to rare winter visitor, recorded widely but typically in small numbers in December–March and, more rarely, April–May. According to M. Özen (*in litt.*) steadily increasing numbers have occurred in the western two-thirds of Anatolia in recent years, although few details are available. Whooper Swans may reach western Turkey across the Black Sea from the Ukraine or via the west coast of the Black Sea and, if the latter, would suggest that birds should appear there in winter too. According to recent Turkey bird reports (1987–96; Kirwan & Martins 1994, 2000), the species has occurred in Black Sea Coastlands, at Kizilirmak Delta, Yenicaga Gölü and Sarikum Gölü, from 10 December (1995) to an exceptionally late 30 May (1993), with a maximum 19 at Kizilirmak Delta on 30 January 1993. In Thrace, birds have been found at Meriç Delta, Igneada and Büyükçekmece in January–February with a maximum 214 on 22 February 1996 at Meriç Delta. In Western Anatolia, Whoopers have appeared at Bafa Gölü, Marmara Gölü, Büyük Menderes Delta, Camalti Tuzlasi and Manyas Gölü in January–February with a maximum 32 on 27 January 1993 at Marmara Gölü. In Southern Coastlands, records are in December–April at Göksu Delta, Burdur Gölü and Çukurova Deltas, with a maximum 67 on 24 January 1993 at Göksu Delta. On the Central Plateau, it has occurred at Kulu Gölü, Eregli Marshes, Hirfanli Baraji and Tuz Gölü in December–April with the largest count being 158 at Kulu Gölü on 8 December 1993. The only record further east was at Kabakli Reservoir, Diyarbakir in February 1993.

CHAPTER 5

Range, habitat and population– Russia and Asia, and vagrancy

INTRODUCTION

Russia not only provides the single largest continuum of swan habitat in the world, but it also contains almost the entire breeding range of the Whooper Swan (see (Fig. 5.1). The enormous land area (17,075,200 km²), low human population density (8.52 per km²) and even lower density of swan biologists has meant that, until recently, relatively little information has been available. With this in mind, and with the considerable help of Jevgeni Shergalin, I have attempted to summarise not only information that became available in English from the Third International Swan Symposium but also as much as possible of the Russian literature published since the 1980s (Brazil & Shergalin 2002). Much of what was known was based on supposition, although some local populations were known to have declined and others to have increased (Earnst 1991). Ohtonen (1992), who has studied Whoopers for some years in Finland, considered that densities were higher in Russia, that they increased eastward, and that during the 20th century the population had probably not declined in Russia as much as it had in Scandinavia. How much of this is fact remains to be seen.

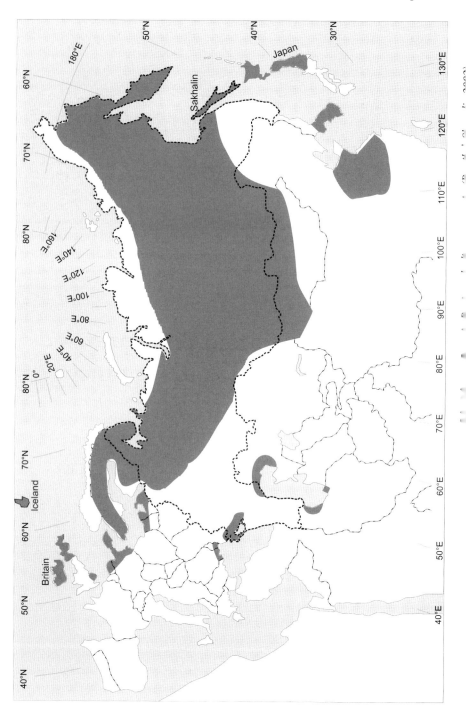

The population in Russia is large, generally secure and seemingly expanding north (Syroechkovski 2002). It suffers from a wide range of anthropogenic influences, but in some areas is also reoccupying breeding haunts where such negative influences have declined (Brazil & Shergalin 2002). It ranges from the Kola Peninsula, in the northwest, east to the Anadyr River of Chukotka, and Kamchatka in the extreme northeast. Although the northern limit to its normal breeding range lies close to 67–68°N, it has occasionally bred north to 70°N and, exceptionally, 72°N. In the south, the species regularly breeds as far south as 62°N in western European Russia and between 55°N and 50°N in Sakhalin and Kamchatka. Towards the southern extent of its range nesting is irregular both in distribution and success. It winters south to *c.*47°N on the northern shore of the Caspian, to 50°N in northern Kazakstan, and 47–50°N at Lake Balkhash, Mongolia, across northeast China, and into southern Primorye around Lake Khanka. Its southernmost regular wintering grounds are in Japan, where it occurs at latitudes as low as 35–40°N in large numbers (Stepanyan 1990; Brazil 1991).

Estimating population size in a region that comprises 80–90% of the species' range is fraught with difficulties and, in the case of the west Siberian population, contradictions. The latter has been placed at anything from 17,000–250,000, while the west *and* central Siberian/Caspian population has variously been put at 20,000–250,000[1] (Ravkin *et al.* 1988; Delany *et al.* 1999), while the central Siberian and east Asian population has been estimated at 60,000, giving a 1% threshold for Ramsar criteria of 600 in that region, and making it very similar to the northwest European population (Laubek *et al.* 1999; Miyabayashi & Mundkur 1999). All three swan species breeding in Russia, the Mute, Tundra and Whooper have been expanding their ranges and increasing in numbers according to Shchadilov & Belousova (2001).

RUSSIA AND ASSOCIATED STATES AND REPUBLICS

Beginning in the west, I have considered Russia and adjacent areas as four large regions: Western Russia, i.e. the region west of the Urals; Western Siberia, from the Urals east to the Yenisei; Central and Eastern Siberia, from the Yenisei east to the Lena; and the Russian Far East from the Lena east to the Bering Sea. Each of these is as large, or larger, than the entire European range.

WESTERN RUSSIA: FROM EUROPE TO THE URALS

Western Russia stretches east from the Finnish border to the Pechora and the western slopes of the Urals (see Fig. 5.2). Whereas this region is rather narrow in the north, to the south it spans from the shores of the Black Sea to the Caspian. A mere fraction of Russia and the republics of the ex-USSR, it is larger than Europe.

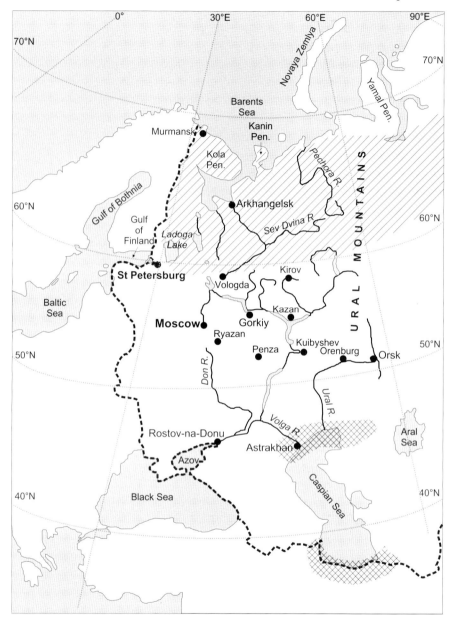

Figure 5.2.   Western Russia showing the main locations mentioned and the approximate breeding and wintering areas of the Whooper Swan (Brazil & Shergalin 2002).

Here, the Whooper's breeding range reaches north to the Barents Sea on the Kanin Peninsula. Further east, its northern limit is Cheshskaya and Korovinskaya bays. It also breeds in the Pechora delta at *c.*67°N, between the Pechora and Kuya rivers and along the middle reaches of the Shapkina, as well as at the northern border of sparse forest on the rivers Kolva and Kolvavis, along the middle Hosedayu and Ad'zva rivers, and the upper Bolshaya Nertseta and Bolshaya Rogovaya rivers. From the latter, its distribution shifts south to the lower Vorkuta and Yun'yakha rivers. Occasionally, it breeds in the southern Russkiy Zavorot Peninsula and in the shrubby subzone of the Bolshezemel'skaya tundra (Mineev 1995).

The status and even integrity of this population, which winters in the east Mediterranean/Black Sea and southwest Asia is in doubt. Is it one population that divides in winter or are they separate populations? If the latter, what is the extent of the populations. Is there any interchange between them? Uncertainties and variation in coverage hamper comparisons, but there were, for example, 2,364 Whoopers wintering in Romania, 574 in the Ukraine and 1,510 in Iran in 1995, but in 1996 just 239 in Romania, and 1,810 in the Ukraine (Delany *et al.* 1999). Assessing population size and trends becomes increasingly difficult as one moves east. Coverage is low and this, combined with the known effects of hard weather, means that figures are unavailable. It is thought, however, that numbers wintering in the Black Sea have declined since the early 1970s, while those in the northern Caspian rose to a mean 13,900 in 1970–80 from just 1,850 in 1955–69 (Delany *et al.* 1999).

Scott & Rose (1996) considered there to be two populations, the 'Western Siberian/Black Sea/East Mediterranean' population, estimated to be 17,000 (based on Rügers *et al.* 1986; though see Chapter Four) and the Western Siberian/Caspian population, provisionally estimated at 20,000. Meanwhile, Rees *et al.* (1997a) put the west Russian wintering population, including that wintering in the Caspian, at 10,000–70,000. However, the boundaries of their populations are not clear, with some from northern parts of western Russia clearly migrating west into Europe to winter (Mineev 1995), while others probably migrate south to the Black and Caspian Seas. Important concentrations were also recorded in Azerbaijan at Aggel Lake (900) and in Turkmenistan in the Amu Darya Valley (mean 300) and at Lake Sarakamysh (mean 850) (Perennou *et al.* 1994).

West of the Urals in summer, non-breeders occur in flocks and individually in the northern Russkiy Zavorot Peninsula, Bolvanskaya Bay, the lower Chernaya River, Haipudyr Bay, along the lower Lymbadayaga and around Karskaya Bay.

Research has shown that birds arrive on the Kola Peninsula via the Gulf of Bothnia (see Figure 10.1; Bianki 1990). In spring many cross the Gulf of Kandalaksha, with the first appearing on 23–24 March, although the mean is 18 April (range 23 March–5 May; N=24). In Lapland Nature Reserve (NR), Whoopers appear even earlier, averaging 13 April (range 25 March–30 April; N=35). However, the apparent later arrival from the Kandalaksha Skerries may be connected with the rarity of observation there. Peak spring migration through Kandalaksha is during the last third of April, and intensity decreases during the first third of May, but it is still comparatively significant (Bianki 1999).

On the Kola, Whoopers prefer richly vegetated lakes amid taiga forest, and nests near water on islands or in bogs. Aerial surveys during July 1986 revealed that there were three areas of concentration, in taiga around the upper Ponoi River and the Umba Basin and in forest-tundra between Kharlovka and Iokanga. Non-breeders arrive later and, whereas early migrants are mostly in pairs, they are usually in flocks. The total population was estimated to be 1,600–1,700 in 1986 (Bianki 1990). In the Lapland NR western Kola, Bragin (1987), and Semenov-Tyan-Shanskiy & Gilyazov (1990) reviewed the period 1930–87. Whoopers arrive on average on 13 April, ranging from 25 March (1935) to 30 April (1940), generally as streams and bogs are thawing and mean air temperatures are –3°C. They arrive 11–14 days earlier 100–130 km to the south and west on the Kovda and Girvas rivers, and depart as shallow lakes freeze, on average on 18 October, although as early as 20 September (1966) and as late as 30 November (1950). Spring flocks are usually smaller, 6–15, than in autumn (10–18 but sometimes 27–50). Numbers fell considerably in 1930–50 but have since recovered. As elsewhere, the proportion of the population actively breeding is *c.*30%, of which only 60% is successful.

Around Murmansk, in northern Karelia, and in coastal Arkhangelsk, Bianki & Shutova (1987) made aerial surveys of swans in late July–early August 1975–77. Ground surveys west of the Pesha River indicated that all were Whoopers, while in the Pechora and Khaipudyr bays, most, if not all, were Bewick's. They found the highest densities among aapa-bogs in Karelia, where they averaged 4.8/100 km². In the southern Kola, the density was highest at 8.3/100 km² in aapa-bogs and swamp forest. As in much of the rest of the species' range, only single pairs occupied large lakes, thus where large lakes were commoner, swan densities were lower, conversely, the majority of broods in the taiga were at small lakes or in bogs. Further north, density was lower at 3.3/100 km². The Umba Basin and western Murmansk support fewer. Breeding occurs only at low density in the Lapland NR, with only 3.3/100 km² in the Notozersko-Tulomskaya depression, and the whole of the Murmansk region has suffered from increased human access and a subsequent decline in swan numbers. Northeast of a line joining Kola Bay with the Pyalitsa River mouth, forest-tundra and tundra form a complicated mosaic, with few nesting swans. The highest density (12.2/100 km²) was in the Mezen River lowland on the Kanin Peninsula. There, in the 1950s, Whoopers nested at considerable densities with pairs as close as 3–5 km. Densities were lower in the southern Kanin. Near its base swan density was less than in its middle regions, with 9.5/100 km² on the west bank of the Kuloi, but only 4.7/100 km² between the Mezen' and Pesha rivers. Whoopers were absent from the tundra region to the east, but in western Pechora, densities rose to *c.*8/100 km². East of Bolvanskaya Bay, and in tundra east of Kolokolkovaya Bay all swans were considered to be Bewick's. In the enormous Pechora delta, Whooper densities reached very high levels, 28.3/100 km². Bianki & Shutova (1987) estimated that there were around 1,250 Whoopers in northern Karelia, 1,510 in Murmansk and 2,470 in the area of the White Sea in Arkhangelsk.

Continuing east from the Kola is the Kanin Peninsula, then to the east but still west of the Pechora is the Indiga River, which reaches the Barents Sea close to 49°E. In this basin, the Whooper breeds at large glacial lakes, in grassy and shrubby vegetation including sedges, sabre-grass, horsetails and willows. The highest density is reached along the river's upper reaches at $0.3/1 \text{ km}^2$ (Mineev 2000). To the north, solitary Whoopers have been recorded as vagrants on the Yugor Peninsula, and on Novaya Zemlya in July–August, although in 1995 breeding was confirmed on the southern island. At approximately 72°N, this is close to the species' current northernmost limit. There are also archaeological remains of two from Vaigach, and in October 1987 non-flying cygnets were captured at Varnek (south of Vaigach) and released on the tundra in May 1988 (Kalyakin 1999).

Human influences in western Russia during the mid 20th century greatly reduced the population and range below 19th and early-20th century levels. During the latter half of the 20th century it began to reoccupy its former range, the southern border of which stretches from 62°10′N at the Russian border, rises to 63°20′N in central Karelia falling to 62°N at the eastern border of Karelia with Arkhangelsk. Continuing east, the southern border is at 61°45′N declining to 61°20′N at the border with Vologda. South of here, it breeds irregularly in southern Karelia, the St Petersburg region[2] and north Vologda (Hoklova & Artemjev 2002). In the southern and middle Komi Republic (the northeastern part of western Russia between Arkhangelsk and Permskaya region[3]) the Whooper is absent in the breeding season. In the southern taiga it is common only on spring migration when flocks appear briefly on the snowmelt floodwaters and swamps.

East of the Pechora, along the northern fringe of the Urals, in the basin of the upper Pechora, it is uncommon and breeding has not been confirmed, but it is numerous on branches of the mid and lower Pechora (Mineev 1995), and the density of pairs in the delta increased considerably in 1980–2000. Some 55.9–73.2% of Whoopers do not defend breeding territories, instead they congregate in non-breeding flocks (Shchadilov *et al.* 2002).

Whooper Swans from the northern part of western Russia, at *c.*30–55°E, migrate mainly west and southwest, towards the Severnaya Dvina and Onega rivers, and may be joined by birds from as far east as the Yamal Peninsula, which is northeast of the Urals. They continue, via the Gulf of Finland to the southern Baltic, returning along the same route (Mineev 1995). Prior to the main exodus, concentrations occur in the Pechora delta, and in Korovinskaya and Kolokolkovaya bays, where thousands gather in flocks of up to 200–300. It is possible that some also migrate south towards the Black and Caspian Seas. In spring, the first birds arrive east of the Pechora in mid April to early May. Even along the upper Pechora, they may arrive as early as 7–20 April, although in the lower Pechora first arrivals may be ten days later, on 17–20 April. Migration is frequently via river valleys, with arrival in spring either in pairs or small flocks of up to ten, although at resting sites they may number up to 50 (Mineev 1995). In autumn, non-breeding pairs and small flocks depart first, sometimes from mid September, moving west and southwest and resting at wetlands en route. Around Urdyzhinskoe Lake pairs and small flocks of 16–18 occur annually until 25–28 September. Weather seems a critical factor, and before the first frosts the number of migrants increases sharply. During the final days of September flocks of 30–100 (without broods) are on the move. While the first wave consists almost entirely of non-breeders, the second includes many breeders, with pairs families and flocks containing broods passing during early and mid October (Mineev 1995).

In European Russia the main breeding range lies north of 63–64°N, but its southern border is unclear. South of the Kola, the species currently breeds commonly south to *c*.63°30′N in Karelia, with birds occasionally summering in southern Karelia. In Europe there has been a spate of breeding records south of the historical range, and such range extension may also be occurring in western Russia, for example in the St Petersburg region (*c*.60°N). According to Mikhaleva (1997), the species bred in the St Petersburg region until the mid 20th century, since when it has only been a passage migrant. Not only did breeding occur prior to the late 1960s, but it also occurred in 1986 and 1997, on Ladoga Lake (Vysotskiy 1998; Golovan & Kondratiev 1999). There have been occasional breeding records elsewhere in recent years, in the Zaonezh'ye (Trans-Onega), at Pudozhskiy (on the border of Karelia and Arkhangelsk) and in Vodlozerskiy National Park (NP). Whoopers also breed in the region of Tolvojarvi (Khokhlova & Artem'yev 2000) and bred regularly on Vyal'ye Lake (Luga) south of St Petersburg until 1956. After a ten-year absence, a pair summered in 1967 and again in 1998. In May 1999, a nest was found and in early June a brood of five cygnets hatched (Golovan & Kondratiev 1999). During 1995–99, a new breeding population was discovered in Kenozero NP (between southeast Karelia and the Arkhangelsk region at *c*.62°N). Here, in the watershed of the White and Baltic Seas, Whoopers breed annually on certain forest lakes, although no population estimate is available. Breeding pairs and young have been seen at numerous lakes within the park, over the five-year period up to 1999, suggesting the presence of a stable population, the southernmost (61°30′N) in northwest Russia (Khokhlova & Artem'yev 2000).

In the upper Volga, Whooper Swan is a spring migrant with small flocks (largest 52) at swampy wetlands and lakes in late March–April. Larger concentrations are known from the peat areas of Sokoliy Mokh in the St Petersburg region. In summer 1982, one was at Kremno Lake. They reappear in autumn from late September until the freeze in November. At this season they are usually seen singly, in small flocks of up to five (presumably families) and sometimes with geese (Nikolaev 1998). The nearest breeding site is Darwin NR, on Rybinsk reservoir in Vologda region, where three pairs have bred since 1983 in peat bogs, and pairs have bred regularly at lakes in Ustyuzhinsk district, Vologda (Nikolaev 1998).

Whooper Swan seems to be a rare passage migrant further south, e.g. in Ivanovo region, in particular on Gorky reservoir on the Volga, *c*.200 km from Moscow, between April and 11 May during 1960–65, but only rarely since (Gerasimov *et al.* 2000). To the east, in Kirov, particularly the Kama drainage and Tatarstan, it is now listed in the local Red Data Book as a rare migrant and former breeder. During the late-19th and early-20th centuries, however, it nested quite commonly across much of Kirov and the southern border of its breeding range crossed Tatarstan. Even by the early 1910s, there were just a few isolated pairs. Wetland drainage and hunting both contributed to its decline, such that in 1926–65 no nests were found in Tatarstan, although in 1965 a pair bred in Rybno-Slobodskiy district. Spring migrants begin to pass through Kirov and Tatarstan before the river ice breaks up, forcing them to rest on polynias[4] in early April, with mature birds moving through very quickly and first-years passing later and sometimes remaining until mid May. Return migration begins in mid September, and by late October all have left, although occasionally individuals winter on lake or river polynias, presumably because of injury or disease (Sotnikov 1999).

In central-western Russia[5], during the mid-19th century, Whooper Swan was a common migrant and winter visitor, with some breeding at large ponds and lakes. In particular it was considered quite common on the middle Volga and Kama. By the late-19th century, however, its range was already contracting and then bred only on the Mesha, Vyatka Volga, and Kama rivers. The species certainly continued to breed until 1926, when two young were trapped for Kazan Zoo. Between then and 1965, however, there were no certain records. A pair was recorded in summer 1935 (19 June); then in 1965 a pair bred, once again in the Kama floodplain (Artem'yev & Popov 1977). Despite annual research, those authors observed birds only in 1956, 1959 and 1960 on spring migration in April, and once in October (1959), indicating that it is now only a rare visitor to the region.

In Mordovia, between Gorky and Penza (*c.*55°N, 45°E), the Whooper is currently rare on migration, although it was apparently commoner in 1900–20. It also bred there in the 19th century, but has not done so since (Lysenko *et al.* 1997). At similar latitudes to the east, in Ul'yanovsk region between Samara and Kazan, the species formerly nested in valleys of larger rivers. Now, however, it is uncommon and only recorded on migration, more often along the Volga River. In 1995, 1997 and 1998, however, 2–4 Whooper Swans summered on Kuibyshev reservoir (Barabashin *et al.* 2001). In Kuibyshev region, Gorelov *et al.* (1987) noted that Whoopers were more widely recorded in the 19th century, but anthropogenic pressures during the early-20th century led to its decline. Nevertheless, during the 1970s–80s they occasionally appeared in two districts, and from 1970 two pairs bred successfully in Avgustovskiy Game Reserve (GR), with six pairs in 1977. In 1979, two pairs were also recorded in Rostashinskiy GR. In Khvorostyanskiy a pair nested at the Chagra River in 1971; in June 1977 a pair nested near Abashevo, and later the same month two more were discovered nearby (Gorelov *et al.* 1987). These and several other summering records indicate that the species may be recolonizing the area.

In nearby Mari[6], which lies northeast of Mordovia and northwest of Orenburg, numbers were so great in the 16th and 17th centuries that hunting was commercial significant. Thereafter, however, they fell, until extirpation in 1919. Then, in June 1976, eight were seen on Palenoe Lake, and in July 1977 eight visited fishponds near Joshkar-Ola. In subsequent years pairs and small groups of up to 11–15 have occurred on various waterbodies in the republic, remaining throughout the spring/ summer. In 1983, a pair bred for the first time in Mari at Cheboksary reservoir. In the years following, the number of pairs increased until by the late 1980s there were *c.*50 adults summering, among which 8–9 pairs bred (Baldaev 1990).

In Ryazan, south of Moscow, Whoopers bred in small numbers until the 1960s, otherwise it is a scarce migrant in early April to early May and from early October to early November (Kotyukov 1990). To the south, in Tula, Whooper Swan formerly nested on lakes in the north, and also occurred on migration. From 1879, however, there was no reference to the species in the region's literature. Then in 1956, and several times during the 1980s, it was seen on migration (Miller & Skalon 1990).

In the extreme southwest, in the Don River region, just east of Ukraine, the Whooper Swan is an uncommon migrant that uses the lower Don as its route. It is considerably less numerous than Mute Swan, occurring occasionally in spring, in small numbers during mid-March and early April, and is even scarcer in autumn. Nevertheless, there have been occasional winter records (e.g. in January 1988). Numbers at this season are subject to sharp fluctuations at most sites where it occurs. It has even been recorded as a scarce breeding bird. Two pairs bred at Novolipetsk metallurgic plant in 1981, and pairs have attempted to breed in Lipetsk and Voronezh regions, but such cases are rare, less than annual and have sometimes been curtailed by poaching (Belik 1990; Sarychev *et al.* 1990). More recently in these regions, Klimov *et al.* (2001) regarded it as a rare passage migrant and breeder, with 10–15 pairs.

In the region between the Volga and Ural rivers (south of Kuibyshev and north of the Caspian), the species has bred but is commoner as a passage migrant. For example, pairs nested in 1966 and 1969 on Kamysh-Samarskie Lakes, and a pair with cygnets was seen in early July 1970 on Ak-Kul' Lake. Individuals have also occurred in summer on Kisyk-Kamysh Lake in 1967 and on the Bolshoi Uzen' River in 1961. Small numbers of migrants (in flocks of up to 42) pass through in late March (from 24 March in Gur'yev region), early and late April, and late May, and between 10 and 28 October (Shevchenko *et al.* 1993). Similarly in Orenburg, straddling the middle and upper reaches of the Ural River, it is most visible on spring migration in late March–April when flocks of up to 60 occur around Orenburg and in the Trans-Uralian steppe (Samigullin 1990). The species apparently bred in the region in the late-19th century. Since then, however, it had only been reported on migration, that is until May 1995, when it was again found breeding, in Svetlinskiy district. The nest was set among reeds on an occupied Muskrat den (a fairly common situation in Russia) *c.*25 m from open water (Kornev & Korshikov 1995).

In the south part of western Russia, in the Volga delta, Whooper Swan is a very common migrant and a common winter visitor in October–April, with numbers stable in recent years (Rusanov 1990; Rusanov *et al.* 1999). Whoopers form mass concentrations each winter in the western delta, particularly favouring areas of lotus *Nelumbo nucifera*, whose seeds and roots they prefer. In the Damchik area of the Astrakhan NR, where such thickets cover *c.*3,000 ha, concentrations in late October–November have reached 10,000–30,000. Rusanov (1987) identified three important areas: the shallow-water delta of the Volga to a depth of 2 m (one million ha); Kizlyar region, and the region of the Kulaly and Morskoi islands. Because low temperatures cause the delta and northern Caspian to freeze for much of the winter, swans mainly winter in the microhabitats offered by patches of ice-free water, mostly in the southwest of the delta. The Kizlyar region consists of Kizlyar Bay and the coast of the Caspian north of the delta, south to Chechen Island, and the northern Agrakhan Gulf, and covers 433,000 ha. This region is also mostly shallow (<1 m) although around Tyuleniy Island depths reach 2 m, and to the east 5–6 m. Ice cover lasts a mean 66 days from late December until late February. Winds help keep some areas ice-free, allowing birds access to submerged vegetation such as pondweed and eelgrass. Around the Kulaly, Morskoi and Rybachiy islands water salinity is very high, averaging 10.9–11.2%, emergent vegetation is absent, and submerged vegetation consists mainly of *Potamogeton pectinatus*, eelgrass, Rock Cress and wigeongrasses. Air temperatures here are somewhat lower than near the western shore, and ice cover remains on average 67 days from late December until early March. Strong easterly winds dominate and keep large areas ice-free. The numbers of swans wintering in these regions is related to winter weather with fewer in colder winters. When conditions worsen suddenly, the swans leave the delta, sometimes to the Dagestan, Azerbaijan and Iranian shores (Rusanov 1987).

Large concentrations occur off the mouths of various channels of the Volga, with numbers ranging from several hundred to several thousand (1955–69), and as many as 33,200 in the Kizlyar region in 1980 (although the average was 7,300). In mild winters more remain in the Volga delta area and numbers are fewer in the Kizlyar Gulf (Rusanov 1987). The most remote and hence least studied region is that near Kulaly, Morskoi and Rybachiy islands. Here too swan numbers reached their highest during mild winters in the 1970s (5,000–10,500). In addition, in some winters Whoopers are observed on the open sea in areas where Caspian Seals concentrate to give birth. Sightings are more frequent in severe winters, when the swans may easily cross from one shore to the other of the Caspian Sea.

The northern Caspian Sea is an important wintering area; in 1976 and 1980 numbers exceeded 30,000, although some Mute Swans may have been included. Mean winter totals for 1970–80 (13,900) greatly exceeded those in 1955–69 (1,850), perhaps as a result of the hunting ban in place since 1956, and birds shifting from the south Caspian Sea wintering grounds. Fewer, and smaller broods are recorded in concentrations in the Volga delta in years of hard weather on the wintering grounds (Rusanov 1987, 1990).

WESTERN SIBERIA: FROM THE URALS TO THE YENISEI

As one travels east, first from western Europe to western Russia, then across the Urals into Siberia, the Whooper's range expands dramatically but, in contrast, the number of swan biologists is fewer and the amount of available information considerably smaller.

Population sizes are difficult to assess as access on the ground is so difficult and from the air so expensive, and information is sometimes contradictory, making it difficult to draw overall conclusions.

TRANS-URALIA

In the steppes of the Southern Urals, research was conducted in 1968–95 in Orenburg region (see Fig. 5.3). There, in spring, Whooper Swans appeared in small flocks of up to 60 from 20 March. They appeared in the middle reaches of the Ilek (in Aktyubinsk region), the middle Ural River (between Orenburg and Orsk), and on the Sarinskiy and Ural-Tobol plateaux. Migration continues throughout April until about 20 May, with its general direction being northeasterly. At this time, flocks sometimes rest for 1–3 days. Samigullin & Parasich (1995) reported as many as 700 on the Ural-Tobol plateau, however, both Whooper and Mutes were involved, and their relative proportions not noted.

Since 1992, Samigullin & Parasich (1995) have also found the species breeding on the Ural-Tobol plateau, with four nests and five broods in 1992–94 in Svetlinskiy district (54°N; 64°E). As is typical, pairs breed separately at shallow-water sites, building their nests amid reeds or bulrush (cattail) thickets, and using reeds, bulrush, and sedges as nesting materials. One was found on old floating vegetation and another was on a Muskrat den. They estimated that 5–9 pairs breed annually on the Ural-Tobol plateau. Clutches of 3–5 eggs are completed between mid May and mid June, with 2–5 cygnets hatching in late June–late July.

Autumn migration occurs through the Ural-Tobol plateau from early October until mid November, with most moving southwest. The peak is in the first half of October when Whoopers occur in flocks of up to 200 on Iriklinskoe reservoir. The largest mixed concentrations with Mute Swans are in mid October on lakes of the plateau, when up to 1,000 are present. Timing varies between years but, in autumn, it appears to be driven by sharp falls in temperature and, in particular, the fall of daytime temperatures from 5°C to 0°C. In 1979–94, 41% of flocks consisted of fewer than ten, flocks of 11–20 comprised 32.3%, 21–50 18.6% and 50–1,000 6.1%. Surveys in mid October in 1983–87 indicated that 67% of swans passing through the region are Whooper, the rest Mutes (Samigullin & Parasich 1995). It is a common migrant through southern trans-Uralia (Samigullin 1990; Gordienko 2001) during spring and autumn, usually in flocks of 20–50 (48%), and often forming mixed flocks with both Mutes and Bewick's in concentrations of 1,000 or more. The majority of Whoopers pass through Chelyabinsk before the main passage of Mute Swans. The largest flocks occur on lakes in the forest-steppe with maxima of *c*.500 at Kurlady, 349 at Tishki, 130 at Sinara and smaller numbers (fewer than 60) elsewhere. Total numbers are estimated at 5,000–7,500 (mean 6,200; Karyakin & Kozlov 1999).

In the 19th century it was quite common on lakes in north Chelyabinsk region, and even in the early-20th century it nested at several places, although it

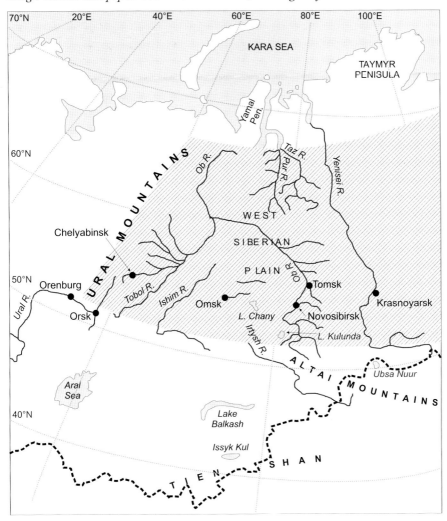

Figure 5.3.   Western Siberia showing main locations mentioned and the approximate breeding range of the Whooper Swan (Brazil & Shergalin 2002).

was declining and by the 1930s–40s only summering birds and migrants were observed. Then, in the 1980s–90s, it began to reappear as a scarce breeding bird at 18 lakes in the region, three in the eastern Urals and the others in the forest-steppe of Trans-Uralia (Karyakin & Kozlov 1999; Gordienko 2001). It also breeds on Travyanoe Lake and also occurs on various other waterbodies not investigated by Karyakin & Kozlov (1999). The total number of pairs estimated to breed in the region is 60–90, with a further 50–120 non-breeders.

In Kurgan Oblast[7] (east of Chelyabinsk and west of Omsk), Braude (1998), who conducted swan surveys in August 1989 and 1998 in its eastern part, found that numbers were considerably greater in 1998 than in 1989, with 6,832 (Mute and Whooper in 1998) as opposed to 1,916 the following year. Whereas in 1989 only four pairs of Whoopers were found, in 1998 56 pairs with broods were located; further investigation indicated that they had increased throughout Kurgan Oblast (Braude 1998).

In the region immediately east of the Urals, Whooper Swan is a common breeder in the northern forest-steppe. In the south it is an early-spring migrant which arrives before the end of March and in early April once daytime air temperatures rise above freezing, despite waterbodies still being ice-covered. The local forest-steppe population arrives early in April, occupying lakes at a density of up to $0.1–0.2/km^2$. Heaviest migration is in the latter half of April along the Tobol River (up to $47/km^2$) on shallow saline lakes, temporary floods and flooded water meadows. Here, they migrate north (65%) or northeast (35%). As elsewhere, many are in pairs (40%), slightly fewer in flocks of 20–50 (36%) and more rarely in small flocks of 3–10 (24%), perhaps comprising families from the previous year, and just 4% migrated alone (Blinova & Blinov 1997). Further investigation might show that, as in Europe, migration occurs in two waves, of potential breeders, and non-breeders. The majority have passed by late April, although small groups occur in May on the inter-stream lakes of the northern forest-steppe, at densities of up to $11/km^2$ (Blinova & Blinov 1997).

While Blinova & Blinov (1997) consider that Whoopers breed at high density in northern forest-steppe between the Tobol and Ishim rivers[8], they have not been found breeding in the southern forest-steppe and are certainly rare in the steppe zone of Kazakhstan. Their description of the breeding habitat includes shallow overgrown lakes and water meadows, where they occur at average densities of $3–4/km^2$ during early summer. Although they also occur on lakes outside the Tobol Valley, breeding has not been confirmed. Departure commences during the first half of September, when they begin to appear on large freshwater lakes in the southern forest-steppe between the Tobol and Ishim. Migration continues throughout September and early October, with some flocks halting at open flood lakes in the southern Tobol Valley and others at slightly saline lakes between the rivers. The majority of flocks consist of 11–60 (68%) and fewer (23%) of 3–10, which presumably represent new family parties, and more rarely they travel alone (9%). In contrast with spring, pairs were not observed. The majority (80%) fly south and the remainder chiefly southwest (19%).

In the taiga of middle Trans-Uralia[9], Antipov (2001b) made aerial surveys of the Whooper population. Of the 2,620 swans counted over seven years, most were along the left bank of the Konda River with about half in Elizarovskiy Federal Game Reserve. Densities ranged from $0.01/km^2$ in some swamps to $0.17/km^2$ in flooded habitats of the Ob River. The past status of both Whooper and Mute Swans in Omsk region, spanning the Irtysh River drainage, is poorly described. Furthermore, some of that published is misleading, contradictory and pertains only to the late-19th century, suggesting that confusion over these species' identities may have added to difficulties in agreeing over status. However, a recent 15-year study by Yakimenko (1997) has helped clarify the current status of both species in the enormous Omsk region. Whooper is a migrant that also breeds commonly in the steppe, forest-steppe and southern taiga forest. Mute Swan also breeds, principally in the southern forest-steppe and steppe north to the latitude of Omsk city (55°00N; 73°30E). The region covers a wide range of habitats, but both swans are found on brackish or slightly mineralised lakes. In the northern forest-steppe, migration commences during the last third of March and continues until *c.*20 April, although dates vary annually. In the central forest-steppe, around Omsk itself, movements occur up to ten days earlier. Whooper Swans heading further north continue to migrate into the second half of May, even early June, by which time local birds are already incubating.

Certain lakes are particularly attractive to waterfowl and shorebirds, and concentrations of Whoopers also occur there on migration. For example, Solenoe and Gorkoe lakes are shallow, highly mineralised lakes in the northern forest-steppe. They warm rapidly in spring, consequently submerged vegetation develops early, thus by May the lakes are already highly productive. Concurrently, above-water vegetation is still absent, so waterfowl find food plentiful within the lakes. Both Anseriformes and Charadriiformes appear in their thousands on lakes such as these during late May and the first third of June, although the Whoopers leave, as a rule, by the end of May, unless unfavourable weather delays their departure, e.g. in 1996, when snow continued until late May forcing over 400 to remain on Solenoe until *c.*10 June. In autumn, Whoopers leave northern forest-steppe on 20 September– 15 October, flying south at heights of 100–400 m in flocks of tens or even hundreds. Small flocks roost on lakes in the taiga zone and may remain for several days (Yakimenko 1997).

Estimation of the breeding population in this region is problematical due to the difficulty in accessing the huge number of lakes that vary considerably not only in size but also in their biotic characteristics. In the southern taiga, where oligotrophic sphagnum bogs are widespread, very few pairs breed. Rather more occupy the deciduous forest zone at densities of 0.8–1.0 pairs/100 km². The estimated population of this habitat was 80–100 pairs. The northern and central forest-steppes are the most densely and most evenly populated, with 0.4–2.0 pairs/100 km² and perhaps 270–300 breeding pairs. In the southern forest-steppe and steppe, divided by the Irtysh River, and in Kurumbel steppe on the east bank of the Irtysh, Whoopers occur at a density of 0.7 pairs/100 km². On the west bank two pairs probably bred at Lake Alabota, but none was found elsewhere. Numbers breeding here are low because highly saline lakes unsuitable for breeding predominate and much of the area has been developed for agriculture. As a result, the numbers breeding in these habitat types is not high—perhaps 50–60 pairs. In total, Yakimenko (1997) considered there to be 400–600 breeding pairs in the entire Omsk region.

Further east, in Novosibirsk region, particularly in the 120,000-ha Kirzinskiy State Game Reserve on the northern shore of Chany Lake, there is a typical forest-steppe landscape, with an abundance of natural waterbodies of different types. During 1968–79, Fedorov & Khodkov (1987) found that Whooper Swan was common on forest-steppe lakes, with breeding pairs on 12 waterbodies within Kirzinskiy GR and on two of the largest islands in Chany Lake. The first birds arrived in early April, when temperatures may still be $-20°C$ with snow still lying. Migration continues through the second third of April, with egg laying in early May and cygnets hatching in early to mid June, fledging late August–early September. Non-breeders gather on Chany Lake. Later families with broods also move to the lake. During the 1960s, apparently as many as 5,000–7,000 gathered there, but in the 1970s numbers within the reserve were no more than 100–130, and no more than 250 at the other principal site (Fedorov & Khodkov 1987).

THE WEST SIBERIAN PLAIN

In this region Whooper habitat is plentiful and it is a common, widespread species in flooded swamps within forest-tundra; densities reach $3/km^2$, although they are considerably lower in simple flood lands $(0.3/km^2)$ and in waterless valleys $(0.2/km^2)$. It is rarer along rivers, occurring at densities of only 0.6/5 km of river. If the forest-tundra zone is examined as a whole, even here it is rare, averaging $0.5/km^2$. In the northern taiga, it occurs at a far lower density on average than in forest-tundra, although the extent of habitat means that more birds actually occur here than in forest-tundra, being commonest on lakes $(2/km^2)$ and rarest in valleys without rivers $(0.005/km^2)$. Along rivers of the northern taiga it is as scarce as along forest-tundra rivers. In the middle taiga, Whooper Swan also occurs at a low density, with the highest densities on lakes $(3/km^2)$, and in the southern taiga it is extremely rare occurring at densities as low as $0.007/km^2$. Considering the forest zone as a whole, Ravkin *et al.* (1990) found its density to be just $0.1/km^2$, but given the vast area of habitat, they considered that perhaps 252,000 were present in the region, although in late summer, when one would expects numbers to be considerably higher, they only estimated a population of 58,000[10]. To the south of the taiga, the species' range extends into forest-steppe and even steppe. In these habitats it occurs less abundantly on average than in northern taiga and forest-tundra, but more commonly than in southern taiga forest. Ravkin *et al.* (1990) estimated numbers in early summer in the forest-steppe zone to be *c.*12,000 birds with another 3,000 in steppe, but only 2,000 in forest-steppe and 2,000 in steppe later in the season[11]. Highest densities were at lakes $(0.6/km^2)$ in forest-steppe and at steppe lakes $(1/km^2)$, while in dry valleys they were very scarce $(0.02/km^2$ in forest-steppe and $0.0004/km^2$ in steppe). Considering the entire West Siberian plain, Ravkin *et al.* (1990) estimated numbers in early summer at an astonishing 380,000, with 70,000 later in the season.

Thus recent surveys in this region suggest a local population several times greater than was previously estimated for the species as a whole! Total figures are difficult to assess, some data is questionable, and Ravkin (1991), an important source for the region, appears to give contradictory information, stating at one point that in Siberia the population is mainly concentrated in northern taiga, but that the highest density was in forest-tundra with slightly lower densities in northern and especially middle taiga, and considerably lower densities in habitats such as steppe. Further information from the region perhaps only confuses the issue more. Vartapetov (1984) observed Whoopers on a large lake within a peat bog, and on pastures among sub-taiga forest at densities as low as 0.02/km². He reported that whereas it was an extremely rare bird of marginal sub-zones of taiga, it was observed considerably more frequently in the Ob Valley. He concluded, however, that everywhere the species is *rare* or *extremely rare*. Could the situation really have changed so much for the better between 1984 and 1991?

The vast expanse of forest tundra and taiga doubtless provides ample habitat for an enormous number of Whoopers, but the total of 316,000 counted, and 380,000 birds estimated, on the west Siberian plain during the first half of summer by Ravkin (1991) does beggar belief, being more than three times the previous estimate for the entire population (Brazil 1981c; Madge & Burn 1988), especially given that he found only 67,000 (estimated 70,000) in the second half of the same summer when one might expect numbers to be inflated by the addition of young of the year. Because the totals are based on sampling and extrapolation, however, the overall degree of error was accepted to be very large and the 95% confidence limits amounted to 151,000–662,000 (Ravkin 1991). Nevertheless, even the lower end of that range means that the west Siberian Plain population is considerably larger than previously estimated and, if at all accurate, requires a rethink of the total world population.

Early in the year, particularly between mid May and June, Whoopers are extremely active and vocal, making them far more readily observed; it was during this period that Ravkin made his earlier estimates, which he considered may have been inflated. Taking all possible reasons for overestimation into account, Ravkin (1991) lowered his estimated population by 45% to 173,800 (95% confidence limits 83,000–364,000). There is a further complication: during the second half of June, swan activity drops as non-breeders, which according to Ravkin (1991) comprise *c.*75% of the population, fly to remote lakes that were not censused, thus counts might possibly be underestimates! Given this, anyone's estimates for this enormous region remain essentially guesstimates, but a figure in the region of 100,000 for this population may be reasonable, and even this requires that the world population is considerably larger than previously estimated. Nevertheless, and in extreme contradiction to Ravkin, the western Siberian population has also been estimated at just 17,000, wintering mainly in the Black Sea, and regarded as probably declining (Scott & Rose 1996).

Krivenko (1989; in Ravkin 1991) had made the only previous estimate of this population, in 1971–86, finding late-summer numbers to be of *just* 52,000! That is still almost as large as the entire northwest European population. Interestingly, their late-summer population estimates did not differ so markedly, and Ravkin considered it likely that Krivenko had underestimated his, for which there is perhaps

Figure 5.4.   Small flock of Whoopers arriving at a taiga forest lake during migration.

good reason, as Whoopers can be extremely wary when nesting and during moult they occupy remote, inaccessible and undisturbed lakes, which were not censused, while others may depart for Kazakhstan.

Further north, aerial surveys of the valley of the Ob River at *c.*61°24′N; 70°20′E confirmed Vartapetov's (1984) observations that they were commoner there than in the taiga inter-fluvial. These surveys also revealed that Whoopers occur densely, though unevenly, throughout the Ob Valley area, with 3–4 broods observed just 100–150 m apart, an unusual concentration. Furthermore, broods reared in the area were large (mean 3.8 in 1994, and 5.0 in 1995, with two broods of eight in September 1995; Poyarkov & Johnson 1996), indicating that the Ob River provides particularly good breeding conditions. Further observations in September 1995 of broods of nine and 11, and an earlier observation of 14 cygnets in semi-natural conditions at Askania-Nova Zoo indicate that Whoopers may either combine broods or that large broods may originate from joint clutches. Brood amalgamation has been reported in Trumpeter Swan, in which 11.5% of broods were involved in amalgamation and, was coincident with high densities of breeders (Mitchell & Rotella 1995, 1997), and for Bewick's (Brazil 2002), but not yet for Whooper Swan.

On the northern taiga of the West Siberian Plain, Whooper Swan is both a migrant and a breeder but, although widespread, it is everywhere very rare according to Vartapetov (1998). Birds arrive along the lower Ob River from early April, with migration continuing until mid May and the most intensive movements in early May (Fig. 5.4). Around the Kazym River none was recorded in June–August, whereas in the Ob-Purov interfluvial pairs and singles were observed on marshland lakes during the second half of June. In the Tazov northern taiga, Whoopers were found along the Taz River only during the first half of June, though they apparently formerly nested there (Vartapetov 1998).

Aerial surveys of the Yamal-Nenets Autonomous District, undertaken in June–July 1977 and 1978 by Krivenko *et al.* (1984) found that highest nesting densities were in the Ob River floodplain, at the Nadym River mouth, and along the east bank of the Nadym Ob. Density was lower in the floodplains of the Taz and Pur, and much lower elsewhere. Based on the extent of the aerial surveys and the available habitat, they estimated a population of up to 13,000 in the area, with *c.*3,000 pairs breeding. They further estimated that the region would hold *c.*35,000 birds by autumn. The following year, the late spring, cold summer and extraordinarily high water levels, greatly affected numbers and dispersal, leading to much lower numbers in the area.

In the Yenisei River valley, which forms the eastern border to this region, Whooper Swans occur in the breeding season at densities of 0.2/km². On the basis of surveys over several years, Vartapetov (1998; *in litt.* 2001) concluded that the highest density is on wetlands and channels in the Ob valley, with fewer in tussock swamps and lakes on the plains, in the Yenisei floodplain and along the Taz River, and lowest densities in the floodplain and swamps of the Ob River. Trends through the summer are for singles to be seen more frequently during June, while in July, although fewer birds are seen, these consist of breeding pairs, and none was seen in August. The total population here is estimated from aerial surveys at 5,000, with most (60%) in floodplain swamps, fewer (28%) on waterbodies and least (12%) on flooded areas (Krivenko *et al.* 1984; Vartapetov 1998). Estimating numbers is extremely difficult as climatic conditions can alter the extent of available habitat dramatically between years. Thus, in the northern taiga in 1977, 55,000 were encountered, whereas in 1978 only 8,200 were found. Conditions varied considerably between years. While 1977 was relatively warm and, with low water levels, the majority (79%) of Whoopers remained in the Ob floodplain, in 1978 spring was later, the summer cold and high water levels prolonged. As a result, larger numbers (30%) remained in flooded areas and in the interfluvial marsh-lake areas. In such years, when floodplain habitats are less suitable for breeding, fewer pairs nest and the success of those that do is reduced (Krivenko 1991 in Vartapetov 1998).

Along the west bank of the Yenisei in Krasnoyarsk territory, from the Kas River to the upper Bolshaya Keta River in the north, is an area of relatively dense nesting of both Whoopers and Bean Goose (*middendorffi* and/or *johanseni*). Here, Whooper Swan is widespread and nests at oligotrophic bog lakes of at least 25–30 ha in tundra and at mesotrophic taiga lakes (Zabelin 1996). Zabelin (1996) estimated that 200–250 pairs bred in the Turukhansk region of Krasnoyarsk in the early 1990s.

The Ob River region north of Tomsk is enormous, covering some 317,000 km², and five large rivers and a considerable number of smaller rivers join the Ob. The interlacing rivers and their floodplains with considerable swampy areas provide a wide range of wetlands suitable for various waterfowl including Whooper Swan, which breeds throughout the area, including in the Elizarovskiy GR in the lower Ob floodplain (Vengerov 1990b). The population is considered rather stable because survey results over a limited area were very similar in 1963–65 (900–1,000) and 1986–87 (940–1,320) (Adam *et al.* 1990). This reserve, which is a Ramsar site, was surveyed in May–September by Antipov (2001a). Distribution in floodplain habitats was affected by climatic and hydrologic conditions. Thus in 1978, low temperatures and strong winds negatively affected numbers, as did high water levels in 1979. Swan densities after the breeding season were the lowest in these years, at 0.2–0.48/km^2 with the lowest densities in wetlands without sors[12]. In milder years, densities in early summer ranged from 1.9 (1981) to 7.8/km^2 (1979) and in late summer from 2.1 (1982) to 4.8/km^2 (1977). Elizarovskiy is one of the main breeding and moulting grounds of the species in western Siberia, and is internationally significant. Mean density is apparently many times higher than in equivalent habitats of the middle and lower Ob and Irtysh (Antipov 2001a). Spring arrival is swift throughout the Ob region with birds appearing on 10–16 April in most years. Autumn migration is from mid October and seemingly almost simultaneous throughout the region, with *c.*6,000–8,000 passing through, compared with *c.*4,000 in spring (Adam *et al.* 1990). Hunting occurs, but is uncommon because of the tradition of protecting swans, and also perhaps because of the difficulty of hunting them.

In the northern part of west Siberia, the enormous Ob River, joined by the Irtysh at 62°N, drains north into Obskaya Bay (as large as Lake Baikal) and thence the Kara Sea. Vengerov (1990a) studied an area of *c.*76,600 km^2 of the Ob drainage, a region strongly intersected by channels, with innumerable lakes and flooded meadows-sors. The shoreline vegetation of these waterbodies typically consists of sedges, but also club-rushes and reeds. In spring, the majority of lakes are flooded by snowmelt and sors and large channels provide excellent feeding, nesting, and moulting and migratory resting sites. While breeding pairs move to the northern parts of this region to nest in April and early May, moulting birds remain the summer, arriving from the first third of May. At first these consist of solitary individuals and small flocks of 3–5, only later do larger groups appear or form. During mid May to early June, these flocks range from the low twenties to as many as 150, with the largest involving over 200 (Vengerov 1990a). Movement around various possible moulting sites continues into mid June, but the larger groups break into pairs and small flocks of fewer than ten among sors. Moult begins in early July and is finished by early September. At the end of summer they aggregate again and remain together until their departure (Vengerov 1990a), with up to 1,000 in Elizarovskiy GR during the late 1980s.

ALTAI KRAI

While the Ob flows for more than a 1,000 km north of Tomsk towards the enormous Obskaya Bay and the Arctic, to the south the topography is vastly different as one enters the Altai Krai[13] where the Ob originates. The Whooper Swan, though rare here is widespread. It has been known since at least the 19th century at lakes in some regions. Whoopers even breed in southeastern Altai, particularly on the plains, at wetlands among forested regions and high mountain lakes. It breeds regularly on lakes in the Kulunda hollow between the lower Kulunda and Suetka rivers, on Julu-Kul' Lake, and on lakes of the Kulu-Kul' hollow. It occurs at low density, with even vast lakes usually inhabited by just one pair. In the east it is rather rare, with widely scattered pairs. It only began to breed at various lakes in Kulunda steppe after the establishment of Blagoveshchensk reserve, where numbers increased slowly from one pair in 1978 to *c*.20 pairs by 1986–87 (Kuchin 1988; Kuchin & Kuchina 1990; Kuchin 2001a,b).

In spring, they arrive quite early at breeding sites, even before most snow has melted and while lakes are still frozen. The first migrants have been reported as early on 10 March, on the Biya River. In the upper Ob River, the first arrived on 12 March. Despite the occasional early arrival, the main migration occurs in mid April. Although they usually arrive in pairs at breeding sites, sometimes flocks of immatures appear at lakes and remain throughout the summer. Disturbance is a significant factor, driving pairs away from lakes that they may occupy briefly in the spring. Whoopers begin to migrate at the end of September and by mid-October they have all gone. On lakes in the upper Ob forests, they remain until consistently low temperatures lead to the formation of ice. Thus, on Bolshoe Kamyshnoe Lake a flock of seven remained as late as 10 November in 1986, although in 1987 a flock of 24 was forced away on 26 October (Kuchin & Kuchina 1990).

The wintering areas of this population are unknown, although it is possible that some winter in the Altai Krai. The coincidental departure of birds from the upper Ob forests and arrival at ice-free spring-fed lakes on the plain of the left bank of the lower Katun River suggest this. Confirmation comes from local people, who have noted that swans regularly winter on three non-freezing lakes and on small rivers emanating from these. Over 1966–86, arrival varied from late September (1966) to the first third of November (several years), and was usually coincident with the first consistently low temperatures, deep snowfalls and ice on other lakes. During the period, the number of swans on these lakes increased steadily, from just ten in 1967 to 35 in 1986, partially as a result of the area being made a game reserve in the mid

1970s. Wintering numbers at these lakes continued to rise, to 98 in 1993, and to maxima of 253 in December 1996 and 232 in December 1997, apparently due to changes in climate (Kuchin 2001a).

TUVA AND KHAKASSIA

To the east of the Altai, in Tuva, the Whooper Swan was once a quite common bird, breeding at Khadyn, Chagytai and Jerjaryk lakes, and it also winters at Ubsu-Nur (across the border in Mongolian Tuva). Unfortunately, it is declining and in some places it has disappeared completely, breeding now occurs only in little-developed regions of eastern Tuva and on Ubsu-Nur. The middle reaches of the slow-flowing Belin River, provided good habitat for waterfowl and Whooper Swans certainly bred there during the 1960s. Ubsu-Nur is an important site that supports not only breeders but also small flocks of non-breeders (Baranov 1990). Whooper Swan is scarce on spring passage or at least rarely recorded and then only in late April–early May. In autumn, it is somewhat commoner, with as many as 500 occurring in some areas (Savchenko *et al.* 1986, in Baranov 1990).

Emel'yanov & Savchenko (1990) made aerial surveys of Tuva and adjacent Khakassia between spring and autumn periods in 1980–87. They found that spring migration began in the south from mid April to early May, and involved individuals, pairs and small flocks, its timing coinciding with the formation of the first ice-free open water on rivers and shallow lakes. Migrants pass through Tere-Khol in southern Tuva, Khadyn Lake in central Tuva, and Ulukh-Kol and Bele in Khakassia over 15–30 days. They are most often recorded at medium-sized and large water-bodies west of the Yenisei. The intensity of passage is very low, with only small numbers involved, although circumstantial evidence suggests that more may pass through Khadyn Lake and Ulukh-Kol at night. Ulukh-Kol supports the largest concentrations in the south of western Siberia in spring, with over 2,500 during the period 23 April–9 May 1987; large numbers do not occur here in autumn because of hunting. Most depart at night and early morning and head north or northeast.

Autumn migration through the region begins in early September and continues into the first third of October. It usually occurs in a single wave of 10–20 days. In Khakassia, the peak was during the first four days of October in 1986, and in Tuva it was on 7–10 October in 1980 and 1982. Whoopers are commonly seen in flocks of 3–20 and they prefer more open steppe lakes as resting sites. After Ulukh-Kol, Bele Lake is the second most important wetland in the region with up to 370 Whoopers in spring and up to 500 in autumn.

CENTRAL AND EASTERN SIBERIA: FROM THE YENISEI TO THE LENA

Central and eastern Siberia are defined as the region from the Yenisei in the west (*c.*85°E) to the Lena in the east (125°E) and from the Taimyr Peninsula in the north (77°N) to the border with Mongolia in the south at *c.*50°N (see Fig. 5.5). This region, itself almost as large as Europe, spans a similar range of habitats to those in western Siberia, and also incorporates not just two of Russia's greatest rivers (the Yenisei and Lena) but also the single largest, deepest and oldest freshwater body on earth, Lake Baikal (Brazil & Shergalin in press).

The Whooper's distribution here is disjunct and highly variable, thus Emel'yanov & Savchenko (2001) considered that in the Yenisei region the number is extremely low and does not exceed 400–500. East of the Yenisei, in the lower Tunguska Basin, it is a passage migrant that also nests at wetlands in the floodplains of tributaries of the lower Tunguska, and also occurs on larger lakes in interfluvial areas. Nevertheless, numbers here were small, at *c.*20 pairs in the 1980s. As a result of a sharp decline in motorised water transport and other factors, however, numbers are increasing even in the southern, more developed part of the lower Tunguska Valley, such that now *c.*150 pairs occur in the region (Mel'nikov 2000).

South of the Tunguska, we reach the Angara, the only outflow of the Lake Baikal drainage system, which flows northwest, eventually joining the Yenisei. Here, in the early 1980s, Mel'nikov *et al.* (1990a,b) studied the waterfowl of the middle part of the upper Angara. A peculiarity of this region is the large number of small lakes with abundant old floating vegetation, aquatic vegetation and well-vegetated shorelines and swampy meadows, or grassy bogs. They found comparatively high densities of breeding Whoopers between the upper Angara and the Kotera River, where there was one pair/20 km². Elsewhere, nesting density is considerably lower at 1–3 pairs/100 km². The greatest density recorded was of three pairs within 25 km². The total number nesting in the main part of the Verkhne-Angara depression was 80–100 pairs. The upper Angara depression is, according to Mel'nikov *et al.* (1990a), the only area in the south of central and eastern Siberia, where the Whooper is common. Following the completion of the Baikal–Amur Railway, development in the region has increased and in order to avoid a decline in the species, Mel'nikov *et al.* (1990b) deemed it important that a game reserve be established in the region of the upper Angara.

In the enormous Irkutsk Region, the availability of suitable nesting habitat is limited by topography. Since the 1950s–60s, when industrial development began, Whooper numbers have declined. It used to occur quite commonly as a breeder mainly in the basins of the lower Tunguska and Kirenga rivers, and at isolated lakes of the Prisayanie, the Lena-Angara plateau, and the Vitim-Patom upland. More recently, numbers were very small (just a few pairs) not exceeding 50 pairs in the whole Irkutsk region (Mel'nikov *et al.* 1990a).

Northern Lake Baikal and the upper Angara depression (3,800 km²), which mostly consists of the floodplain of the upper Angara River and its tributaries was

Figure 5.5.   Central and Eastern Siberia showing main locations mentioned and approximate breeding range of Whooper Swan (after Brazil & Shergalin in press).

also studied by Sadkov & Safronov (1990). In the middle reaches of the upper Angara the depression reaches 60 km in width, and is covered with sedge/grass meadows and sedge/horsetail swamps, with a number of lakes connected by channels. Whoopers begin to arrive here during the first half of April with migration continuing into May and early June, although total numbers are rather small. They arrive in pairs and small flocks of 6–30, some at night, others in early morning from the southeast having traversed the North-Mui ridge and northern spurs of the Barguzinskiy ridge, or passes in the upper tributaries of the upper Angara. Flocks of 30–100 may pause in river valleys, and on lakes in Kumoro-Uoyan, dependent on weather conditions. In autumn small numbers appear on passage during the first half of September, but migration ends during October because all water in the region freezes.

THE BAIKAL REGION

Until the 1950s–60s, the species was a widespread breeder in the Lake Baikal region. Its range included practically the whole shore of Lake Baikal, but since the 1960s its numbers and distribution have declined sharply and it is now considered rare with as few as 45–50 pairs in the entire Lake Baikal basin (Dorjiev & Elaev 2001). Pairs breed regularly only on suitable waterbodies of the Svyatoi Nos Peninsula, in Chivyrkui Bay, in the upper Angara Valley, and in the Barguzin Valley[14], although Yumov (2001) considered it to be only rare there.

On migration, they appear along both shores of Lake Baikal, and in western Trans-Baikalia along the Selenga River and its tributaries, and on Gusinoe Lake, Taglei Lake, and in other depressions. Current migrant numbers are also considerably lower than 50 years ago. In the 1950s–60s, polynias in the Selenga delta and Chivyrkui Bay held up to several hundreds, and flocks of up to 500 were recorded annually in spring in the Borgoiskaya and Barguzinskaya depressions (Dorjiev & Elaev 2001). In southern Baikal and in southeast Trans-Baikalia, it is an uncommon, even scarce, migrant from mid April until late May, more rarely in June, and is also a rare breeder here. In middle Baikal spring migration starts about ten days later and nomadic non-breeders have occasionally been sighted in June–July. Simultaneous arrival at sites far removed from each other has also been noted in this region (Shinkarenko *et al.* 1990; Popov *et al.* 1998; Elaev *et al.* 2000). In autumn, the first reappear in southern Baikal from 13 September and the last have passed by 23 October. Although they may occur widely in autumn, numbers are low, with small flocks of no more than 15–30 and usually much smaller groups of just 4–8, which are presumably families.

In the 1970s–80s, the species bred only once in the Selenga delta, in 1987; since the mid 1980s however one or two pairs have bred regularly in the western delta (Fevelov *et al.* 2001). The decline in numbers visiting southern Baikal during the mid 20th century has been attributed to loss of habitat, particularly at stopover sites, caused by the raising of the water level in Lake Baikal following the construction of the Irkutsk dam. Now only small numbers up to 220 are recorded (Shinkarenko *et al.* 1990).

More generally in central and eastern Siberia, the Whooper is considered rare in taiga and forest-tundra habitats. Along the Yenisei, it occurs as far as 70°N, although breeding has only been confirmed to 68°15′N in southern forest-tundra around Ust-Khantaika. The southern limit of its modern breeding range in this region is less clear. In the early 1960s, it occasionally nested in swamps and lakes along the upper Kamenka and Gorbilok rivers (58°N; 105°E), and it has also bred south to Sayany and in Khakassia, although it is rare there, while in the Minusinsk depression the Whooper is a rare nomadic species (Rogacheva 1988, 1992). Its natural breeding range may continue south beyond the border of Russia here, as it also breeds in Mongolia (Batbayar 2000).

Along the east bank of the Yenisei, Whoopers probably breed on most suitable waterbodies in the Putorana Mountains and around the Norilsk lakes. Near Pyasino Lake its breeding range abuts that of its smaller relative the tundra-breeding Bewick's Swan. The richest region for Whoopers in all of central and eastern Siberia is deemed to be the northern taiga and forest-tundra along the west bank of the Yenisei. In the middle taiga region, numbers used to be high along the upper reaches of the Taz and Elogui basins, but although they still occur, they are now rare. Particularly in the 1950s–60s, they were often killed by hunters and fishermen, although by the late 1970s–80s such persecution was reduced, and the prognosis for the population was more favourable (Rogacheva 1988, 1992).

THE RUSSIAN FAR EAST: FROM THE LENA TO THE BERING SEA

From the Lena River east lies yet another enormous region variously known as the Russian Far East or Yakutia, Chukotka and Kamchatka. The Russian Far East includes the region from the Lena (*c.*125°E) to the Bering Strait, the Kamchatka Peninsula and Anadyr region facing the Bering Sea at 155–180°E. It stretches from the shores of the east Siberian Sea at *c.*72°N as far south as the narrow finger of Russian territory known as Ussuriland, to approximately 43°N, and including the Amur River, Sakhalin, the Kurile Islands and the Sea of Okhotsk (Brazil & Shergalin in press; Fig. 5.6).

Only one population of Whoopers is currently recognised east of the Urals. Variously described as the central Siberian and/or eastern Asian population, it was estimated to comprise 30,000 birds by Perennou *et al.* (1994) and Rose & Scott (1997). I feel, however, that this is an over-simplification based on limited information, as migration data suggest that birds from this vast area winter in quite different regions. Some east of the Urals may migrate west or southwest, while others move south or southeast. Furthermore, those breeding in central Siberia may migrate to Central Asia, Mongolia and China via Lake Baikal, whereas those breeding in Yakutia and Chukotka migrate via the east Russian coast and Sakhalin to Japan. More recent estimates are of 60,000 for this population, but this figure may in fact be appropriate for the Russian Far East alone. Approximately half of this total winters in Japan, with a further 10,000–15,000 in China, 7,500–8,000 in Kamchatka and 4,000 in the Koreas (Miyabayashi & Mundkur 1999), although their figures for China are very much higher than those of Li (1996) and Ma (1996). A small number, fewer than 50, winters on the Aleutians and Pribilofs of Alaska and breeding has also occurred (Sykes & Sonneborn 1998; Mitchell

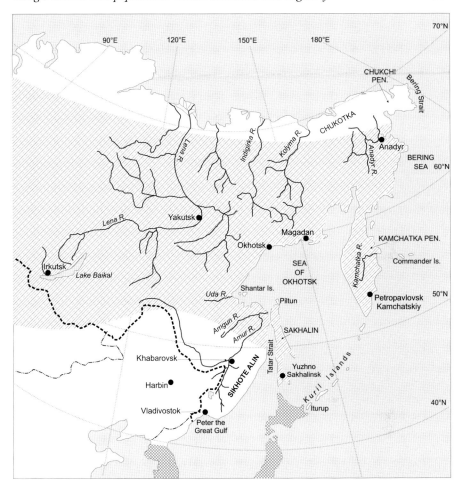

Figure 5.6.   The Russian Far East showing the main locations mentioned and the approximate breeding and wintering areas of the Whooper Swan. ▨ = *Breeding* ▨ = *Wintering.*

1998). Though politically and geographically part of the New World, biogeographically the links to Asia are so great that this cannot be regarded as a separate population.

Starting in the west, we must first consider the Lena River. In the lower Lena Valley, Whoopers occur widely in the taiga zone and even penetrate the tundra, although human activity has greatly reduced availability of nesting sites within the main river floodplains. For example, compared with the 1950s, breeding on the large islands of the lower Vilyui has dropped by a factor of 7–10. A similar decrease has been noted from the Lena River; declines attributed to losses of both nesting and moulting habitat. Nevertheless, Labutin *et al.* (1987, 1988) found pairs and small flocks many times in summer on tributaries of the Lena, though more commonly closer to the foothills. And on migration, during May and September it is by no means rare. Further north, towards the Lena delta, it was rare although it has bred occasionally as far north as the southern delta (72°N) and especially so since the 1970s (Perfil`yev 1972, 1976, 1979, 1987). It also strays to the subarctic tundra where, for example, a single was recorded on 9 August near the lower Kengdei River (Sofronov 2000).

At the opposite extreme, in the far south of the Russian Far East, between the Sikhote Alin mountains and the border with China, the Lake Khanka area is important. Glushenko (1990) considered that the Khanka-Sungacha Lowland to be the only reliably reported breeding site in the Primorskiy Krai, although Vorob'yov (1954) was unaware of it breeding there, so perhaps it is a recent colonist here, one of the southernmost breeding sites in its entire range at *c.*45°N. In 1987, censuses on 23–25 May and 11–12 September found five breeding pairs in eastern Khanka, between the Spasovka mouth and the Sungacha mouth, and one pair in the Sungacha Valley. In addition, 25 summering Whooper Swans were found. During the September census only four solitary adults and five pairs were relocated, only two of which had chicks. Numbers breeding are small, just 1–6 pairs in 1980–88. As the water level in Lake Khanka rises in late summer swan numbers decline with deteriorating feeding conditions (Glushenko 1990; Glushenko & Bocharnikov 1991).

Although breeding is scarce, it is a common migrant through Ussuriland northeast of Lake Khanka. In spring it appears very early in southern Primorye, sometimes by 8 March, although more typically from mid March and the last have passed by mid April. Large flocks were observed flying north over Lake Khanka in early April 1945 presumably en route from Korean or perhaps Chinese wintering grounds. In autumn, passage occurs during October and early November, and although, according to Przewalski and Jankowskii, some used to winter in coastal southern Primorye, no recent data are available from the Gulf of Peter the Great (Vorob'yov 1954). In southern Ussuriland, it is not especially common, though it appears on migration at both seasons in flocks of up to 80, in spring during the last third of March and early April, occasionally until 22 April. Autumn migration is in October–November and is far more concentrated, with several thousand at sea, river mouths and freshwater lakes. The first birds appear in early October, with numbers peaking in late October–early November, e.g. on 22 October 1962, a flock of 1,220 was reported on Khasan Lake (Panov 1973).

During early spring (March), they pass through south Primorye in small flocks of 10–15 birds, but as the intensity of migration increases, so flock size also increases. The majority (57%) passed through in flocks of 30–70, and the largest flock recorded was of 122 on 29 March 1989. Although some were noted at night, the majority (51%) passed through between 18:00 and 20:00. Although weather conditions seemed not to affect passage intensity, the height of passage (normally 150–300 m) was lower in windy weather and higher (500 m or above) in clear weather with light or no wind (Gorchakov 1996).

In the Khabarovsk Krai, which includes the valleys of the Ussuri and Amur rivers, the Tatar Strait separating the island of Sakhalin from the mainland, the mouth of the Amur, and the Shantar Islands in the westernmost Okhotsk Sea (Fig. 5.6), the Whooper is a common migrant but only a scarce breeder. It formerly bred and may still breed at lakes in the lower Amur (Vorob'yov 1954) and in Khabarovsk region it continues to breed in the north and centre, but is declining and rare, being added to the region's Red Data list in 1999. Until the mid 1980s it nested on lakes within 90 km of Khabarovsk City, however the southern border of its range retreated north to Bolon Lake as a result of forest fires, poaching and other disturbance, and it has declined from 20–25 pairs to just three in 1999 (Roslyakov & Voronov 2001). The most important site for breeding is currently Mukhtel' Lake, and for migrants the Amur estuary. Neither, unfortunately, has protected status, and disturbance at Mukhtel' has increased considerably since a recreational site was developed, and poaching continues in the Amur estuary.

The first migrants appear along the lower Amur in the last third of April, with mass migration occurring in early May. In 1986–87, Roslyakov (1990) observed flocks of 75–180 at very low altitudes (50–150 m), crossing the Amur towards Bolon Lake. In springs 1986–87 *c.*2,000 were seen between 28 April and 12 May. In the Bolon Lake basin on 1–14 April 1986 and 1987 at least 2,000–2,500 were seen. That figure is, however, a considerable reduction from numbers formerly recorded. In 1976–80, 5,000–6,000 were counted, although Roslyakov (1990) considered that in earlier counts migrating Bewick's Swans might have been mistakenly included. Spring passage over the Amur occurs in mid and late April until late May. The Amur estuary is a very important stopover for both Whoopers and Bewick's, especially during May as coastal ice is breaking up then (Babenko 2000). During spring, flocks ranging in size from a few hundred to *c.*1,500 Whoopers have been found on Orlik Lake, on the My, Chomi and Amur rivers, and the Gulf of Schast'ia, with 5,000 or more passing through the lower Amur area. Babenko (2000) reported as many as 2,000–3,000 on the Shantar Islands, and on 17 May 1991 on the Amur estuary, a total of 4,150 was found.

Autumn migration is best witnessed along the shore and the Tatar Strait, where at least 8,000–10,000 pass through, whereas inland fewer birds, perhaps 2,000–2,500, occur from early September. Low-altitude migration is typical, usually no higher than 200–300 m. Departure from the Shantar Islands occurs from mid October and great flocks, numbering thousands, occur at the Amur estuary during the same period, then along the coast and through the Tatar Strait. Large flocks of 5,000–6,000 may occur in the Gulf of Schast'ia, at the Amgun' River mouth and on Bolon Lake; and mixed flocks (with Bewick's) of 3,000–5,000 gather on the western shore of Sakhalin. Flocks of up to 100 early migrants (presumably non-breeders) have been seen on August (1986) on Mukhtel' Lake (Babenko 2000).

Before 1970, Whooper Swan was a rather scarce breeder in the lower Amur area and the Shantar Islands, with no more than 50 pairs nesting, mainly at inaccessible northern sites. During the late 1970s, numbers increased considerably, with at least 350–400 pairs in 1987 in all Khabarovsk territory; furthermore they began nesting closer to human dwellings, and even in the vicinity of Khabarovsk. Since then, further increases have been noted in some areas, but decreases in others (Roslyakov 1987, 1990; Babenko 2000; Brazil & Shergalin in press). In 1997–2000, however, Poyarkov (2001) reported a sharp decrease in the number of several waterfowl including Whooper Swan. Babenko (2000) also considered that the breeding population of this area has been declining since the mid 1980s, perhaps because of increased disturbance and poaching.

The only information on Whoopers wintering in the Khabarovsk Krai is old, with reports in the 1930s from the Ul'ya River, which flows into the Sea of Okhotsk (see Babenko 2000). A little to the east, on Sakhalin, small numbers of Whoopers have been reported in winter, but as all inland waters freeze for 4–6 months, the few wintering birds are restricted to polynias and springs. Numbers are small, perhaps no more than 50, and are not infrequently shot by hunters or caught by predators. On the volcanic island of Iturup, in the southern Kuril Islands, 30–40 Whoopers winter on polynias and near various hot springs, although they are unfortunately the target for local hunters here too. In addition, Ben'kovskiy & Ben'kovskaya (1990) considered that small numbers might also breed here, given the presence of many remote, suitable sites.

Continuing north through the Russian Far East, in the Kava River drainage of Magadan region, small numbers of Whooper summer and breed, and 40–50 moult on the Kava itself. Spring migrants, heading north or northeast, arrive from the direction of the Ushki Gulf. They fly mostly in pairs or small flocks of up to 60–70, most commonly 15–30, at heights of 70–150 m and commonly follow river valleys. Arrival here is during late April and early May, usually coinciding with the first breaks in the ice. In some early and extensive polynias, spring migrants may arrive earlier such as on the upper Kava. Mass spring migration occurs in May, ending before June; migration recommences in late September and peaks during October (Krechmar & Krechmar 1997).

In northern Yakutia, east of the Lena, the species is distributed in the taiga forest zone being replaced on the tundra by Bewick's Swan. Further east, however, to the east of the Kolyma River in Chukotka, the Whooper's range extends well beyond the border of the taiga and it inhabits the entire Beringian forest-tundra, but it is Bewick's that breeds further north on the Chukotka Peninsula. Overwintering by Whoopers has even been recorded here, on polynias in the Apuka and Anadyr rivers, by far the northernmost wintering observations on record in Asia, at c.65°N (Kishchinskiy 1988). Along the Lena River it breeds as far north as 72°N, whereas in the Vilyui River basin it breeds only to c.66°N (Degtyarev 1987). In the drainages of the Indigirka and Kolyma, it occurs widely in the northern taiga and, in certain places in the forest-tundra, to 69–70°N. Syroechkovski (2002) considers that Whoopers have expanded their range north on the Taimyr (central Siberia) and in Yakutia in recent decades, and this pattern can be expected to continue as global warming influences spring thaw and autumn freeze across northern Russia (Fig. 5.7).

Figure 5.7.   Global warming may be encouraging Whoopers to expand their range northwards in northeastern Russia.

Numbers appear to be lower in western Yakutia, with small numbers in the middle and lower Lena basin, but considerably higher numbers in the Indigirka and Kolyma lowlands, with as many as 2,000–8,000 there in autumn. Densities in the middle Kolyma have been unstable, sometimes rare sometimes a common moulting visitor, but they have declined considerably, however, from 21 birds/100 km^2 in 1966, to 8 in 1970, and just 2 in 1975. Aerial surveys conducted in 1983 indicated that in the Indigirka River basin density was 12.7/100 km^2, 10.2 in the upper Kolyma, and 6.5 in the middle Kolyma. Declines in densities, particularly in central Yakutia, have been attributed to their nesting and moulting habitat becoming degraded through agricultural development, and intensified disturbance by sport hunters and fishermen (Krivosheev 1963; Degtyarev 1987).

In northeast Russia, in the lowlands between the Indigirka and the Kolyma rivers, the considerable number of lakes provide ideal summer habitat and, unsurprisingly, it is in this region that it is most common in Yakutia. Within the limits of the taiga, it can be found everywhere, and along river valleys it even penetrates massifs. Along the Indigirka River, it occurs as far north as 70°N, 30 km from Olenegorsk. The northern border of its range deviates south to 69°30′N through the foothills of the Ulakhan-Sis and Suor-Uit ranges. To the east, in the Kolyma Valley, singles have been observed almost to the river mouth at 68–69°N. Research in 1978–87 between the Indigirka and Kolyma rivers provided information on migration, population and ecology. The highest density was 25.3/100 km^2 east of the Indigirka. Northern regions, and also the southeastern lowland basins of the Ozhogin and Yasachnaya rivers are also quite densely inhabited with 9–16/100 km^2. These regions all have many suitable lakes, but the latter two are also used for agriculture and water transport, thus disturbance is greater. In higher regions, lakes are considerably fewer and numbers of Whoopers lower. In the Sedema River basin, for example, where altitudes reach 230 m, density was 6.8km^2, while on the Alazeia and Yukagir plateaux, which reach 500–900 m, Whoopers occur only in river valleys with floodplains and at densities of just 0.01–0.5/100 km^2 (Degtyarev 1990).

Unfortunately, Degtyarev's (1990) paper is unclear on the extrapolation to the Indigirka area (127,200 km^2) and he may have found them at densities of 6.2/100 km^2 throughout, or perhaps numbering 6,200 in the region. It is clear, however, that he estimated the population in the basins of the Alazeya and Kolyma rivers to be 5,800. The population was considered to have increased until 1966, but thereafter, as a result of the loss of both breeding and moulting habitat due to lake drainage, it decreased dramatically. In 1983–84, fresh surveys of the same region indicated that the population had stabilised and might be increasing again, a hypothesis supported by observations of migrants from a fixed point in the middle Kolyma, where numbers also increased during the 1980s.

In Yakutia, breeding Whoopers have a distinct preference for small wetlands (68.6–77.4% of pairs at small lakes 0.01–1 km^2), whereas moulters prefer large (6–10 km^2) lakes. Whoopers are typically found here in small numbers, in pairs (53.2–74%), alone (14–34.4%) or rarely in small groups of 3–15 (1–9.3%) (N=368). By the last third of August pre-departure concentrations develop and birds may gather into flocks of up to 35. In Zapolyar'ye (Trans-Polaria), reproductive capacity appeared rather low. Although as many as 49–65% in peripheral areas and as many as 78% of those in central and southern areas formed territorial pairs, few actually bred (Degtyarev 1990). In the sparse forests of the tundra by the Alazeya and Kolyma rivers, the proportion of young was only 4.3% (in 1984), in the lower Indigirka, further west, that figure was 7.8% (in 1980) and 7.1% (1985), while in the middle reaches of the Indigirka and Kolyma rivers it was 7.7% (in 1983). Not only did few pairs breed but also output was low, with the average brood size just 2.6–3.3 (N=25). In marginal areas output was even lower at just 1.7 (N=6). In northeast Yakutia, however, the Whooper appears to have a healthy population that is at a higher density than in adjoining regions, and it has probably reached its maximum density, as indicated by its stable numbers and low rates of reproduction (Degtyarev 1990).

The extreme northeastern part of Russia, Chukotka[15], is also within the Whooper's breeding range. Although little is known of this population, wherever taiga and forest-tundra habitat exists in a relatively undisturbed state, then it probably remains common. It is widely distributed in the taiga, with the northern limit of its breeding range coincident with the distribution of coniferous forest over most of the region. Only in the east, apparently, does it extend beyond the northern limits of the taiga zone treeline (Kondratiev 1991). It is a common breeder in the Anadyr forest-tundra, absent only from the lower river. Population data and information on its breeding status in this large region, are, however, very limited. Preliminary estimates suggest that a modest 2,200 individuals inhabit the region.

The populations of the Penzhin-Parapol Valley and Anadyr were considered quite stable. Birds banded in Japan have not been found in this northeastern portion of the Russian Far East prompting Kondratiev (1990) to speculate that those breeding there winter in Kamchatka, while those reaching Japan do so presumably from Yakutia and possibly further west.

According to Krechmar (1982a, 1990) and Krechmar & Kondratiev (1986), there is a small but very stable population of 100–200 breeding in small- and medium-sized lakes in depressions along the middle Anadyr in forest-tundra and shrubby tundra landscapes, and 1,100 in the Penzhina-Parapol Valley further south. Of about 184 Whoopers counted in 500 km^2 of the Magadan State Reserve, only *c.*30% were breeding pairs, and the rest non-breeders (Kondratiev 1991), a figure remarkably coincident with estimates from northwest Europe, and presumably also true of northern Yakutia and Chukotka, where it ranges almost to the shore of the Bering Sea, the extreme eastern Asian outpost for this species. Here, its range does not meet that of Bewick's Swan, which is completely absent from the middle reaches of the Anadyr, even on migration.

In spring, Whoopers begin to arrive in Anadyr during mid or even early April when wintry conditions may persist and nocturnal temperatures often fall to −30°C and, occasionally, even −40°C. The only habitat available to early arrivals are the polynias, open patches of unfrozen water, amid the still icebound Anadyr and its tributaries (Fig. 5.8). Both on the polynias and at the breeding sites in the floodplain landscapes of the Anadyr, Whoopers usually appear in pairs or small groups of 3–5, presumably pairs and their previous year's offspring (Krechmar 1990). They remain until intensive snowmelt exposes their nesting habitat, which may be up to several tens of km from the polynias. Waiting at the polynias allows

Figure 5.8. Polynias are vital for wintering birds and early migrants.

that portion of the population that will attempt to breed the possibility of reaching the breeding sites almost immediately upon conditions becoming favourable. This movement, from the polynias to the breeding territories, occurs in mid May or in the final third of the month, making these among the latest Whoopers to commence breeding. Just after their appearance in the nesting range, they visit and remain at 'oases of spring', i.e. sites where snow melts early, and wind-blown silt and sand promote rapid melting. At such sites, they gather in small parties of up to ten with other waterfowl and Sandhill Cranes and, as soon as small shallow-water grassy lakes are released from ice, they visit them to feed. During this same period (mid to late May), considerable numbers of Whoopers continue to arrive from the wintering areas, but most of these later arrivals probably do not breed (Krechmar 1990).

Northeast of the Anadyr, towards the Bering Strait, Whoopers become scarce. Portenko (1981) considered that their rarity in the central and eastern Chukchi Peninsula made it more likely that they only visited there to moult.

To the south of Chukotka lies Kamchatka, which separates the Okhotsk Sea from the Pacific. In the north lies the Penzhina-Parapol Region[16], a forest-tundra of lake–alas lowlands, with weakened permafrost, summer thawing and many lakes. There, Gusakov (1987) studied the population by means of aerial and ground surveys in 1977–83 covering as much as 17,000 km^2 in some years. The habitat of flooded hollows, shallow lakes and their islands provide ideal habitat for Whoopers, which are common, and a wide range of other waterfowl. Bewick's Swan is only a rare migrant through this region, the majority passing further west. During the first aerial survey (20–23 September 1979), when some lakes were already freezing, the population consisted of non-breeding pairs (lacking broods 50%), cygnets (27%), breeding pairs (18%), and singles (5%). The average size of fledged broods was 2.9 (somewhat higher than Krechmar 1982a,b) found just to the north along the Anadyr). Average density in Penzhina-Parapol was of 7–9/100 km^2 in 1979–81, during which time up to 60% of the population consisted of non-breeding pairs.

Kamchatka is both an important breeding ground and a surprisingly important wintering area. Given the generally extremely severe winter climate of northeast Russia, a surprisingly large number of Whoopers winter here because of the peninsula's comparatively mild maritime climate, its considerable number of ice-free rivers, numerous springs (many of them hot), and the extensive aquatic vegetation that is available year-round. In addition, the extensive plains, the density of wetland habitats (10,000 km^2 of swamps, and more than 10,000 lakes) provides favourable resting areas for migrants and a considerable extent of suitable breeding habitat. In the north, the limit of breeding in Kamchatka is reached at the narrow base of the peninsula between the Penzhinskaya Gulf and the Parapol depression (Gerasimov & Alekseev 1990). With the exception of the small population in Anadyr, and the exceptional record on Attu in the Aleutians, Kamchatka represents the northeastern limit of the species' breeding range. Nesting occurs as far south as the Pravaya Khodutka River in the southeast, where in early July 1980, 4–5 pairs and two broods each of two cygnets, were sighted. Aerial surveys of the upper and middle

Kamchatka River, in the central peninsula, on 25–26 May 1986, found 21 pairs with clutches and cygnets, and two pairs without nests in the upper third of the valley. Another survey, on 22–25 May 1987, found seven pairs on nests, 11 pairs without nests and a further 30 alone or in flocks mainly in the middle river basin (Gerasimov & Alekseev 1990). Extrapolation of observed data to the area of habitat suitable for breeding in the entire Kamchatka Valley suggests that at least 90–100 pairs may breed in the central peninsula (Brazil & Shergalin in press).

The importance of migration along the peninsula was confirmed by Gerasimov (2000), who found that 2,200 Whooper Swans passed through the Ozero Kharchinskoe reserve in central Kamchatka between 27 April and 27 May 1999.

An initial survey of the whole of Kamchatka in 1966–68 by questionnaire indicated that a considerable number of Whoopers, some 5,000–5,500, winter there (Gerasimov & Alekseev 1990). During late February and early March 1980, the first aerial survey was conducted. The region being far too extensive to cover adequately with the single aircraft available, a questionnaire survey was made prior to the flights in order to help locate wintering concentrations. To give some indication of the scope of the undertaking, and to help envisage both the landscape and distribution of the swans, this highly volcanic peninsula covers c.472,300 km^2, whereas the entire UK amounts to only 244,020 km^2, and Iceland just 102,820 km^2.

From the aerial survey of 1,977 km of non-frozen rivers, and the visual counts of 1,287 Whoopers, combined with additional information from hunters and responses to questionnaires, it was estimated that the minimum wintering population in 1980 was a considerable 7,500. That is more than the wintering population of Scotland, England and Wales combined at the latest census in January 2000 (Cranswick *et al.* 2002). The Kamchatka wintering population was distributed as follows: in the south there was a minimum of 1,300; along the southeastern shore 1,200; in the southwest 800; in the middle region of the west shore 1,000; along the northwestern shore 500, and in the southern part of the central peninsula another 1,200. Regions in the northeast of Kamchatka, including several river basins, could not be investigated from the air, but questionnaires indicated that only about another 20 Whoopers would have been missed from this region (Gerasimov & Alekseev 1990).

A repeat survey, in 1984, of the richest parts of the southern and southeastern regions of Kamchatka, found 1,553 birds along 1,750 km of ice-free rivers (i.e. more birds in a shorter distance than during the previous survey) confirming that the lower limit of the previous survey had probably been correct. The new survey also revealed the impracticality of complete coverage by aerial survey of all non-frozen water channels. Even small channels and springs may hold considerable numbers of swans, and these numbers vary. For example, on 11 December 1984, at a spot on the Pinacheva River used for resting and roosting, 60 were counted, but two days later 100 were seen at the same site from the air, and 216 were found there on 15 December 1984 during a count from the ground. Given this extent of variability, Gerasimov & Alekseev (1990) concluded that in the many branching water channels, amounting to thousands of km, the likelihood of missing flocks was very high. Thus, numbers wintering in Kamchatka are likely to be even higher than indicated, making Kamchatka one of the most important regions for wintering Whoopers in the Far East, outside Japan.

To the east of Kamchatka lie the Commander Islands, the westernmost part of the chain that stretches east to Alaska. As Whoopers both breed and winter in Kamchatka, it is not surprising that they are also regular migrants, more frequent in autumn, on the Commanders, and winter there from early October to mid April (Johansen 1961).

Across the Okhotsk Sea, to the southwest of Kamchatka, lies the island of Sakhalin, which lies so close to the east coast of the Amur region and to the northern tip of Hokkaido that it is visible from both. It is a major stepping-stone between far eastern Russia and northern Japan. On this Russian island, the Whooper Swan formerly bred on lakes and rivers in both the south and north until the 1930s. Now, however, it appears to be rare as a breeder, and restricted to the north, on lakes near Tungusskiy Cape, the northern Chaivo Gulf, the surroundings of Baikal Gulf and the Shmidta Peninsula. Nechaev (1991) has also seen birds during summer at sites in the south, e.g. Svobodnoe Lake, where they bred as recently as 1983, and in the latter half of the 19th century, small numbers were also reported wintering in the south of the island.

Sakhalin is an important and traditional flyway for both Whooper and Bewick's Swans. The former is abundant on migration, with many sites being used as transit stops by birds as they head north to breed or south to winter in Japan. In spring, the first flocks appear in the south of the island on shallow bays and lakes, especially in Aniva Gulf, as early as the last third of March (Gizenko 1955; Nechaev 1991). Despite such early arrivals, the main mass actually appear during the second half of April, when they concentrate mainly in Aniva Gulf. Both fresh and saltwater are still largely frozen at this time and swans rest at polynias and ice-free river mouths (Zykov & Revyakina 1996). Numbers of migrants decline rapidly in the south during the first third of May, as by then all have left Japan, but migration continues in the north until mid May (Nechaev 1991). Flocks vary considerably in size from a few tens to several hundreds, but individual flock sizes give little indication of the importance of migration sites, which may see a steady or even rapid turnover. Thus, between 09:00 and 11:00 on 8 May 1984, Nechaev (1991) observed 1,200 migrating north, although none of the flocks held more than 100. Further south, in Honshu and Hokkaido, Bewick's and Whooper Swans mainly follow different migration routes which converge on Sakhalin. Direct observation and satellite telemetry of banded birds has shown that there they use the same sites (e.g. Nechaev 1991; Kanai *et al.* 1997).

The largest reported concentration of Whoopers on Sakhalin appears to have been that by Nechaev (1991) who observed 12,500 at Svobodnoe on 12–14 May 1984, a number larger than the known Japanese wintering population of the time (9,476 in January 1984; WBSJ 1992a) indicating that perhaps the island is also an important transit route for Whoopers wintering elsewhere in east Asia, or that the flock also contained many Bewick's. Zykov & Revyakina (1996) counted as many as 16,000 swans in southern Sakhalin (e.g. on 2 May 1993), the majority Whoopers (*c.*1,000 Bewick's). They also observed intensive passage, e.g. over the Susuya River, where 646 swans passed in just two hours on 5 May 1993. Flock sizes were 11–200 (mean 58). Flocks also rest in southern Sakhalin on the channel connecting Tunaicha Lake to the sea, where as many as 1,000–2,000 can occur, and on Tunaicha Lake where up to 5,000 swans gather (Zykov & Revyakina 1996). While the majority pass through during April, occasional individuals are reported until early June (e.g. near Aniva Cape on 1 and 6 June 1980; Nechaev 1991). In northeast Sakhalin, the distribution of swans in spring is entirely dependent on annually variable ice conditions. Consequently, they may be widely dispersed or greatly concentrated, depending on the availability of open water; for example hunters reported a concentration of several thousand in the southern Nabilskiy Gulf in spring 1992. In autumn, swans concentrate in the northeastern Pil'tun Gulf, where as many as 15,000–16,000 may rest (Zykov & Revyakina 1996). Again, both Bewick's and Whoopers occur together, and their proportions at this season are unknown. On the whole, however, birds pass through quickly in autumn, resting only a short time, whereas in spring they may stay for prolonged periods.

Given the immense scale of the four regions of Russia addressed in this chapter, and given the enormous extent of suitable habitat for the species, it is clear that it is to Russia that we must look in the future to clarify the true status of the Whooper Swan.

Figure 5.9.   In east Asia Whooper Swans frequent volcanic landscapes.

ASIA

Bordering Russia, from Armenia in the southwest to China and Japan in the southeast are a range of countries that are themselves many times greater in area than Europe. In some of these, such as Kazakhstan, Mongolia and China, the Whooper Swan is a breeding species, while from the shores of the Caspian Sea to those of the Yellow Sea and the Sea of Japan it is a significant migrant and winter visitor. Here on, therefore, I will address this further enormous portion of the west, central and east Asian range, before discussing the subject of vagrancy.

ARMENIA

In Armenia, situated in that broad saddle of land between the Black and Caspian Seas, and adjacent to easternmost Turkey and northwest Iran, the Whooper Swan was considered, historically, to be a rare winter visitor and only to Lake Sevan. Since about 1980, however, despite a fall in water level at Lake Sevan and the complete disappearance of Lake Gilli, numbers have increased sufficiently for the species to be regarded now as an uncommon winter visitor, primarily to the Lake Sevan basin and the Araks Valley. There, it occurs on natural and man-made lakes of 50–60 ha; Whoopers occur only rarely at river mouths and floodplains. The earliest autumn arrival was 3 November 1989, and the latest departure on 18 May 1990. Unfortunately, no overall figures are available for the population wintering in Armenia, but birds seem to occur in small flocks of up to 39 (Adamian & Klem 1999).

AZERBAIJAN

Swans appear to be locally common in Azerbaijan, on the west shore of the Caspian, with numbers occasionally reaching 1,000–1,300 at Kyzylagach Nature

Reserve for a short period mainly during January and increasing in February (Tkachenko 1990). Land-based surveys found 13 at Sarysu Lake, 102 at Shorgyol Lake and 104 at Mahmudchala Lake in 1996, while aerial surveys in 1993 discovered mixed flocks of 742 Whooper and Mute Swans (3:1) at Shorgyol Lake, 1,193 (2:1) at Mahmudchala Lake and 1,383 (1:10) at Kyzylagach (Soultanov *et al.* 1998), indicating that numbers may vary considerably.

Kazakhstan

North and east of the Caspian, and stretching as far east as Mongolia lies Kazakhstan. Once a widespread winter visitor here, Whoopers can be found now only in a handful of places in the centre, north and southeast. Pridatko (2001) described its range as forest-steppe, steppe and even desert-zone habitats, but excluding high mountains. In Kazakhstan and elsewhere in Central Asia it has declined. They were formerly numerous on migration in northern Kazakhstan and thousands wintered on the Caspian into the 1970s. It is thought to have decreased since 1983 because of its sensitivity to human activity, habitat and climate changes, and poaching. Nevertheless, large numbers still winter on the coastal plain of the Caspian (Kovshar 1996).

In 1985, Andrusenko (1987, 1990) found a pair breeding on Bolshie Shabury Lake (at *c.*51°N, thus equivalent to southern Belgium and southern Poland). According to Kovshar & Levin (1982) and Egorov *et al.* (2000), Whooper Swan has bred in Kazakstan since at least the 1940s, e.g. in Taldy-Kurgan region in April 1947, and again in 1953 and 1954. In 1956, they nested at Zharkol Lake; in 1957 at Kamyshykol Lake in central Kazakstan, and also at Kozhakol Lake; in 1958 (and perhaps in other years) they bred at Monastyrskie Lakes in Kalba region, although development of a tin mine put paid to summering there. More recently, Zhatkanbayev (1996) studied both Mutes and Whoopers in 1984–93 in southeast Kazakstan in the Ili River delta where it enters Lake Balkhash (*c.*45°N). In addition to 100–120 pairs of the former, 10–15 pairs of Whoopers nested irregularly, with more on spring and autumn migration. As well as breeding there, small numbers occurred at ice-free spots in the 1990–91 and 1992–93 winters. Confirming that extralimital breeding is not confined to Europe, in 1999, nesting was confirmed in the central Tien Shan, in eastern Kazakhstan at Tekesskoye reservoir, where a female raised four young (Berezovikov & Belialov 1999).

An example of the pitfalls to be found when attempting to review work in this region is available from central Kazakhstan. There, Vinogradov & Auezov (1991) worked on breeding swans, but unfortunately many of their figures relate to both Whooper and Mute Swans combined, making their information unusable for many purposes, as they cannot even be reconciled with the brief comments or map in the Red Data Book (Kovshar 1996).

Studies from the late 1950s to late 1970s in northern Kazakhstan, mainly on waterbodies of the Ishim-Tobol interfluvial, showed that spring arrival was associated with daytime temperatures rising above 0°C, but while lakes and meadows are still frozen. Mean arrival was on 8 April (28 March–15 April), continuing into late April–early May. Migration is on a broad front, in small flocks, generally fewer than 30, and on a northeast or north-northeast heading (58.4%). Drobovtsev & Zaborskaya (1987) noted that in autumn, there are different waves to the migration with the first birds passing in mid September, but with peak migration occurring 10–14 days later, typically in October. Mass migration appears to follow a sharp fall in temperature and the transition of average temperatures from about 5°C to 0°C. During fine weather with weak winds they fly at heights of 600 m or more, whereas in dull or rainy weather with strong winds they fly at 70–100 m. In autumn, flocks of 15 (presumably families) comprised 27.5% of flocks, with flocks of 6–10 contributing another 30.0%. Larger flocks, of more than 30 made up only 7.5%, and the largest flock was only 35. Autumn flocks, flying southwest, typically stop at shallow, open-water sites and avoid rivers and swamps. Autumn flocks consist largely of families.

Rapid human settlement of previously wild areas and associated disturbance has caused local declines in breeding numbers. (Auezov & Grachev 1987; Drobovtsev & Zaborskaya 1987).

Observations in southern Mangyshlak, on the eastern shore Caspian, near the Kazakhstan/Turkmenistan border, by Molodovskiy (1965) in the mid 1950s demonstrated that the Whooper Swan was formerly a common, though not numerous, migrant and winterer, even as far south as Aleksandr-Bai Gulf and Kazakh Gulf, with small flocks following the shore and passing mainly at night. Spring passage, in March–June indicates that birds wintered even further south still. Although numbers were generally low, in cold winters hundreds concentrated offshore (Molodovskiy 1965).

The northern and northeastern Caspian Sea was first recognised as a mass-moulting area for Whoopers more than a century ago by Karelin (1883), however since that time the situation has changed considerably and now very few migrate there to moult. Records from the 1960s indicate that even small flocks of 5–10 are rare (Poslavskiy 1972).

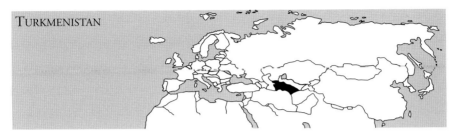

TURKMENISTAN

Turkmenistan, situated east of the Caspian and bordering Iran and Afghanistan, lies at the southern limit of the species' wintering range. Although birds breed to the north in Kazakhstan, and have been reported as vagrants in Iran, little is known of its status in Turkmenistan. The few available publications provide only rather vague information indicating that it prefers the eastern Caspian Sea, occurring only sporadically on inland waters. It has been reported from the Osushnye Islands (several) and a flock of nine in January 1976 at the Slipovskie Islands. It was reported to occur in Chikishlyar region in January, but that numbers were gradually decreasing (Rustamov 1994; Khokhlov 1995).

UZBEKISTAN

In Uzbekistan, the Whooper Swan is a migrant and winter visitor. On migration it can occur at open-water sites throughout the republic, in particular along the lower Amu-Daria River, on lakes in western Uzboi, along the lower and middle Zarafshan River and on the Surkhandar'ya, while older reports (19th and early-20th centuries) are from the middle Syr-Daria River, Golodnaya steppe and the Aidar-Arnasai lakes (Kashkarov 1987). Until the early-20th century at least, the species was quite common in winter on the unfrozen southern Aral Sea. With the astonishing demise of that once enormous lake as a result of river redirection and drainage engineering, the birds presumably no longer winter there. Regular wintering has been noted elsewhere, particularly adjoining the Syr-Daria River basin, but numbers are low and appearances sporadic (Kashkarov 1987).

The timing of the migration depends on latitude, with birds in spring observed as early as late February near Bukhara, although more typically it occurs in March to mid or late April. In spring, they are most often seen in pairs and more rarely in small groups, whereas in winter singles and pairs are more common. Those along the lower Amu-Daria River and the shore of the Aral Sea were observed moving during late November, and it was speculated that they were en route to wintering sites on the Caspian (Kashkarov 1987).

Kashkarov (1987) regarded the species as less common now in Central Asia than previously and, because hunting is prohibited, he considered the reason for the decline to be human activity within the breeding range.

KYRGYZSTAN

In Kyrgyzstan, Whoopers winter close to the southern limits of its western range. Several hundred (800–1,350) winter on Lake Issyk-Kul in the east (Kasybekov 1993). They usually occur in small flocks of 5–10, sometimes 30–40, mainly in Tyup Bay at the east end of Issyk-Kul, and also along the western shore; these two areas hold up to 70–80% of the wintering Whoopers (Kydyraliev 1990).

The first birds may begin to arrive surprisingly early, given that the country is at a similar latitude to southern France or Greece, sometimes before the end of September, though the majority typically arrive in mid October. In some years, however, they do not arrive until November. It is assumed that the extreme variation in arrival dates reflects climatic conditions in the breeding range. Whoopers typically remain until the middle or latter third of March, but sometimes individuals and even small groups oversummer. These include sick and injured birds, numbering 8–10 individuals. In 1972, however, on Issyk-Kul, a surprising total of 57 summered, with 33 there in 1986 (Kydyraliev 1990).

West and southwest of Issyk-Kul, Whoopers are sometimes recorded on spring and autumn migration at the high-mountain lakes of Son-Kul and Chatyr-Kul (Kydyraliev 1990), begging the question 'where are they going to and coming from?'. In 1968, a pair was recorded during July on Son-Kul and, in July 1974, breeding was confirmed when a pair produced two young there (perhaps the southernmost breeding record ever, as Son Kul is at *c*.41°50′N). They bred again in 1975, when a nest was found with five eggs. Only two hatched, however, and the young were subsequently killed by poachers. Although not known to have bred since, two summering pairs were found in June 1984. Summering has also occurred on Chatyr-Kul, and a moulting bird was ringed there on 19 July 1973 (Kydyraliev 1990). At *c*.40°40′N, Chatyr-Kul appears to be the southernmost site where wild Whoopers have summered.

MONGOLIA

The extreme climate, particularly the prolonged, very cold winters, cause most wetlands in the Central Asian highlands to freeze for several months. Consequently relatively few waterfowl winter in the region. The winter climate of the Great Lakes Depression in western Mongolia is severe, with average air temperatures in January falling to between −20°C and −25°C. Most lakes and rivers are frozen in late November–late April, though some outflows and inflows remain ice-free throughout, permitting certain waterfowl species, including Whooper Swans, to survive the long winter (Batbayar 2000; Batbayar *in litt.* 2001).

Until recently, the Whooper was little known in Mongolia; unfortunately, limited ornithological activities and funds have precluded study of its breeding biology, ecology and status there. Recent work, particularly a survey on 19–21 January 2000 of wintering waterfowl in Khar Us Nuur National Park, western Mongolia (a Ramsar site), has greatly clarified the situation, confirming that as many as 397 Whoopers occur at this season. Earlier, in December 1994, there were 370–400 Whoopers in the northern Khar Us, Khar lakes, and Teel and Zavkhan rivers (Munhtogtoh 1995; Batbayar 2000). They also winter annually on the Khomyn-Kholoi channel in the hollow of the Great Lakes (Bold & Fomin 1996).

Whooper Swan is a common and widespread breeder in Mongolia, where it occurs in early April–late September at most lakes (fresh and saline) and river systems. As many as 2,000 pairs are estimated to breed according to the Mongolian Red Data Book, although no methodology for this estimate was given (Shiirevdamba 1997). Bold & Fomin (1996), however, considered the total population to be *c.*3,500 (many of which would be non-breeders) and that it was in a human-induced steady decline. Its breeding range stretches from the Mongolian Altai to the Khalkhin-Hol River, and from the national border in the north to the Barun-Khurai depression in the Valley of Lakes and the Moltsog-els sands (Bold & Fomin 1996).

In the north, where summer temperatures reach 10–20°C, Whoopers breed at lakes in the Darkhat Depression and in the Khovsgol region; with small numbers nesting in the northern Khangaii lakes plateau. In eastern Mongolia, the sparsely dispersed steppe lakes (mostly saline) and rivers also hold 100–500 Whooper Swans during the breeding and moulting seasons. In 1990–94, for example, 342 were recorded on Khokh Lake, and 106 on Khaichiin Tsagaan Lake (Shiirevdamba 1997). In the west and northwest, it breeds commonly at several lakes and rivers in the Depression of the Great Lakes region. In the far west, it is less common, although a few pairs breed in lakes of the Mongol Altai. The far southwest of the country has few rivers or lakes with the exception of the Bulgan River, where Whoopers occur very rarely.

The Great Lakes Depression of northwest Mongolia contains a number of the country's largest lake in an arid semi-desert area, and attract many migratory water-birds on both spring and autumn migration, including both Whooper and Mute Swans. Here, in May 1995, 273 were recorded at Dalai Nur (A. Bräunlich *in litt.* 2001); in June 1998, over 100 were at Khar Us Nuur, 21 at Khar Nuur and 24 at Airag Nuur, with 150, 50 and 43 at the same sites in September that year (Liegl 1998). Very few data for this area are available, although Batbayar (*in litt.* 2001) suspects that many more Whoopers occur here because of the ready availability of suitable breeding habitat. Further southeast, it breeds commonly in the Valley of the Lakes, although no numbers are available.

Large numbers in the breeding season are of interest, though whether they refer to late migrant flocks or non-breeders is unknown. Along the east side of Ayrag Nuur, in northwest Mongolia, 300 were seen on 15–16 June 1996 (M. Köppman per A. Bräunlich *in litt.* 2001). At Erchel Nuur, north-central Mongolia, 156 adults were present on 13 June 1996. The largest flock reported was of 850 at Ich Nuur in central-south Mongolia on 13 September 1998. Flocks of 90 adults at Ayrag Nuur, Uvs Province, on 2 July 1995 and of 115 adults there on 13 July 1995 may possibly have involved non-breeders gathering to moult (A. Bräunlich *in litt.* 2001).

In general, the Whooper Swan appears secure in Mongolia, although it was con-sidered to be in decline by Bold & Fomin (1996) as a result of anthropogenic influ-ences. The extreme cold of winter is known to take its toll; on at least two occasions several dozen to more than 100 have died, although the major risk they appear to face is the drying out of small steppe lakes in east, south and central Mongolia (Bold & Fomin 1996; Batbayar 2000; Batbayar *in litt.* 2001).

CHINA

The Whooper Swan has generally been described as occurring in winter along the coast (e.g. Madge & Burns 1988), but otherwise its status, distribution and ecology in China were very poorly known until Li (1996) and Ma (1996). As access to Chinese literature has proven difficult, despite an increase in swan-related research during the 1980s–90s, it is very fortunate that an excellent summary is available in the form of Ma & Cai (2000), which I have relied heavily upon.

Other than as a winterer, the Whooper Swan is now a rare breeder, and confined to the north, including the Inner Mongol Autonomous Region, and Heilongjiang and Gansu provinces, with most in Xinjiang, in the northwest. There, 3,000–5,000 occur in Bayinbuluke National Nature Reserve, which covers 23,835 km², most of which is grassland and thus extremely important for grazing livestock, in the central Tien Shan. This is the most populous breeding areas in China, and perhaps the highest altitude breeding area in the world, at 2,400 m. In the past as many as 20,000 Whoopers were estimated to summer or breed there (Liang 1982).

In contrast to Europe, Whooper Swans in China are in decline and threatened by direct over-harvesting, hunting, disturbance of cygnets, removal of eggs, destruction of nests, and habitat loss resulting from intensive economic development. They have been extirpated from some regions and are undergoing a major decline in others. Without conservation efforts the species is even considered likely to disappear (Ma 1996), although prior to the 1970s it was a common breeder. Losses in Heilongjiang have been equally as dramatic, falling from more than 4,000 prior to the 1970s to 212 in May 1984 and only c.50 in 1990–92 (Li 1996). A similar decline has apparently occurred in Inner Mongolia and Gansu, although there are no specific data.

In Xinjiang, birds arrive early, in late March–April, and breed mid April to June, with peak egg laying in early May and hatching early to mid June. Severe snowstorms in spring 1996 killed nearly 460 Whoopers in April at Bayinbuluke (Ma 1997). Many nests were abandoned as a result of cold and breeding that year was, as a result, poor. They commence autumn migration in October–November, but since the 1980s with the provision of food in the reserve, 200–350 remain into December, declining to c.100 in midwinter (January–March) when temperatures reach −30°C, snow covers the region and marshes and swamps freeze to 2–4 metres. With the exception of a few springs and their vicinity all water freezes. Family parties occupy the springs as if a feeding territory, usually of 3–7 birds, but invariably fewer than 20 (Ma & Cai 2000).

Approximately 10,000–15,000 Whoopers winter in China (Li 1996; Ma & Cai 2000) at inland lakes, rivers and marshes or, more commonly, along the coast, in Qinghai, Shandong and Henan provinces, as well as along the lower reaches of the Yangtze River. At this season they can be found at Yungzheng, the Huang He (Yellow) River delta and at Zhengko (Shandong); along the Huang He river valley, at Xinxiang (Henan); and at Qinghai Hu and Shihhetze (Xinjiang). In particularly cold winters, they also appear in northern Sichuan, in Hubei, Dongting in Hunan, several sites in Jiangsu, and at Shenggin Lake in Anhui. More occasionally still they reach Yunnan and Fujian provinces and Taiwan (Ma & Cai 2000).

Three main wintering populations are known. The first is in Rongcheng Natural Reserve (Shandong). Conservation measures and the provision of food in winter have led to a considerable increase in numbers from 300–500 in the 1980s to 5,000 by 1992 (Li 1996). They also winter in Bohai Bay (Liu 1987). The second wintering population is in Yangcheng reserve (Jiangsu), though numbers vary widely (more than 3,000 in January 1990 but just 17 in January 1992) suggesting that the birds there are either transient, dependent on weather conditions, perhaps subject to considerable disturbance, or that varying coverage/survey methods have been employed. The third concentration is at Poyang Lake Natural Reserve (Jiangxi). There, more than 1,000 were seen in October 1989 and 2,104 in January 1992 (Li 1996). It should be noted, however, that there has been a 90% decrease in water-fowl populations at Poyang since then as a result of massive systematic and well-organised hunting for restaurants in Guangzhou and Shanghai. It seems likely that fewer than 1,000 swans now occur (G. Carey *in litt.* 2001)

In addition, smaller numbers have been reported in Qinghai and Henan provinces. Early specimen records are mentioned from Zaidam/Qaidam and Koko Nor/Qinghai Hu in what was then Tibet (Vaurie 1972). As many as 1,200 were observed at Qinghai Hu Nature Reserve, Qinghai, in December 1982 (Li 1996).

From Li's (1996) observations, it is clear that the population in China is unnaturally low (Table 5.1) and furthermore that it is in decline largely, if not entirely, as a result of human activities, most particularly the intensive reclamation of wetlands since the 1970s. In the Three River Plain areas in Heilongjiang Province more than 60% of wetlands have disappeared since the 1970s. Furthermore, nesting swans are persecuted by fishermen who collect their eggs and young during the breeding season, and by hunters who shoot them in winter, in addition to being affected by pollution. As countermeasures to this decline, the Whooper Swan is now listed as category two under the Chinese Wildlife Protection Law and by the end of 1993, 34 nature reserves where Whooper Swans are protected had been established. Other sites of potential international significance include Qinghai Hu, and Xiaochi in Shanxi, both with more than 500 in winter (Perennou *et al.* 1994).

Table 5.1.   Whooper Swans wintering in China (after Li 1996).

Site	Location/Province	Year/Month	Number
Bayan Bulag NR	Xinjiang	1992/1	74
Beidagang	Beijing	1990/1	24
Tanyang NR	Shandong	1984/12	202
Old Course of Yellow River	Shandong	1988/12	26
South Four Lake NR	Shandong	1988/12	40
Laizhou Bay	Shandong	1988/12	30
Jiaozhou Bay	Shandong	1990/1	6
Rongcheng Sea Beach	Shandong	1990/1	5,000
Pangzai	Henan	1993/1	171
Heigangkou	Henan	1993/1	33
Shihezi	Xinjiang	1992/1	95
Qinghaihu NR	Qinghai	1982/12	1,200
Caidamu River Basin	Qinghai	1991/1	83
Yangcheng NR	Jiangsu	1990/1	3,000
Shaobo Lake	Jiangsu	1990/1	42
Poyanghu	Jiangxi	1989/10	1,000

Despite there being generally little inter-population variation among Whoopers across their range, there is some genetic evidence that eastern birds, those occurring in China and Japan (and by implication in Russia) are distinct from those in the far west (Harvey 1998).

THE KOREAN
PENINSULA

Sandwiched between the huge expanse of swan-poor China and the relatively small, but swan-rich Japanese archipelago, lies the enigma of east Asian ornithology, the Korean Peninsula. Lacking the distinctive biodiversity of either of its neighbours, and with much of it inaccessible, it has been largely overlooked. Nevertheless, the Koreas lie within the species' wintering range.

NORTH KOREA

Data from this ornithologically little-visited country are scarce, though Tomek (1999) cites published records of Whoopers from October, November, January, March and April indicating that they winter there. Until the 1960s, however, available data suggested that Whooper Swan was merely a spring and autumn migrant along coasts of North Korea en route to well-known wintering areas in the southern peninsula (Gore & Won 1971). More recently, however, it has been known to winter in the north, with flocks of several dozen to 300–500 wintering off coasts of North Korea. The largest number recorded was 1,000 in bays west of Khechzhu (28–30 November 1989) comprising large flocks of 200–300 adults, with families seemingly separate or in small groups. Some have even been observed wintering on unfrozen waterbodies to the north of North Korea (Fiebig 1993; Stepanyan 1998; Tomek 1999) where wintering was unknown prior to the 1970s, indicating perhaps that the limit of its wintering range has been shifting north.

SOUTH KOREA

Wetlands of the Korean Peninsula have largely disappeared over the last 50 years, with Won (1981) noting how rapidly swan habitats have been lost. Prior to 1939, Whoopers arrived in the north and moved along both coasts as waters froze, arriving in the south by late October and remaining until late February–early March. The Korean War of the 1950s, rapid government-sponsored farm expansion in the 1960s, and industrial development since the 1970s, have mirrored not only Japan's rapid economic expansion but also her development at the expense of the natural environment.

A small number of swans (c200 Whoopers and Bewick's) formerly wintered on Jido Island, and up to 2,000–3,000 on the Nakdong Delta in the extreme southeast (Won 1981). A major barrage constructed across the Nakdong River has, however, greatly disrupted the site, and has doubtless led to reduced numbers wintering there, although I was able to observe several hundred on the river in February 1995, and smaller numbers at Chunam reservoir, west of Pusan in the same month.

The most recent information suggests that fewer than 2,000 now winter in the country, with the Nakdong estuary accounting for 1,396 of the 1,955 recorded by the Korean Department of the Environment on 7 February 1999 (Cho *in litt.*), although Park (in Miyabayashi & Mundkur 1999) regarded the wintering population here to be larger, in the region of 3,500. A further 82 were recorded in the extreme south. Thus, 1,478 of 1,955 were at just four sites in the extreme south, with the rest of the country contributing only 477, all but one of which were along the west coast or at inland sites closer to the west coast. The critical factor affecting wintering on the Korean Peninsula is no doubt the very low winter temperatures, which cause many waterbodies to freeze. In addition to the Nakdong estuary, the only site of international importance is the Kum River (Perennou *et al.* 1994).

JAPAN

First noted by Temminck & Schlegel (1845–50) the species has subsequently been recorded widely: on Hokkaido, Honshu, Sado, Shikoku, Kyushu, and Tsushima, and even on Chichi-jima, Ishigaki-jima and Kohama-jima. It is a regular and locally common/abundant winter visitor in October–April, with some occasionally arriving in September and late migrants leaving Hokkaido during early May (Fig. 5.10). At a number of sites injured birds oversummer. It occurs at several well-known sites in Hokkaido and northern Honshu, particularly those protected as natural monuments (OSJ 1974; Yamashina 1982; Brazil 1984b, 1991).

Generally, Whoopers predominate in Hokkaido, Aomori-ken and recently also in Miyagi-ken, while Bewick's predominate south of Aomori-ken (Kuroda 1968; Horiuchi 1981; Ohmori 1981b; Brazil 1983c; WBSJ Research Division 1988). Horiuchi (1981) found 96% of all Whoopers in just five prefectures, 53% in Hokkaido, 18% in Aomori, 13% in Miyagi, 7% in Niigata and 5% in Yamagata. Whoopers are generally less common in central Honshu and uncommon in the southwest where Bewick's replaces it (Brazil 1983c), and it is a rarity further south. The chief concentrations are at Toro-ko, Furen-ko, Akkeshi, Odaito, Toufutsu-ko

Figure 5.10.   Odaito. A regular coastal wintering site in east Hokkaido.

(about 10,000 on migration), and Utonai-ko in Hokkaido, at Kominato, Ominato and Jusanko in Aomori-ken, Izunuma in Miyagi-ken, and Hyoko in Niigata-ken (Wada 1961; OSJ 1974; EAJ 1999). The concentration recorded from Shinjiko in Shimane-ken (OSJ 1974) is considered to be in error as the great majority of swans there are Bewick's (Brazil 1991). It is less common in Iwate-ken, Fukushima-ken, Fukuoka-ken, Miyazaki-ken (Kuranari 1961; Onami 1961; Yamamoto 1961), and tiny numbers occasionally reach Kyushu. It formerly wintered in Tokyo, but all known sites have been reclaimed (Kuroda 1931a, 1931b; Brazil 1991).

Odaito in eastern Hokkaido was probably the single most important site with several thousand, reputedly 10,000, Whoopers spending all or part of the winter there until the early 1980s, although this flock is susceptible to periodic winter mortality. Since the mid 1980s numbers wintering in Hokkaido have declined and more have wintered in Honshu. Nevertheless, 2,000–3,000 still winter at Lake Akkeshi, on the southeast coast, and as many as 8,000 occur on autumn migration, making it the most important site in Hokkaido now (Albertsen 1997; EAJ 1999). Family parties remain together all winter, with broods of 1–5 cygnets, 1–3 commonest. The proportion of cygnets appears to decline as winter progresses and it has been suggested that families with cygnets may prefer to move to milder regions where their offspring's survival chances might be higher (Brazil 1983c).

The first migrants reach Hokkaido in October; in the southeast 5,000–8,000 gather in mid November, remaining until coastal waters freeze in December, then moving south along the Pacific coast. In southwest Hokkaido, Whoopers arrive in late October–early November. Numbers peak from late February to mid March, with northward movement commencing in late March of birds that left Honshu in mid March. By mid to late April the majority have departed Japan, although I have seen individuals as late as 4 June, and an apparently uninjured wild bird summered in 1986 (Brazil 1991). Subsequently, injured birds, mostly with broken wings, have summered in Hokkaido during the 1990s and early 2000s, although an apparently healthy pair summering in east Hokkaido in 2001 was very unusual.

The wintering population was around 11,000 by 1970 (Scott & The Wildfowl Trust 1972). Although Ohmori (1981b) reported only 8,416 in 1979, it has increased considerably since, with 13,777 recorded during the nationwide count in January 1988 (WBSJ Research Division 1988), and 32,423 in mid January 1999 (EAJ 1999). This apparent tripling of the winter population since 1970 may result from a combination of improved counting coverage, actual growth in the population, and concentration of birds attracted from other areas of east Asia. Approximately 8,000 wintered in northern Honshu, arriving in early November and peaking in late November–December. A few wintered regularly south to Niigata-ken, where sometimes 1,900 have been known to gather (WBSJ Research Division 1984). More recently, of the 32,000 wintering in Japan, only 15% remain in Hokkaido, the rest moving to Honshu (EAJ 1999), a considerable decline from the 53% considered to winter in Hokkaido by Horiuchi (1981). It is unusual south of northern Honshu and rare elsewhere in Japan, except during severe winters.

Annual waterfowl counts provide the most accessible data on the species' recent status and indicate that it continued to increase into the late 1990s (see Fig. 5.11). It should be noted, however, that not all Whooper sites are counted, thus an unknown and variable number are missed each year. Total numbers of swans (Whooper, Bewick's and Mute) rose from *c.*14,000 in 1982 to 37,000 in 1991, with 17,947 Whoopers nationwide in January 1990. In 1991, the number of Whoopers had risen to 18,061 but in 1992 numbers recorded fell somewhat, to 16,840 including one in Fukuoka, northern Kyushu, which was unusually far south (WBSJ 1990, 1991, 1992b). By 1999, in addition to 32,423 Whoopers (a mean growth of 300 per annum in the 1990s), there were also 26,684 Bewick's and *c.*150 Mutes (EAJ 1999; Albertsen & Kanazawa 2002). Their distribution remains essentially the same as formerly, with the majority in Hokkaido and in the northernmost prefectures of Honshu.

Supplementary feeding of swans is particularly common in Japan, there were 294 such sites in 1999, with 62% of Whoopers wintering at such sites (EAJ 1999; Albertsen & Kanazawa 2002). The practice presumably contributing to the rapidly rising population.

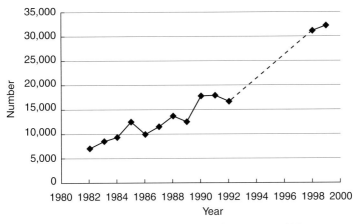

Figure 5.11.   Numbers of Whooper Swans wintering in Japan 1982–1999.

VAGRANCY

This long-distant migrant has, not surprisingly, appeared well outside its normal range as an accidental or vagrant, and will doubtless continue to do so. Tracing original records of vagrants has proven difficult, and some, if not many, have probably been overlooked in this analysis. Although there are rarely difficulties in confirming the identification, that the species is also kept in waterfowl collections, both within and outside its normal range, means that assessing the validity of extralimital records is difficult. Distinguishing escapees from wild vagrants may prove impossible (if only all captive birds were conspicuously ringed, then ornithologists would not face this dilemma!).

OLD WORLD VAGRANCY

In the Old World, the Whooper Swan has been reported as far afield as on Jan Mayen, Svalbard and Bear islands, presumably having overshot on spring migration. Conversely they have also reached well south, visiting at times most of southern Europe and also North Africa (Fig. 5.12).

SLOVENIA

In Slovenia the Whooper Swan seems to be only a vagrant, with for example three in December 1994 and two in January 1995 (Anon 1995b).

THE MEDITERRANEAN REGION

The Whooper Swan has been reported in winter from Italy (Castellani *et al.* 1985), while on Sicily, it is an accidental visitor with 11 records in the early 1800s–1989 (Iapichino & Massa 1989). On the Balearics, it was recorded in February 1938 and January 1964 on Mallorca (Bannerman & Bannerman 1983). It is rare or occasional in Albania (Peja & Bino 1996), while in Greece, it is mainly a hard-weather visitor, reaching as far south as northern Crete, Rhodes, Kos and the Ionian islands (Handrinos 1996; Handrinos & Akriotis 1997). On Cyprus, it is accidental. A young male was shot on 28 December 1910, and three were present on 27 January–2 February 1963 (Flint & Stewart 1992). More recently, three were at Spiro's Pool on 5 December 1994, and two were at Larnaca on 18–28 December 1994 (Anon 1995b). There is a questionable record of one from northern Morocco in the late-19th century (Heim de Balsac & Mayaud 1962). More recently it has been confirmed there, with a reliable record of one at the Moulouya Estuary, in January 1982 (Thévenot *et al.* 1982, 2002). It has been reported once from Tunisia (three in January 1966 at Lac Ichkeul; Thomsen & Jacobsen 1979), and elsewhere in northern Africa it has been reported from Algeria in the late-19th century (Heim de Balsac & Mayaud 1962). It is also a rare winter visitor to lower Egypt, where it was recorded in November and December during the late-19th century, and more recently in autumn 1948, and along the Nile in February 1976 (Brown *et al.* 1982; Goodman & Meininger 1989). It has also been reported from Israel, where it was first recorded near Jerusalem in December 1863. One was recorded at Ma'agan Mikhael in winter 1984–85, three were observed migrating north in March 1985, with one remaining until the month's end (Shirihai 1996) and most recently two

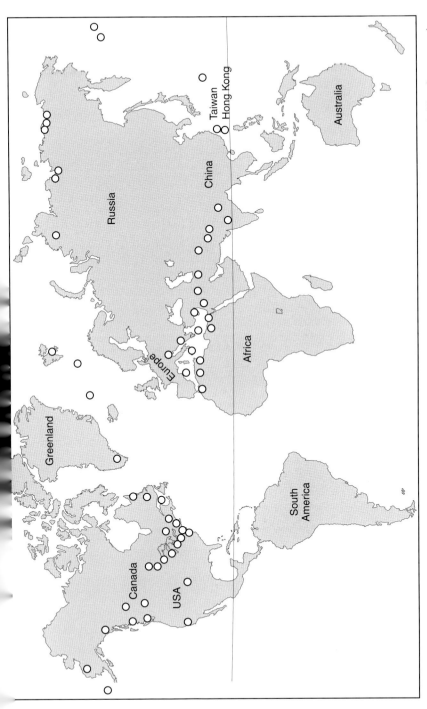

Figure 5.12. Records of Whooper Swans outside their normal breeding and wintering ranges (those in eastern North America probably refer to escaped or released birds).

were at reservoirs in central Judaea and the coastal plains in January and February 2002.

ASIA

Until the mid 1960s Whoopers were reported as occasional winter visitors to Seistan in eastern Iran, and six were in the southern Caspian in January 1977, and they have also been reported from northern Iran, Iraq and Afghanistan (five in January 1976). It is a very rare straggler or vagrant in severe winters to Pakistan (Sind), India (e.g. Kashmir, Punjab, Baluchistan and Rajasthan) and Nepal (Jeacock 1945; Ludlow 1945; Bivar 1947; Ali & Ripley 1968, 1983; Ahmad 1985; Inskipp & Inskipp 1985; Roberts 1991; Al-Robaae & Salem 1996; Porter *et al.* 1996; Grimmett *et al.* 1999). Further east, it has reached Taiwan, the Nansei Shoto and Kyushu, Japan (WBSY 1982; Brazil 1991), and has also strayed more than 1,000 km into the central Pacific to reach the Ogasawara Islands, Japan (Vaurie 1965).

NEW WORLD VAGRANCY

The Whooper Swan has also appeared in the USA and Canada, particularly on the west coast. There, the high density of observers and interest in rarities has guaranteed it considerable attention. However, its presence also generates considerable controversy. Widely regarded as the first documented record is that of three on St Paul in the Pribilofs, in December 1941 (Wilke 1944), but this is well pre-dated by one that reached Maine in 1903 (AOU 1998). As one might expect, the Whooper Swan has strayed particularly to Alaska, where it is regular in October–May, in the outer Aleutians (Kenyon 1963; Mitchell 1998). It is now considered to be a regular winterer there in small numbers, family groups and small flocks of fewer than 50 (Mitchell 1998; Nehls in press; Fig. 5.12)

Birds may overshoot their normal range in spring, migrate southeast rather than southwest in autumn, or even be pushed by severe weather. For example, more were seen in the unusually stormy season of 1962 than in the two previous years, with up to 23 on one day, and 122 on 19 January–4 April (Kenyon 1963). It has even nested in the USA, on the Aleutian island of Attu.

It has also wandered as far south as Oregon and California (Tweit & Johnson 1992; Yee *et al.* 1992; Tobish 1996). It has also been recorded in New England (Palmer 1976); records on the eastern seaboard (and Greenland) presumably relate to arrivals from Europe. Examination of bill patterns of vagrant Whoopers in North America, though not conclusive, might provide some indication of their origin.

Of more difficulty is the assessment and management of released or escaped birds. There have been birds in Massachusetts since the arrival of three in Essex

County in spring 1993, which are now confirmed to have been deliberately released by waterfowl breeders (Berry 1997; French 1997; Patten 2000; French *in litt.*). In spring 1994, the original three were joined by three immatures, two of which disappeared that summer, but one may have remained to become the fourth in a quartet of adults seen widely in the area for several years. In October 1996 two additional adults appeared accompanied by four young, though whether they had bred elsewhere in Essex or somewhere further north is unknown. Coincidentally the same year wild Whoopers bred for the first time on Attu, thus in the same year North America experienced its first wild and feral breeding. The family retired to the Parker River National Wildlife Refuge where both adults and two offspring were present at the end of July 1997, thus confirming that an adult with three yearlings, in June 1997 in New Hampshire, were different and represented probably a second instance of feral breeding. Neither pair attempted to breed in 1997 (Berry 1997).

Berry (1997) also recorded that Whoopers were reported from North Carolina, Pennsylvania, New York, Massachusetts, New Brunswick, Québec, and most northerly of all, Labrador. With none of the Essex birds individually identifiable we shall never know whether these records related to the winter wanderings of those birds or to genuine vagrants. The records from Québec and Labrador (each of three) are the most likely to have been wild birds, as they appear to have been very wary of people (unlike the Essex birds). Given that the Whooper is a common breeder in Iceland, and that it regularly reaches Greenland, even having bred there (Hagerup 1891), it does not appear surprising that wild birds should reach eastern Canada, from where it is a relatively short journey to the USA. In contrast, individuals reported from the mid-west prairies, Iowa, Ohio, Indiana and North Dakota, were all deemed likely to have been escapes. There have also been reports from Minnesota, Missouri, a captive pair that raised a cygnet in Illinois, and an adult in Saskatchewan (Patten 2000).

VAGRANCY IN WESTERN NORTH AMERICA (CANADA AND USA)

In general, birds seen in western North America are assumed to be wild, whereas those on the east coast are more suspect. On the whole, Eurasian waterfowl are much more numerous on the west coast, even outside Alaska. Interestingly, south of Alaska, with the exception of records from Oregon, and California, which were close to the coast and may have involved the same returning individual over several years, west coast USA records have all been well inland, where they have mixed with Tundra Swans, not Trumpeters (Mlodinow *in litt.*). My inclusion of records here does not imply that they are accepted records of wild birds, only that their identity is beyond doubt.

ALASKA

In the western and central Aleutians, Whoopers are uncommon and local winter visitors, usually from early November until mid April, although it has been recorded as early as 26 October and as late as 8 May. It most often appears in family groups or small flocks, but as many as 17 have been reported together on Atka (11 February 1963) and 31 on Amchitka (10 April 1970) (Kessel & Gibson 1978). Small flocks of 5–15 occasionally winter on Adak, from as early as 9 November to 7 April (Byrd *et al.* 1974). On remote Attu, *c.*700 km east of their nearest Palearctic breeding grounds, a pair first located on 20 May 1996 had five cygnets from 5 June. In 1997 they nested again and laid five eggs at Lake Nicholas, one of the largest freshwater bodies in the western Aleutians and surrounded by extensive willow thickets Tobish 1996; Sykes & Sonneborn 1998; Mitchell 1998). It seems, unfortunately, that three were shot during 2000 (Bakewell *in litt.*) and so this tiny foothold of genuinely wild birds in North America may have been lost.

Elsewhere in Alaska, its status is quite different. It is very rare in the Pribilofs, where it has been seen in winter, late spring and summer. In western Alaska it is a very rare spring/summer visitor, with four records (June 1967, June 1975, June 1976 and June 1965). In south-coastal Alaska it is an accidental, there is a record of two in October 1977 (Kessel & Gibson 1978).

Despite a number of records from the continental USA, it is surprising that the Whooper Swan has not yet been reliably reported from Canada. Godfrey (1986) considers it hypothetical, and mentions only one record of supposed escapees in Ontario. A number of records, however, have surfaced subsequently of extremely problematical status. Records have been assessed by regional rarities committees and in most cases rejected, while occasionally being accepted in other publications!

BRITISH COLUMBIA

A family group, widely considered to be wild, was observed north of Merritt in November 1999, the first provincial record if accepted (Marven 1999); however the occurrence is made questionable by the possibility that they were the same family, in fact a hybrid family, that bred in Alberta in 1999 (see below), and may have travelled as far south as northern California by January 2001 (Various 2001). Of two previous records, from Vancouver Island, one, in November 1998, was not officially reported, and the other, in July 1996, though widely seen was rejected on the basis of its unknown origin (Various 2001).

ALBERTA

During summer 1999, a Whooper bred with a Tundra Swan (both escapes from a local waterfowl collection) producing two young near Calgary. Their subsequent movements are unclear but may have included Mamit Lake, calling into question the validity of the British Columbia record (Marven 1999), and most recently from northern California in January 2001, though surely different from a Whooper seen there in the early 1990s in the company of a Tundra Swan (Various 2001).

ONTARIO

Three immature Whoopers wintered in Ontario in 1978–79 at the west end of Lake Ontario. Two (considered to be the same) returned in February 1980 to Bronte, and a further record occurred in 1981. However, all of these were considered to be escapees (Godfrey 1986; James 1991).

OREGON

Since autumn 1991, individuals have been found wintering in the Lower Klamath Basin or among swan flocks in northern California, with up to five at Lower Klamath National Wildlife Refuge in 2000–01. In 1991, one previously at White Lake, California, was seen just across the border in Oregon on 13 December (Gilligan *et al.* 1994). In 1992–94 singles were seen in Lake County on three occasions, possibly a returning bird (Tweit & Gilligan 1995; St Louis 1995; Bailey & Singer 1996). One of the same or another returned in November 2000, with three possible hybrid cygnets (Mlodinow & Tweit 2001) indicating that perhaps the mixed Albertan pair bred twice. This is presumed to have been the same family group that appeared shortly afterwards at Klamath. An individual present near Airlie on 27 November–1 December 1997 is the only Oregon record west of the Cascades (Tweit & Gilligan 1998; Korpi 1998; Tice 1998). Most observers seem of the opinion that eastern Oregon records mostly pertain to the same individual that has also appeared repeatedly in northeast California.

CALIFORNIA

In heavily birded California, Whooper Swans have been recorded several times, in January 1984 in Colusa County, with a flock of c750 Tundra Swans (Roberson 1986), then in December 1988 in San Joaquin County. One wintered from late November 1991 until mid January 1992 at Lower Klamath, Siskiyou County, where one was present on 26–27 February 1994. On 17 January–7 February 1995 one was in Colusa County, returning on 3–4 January 1996 (Howell & Pyle 1997; McCaskie & San Miguel 1999). On 28 January 1998 an adult was again at Lower Klamath (Erickson & Hamilton 2001), and one accepted as wild was present on 1 January 1999 (Rogers *in litt.*). An adult with two cygnets was reported at the same place in January 2001[17] (Various 2001) with the adult present on 1 January– 10 March 2001, and another adult with three cygnets on 10 January–10 March, making five in total (Glover *et al.* 2001).

Montana

An adult was photographed at Somers, western Montana, on 5–11 March 2001 (Trochlell 2001). It might have been one of the Albertan birds.

North Dakota

A single was recorded in North Dakota in 1992 (Bergman & Homan 1995), but I have been unable to trace the location.

Vagrancy in Eastern North America (Canada and USA)

The origins of Whoopers in eastern North America are difficult to assess, although one could argue that groups of subadults in summer could have been wandering non-breeders. However, some have bred ferally in Massachusetts, and these are perhaps responsible for additional records in Maine and Québec. It will become increasingly difficult to assess records of this species in the northeast, though some argue that all records from the USA are of captive origin (Patten 2000).

Québec

The first records from Québec appear to have been of three in Gaspé Bay on 21 July–11 August 1994, then of two subadults on Anticosti Island, on 11–16 June 1995 (Aubry & Bannon 1995; David 1996). It is possible that these derived from a group of escapees in New York, in 1993, although it is also possible that genuine vagrants from the Icelandic population could reach Québec. Subsequent records involved single adults at Saint-Barthelemy on 2 May 1999 and Cowansville on 14 May 1999 (Bannon & David 1999). Most recently, a group of five, all showing some signs of immaturity though largely white, were at a number of locations on 8 June–29 July 1999 (David *et al.* 1999), and in summer 2000, a single was seen several times in Granby and Yamaska Park.

MAINE, NEW HAMPSHIRE, MASSACHUSETTS, NEW YORK, NEW BRUNSWICK

Whoopers may be in the process of establishing themselves in Massachusetts, having bred ferally several times in Essex County in the late 1990s. They seem to have first appeared there in 1992, when three were presumed to have been the same three reported on Long Island, New York, in 1991. These were joined by a further three in Essex County in April 1994 (Perkins 1994). At least seven were in Essex County into early May 1999 (Perkins 1999). Birds were certainly present at Plum Island, Massachusetts, prior to May 1995, as birds presumed to be from there were reported at several localities in New Brunswick during late May 1995 (Maybank 1995). Four adults summered at Plum Island in 1997 (Petersen 1997). In November 1998 a total of 16 birds were at four localities in Massachusetts and New Hampshire (French *in litt.*). An old record of one in Washington County, Maine, in 1903 presumably relates to a wild bird (AOU 1998). However because of the number of feral birds in the region, one in Maine, on 12 October 1999, must now be considered of uncertain origin (Ellison & Martin 1999).

OHIO

In January 2000, a Whooper was seen accompanying Tundra Swans in the Cincinnati region (Ohio Bird News 2000).

CHAPTER 6

Food, feeding and flocking behaviour

THE WHOOPER SWAN'S DIET

The Whooper Swan is essentially a vegetarian; its diet ranges from freshwater and marine algae and higher plants on the one hand, to terrestrial plants, agricultural crops and even invertebrates, freshwater molluscs and aquatic insects, refuse, carrion, prey and fish eggs on the other though, of course, the vegetarian component greatly predominates. The range of foods recorded is surprisingly diverse (Appendix 3). It feeds by a combination of sight and touch. Its predominantly vegetarian diet is reflected in the great length and morphology of its alimentary canal, with in particular a powerfully muscular gizzard containing considerable amounts of sand grains, and its long caeca (25–26 cm; Shaw 1938).

Whoopers feed at a wide range of wetlands, preferring shallow lakes and coasts, but also on marshes, grazing from riverbanks and also rivers, and on agricultural land. In winter, in particular, they consume various emergent and submerged aquatic plants as well as grasses, spilled grain from stubble fields and winter cereals. In some areas they also feed on root crops, notably waste potatoes, turnips and sugar beet left after the harvest, and in intertidal and subtidal areas eelgrass is a common food item (Scott & The Wildfowl Trust 1972; Brazil 1981c; Sheppard 1981; Owen *et al.* 1986; Thom 1986; Batten *et al.* 1990; Kirby *et al.* 1992; Laubck 1996). Food plants may be pulled, twisted and broken off, clipped, pecked or nibbled with the bill; they may be taken from above or below water; paddling may be used to expose or loosen submerged foods, and food is swallowed with no further 'handling'.

Certain species are notable because they are commonly reported in the diet from widely separated localities, such as: eelgrass, wigeon grass (beaked tasselweed) *Ruppia maritima*, pondweed *Potamogeton* spp., Canadian pondweed *Elodea canadensis*, various sweet-grasses (*Glyceria maxima, G. fluitans* and *G. aquatica*) and stoneworts (*Characeae*; Witherby *et al.* 1940; Sparck 1958; Scott & The Wildfowl Trust 1972; Wilmore 1974; Charman 1977; Owen 1977; Bianki 1988). Owen & Cadbury (1975) found that when feeding on meadows, Whoopers preferred soft grasses such as *Agrostis stolonifera, Alopecurus geniculatus* and sweet-grass *Glyceria fluitans*, but that reed sweet-grass *G. maxima* was also important in late winter.

Other species are notable because they appear only infrequently or only locally in the diet, such as fruits, berries, roots, aquatic insects and larvae, worms, small shellfish, small frogs and small fish (Naumann 1897–1905). Witherby *et al.* (1940) recorded them as taking not only the leaves of thread-leaved water crowfoot *Ranunculus trichophyllus* in Iceland, but also the seeds of cotton grass *Eriophorum scheuzeri*, while Hilprecht (1956) refers to acorns and plums, but as Scott & The Wildfowl Trust (1972) concluded, 'this must be exceptional', as is that Whoopers wintering in southern Kamchatka will take a range of animal matter from caddisflies *Trichoptera* to fish, and late-spawning *Oncorynchus nerka* salmon eggs (Gerasimov & Alekseev 1990; Ladygin 1991). The taking of fish carcasses, such as those of Coho Salmon *O. kisutch* and small Rainbow Trout *O. mykiss* and salmonid eggs has also been noted in North America among Trumpeter Swans (Mitchell 1994). In Hokkaido, Japan, where Whoopers rarely feed on land (because of snow cover), I have once seen them browsing on the somewhat tough and leathery, but still green, leaves of dwarf bamboo *Sasa* spp. emerging beside a lake shore as the winter's snow cover melted early in March (Brazil 2002b; Fig. 6.1). During spring and summer, breeding adults will, occasionally, take emerging Chironomid flies, as will young cygnets (Blomgren 1974; Haapanen *et al.* 1977; Rees *et al.* 1997a); however, animal matter is far less common in the diet than among Tundra Swans.

Black & Rees (1984), studying Whoopers in the Caerlaverock area of southwest Scotland, analysed faeces and found that the Islesteps flock, when feeding primarily from grass fields, relied almost entirely on rye grass *Lolium perenne* but also took timothy grass *Phleum pratense* and marsh foxtail *Alopecurus geniculatus*, and a number of other grasses to a lesser extent. They also visited 'merse' areas where clover stolons *Trifolium repens* comprised over 75% of the diet, along with smaller amounts of *Festuca* sp. and other merse grasses. When feeding near the River Lochar they also fed on the roots, tubers and leaves of: water starworts *Callitriche* sp., water forget-me-not *Myosotis scorpioides*, celery-leaved buttercup *Ranunculus sceleratus* and the bur-reed *Sparganium emersum*.

Figure 6.1.   Dwarf bamboo emerging from beneath snow is a rare food for Whooper Swans.

The summer diet differs somewhat from that in winter because of the different floras in the widely separated ranges. Thus, in Iceland during summer they often take filamentous green algae, most notably *Cladophora aegagropila*, as well as water crowfoot leaves, water buttercup and, later in the season, cotton grass seeds (*E. scheuzeri* and *E. angustifolium*) and even the sugar-rich berries of the crowberry *Empetrum nigrum* and blueberry *Vaccinium myrtillus* (in Iceland and Finland). Blueberries and mountain cranberries *V. vitis-idaea* are also eaten occasionally by Trumpeters in North America (Mitchell 1994), while in both northern Norway, Finland and Iceland, the water horsetail *Equisetum fluviatile* is very important, if not the preferred, summer food plant (Roberts 1934; Witherby *et al.* 1940; Scott & The Wildfowl Trust 1972; Cramp & Simmons 1977; Haapanen *et al.* 1977; Brazil 1981c; Myrberget 1981; Einarsson 1996). Various sedges (including *Carex lyngbyei*, *C. rostrata* and *C. rariflora*) also occur commonly in the diet, as does the spike-rush *Eleocharis palustris*. Alpine bistort *Polygonum viviparum*, dwarf *Salix herbacea* and woolly willow *S. lanata* leaves are taken less frequently (Gardarsson & Sigurdsson 1972). In China, the primary foods are the leaves and stems of aquatic plants such as pondweed and crowfoot, as well as roots, tubers and seeds, while the cygnets reportedly forage on invertebrates, especially aquatic insects, during their first summer (Ma 1996).

Eelgrass and wigeon grass used to be very important estuarine food sources, but increasingly since the 1940s in the UK at least, they have taken not only more seeds, rhizomes and tubers, but also crop plants and relied less on such marine species; although they remain common in the diet elsewhere. Eelgrass is, for example, an important food for several thousands passing through east Hokkaido in autumn (see Fig. 6.2).

At Loch Spiggie, Shetland (UK), Whoopers usually feed in the accepted 'traditional' manner, on aquatic plants, particularly at the shallow northern end, where they can reach the bottom by upending. When the water level rises, making the submerged food supply unreachable or as it becomes exhausted as winter progresses, they move out to graze on the adjacent fields or fly 1.6 km to Loch of Hillwell (Dymond 1981). This example may illustrate in practical terms how marshland grazing could have been naturally linked with aquatic feeding (when marshes would have been a common peripheral aspect of many natural wetlands), and how from marshland, birds may easily have explored adjacent water meadows and fields for suitable food.

Though once an uncommon habit, many of Britain's wintering Whoopers now rely on farmland for foraging. Carbohydrate-rich foods, such as potatoes, turnips and sugar beet and, to a lesser extent, grasses, and waste grain spilt during harvest, are favourite foods in autumn and early winter. In mid and late winter Oilseed

Figure 6.2.   Whooper Swans foraging on abundant eelgrass, Hokkaido.

Rape is also popular. In late winter and early spring, protein-rich grass and newly shooting winter wheat form an important part of the diet. Field feeding also occurs commonly in Ireland, Sweden, Denmark and Germany, and even at the opposite end of the range in Japan, though less commonly and its origins there have not been traced. As lowland forest clearance for cattle pastures in Hokkaido began as recently as the late 1800s, the habit there can be no older than a century, though in Honshu their occasional use of rice stubble fields may have begun at any time in the last 2,000 years.

Some authors, notably Owen & Kear (1972), Cramp & Simmons (1977) and Owen *et al.* (1986), in referring to grain or potato eating, have given the impression that such feeding is essentially incidental to other foraging and that it occurs when preferred aquatic plants are lacking; that it is resorted to as a consequence of severe winter weather; occurs as a result of an artificial feeding situation such as exists at WWT centres, or that it is a developing habit. This is, however, something of a misrepresentation. As Brazil (1981c), Sheppard (1981), Reynolds (1982) and Colhoun *et al.* (1996) have all indicated, in parts of both Scotland and Ireland grain and root consumption form the basis of the winter diet. In some areas, Whoopers feed in stubble only a few km from quite suitable freshwater habitats, which support other waterfowl including Whoopers at different times. The use of agricultural land, although a relatively recent phenomenon in the West, is not just a second choice, as implied by Cramp & Simmons (1977), but even preferred in some places or at certain seasons. The Orkney birds are a prime example. As Reynolds (1982) observed there during October–December they feed exclusively on spilled grain among cereal stubble, and visit lochs only for roosting and preening, not for feeding. Colhoun *et al.* (1996) too, found that in northwest Ireland, stubble, potatoes, winter cereals and grasslands were all important foraging areas, in addition to aquatic sites, and that these habits were leading to conflicts with farmers.

In at least Orkney, central Scotland and northwest Ireland (and no doubt far more widely), the Whooper's principal foods, consisting of spilt grain in stubble fields, potatoes and grasses, vary seasonally. From arrival on the wintering grounds until the year's end at least, spilt grain from stubble fields is favoured. In mid winter, when food is scarce, all three are taken as and when available, and particularly in January–February they rely partly on stubbles and partly on loch vegetation, grass or floodwater areas, shifting increasingly to grasslands from March until their departure, primarily in April (Brazil 1981a,c; Sheppard 1981; Reynolds 1982). By February most stubble and potato fields have been ploughed and only grass is available to the swans. By then, however, early-spring growth has usually begun, making it more than adequate (Brazil 1981c; Sheppard 1981). The only other significant food is in fields of winter cereals. The cereals themselves are grazed, but it seems to be often leftover potatoes from the previous crop, that are the attraction (Sheppard 1981).

That Orkney birds switch only during January from feeding on stubble fields to more than 50% feeding at lochs (Reynolds 1982) is evidence that, where available, it is spilled grain that is preferred; it presumably serves as a concentrated protein and carbohydrate source with which to replenish depleted fat reserves after autumn migration. As the availability of spilled grain declines, submerged aquatic vegetation in lochs becomes an attractive alternative. Although such aquatic food is available, the departure of many birds from Orkney during late December and January is coincident with the depletion of spilled grain, and further circumstantial evidence that spilled grain is preferred. A similar pattern of switching between foods has been reported from northwest Ireland, where Colhoun (1997) found that Whoopers selected cereal stubbles in autumn, waste potatoes in midwinter, followed by an increasing dependence on winter cereals and grasslands in midwinter and spring, with local preferences for oilseed rape and beet, where available.

Early-spring growth of grass is particularly nutritious and, again using Orkney as an example, Reynolds (1982) considered that growth there was more likely to resume early than in mainland regions with more extreme temperatures. The observed spring influx of Whoopers to Orkney is presumably linked to the availability of this highly nutritious grass, which would serve to provide the energy reserves for continued northward spring migration and for breeding.

The WWT centre at Welney (UK) is a focal roosting point for a large number of swans, more than a 1,000 Whoopers among them. The provision of food there, and the style of agriculture in surrounding areas, clearly has affected foraging behaviour. There too, contrary to Cramp & Simmons (1977) and Owen *et al.* (1986), the major part of their preferred diet in the first half of winter at least, is arable waste either tipped out on the reserve for them or gleaned from the surrounding farmland (Kemp 1991). Whoopers have learned to exploit four distinct food types: the natural vegetation of the Ouse Washes (fen feeding), grain specially provided for them, waste arable crops, particularly potatoes, on surrounding farmland (Owen & Cadbury 1975), and also newly growing winter wheat (Dafila Scott *in litt.*).

Further evidence as to the significance of waste potatoes for Whoopers came in the 1989–90 winter when a shortage of these coincided with extreme flood conditions. This led to a temporary change in the habits of many Whoopers, which spread out along the entire Washes, instead of concentrating on the Welney reserve. Some fed on aquatic vegetation. Many more, however, joined Bewick's Swans on the surrounding arable land taking a mixture of post-harvest potato remnants, sugar beet tops and winter cereals. There was also a tendency for some wintering at Welney to remain less dependent on the reserve during the 1990–91 winter with more birds regularly field feeding and/or spreading out along the Washes. Whereas Bewick's Swans regularly fly 15+ km to feed on arable land, most Whoopers are typically within 2 km of the Washes (Kemp 1991).

Soon after their arrival, the Whoopers settle into a daily routine of roosting at Welney then flying out at first light to the surrounding arable countryside to forage. In some years they feed first on spilt grain in stubble fields, then move to sugar beet and potatoes after these have been harvested (Bowler *et al.* 1992). In other years, it seems that root crops are generally preferred; the potato and sugar beet remnants left after the harvest form the dominant element of the diet with only some birds frequenting stubble fields (Rees *et al.* 1991b). After foraging the swans return at dusk to roost on the reserve.

During the 1990–91 winter, both Whooper and Bewick's from Welney developed a distinct preference for sugar beet. Waste potatoes appeared to be fewer, and the beets were believed to have higher sugar content than in other years (Rees *et al.* 1991b), which presumably accounted for this. The habit was again popular in 1993, although, after their arrival, the swans fed mainly on set-aside stubble, they also fed on the remains of the sugar beet crop, which had not been ploughed due to the unusually wet autumn, and continued to do so throughout the winter. So popular did this food prove that sometimes mixed flocks of more than 2,000 Whooper and Bewick's were found on a single beet field close to the reserve (Bowler *et al.* 1994).

At times, Whoopers at Welney resort to or are forced (as a result of disturbance by farmers) to use more traditional styles of feeding, on underwater vegetation or grazing on wet marshland (Kemp 1991). Particularly in late winter/early spring they especially favour areas containing a mixture of creeping bent *Agrostis stolonifera* and reed sweet-grass (Rees *et al.* 1991b). They appear to grub the roots and/or shoots of sweet-grass in particular (Kemp 1991).

At another WWT centre, at Martin Mere, some Whoopers fly to fields adjacent to the reserve, where they feed during the day alongside Bewick's Swans, while others visit fields near the Ribble Estuary. They favour barley stubbles, where they feed both on the grain left after harvesting and on sprouting cereal shoots. They also forage on newly sown winter wheat and the remains of the potato and carrot harvests during early winter, but move to improved pastures in mid to late December (Bowler *et al.* 1992, 1994).

In Scotland, at WWT Caerlaverock, heavy rainfall in early winter may cause flooding, in which case the swans disperse widely to feed. Also here, large numbers of Whoopers have been observed feeding from an oilseed rape crop, a habit common in Denmark but still scarce in the UK, though perhaps on the increase as recent research by Chisholm & Spray (2002) suggests it is the dominant habitat of Whoopers in the Tweed Valley (Scotland) from December to March. Even more unusual is that the swans have even been known to leave the roost at night to feed on rape by moonlight, returning later to the refuge to feed on the grain distributed at midday (Bowler *et al.* 1994), a rare case of feeding on land after dark.

Schneider-Jacoby *et al.* (1991), who observed Whoopers on Lake Constance on the borders of Austria, Germany and Switzerland, noted that they fed on the tubers of fennel (sago) pondweed *Potamogeton pectinatus*, which they dig from the sediment at depths of up to 24 cm. Apparently no other species is able to exploit these food resources, and a number of diving ducks and Eurasian Coots, forage around the swans benefiting from their waste.

Generally it is the leaves of plants that are eaten, but some seeds and berries are also taken; I have seen, for example, Whoopers stripping seed heads from thistles *Cirsium* spp. in Scotland and taking berries such as those of crowberry in Iceland.

The large bill enables it to eat coarser vegetation and harder materials, such as root crops, than many of the smaller billed geese. Given that they eat both terrestrial and aquatic plants, as well as crops and such things as refuse, the range of foods is probably broader than that of any goose, although in their terrestrial feeding ecology they are very goose-like (Brazil 1981c).

Whereas throughout the majority of the range they feed in relatively natural environments, in some regions the rich crops and crop waste of agricultural land have become important foods. At some localities they are provisioned with grain and root foods. Usually, at sites such as Caerlaverock, Martin Mere and Welney, official carers do this. In the Far East though, particularly in northern Japan, where Whoopers are extremely popular for sightseeing, feeding them plays an important role in this interest and birds have become very confiding (Anon 1955). At many such localities they have become extremely approachable, even to the point where at Odaito part of the flock frequently loafs on a snowbound car park adjacent to the shore waiting to be fed by hand. As a result of the popularity and prevalence of feeding wintering swans in Japan, Whoopers that feed on an entirely natural diet are now something of a rarity (Brazil 1984b). The food commonly given consists mainly of cereals, rice, waste tea and bread (Ohmori 1981b), and brands of rice-crackers flavoured, amazingly, with Antarctic krill! At more natural sites the birds take plants such as eelgrass and other marine species (Brazil 1991). Ohmori (1981a,b) noted that local agricultural products are the main foods fed to Whoopers in Japan. On the assumption that the daily minimum food requirement is 10% of body weight, 400–500 g (dry weight) is given per swan, except when cold weather prevents access to any natural foods, such as at Inawashiro-ko, where Ohmori (1981a,b) found 700 g/day reduced mortality rate.

Because of their predominantly vegetarian diet, the gizzard plays a vital role in grinding and processing vegetable matter. In order to maintain its effective functioning, Whoopers are dependent on grit collected from their foraging environment, and substantial amounts may be ingested. Roberts (1934) found that one stomach contained 60% black volcanic sand and lava pebbles, and only 40% leaves and seeds, and all of those that I have examined have been filled with various types of sands or grits. This requirement unfortunately predisposes them to lead poisoning contracted when ingesting hard pellets or weights instead of grit (Chapter Eleven).

Though the vast majority of the Whooper's food is vegetable matter, and although Sparck (1958) found no animal remains in Whooper stomachs, animal matter has also been recorded in various parts of the range. Such items as worms, freshwater molluscs and various aquatic insects have been taken, though many may merely have been ingested incidentally, because they are common inhabitants of the main food plants of swans. On their Danish wintering grounds, however, ingestion of molluscs was not incidental, there hundreds of swans apparently took to feeding on both marine and freshwater mussels (*Mytilus edulis* and *Unio pictorum*) during cold weather (Rees *et al.* 1997a). Whoopers have even been observed visiting rubbish dumps rummaging in and, presumably, eating rubbish, in Ireland (Ruttledge 1963) and waste potatoes from rubbish dumps in Scotland (Baxter & Rintoul 1953). At such sites, one wonders what else they might eat!

Animal matter does not merely occur incidentally in the diet, nor does it only include carrion. There have been a number of cases where Whoopers have attacked and even eaten other birds and mammals. Mayfield (1952) described a captive Whooper killing a Mallard duckling and a Canada Goose, though whether this was for food is unclear. Wild Whoopers have been seen to eat a dead rat (Honda 1979). I watched with astonishment as a Whooper Swan attacked a family of Eurasian Wigeon during the breeding season in Iceland killing, though not eating, one of the young (Brazil 1983b); and in Japan, Iijima (1984) observed young Whoopers eating two Mallards that had become entangled in fishing nets. Though by no means common, meat eating has been reported at both the western and eastern extremes of the range and doubtless occurs elsewhere too.

During the breeding season, Whoopers at freshwater sites feed almost exclusively on roots and shoots of aquatic plants, on algae and incidentally on aquatic invertebrates, while those in bogs, fens or marshes take sedges, horsetails, cotton grass, grasses, willow and bistort leaves. Blomgren (1974) and Haapanen *et al.* (1977) noted that aquatic insects formed the main food of cygnets during the first few days of life, and Ma & Cai (2000) apparently observed this throughout the first summer. Although Myrberget (1981) considered poor insect productivity in a cold summer as a possible factor affecting cygnet survival, I have not observed this dependency on animal matter myself, nor has it been widely reported.

Various feeding methods

METHODS OF FEEDING

The habitats that the swans usually feed from include open saltwater (both marine and inland; the latter primarily in Central Asia), saltmarshes, tidal estuaries, open fresh water, running fresh water and agricultural land. The Whooper has a high, wide skull, powerful jaw muscles, a spiny tongue, a thick horny nail at the bill tip and serrated edges to both mandibles, combining to give it a powerful grip and the ability to tear tough vegetation, aided by the leverage afforded by its long neck. Feeding methods are directly related to the habitat being used and the type of food being gathered, and therefore vary greatly between foods. Martin (1993) considered that upending and dipping were the two main feeding methods, yet acknowledged that in some areas most winter feeding is on farmland, where of course neither is possible. While dipping and upending are important on water, they are only two of eight distinctly different methods, three of which are usual on water and five on land (Brazil 1981c; 1984a).

Dabbling

In waters of all depths, feeding may be either from the surface, on emergent or floating vegetation, or just below it on submerged plants. In both the body lies horizontally with the head parallel to the water's surface, whether on it or just below it. In these situations the head is normally moved both sideways and backwards and forward in a sweeping motion, and the neck is bent upwards into a characteristic kink (Fig. 6.3). This method is referred to as *dabbling* because of its obvious similarity to the feeding method of various dabbling ducks *Anas* spp.; food is obtained either by sieving or by grasping and tearing.

Dipping

In shallow water, Whoopers feed by *dipping*, that is they submerge their head and neck while the body remains horizontal on the surface (mean dip duration 8.6s). At times, only the head is submerged, while at others the entire neck is submerged, nevertheless the body remains horizontal throughout (Fig. 6.4). Such actions may be accompanied by powerful foot-paddling movements to loosen the roots of underwater plants (Venables & Venables 1950) and stir food to a depth where it is easily reached. I observed craters up to 10 cm deep and 50–100 cm across in the bottom of a shallow lake in Iceland, where Whoopers regularly foraged by dipping and foot paddling, and it seemed very likely that the latter had loosened plant roots and shifted silt on the bottom, thus forming the craters (Brazil 1981c). Foot paddling has also been noted in captivity at Moscow Zoo, where it only lasted up to four seconds, and was immediately followed by *dipping* (Rezanov

Figure 6.3.   The posture while drinking closely resembles 'dabbling'.

Figure 6.4.   Dipping for eelgrass.

1990). Dipping enables them to reach nearly 45 cm (Ogilvie & Pearson 1994). In flowing water, feeding methods are again dependent on depth, but when *dipping* they usually maintain their position against the flow by paddling, and thus face upstream.

Upending

In deeper water, that is 100–120 cm (the reach of the Whooper Swan according to Wilmore 1974 and Owen *et al.* 1986), Whoopers employ the *upending* method (mean duration 10.3s) (see Fig. 6.5). A bird first submerges its head, then its neck (as if dipping), and finally swings forward and down. During this manoeuvre, it will paddle slowly, submerging the whole of the foreparts, leaving only the hindquarters, tail and wings protruding vertically (remaining thus for up to 20 seconds; Airey 1955; Brazil 1984a; Rezanov 1988; Ogilvie & Pearson 1994). The bird may paddle in order to help maintain its balance and will usually be forced to do so if upending in flowing water. The second or most extreme stage of upending, presumably in relation to food in slightly deeper water, involves them in upending with both feet in the air.

Although Cleland (1906) considered upending effortless, I have seen birds make several attempts before managing to upend successfully, indicating that energy and skill are required to reach the point of vertical balance. It takes longer to reach the final upending position than dipping, because the bird must dip first, then rock forward into the upend position. Upending, therefore, appears to be more energy consuming since paddling is often required to maintain balance and position. The position does, however, have the advantage of allowing feeding at greater depth. Casual observations suggest that upending is weather dependent, being rarely used on windy days; presumably because the position is unstable, as a larger surface area is presented to the wind, than when dipping.

Figure 6.5. A young Whooper upending.

Digging

Digging, using the bill, is restricted to areas of soft ground. In marginal aquatic habitats, such as marshy areas, including waterlogged pastures, Whoopers occasionally use their strong bill to dig for roots and rhizomes of wetland plants. This is done vigorously and produces characteristic holes surrounded by piles of discarded mud and vegetation (Brazil 1984a).

Grazing

Although dabbling, upending and dipping are typically used in aquatic habitats throughout the year and digging may occur in wetlands, many do not habitually use any of these four techniques as major components of their typical routine. Feeding behaviour is flexible and in some areas they *graze* on agricultural land. In such areas Whoopers are extremely goose-like in their general behaviour, making morning and evening flights to and from feeding grounds and having a distinctive diurnal feeding rhythm with morning and afternoon peaks in feeding separated by a lull (Brazil 1981a,c, 1984a). At some locations grazing may also be seen as an adjunct to more typical aquatic feeding during winter, and on breeding and moulting grounds, where birds leave the water to take advantage of available vegetation.

Because they have larger bills and grosser serrations to the mandible edges and spines of the tongue, grazing Whoopers have a much slower pecking rate than geese. Myrfyn Owen (1977) has recorded peck rates of up to 250/minute for the Barnacle Goose, which typically grazes short vegetation while Whoopers peck rates reach only 75/minute and they feed on taller vegetation (Brazil 1981c, 1984a).

Owen (1980), in his extensive studies of geese, found a negative correlation between bill (culmen) length and pecks per minute. Extrapolating from this and assuming, merely for the sake of argument, that a grazing Whooper is in fact a goose, with a 92–116-mm culmen (Cramp & Simmons 1977) then its predicted peck rate would be in the range of 60–80/minute. Grazing Whoopers, with an actual peck rate of 75/minute, obviously fit very closely to Owen's (1980) goose model.

Although in some respects they are very goose-like, there are, however, differences. Whereas grazing geese pace and peck regularly, moving rather steadily across their habitat, pecking down straight in front of them, Whooper progress is periodic and interrupted. They may move several paces but then stop, as they usually peck while stationary. Before walking again, they move their head backwards and forwards and sideways to peck, pecking as they move their head, thereby fully utilising the great mobility conferred by such a long flexible neck. With 25 cervical vertebrae, they have a greater number than any other warm-blooded creature (Wilmore 1974; Todd 1979). Consequently, they cover the ground in a stop-start fashion rather different from the relentlessly busy manner of geese (Brazil 1984a).

While grazing, it usually holds its neck quite tightly curved into an inverted U.

If the vegetation is short, the head is held pointing vertically downward and only the tip of the bill is used for pecking. When feeding among longer, coarser vegetation the head is held horizontally and angled sideways so that the serrations along the bill sides can be used for shearing and tearing, accompanied by strong backward jerks of the head (Brazil 1984a).

PICKING

When foraging across stubble fields Whoopers make many minute movements of the head and bill while picking unharvested or spilled grain; in other respects, however, they move in much the same way as when grazing on grass. Observations at close quarters at sites where they are provisioned have shown that they are quite adept at using the tip of the tongue and bill, and are able to pick two or more individual cereal grains simultaneously. The minute movements of the head seen among birds foraging on stubble probably relate directly to this method of feeding and are extremely difficult to count (Brazil 1984a).

SHOVELLING

Whoopers feeding on harvested grain sometimes employ another technique that I term *shovelling*, as it involves lowering the head until it is horizontal and close to the ground, then pushing it forward using the lower mandible as a scoop. This may be a further method for picking individual grains, but it is also very suitable for lifting entire seed heads lying on or embedded in the soil (Brazil 1984a).

PECKING

In addition to grasses and grains, Whoopers foraging on land also utilise root crops, especially, as noted by Kear (1963), Pilcher & Kear (1966) and Bowler *et al.* (1992), potatoes and sugar beet, just after lifting when waste remains on the fields. Then, using the strong nail at the tip of the upper mandible or the sharp serrated cutting edges of both mandibles, they *peck*, gouge or chop off chunks to be swallowed (Brazil 1984a).

The commonest position while feeding on fields, whether by grazing, picking, shovelling or pecking, is standing. Whoopers do sometimes sit down to feed (Ohlsen 1972; Brazil 1984a), though in my experience this usually occurs during strong winds when standing is perhaps made more difficult.

Whooper Swans are by no means confined to aquatic habitats/foods. They are capable of utilising a wide range of habitats and adapting their diet and mode of feeding to suit them. In general, when feeding on agricultural land, they have a distinctive bimodal feeding pattern with quite separate feeding and roosting areas, but still prefer to return to the security of water to sleep, and thus closely resemble geese. When using aquatic habitats, however, Whoopers usually feed and roost at the same place and then their behaviour is similar to that of many dabbling duck or Mute Swan. This behavioural flexibility has enabled them to take advantage of new or improved feeding situations, such as the more nutritious fertilised crops now available on land, and hand-feeding sites (Brazil 1984a).

In view of the fact that the development of field feeding has taken place in the West largely during the latter half of the 20th century, and because it has led to considerable changes in behaviour, some of which bring it into conflict with man, I have chosen to discuss it in more detail, particularly in comparison with feeding methods in aquatic habitats.

THE HISTORICAL DEVELOPMENT OF FIELD FEEDING AND ITS INFLUENCE ON BEHAVIOUR

During spring and summer Whooper Swans will graze on shores from lush waterside vegetation. A similar habit can be observed around wetlands of various types even in winter. In one sense at least the step to field feeding, as a common habit in winter, was perhaps not such a large one. In the British Isles, Whoopers are known to have flown to fields specifically to feed on potatoes only since the 1930s–40s and, like Mallard, Pink-footed Goose and Greylag Goose, which began the habit somewhat earlier, it was probably due to changes in farming practices (Kear 1963; Pilcher & Kear 1966) that perhaps coincided with a serious decline in the eelgrass stocks which formed their favoured food around the British coasts. At Rugozerskaya Bay, on the Kola Peninsula, Russia, Bianki (1988) noted that following a mass die off of eelgrass in the 1950s the flock of up to 100 Whoopers that had previously stopped there on migration no longer did so. A similar die-off around Britain in the 1940s may well have prompted them to change their foraging habits.

Use of stubble and grass, a behaviour which has become more prevalent in recent decades, and which I studied in some detail in Scotland (Brazil 1981a,c; Fig. 6.6), may have developed independently, from flights to potato fields, from the use of flooded stubble or pasture, or as a response to severe winter weather. Thus, for example, Anderson (1944) noted that when lake levels in Britain rose in heavy rain, Whoopers began grazing on areas of submerged grassland, and this appears to have been the first published account of the habit. Even after the waters had subsided, however, Anderson (1944) observed that several Whoopers continued to graze, perhaps initiating what has now become a fairly widespread habit in certain parts of the West. In Scotland and Ireland, although many still feed in shallow water in the traditional manner, an increasing number forage over farmland, particularly improved grassland (Merne & Murphy 1986; Thom 1986). Such feeding has also increased in southern Sweden where, Whoopers regularly feed on croplands during cold periods (Nilsson 1979). In Germany, they readily forage on farmland (Degen

Figure 6.6.   Whooper Swans on stubble field in central Scotland.

et al. 1996), while in Denmark more than 75% of a considerable wintering population forages on farmland (Laubek 1995c, 1996), and in the Fenlands of England,
Bewick's Swans moved to agricultural land after the freezing of deep and extensive
floods. In the latter case, the habit continued and became more widespread
(Cadbury 1975; Scott 1980a). Where agricultural crops or the remains of such
crops offer an accessible and energy-rich food source, it is not surprising that a flexible species such as the Whooper Swan has taken advantage of them. Whistling and
Trumpeter Swans have also made a similar recent move to feeding on agricultural
land (Nagel 1965; Tate & Tate 1966; Gunn 1973; McKelvey 1979).

In Japan, field feeding is a relatively rare phenomenon occurring either near
major waterbodies where birds have traditionally roosted and fed (and then usually
on wet rice stubble fields) or more rarely among migrants pausing on their way
north, e.g. on grass pastures blown clear of snow by strong winds in Hokkaido
during the late 1980s and early 1990s (Brazil pers. obs.).

It is apparent from the little information available on field feeding by Whoopers
that it is not only relatively recent in Britain and several other European countries,
including Denmark and Germany, but it is also variable in occurrence, in some
regions being considered rare, in others typical (Brazil 1981c; Nilsson 1979;
Reynolds 1982; Laubek 1995c, 1996). This recent development in behaviour is not
confined to swans, it has occurred in Brent Geese in Europe and Snow Geese in
North America (Ogilvie 1978).

Owen *et al.* (1986) considered that when feeding on land Whooper Swans make
one or two flights each day, but that they do not normally wander further than five
kilometres from their roost site. Though this may be true of those flocks that concentrate at and around Wildfowl and Wetland Trust reserves, where an additional

food source is generally provided at or close to a potential roost site, this is not nec-essarily the case elsewhere. In northeast Scotland for example Whooper Swans roosting at the Loch of Strathbeg regularly fly out up to 30 km in order to reach attractive foraging grounds (Buckland *et al.* 1990).

In Denmark, studies of habitat selection in 1991–93 have shown an extreme shift, with more than 75% of those wintering there (up to 15,000 in midwinter) foraging on farmland. On first arrival they fed mainly in shallow water or on stubble fields before moving to wintergreen fields, mainly grain crops, during November, usually with Bewick's Swans. From late November, they shift their attentions to grassland and *Brassica* fields, remaining there throughout the rest of the winter. In spring 1994, following severe weather in February–March that devastated wintergreen crops, both Whoopers and Bewick's switched to foraging on potato fields. Their main foods during winter were otherwise wheat *Triticum aestivum*, barley *Hordeum vulgare*, rye *Secale cereale* and rape *Brassica napus* (Laubek 1995c, 1996).

Degen *et al.* (1996) found that, in Germany too, Whoopers readily use fields for foraging and, rather like those in Scotland and Denmark, they change their habi-tat selection during the winter. In Germany 63.3% foraged on rape fields in January, whereas by March 83.3% had switched to grasslands. In northern Germany, both Whoopers and Mutes feed in high concentrations almost exclu-sively on rape fields throughout the winter and, unsurprisingly, are increasingly considered a pest by farmers (Spilling *et al.* 1998).

From apparently a new habit just 50 years ago, field feeding has become regular, if variable, in several countries. In Denmark, for example, although some 80% feed on submerged vegetation after their arrival in October, by December the propor-tion on water has declined to *c.*30%, and by late winter fewer than 15% occur on water (Laubek 1995c). Both in Denmark and southern Sweden, Whoopers use winter cereals and oilseed rape. However, whereas sugar beet spill is important in Scania, and in the English Fens in November–December, it is not mentioned for Denmark. A much higher proportion of the Whoopers in Scania feed on rape com-pared with Denmark, where more use winter cereals; perhaps because of differing amounts of snow cover, as rape is easier to feed on than waste cereals where there is snow on the fields (Laubek 1995c; Nilsson 1997).

Laubek (1995c), who studied the habitat choice of Bewick's Swans in Denmark, found marked differences between them and Whoopers. There, it seems that although *c.*80% of both swans were feeding on land, no fewer than 75% of the Whoopers were on arable fields, compared to just 25% of the Bewick's. The latter, it seemed, preferred to feed on grassland and stubble.

Just like the Whooper Swans arriving in Denmark, Bewick's arriving in the Netherlands feed first on water, especially on *Potamogeton* tubers, and also like the Whoopers they change later in the winter to winter cereals, rape and root crops (Dirksen *et al.* 1991).

The habitat choice of wintering Whooper and Bewick's in the UK differs some-what from the situation in Scania and Denmark. During the 1990–91 winter, the majority of Whoopers in the UK were in aquatic habitats, and fewer than 15% were on arable, while in January 1995 that figure was as low as 7%. The majority of Bewick's, however, were on arable (Cranswick *et al.* 1996; Rees *et al.* 1997b). Although the proportions differed the foods taken by those on arable land were similar; root crops were most important in November–January, and cereal and stubble fields were also used (Nilsson 1997).

It seems that field feeding by swans is an even more recent habit in Sweden than in Britain. It was first noted in Sweden during the hard winter of 1963, when some flocks were seen feeding on winter cereals close to their normal aquatic feeding areas. The habit is still mainly restricted to the southernmost provinces, where the lighter snow coverage allows swans to find and reach the crops. As in Britain, the change from aquatic to terrestrial feeding habits may have been triggered during periods of severe freezing weather when aquatic vegetation could not be reached, but when fields remained snow-free this enabled the swans to detect new feeding opportunities (Nilsson 1997). In Denmark, field feeding began among Bewick's Swans in the early 1960s, but the habit was slow to spread and even as recently as 1980 both species were still feeding almost exclusively in natural habitats (Laubek 1995c).

Similarly marked changes in feeding habits have also been noticed for Bewick's Swans in the Netherlands. There, until the late 1960s, they fed mainly on aquatic vegetation before switching to arable crops and grasses. However, although they still prefer *Potamogeton* tubers, since 1975 they have depleted this food source in just 1–2 months, after which they have quickly moved to other habitats (Dirksen *et al.* 1991; Nilsson 1997).

The change to field feeding among Whoopers in southern Sweden may be due to the depletion of food in aquatic habitats, which in turn is due to the increasing swan population. While the eutrophication of many lakes has led to a shift from macrophytes to planktonic production, thus reducing the food resources available for swans, there are no indications of such changes in the shallow coastal areas of the south, where extensive submerged meadows of eelgrass and wigeon grass remain available for both Whoopers and Mutes during all but the very coldest winters. While food depletion might be one cause during cold winters, why then do they also shift in mild winters? The situation among Whoopers in south Sweden seems far less clear than among Bewick's in the Netherlands that deplete their aquatic food supply then shift to field feeding (Dirksen *et al.* 1991; Nilsson 1997).

Perhaps the most significant factor influencing field feeding in recent years in several countries has been considerable changes in agriculture, involving not only the choice of crops but also the methods and timing of harvesting (this combined with a decline in the quality of aquatic habitats as a result of eutrophication; Laubek 1998). In the 1950s, for example, sugar beet was rarely available to swans or geese wintering in southern Sweden. After the introduction of mechanical harvesting, however, sugar beet became an important autumn food item from the 1960s. The increasing acreage of autumn-sown cereals, and the recent introduction of new varieties of some crops, in particular oilseed rape, is also considered important in influencing this behaviour (Nilsson 1997).

During the January 1995 census of the northwest European population, data on habitat use was collected from more than 34,000 birds (Laubek *et al.* 1999). Of these, 60% were feeding on arable (mainly oilseed rape, winter cereals and grass) and just 40% at aquatic habitats, confirming the remarkable prevalence of the use of arable in parts of Europe. More interesting still was the regional variation in behaviour. Laubek *et al.* (1999) found that two distinct regions where feeding differed, one where terrestrial feeding dominated, in Denmark, southern Sweden, the Netherlands and northern Germany, where more than 75% were feeding on arable, and a second region dominated by the use of aquatic habitats, namely Norway, Sweden (except Skåne), Finland, the Baltic countries, eastern and southern Poland, Czech Republic, Austria, Switzerland and southern Germany, where 91% fed in aquatic habitats, perhaps partly related to snow-cover and partly to the topography of these regions.

THE IMPLICATIONS OF FIELD FEEDING

It is only to be expected that, where swans have changed to foraging over fields, other aspects of their behaviour will also have changed or new behavioural aspects will have become evident. Field feeders, unlike aquatic feeders, which may choose to roost at their feeding site, must contend with having to fly on a daily basis between safe aquatic roosts and (relatively unsafe) terrestrial foraging sites, a pattern that may be influenced by many factors. Thus their feeding strategies may become adapted to different food types and their daily activity patterns to different aspects of disturbance.

Whoopers using agricultural land have adopted an essentially bimodal daily activity pattern similar to that traditionally shown by geese. They commute between night-time roosts and daytime foraging sites and feed intensively in the morning and afternoon, tending to rest during the middle of the day (Brazil 1981a,c). The same pattern has also been found among Bewick's Swan wintering in the English Fens, where fenland feeding was first recorded as recently as 1970 (Owen & Cadbury 1975). Unlike various geese species, however, which maximise their foraging by sometimes feeding at night (Ogilvie 1978), it is apparently atypical of Whoopers to feed on land by moonlight. None was observed feeding by moonlight by Brazil (1981c), although Owen *et al.* (1986) referred to probable feeding at night as well as by day while on water, and Black & Rees (1984) refer to feeding on moonlit nights in ponds and fields at Caerlaverock, whereas on dark nights they roosted on the water or pond banks. Thus, while Whoopers may be willing to feed at night, where feeding is possible at or adjacent to their normal roosting sites, they seem less willing than geese to risk having to fly at night and the habit has not yet spread. Buckland *et al.* (1990) noted that in the Buchan plain of northeast Scotland, Whoopers often do not fly to roost at all but remain on farmland fields all night. This is a remarkable observation, but unfortunately it is unclear whether the birds were roosting there or whether they were also engaged in nocturnal feeding.

The development, increase and spread of field feeding have brought the swans, not unexpectedly, into conflict with farmers. In Denmark, the number of complaints about swans causing damage to crops increased rapidly during the mid 1980s (Laubek 1995c, 1998) and they are also considered a pest by oilseed rape farmers in northern Germany (Spilling *et al.* 1998). Their direct impact can be through grazing or puddling of cereal fields, where swans gather during wet periods. Such conflict may lead to changes in human attitudes towards swans and in turn this could lead to further changes in the behaviour of the birds. Grazing on agricultural land by waterfowl inevitably leads to accusations of damage caused directly by grazing pressure, from puddling, 'burning' by their droppings or through competition with domestic livestock, and indirectly by causing stock to avoid areas previously used by waterfowl. Attention was first drawn to crop damage caused by Whoopers as early as the late 1950s, resulting from a growing habit of feeding on farmland in central Scotland, and in Iceland (Kear 1964; Harrison 1973). This habit continues to become more widespread; in Denmark, both Whooper and Bewick's have shifted from wintering and staging on lakes, fjords and flooded meadows, to foraging over farmland since about the late 1970s or early 1980s, and there they have caused an increasing number of problems to farmers, especially among those growing *Brassica*, with the problem further increasing rapidly since the mid 1980s because of changes in farming practices (Laubek 1996).

Several experimental studies have examined the effects of goose grazing on grass crops and the findings apply equally well to Whoopers. Geese and swans usually arrive on the wintering grounds in autumn, after most crops have traditionally been harvested (although winter wheat must be particularly vulnerable). Therefore, they could cause damage only when competing directly with livestock for grazing, when eating un-harvested root crops or where they remain late in the spring. Kear (1970a) showed that loss of winter foliage due to grazing was not detrimental to harvest yields of silage grass, winter wheat or spring barley, but lowered silage yields occurred if grass was grazed in May. Wright & Isaacson (1978) considered that the effect of goose grazing on yields depended on the stage of growth of the crops, grazing intensity, soil type and weather, and that grazing does little harm and may even be beneficial, provided the grazed plant's meristem is undamaged. In such cases, yields were actually greater in grazed than in ungrazed plots, indicating perhaps that some manurial effect of the birds' droppings may have been involved (Clark & Jarvis 1978). Patton & Frame (1981) however reported appreciable economic loss where geese grazed improved grassland, because of the need to supply alternative food to domestic livestock.

Laubek (1998) found that although winter oilseed rape experienced delayed growth by swan grazing, the plants were able to compensate in response, with grazed plants producing the same number and sizes of pods as ungrazed plants; more importantly oil content was also independent of grazing pressure. In the case of oilseed rape then, it seems that no negative effects occurred through grazing. Conversely, where Colhoun (1997) studied grazing pressure on grasslands in Ireland, significant yield losses (up to 65%) were apparently attributable to swan grazing. The amount of crop loss experienced by spring seemed related to the timing of grazing in relation to the growth of the sward. Up to 80% of variance in yield loss was attributable to grazing pressure alone. Colhoun & Day (2002) found that losses attributable to Whooper grazing in January, and early and late spring constituted serious depletion by up to 1.4t/ha of 'spring bite' feed for livestock, and of 4.1t/ha of first-cut silage.

Marriott (1973) showed that the experimental application of bird droppings actually increased yields of rye grass and clover, except in those trials where small applications were used. Moreover, sheep were not seen to avoid areas used by geese, nor was 'burning' of vegetation by droppings observed. The only conclusive damage, other than by late-spring grazing, was retardation of crop growth by puddling, caused by tight concentrations of waterfowl feeding in wet conditions on heavy soils (see Wright & Isaacson 1978).

The actual impact of Whoopers on crops, the balance between any potential positive manurial effect of their droppings and any negative impact of their grazing pressure, is yet to be quantified. Although Owen & Cadbury (1975), and Black & Rees (1984) asserted that the extent of damage was minimal, irate anti-swan farmers are, unsurprisingly, difficult to convince! Inevitably, farmers view any damage as significant and there have been increasing complaints about swan damage. Although conflict may be limited to areas where Whooper numbers are large, the problem may be perceived as a serious one, and in particular given the recent discovery that Ireland hosts the majority of the Icelandic breeding population in winter, amounting to some 10,000+ birds, and given a tendency towards increased field feeding there too, the matter of agricultural conflict will inevitably need to be addressed (Merne & Murphy 1986). With the Whooper population apparently increasing over large parts of its wintering range and the considerable shift to feeding on land, the potential for conflict is rising.

From experimental studies of geese, it is apparent that the behaviourally similar Whooper cannot really be regarded as an agricultural pest, occurring as it does in most areas in much smaller concentrations than geese. Convincing farmers of this, however, is another matter. To the farmer any loss of grazing is seen as significant and they are very likely to try and drive birds from their land. Disturbance by farmers and their dogs proved to be a significant factor in the movements of Whoopers between foraging sites in central Scotland (Brazil 1981c) and may have also contributed to their complete absence at times. Such disturbance is likely in other areas where swans regularly feed on land. Should Whooper numbers, or the perception of the problems they cause, continue to rise, then disturbance by farmers will probably also increase and perhaps lead to yet another shift in habitat selection or feeding rhythm. Ireland could be *the* place to examine changing attitudes towards them. There, preliminary data show that swans can have a significant impact, leading to a 90% reduction in spring bite and 50% reduction in pre-silage cut. Colhoun *et al.* (1996) found a significant positive relationship between grazing intensity and the degree of loss, with up to 8% losses of cereal grain yields attributable to Whoopers. More significantly, they also reported that grazing on agricultural land by swans is increasingly being perceived as a problem in Ireland.

Changes in farming practices in recent decades have provided Whoopers with attractive alternative food supplies. Furthermore, the decision in some countries that the area under wintergreen crops be increased to help protect ground water from the washout of pesticides and nitrates (e.g. in Denmark) will increase the alternative food supply, while EU set-aside schemes also enforcing the removal of 10–15% of arable land from food production means that other crops (such as rape for fuel), also popular with swans, will increase in area and as a result conflict between farmers and swans is also likely to increase (Laubek 1995c).

FEEDING ASSOCIATIONS

FLOCKING AND THE COMPOSITION OF FLOCKS

The occurrence of flocks, sometimes referred to as herds, is a particularly noticeable aspect of Whooper Swan winter behaviour. Gatherings range in size from family groups to herds of 100 or more (Fig. 6.7). Although the factors affecting flocking doubtless vary regionally and between countries it is, nevertheless, of interest to examine this behaviour in detail.

At wetlands, especially where these are provisioned, large flocks of several hundred may accumulate in autumn, remain relatively stable over the winter, and then decline in spring. This simple pattern does not occur everywhere. The majority of wintering flocks throughout the range are small (several tens), and as an indication of how extremely different matters may be, let me describe the situation in central Scotland, where Henty (1977) and I (Brazil 1981a,c) examined their movements, relative use of different localities, variation in flock size and proportion of cygnets and brood sizes. Because marked individuals were not present it was impossible to assess how long individuals remained in the flocks, but it was possible to calculate the use of an area in bird days (bd) for comparison between years and localities. It was quite clear, however, from the almost daily changes in flock size at any site that, in central Scotland at least, flocks were not stable units; there was a fairly high turnover, with birds either passing through en route to other regions or absent from a particular site for a short period before returning (or being replaced). This implies a considerable degree of flexibility among the birds, which were able to find alternative roosting and foraging sites.

The lack of marked birds and the extent of movement of individuals independent of flocks meant that birds might have been recorded more than once. Nevertheless, despite the obvious pitfalls, comparisons between years and with other studies are of interest (see Brazil 1981a,c). I logged a total of 6,932 bd in the study area from a total of 191 flocks, over three winters 1977–80. There was considerable annual variation in the extent to which Whoopers were present, which could not be put down to any variation in observation intensity. Without individually marked birds it was not of course possible to know whether an increase in bd

Figure 6.7.   Whoopers are very sociable in winter.

between winters was due to more birds passing through the area or to more individuals staying in the area for longer. Not only did numbers vary considerably between winters, but their movements also differed. Thus, of the 28 localities used within the study area, different sites were used to greatly differing degrees from one winter to the next. The nine most popular localities received more than 250 bd of use each, whereas the remaining 19 localities each received fewer than 100 bd. Yet, among the top nine sites, only four were used consistently in all three winters, two were used in two winters and three were used in only one. Thus, in some parts of the winter range, they show considerable variability in their use of particular locations. At a relatively small number of other wintering sites, particularly those wetlands where they are provisioned (a mere handful of sites in Western Europe), a high degree of site fidelity and consistency of use might be expected, yet even at those there is also a considerable turnover of birds between years. It appears that although Whoopers are very sociable in winter, and clearly prefer to be in a flock, they are also very flexible, altering their feeding and roosting sites in response to seasonal and perhaps even short-term variations in food supply.

In the study area, the Whooper flocks were of a size that is typically reported, that is usually small. In my case they ranged in size up to 134, with little annual variation in mean flock size (35–36). The overall pattern, however, masked a secondary one, which was that if the nine heavily used localities were considered separately from the rest, then mean flock sizes were quite different. The more frequently used areas attracted flocks averaging 39.9–40.1 birds, whereas the less frequently used localities attracted on average only 13.5–18.1 birds. The locations did not differ in holding similarly sized flocks for longer or shorter periods; instead certain sites actually attracted larger flocks than others. It appears that there was something intrinsically different between the sites and certain sites were more, or less, attractive to Whoopers than others. We can see from the great concentrations occurring at certain winter refuges in Britain the extreme variability in flock sizes depending on a combination of food availability and safe roosting sites. Whereas some sites may only attract a few tens others hold more than 1,000.

Boyd & Eltringham (1962), Hewson (1964) Brazil (1981c), Merne & Murphy (1986) and Kirby *et al.* (1992) have clearly demonstrated that smaller flocks are by far the commoner in winter in Britain. Most of Boyd & Eltringham's (1962) flocks were smaller than ten birds, and 95% of all flocks they found were of fewer than 50. That pattern was clearly supported in central Scotland though flock sizes there were slightly larger, with 75% holding fewer than 60, but 75% of all birds were in flocks of larger than 37 (Brazil 1981c). Haapanen & Hautala (1991) found that small flocks predominated in Finland too, confirming that this pattern is not something unique to Britain. Forty percent of all flocks in Finland consisted of pairs, while *c.*75% of all birds were recorded in flocks of five or more.

Another interesting facet of the situation in central Scotland was that large and small flocks did not occur randomly throughout the winter. Larger flocks, especially of more than 40, were commoner between October and the year-end. Before 31 December 75% of all Whoopers were in flocks of 52 or more. Similarly, Boyd & Eltringham (1962) also found that in October–November large groups contained nearly half of all the swans recorded. In contrast, small flocks, especially those of fewer than 30, were commoner between the beginning of the year and spring migration. From January, 75% of birds were in flocks of 19 or more (Brazil 1981c). It seems therefore, that in the Whooper Swan (away from provisioned sites) flock cohesion weakens as winter progresses, a phenomenon certainly not confined to it, as Thompson & Lyons (1964) had previously found among Whistling Swans in Wisconsin where small flocks, isolated pairs or family groups were commoner in spring.

The more common occurrence of smaller flocks in the latter half of winter in central Scotland may be indicative of family parties and pairs leaving larger flocks and moving north independently and perhaps earlier than those birds which will not breed. Sightings of apparently intact family groups in Iceland in spring, and of different waves moving north through Fennoscandia and other regions in spring, support this view.

Despite the highly conspicuous nature of large flocks, especially those at provisioned sites, such flocks are actually highly labile, and it seems that the largest stable unit is the family (Airey 1955; Brazil 1981c). I found no indication among wintering Whoopers that young of previous years were associating with their parents and younger siblings, and it seems not to have been observed elsewhere either, although Scott (1980c) has shown this to be common among Bewick's Swans. The commonest consistent unit may actually be the adult pair, because relatively few Whoopers breed successfully. Haapanen & Hautala (1991) found that *c.*40% of all their flock observations in Finland were of pairs (the remaining 60% being of groups larger than two).

The flocks frequenting central Scotland differed from those at provisioned refuges in that they seemed particularly unstable, often changing dramatically in size or structure even overnight, and with a rapid turnover. In many respects they are perhaps best described as aggregations (Wilson 1975). Nevertheless, belonging to one of these aggregations does appear to confer certain advantages that may be a result of either the physical nature of the flock or of cooperative behaviour; they should therefore perhaps be regarded as social groups rather than as aggregations (Morse 1977).

The number of swans using or passing through an area is usually only a small proportion of a region's wintering population (unless that site is also an important staging area). Central Scotland was no exception with only a small proportion of the Scottish wintering population, estimated at the time to be *c*.4,100 (Brazil & Kirk 1979), and currently at 2,700 (Cranswick *et al.* 2002).

Brazil (1983a) had shown through the use of neck-collars in Iceland that moulting flocks of non-breeders are mere aggregations, there being no stable links between birds unless they should happen to pair there, and in winter the birds that had moulted together became widely dispersed. What was not known then and is still unclear is whether discrete breeding populations, that is birds breeding in one region of Iceland for example, consistently winter as discrete units in a particular area of Ireland or Scotland, or whether they also disperse widely. Just what is the catchment area for a particular wintering area? If breeding birds disperse widely and mix, then the proportion of cygnets in a given wintering area should reflect that in the population at large. If, however, birds from one breeding area do tend to remain together, then as breeding success may vary regionally so that should be reflected regionally on the wintering grounds. That there is some traditional usage of wintering sites could actually support either alternative.

The production of young by Arctic and other high-latitude breeding birds can vary dramatically. Often several reasonably successful breeding seasons occur consecutively, followed by a year of almost complete failure. Detailed studies of the population dynamics of Whoopers have not been made, but existing estimates of breeding success suggest that they follow the same general pattern. The mean percentage of cygnets in most winters ranges from 13% to 21.6%. Only infrequently is the percentage higher or much lower.

In the absence of evidence to the contrary, it was assumed that those in central Scotland during winter were a representative sample of the whole Scottish wintering population. Certainly the very bad breeding season of 1979 was reflected within the study area the following winter, and data from other years agree well with reports from other areas (Bell 1979). Whenever the number of cygnets could be accurately determined, the percentage was recorded for each flock (N=110), and in order to facilitate comparisons with previous studies a mean percentage was calculated each winter (Brazil 1981c). These mean percentages were very similar in 1977–78 and 1978–79 (13.0 and 12.7), but in 1979–80 it was just 3.7%. This extremely low figure was supported by that of the national census undertaken the same winter, which found just 5.1% (Brazil &

Kirk 1979; Brazil 1981c), indicating that the previous summer had been a particularly poor breeding season.

Airey (1955) had suggested that family parties are the basic coherent units during winter and that for some reason they tend to stay away from larger flocks. If true, this would lead to there being higher proportions of cygnets in smaller flocks than in larger flocks. Hewson (1964) found exactly that, with a mean 26% in flocks of fewer than 20, compared with just 15% in larger flocks. I found a similar pattern, with cygnets amounting to 12% of small flocks (fewer than 20 birds), as opposed to 6% in larger flocks (more than 20), although this pattern was not clear in all years (Brazil 1981a,c) (Fig. 6.8). Variation between winters may well result from the considerable effect of poor breeding success in the previous summer season because then there are so few cygnets in any flocks. Furthermore, the proportion of cygnets in small flocks can be much more variable than in larger flocks. This is because small flocks may consist entirely of families (in which case there will be a high percentage of cygnets) or of just non-breeders (in which case there will be no cygnets at all), or small flocks may contain a mixture of families and non-breeders. Large flocks, on the other hand, are likely to be a more representative sample of the total population, only a small proportion of which breeds. Consequently, the proportion of cygnets in large flocks is large only if there are factors affecting the distribution of the birds, such that families specifically choose to join large flocks. Although such flock-joining behaviour by families is not implausible, it seems not to occur and the opposite appears true, that in general families choose not to join large flocks (Airey 1955; Hewson 1964; Brazil 1981a,c). It is widely assumed that individuals in large flocks incur some advantage(s) in terms of more effective feeding

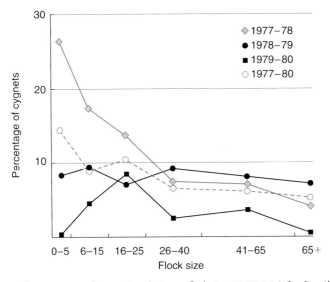

Figure 6.8. The percentage of cygnets in relation to flock size; 1977–80 (after Brazil 1981c).

or predator detection than when in smaller flocks or alone; parents might then be expected to make an effort to join larger flocks so that they and their offspring may benefit. However, although there may be benefits for individuals, perhaps there are also hidden costs for families. There might be some disadvantages for families in large flocks such that they could suffer increased feeding competition, for example (Scott 1980a), and this may make them tend towards remaining in smaller flocks. At provisioned sites, this pattern is likely to breakdown if the amount of food available outweighs any cost of increased feeding competition.

ASSOCIATIONS WITH OTHER SPECIES

Just as grazing herds of large mammals in Africa, the Americas and Asia are followed by Cattle Egrets because of the food they stir up, Whoopers, as large aquatic grazers and rather wasteful feeders, are commonly followed by other species for the food brought to the surface. It is not uncommon, for example, to see groups of dabbling or even diving ducks following behind or even surrounding feeding swans. These gatherings can include tens of birds often of several species. It is sometimes difficult to distinguish between species congregating at a common food supply and those benefiting from the feeding actions of the Whoopers. Sometimes both may be occurring. It is common for two or more swans (Whooper, Bewick's and Mute) to congregate at provisioned sites in Europe and Japan, where they are clearly competing with each other at a concentrated food source, and they may forage together in winter on agricultural land or wetlands. The aquatic feeding activities of all three swans are such that they create opportunities for both dabbling and diving ducks. The dabblers take food from the surface just behind the swans, whereas the diving duck dive around the swans, presumably for food dislodged from the bottom or for invertebrates disturbed by the large feeding birds.

In Iceland, the UK, Japan and Korea, I have, on the whole, observed only dabbling duck, particularly Eurasian Wigeon, Mallard and Northern Pintail, feeding around or among Whoopers. In Hokkaido, I have also seen Goosander feeding actively around them. On one occasion, in eastern Hokkaido, I found a vagrant Canvasback clearly associating with them, and a Bufflehead more loosely so. Venables & Venables (1950) also noted Tufted Duck and Common Goldeneye in Shetland, and Merila & Ohtonen (1987) have found that Whoopers on the Gulf of Bothnia, Finland, in winter are usually followed by 1–6 dabbling and diving ducks. Merila & Ohtonen (1987) found the same dabbling species, but noted that, unlike Eurasian Wigeon, Mallard and Northern Pintail, the smaller Common Teal never feed near Whoopers. They also noted that Common Goldeneye, Tufted Duck, Greater Scaup, Long-tailed Duck and Smew would all escort Whoopers, diving around them for disturbed food. By far the commonest escort in Finland is the Common Goldeneye, 50% of which dive around feeding Whoopers (Merila & Ohtonen 1987). Schneider-Jacoby et al. (1991) and Ogilvie & Pearson (1994) also noted that Eurasian Coot will forage around Whoopers, and doubtless the list of species found in feeding association will continue to grow. In North America, a similar suite of species commonly forages in association with Trumpeters, including Canada Goose, Mallard, Northern Pintail, Common and Barrow's Goldeneyes, Wood Duck and American Coot (Mitchell 1994).

Feeding in such mixed flocks, which is most common in late autumn/winter when food resources are poor and some time remains until the resources are replenished, appears to benefit the ducks, without being detrimental to the swans. As a result of the association, both dabbling and diving ducks are able to expand their feeding areas to water depths or regions normally beyond their reach, leading to more effective use of the available food resources. Interactions under these and other conditions between swans and ducks are usually limited to brief pecking and displacement, with the smaller species invariably losing out, though occasionally it can lead to extreme aggression and even the death of the duck (Brazil 1983b).

Whoopers also occur regularly in association with other swans, especially on migration. They most often winter with Bewick's and Mutes in the West, and in the Far East they are also, very occasionally, seen with vagrant Whistling Swans. Whoopers most often associate with other waterfowl, but in extreme northeast Russia they even associate with Sandhill Cranes (Krechmar 1990). Where Whoopers themselves are vagrants, e.g. North America, they have not only joined Whistling Swans but also even hybridised with them. Associations between swans appear to occur most frequently at a specific and abundant food source. The WWT reserves in the UK are particular examples, where unnaturally dense concentrations of Whooper, Bewick's and Mute Swans form in response to the availability of grain and potatoes (Owen *et al.* 1986). At the Ythan Estuary in Scotland, where dumping of waste potatoes formerly occurred in some winters, large number of Whoopers were attracted, especially in hard weather and this resource was shared with Mute Swans, Greylag Geese, Eurasian Wigeon and Eurasian Coot (Buckland *et al.* 1990).

Although at least ten species of duck have been observed feeding with Whoopers at aquatic habitats, such associations with other waterfowl have not been noted on land, unless those with Bewick's Swans are considered. Most of the ducks associating with them on water, with the exception of Eurasian Wigeon, do not normally feed on land, and feeding swans on land do not stir up or disturb other food items, so the same benefits would not accrue.

WINTER FEEDING AND HABITAT SELECTION

My own studies of Whoopers began with observations for the most part in central Scotland where a population both fed and roosted. They were terrestrial feeders. The history of the development of this habit and its effects on behaviour are interesting and were described above.

I observed swans in two areas close to Stirling; that to the east (*c.*60 km²) was traversed by the tidal River Forth and that to the west (*c.*25 km²) by its two tributaries, the Forth and Teith (Fig 6.9). Both were in flat farmland with a mixture of arable and ley fields. Agriculture depended mainly on winter wheat, barley and sheep. Other cereals, root crops, and livestock were much less common. Two main field types were available to the Whoopers—stubble and grass. Ploughing reduced the availability of stubble as winter proceeded, although it occurred later in cold winters than in milder ones. While the availability of grass decreased only slightly in October–April, stubble decreased in both areas by over 50%. There was, however, always more stubble available to the west than to the east (Brazil 1981c). Thom (1986) noted that farming patterns were changing and that the new tendency was to plough stubbles increasingly early in Scotland. Such changes affected habitat availability, which in turn affected the feeding patterns and habitat selection of terrestrial-feeding Whoopers.

HABITAT SELECTION

Whoopers in the Stirling area did not use crop types in proportion to their availability. In October–December over 40% of both areas consisted of grass, whereas stubble at its maximum in October contributed only 24.3% of the western area and 15.3% of the eastern area. Nevertheless, the swans fed predominantly on the steadily disappearing stubble until January–February, when they changed to foraging on grass (Brazil 1981a,c). In 1977–78, 'autumn' use of stubble overlapped with 'spring' use of grass in January–February. There were, however 151.5 bd spent on stubble in January and 65 bd on grass. In February the situation was reversed, with 71 bd on grass and only 31 bd from stubble. Birds were not recorded on stubble after February. In 1978–79, use of stubble and grass overlapped in December and February. In December, 284 bd were recorded on stubble and 317 bd from grass. In 1978–79 none was recorded from grass in January, although in February 412.5 bd from grass were recorded, but only 69 bd on stubble. Again, none was recorded on stubble after February. In 1979–80, no overlap in use occurred at all. Birds last fed on stubble in December and first visited grass for foraging in January. What can we learn from such details?

The availability of grass declines only slightly during the winter as very little ploughing of such fields occurs. New grass, such as winter wheat and the early-spring growth of leys is of greater nutritional value than old grass, which becomes less palatable as it gets longer (Ogilvie 1978). In autumn, grass is presumably a poorer food source than grain, which is higher in carbohydrates. Later in winter there were fewer areas providing grain for food than grass, which begins to increase in protein content as the growing season recommences. 'Root' crops were uncommon in my study area and their use by Whoopers rare and confined to potatoes (Brazil 1981c). In other areas of Scotland, Ireland, and in the English Fens, potatoes are an important or even very important food (Kear 1963; Kemp 1991; Colhoun *et al.* 1996).

For part of winter 1977–78, some of the eastern study area close to the River Forth was flooded. Swans visited the area briefly in November 1977, but in January–February 1978 it was used consistently for feeding. In February, more bird days were spent there than at either stubble or grass fields, providing further evidence of the species' flexibility and ability to take advantage of short-term changes within its wintering habitat.

My observations showed that Whoopers fed on waste grain in stubble fields from late October until late January/early February, when they changed to feeding on grass (Brazil 1981a,c). There seemed to be a number of possible reasons for this: decreasing seed densities, loss of stubble area and renewed growth of grasses. The situation, however, was more complex as, for the most part, when feeding on stubble they used the area east of Stirling and roosted on the tidal Forth near Manor Neuk, whereas when foraging on grass they moved west of Stirling and roosted on the Forth near Kildean (Fig. 6.9). The reasons for this are unknown. A similar preference for stubble in the autumn and grass in spring has been observed in Aberdeenshire and in Dumfries (J. Kirk pers. comm.). Furthermore, this pattern is not merely confined to a small population of Whoopers in central Scotland. Similar shifts in foraging behaviour have now also been observed in Denmark and Germany (Degen *et al.* 1996; Laubek 1996).

Feare *et al.* (1974) found that in stubble fields used by Rooks, grain density declined rapidly in September–November, and thereafter more slowly until March. This is likely to be a general pattern for stubble fields as grain is utilised by many species as winter food. Murton (1965) found that as stubble fields became exhausted in November–early January European Wood Pigeons moved to clovers *Trifolium* sp. and Brussels sprouts *Brassica* sp., a move more recently found to be echoed by Whoopers in Denmark, which like the pigeons change from stubble fields to *Brassica* fields as winter progresses (Laubek 1996). Thus, although stubble is available throughout the winter, studies of other birds suggest that the grain density in stubble fields is likely to fall to a level too low for swans to feed satisfactorily, hence their change to grass. Although the availability of stubble also declines as winter progresses, the extent of the change in use in Scotland was much greater than the relative change in habitat availability, indicating that the Whoopers were very actively selecting the field types they preferred (Brazil 1981a,c), and essentially switching from carbohydrate-rich to protein-rich food. Whooper Swans have, apparently as a result of feeding terrestrially in certain parts of their west European range, developed a feeding regime and commuting system that closely resembles that of geese.

FEEDING BEHAVIOUR

Whooper Swans may still be regarded as typically wintering on 'shallow lakes, brackish lagoons and coastal bays' (Owen *et al.* 1986) where they consume aquatic plants (Fig. 6.10). Already, by the early 1970s, however, they were considered to

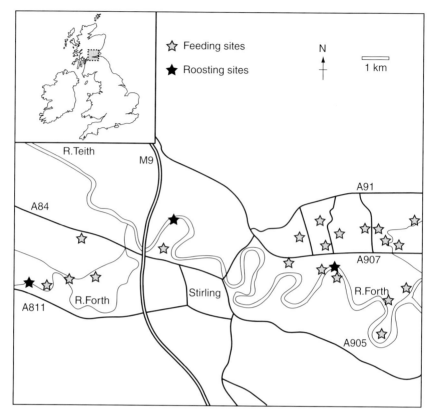

Figure 6.9.   Map of the study areas east and west of Stirling, Scotland (after Brazil 1981c).

forage readily on stubble and arable land when those supposedly preferred aquatic plants were lacking, and it became clear that, in some areas, Whoopers actively select terrestrial habitats. Owen & Kear (1972) noted that feeding on agricultural land, particularly on potatoes and turnips, in Britain seemed to have developed since the hard winters of the 1940s (though it is also likely to have related to changes in food availability in coastal habitats). Certainly, many flocks now feed on agricultural land even when freshwater habitats are available, and despite being less agile on land than Bewick's Swans this habit is increasing. In the Stirling area, despite stretches of the rivers Devon and Forth being available and although Lochs Leven and Lomond are within 40 km, flocks invariably foraged on fields. Furthermore, even those flocks focused on aquatic habitats, such as those at WWT reserves, may spend some time *also* foraging on fields, indicating that the situation is rather more complex than hitherto reported. Terrestrial food plants may be higher in nutrients, especially if fertilised, hence they probably form a very attractive food (Brazil 1981c).

Figure 6.10.   Lakes are important winter habitat.

Whereas in Britain and Europe one would tend to look first for wintering Whooper Swans either on fresh water or on agricultural land, in east Asia the majority occur on saltwater in tidal environments or on brackish coastal lagoons and then, in lesser numbers, on fresh water. In the Far East, foraging on land is unusual, although I have on occasions observed it during winter and during spring migration in northern Japan. In particular in early spring 1984, a large herd was on snow-free meadows in central-east Hokkaido far from the nearest wintering sites, and in March 1997 when birds were feeding on woodland dwarf bamboo (Brazil 1984a,b; Brazil 2002b).

FEEDING BEHAVIOUR IN TIDAL AREAS

With large numbers taking to grazing on land in Britain, Ireland, Denmark and Germany, and with increasing visitor attention focusing on provisioned flocks, it is increasingly difficult for either the casual observer or even the swan biologist, to watch wintering Whoopers following a natural behavioural regime. Studies have tended to concentrate on birds using a mixture of freshwater and terrestrial habitats in winter or fresh water in summer, yet in some regions they frequent primarily saltwater tidal environments. Even in those parts of Scotland where Whoopers have been observed at tidal bays, the same birds have also tended to visit nearby agricultural land and/or freshwater lakes. The increasing use of terrestrial habitats in Europe has markedly affected the Whooper's behaviour, and the only way of observing a more natural behavioural regime is to examine it where supplementary feeding does not occur or is minimal, and where birds never forage on land. Though there may be localities in some parts of Europe or many such localities in Russia where such observations are possible, this opportunity arose for me in Japan.

The behaviour of those wintering at tidal sites can be expected to differ from birds elsewhere, which either commute from aquatic roost sites to feed on agricultural land, or roost and forage at freshwater sites, as on the sea they will be influenced by both diurnal and tidal cycles. Even in Japan, a country presenting apparently ideal opportunities for studying the effect of tide on behaviour, with several significant marine wintering sites, it is increasingly difficult to find sites where some food at least is not provided (Brazil 1984b; Albertsen & Kanazawa 2002). Except at sites with resident 'swan-keepers', the provision of food at sightseeing locations does not usually follow a regular daily pattern, as the number of visitors and amount of food varies greatly during the day and between days. Visitor numbers, and hence food, are considerably higher on weekends, particularly Sundays (Brazil pers. obs.). Even where 'swan-keepers' do provide a fixed amount of food at a given time each day, the behaviour of the birds follows no obvious pattern (other than visiting near feeding time), with birds moving between natural and artificial feeding areas throughout the day (Brazil pers. obs.).

The complex interaction between time of day, tidal state and Whooper behaviour has not been fully examined and explained, although a brief study in a tidal area of northern Honshu, revealed clear differences from both terrestrial and freshwater situations (Brazil 1984b). There, the birds showed no evidence of increasing their feeding activity towards the afternoon plateau observed among Mute, Bewick's and Whooper Swans on fresh water in Britain (Owen & Cadbury 1975; Brazil 1981c, 1984b). Nor did they follow the distinctly bimodal pattern of behaviour typical on land, with morning and afternoon feeding peaks separated by a midday lull. In fact, the reverse was apparent. Feeding tended to peak during the middle of the day or at low tide, decreasing as the tide rose, while roosting was most prevalent in morning and afternoon, lowest at midday and occurred more frequently over high tides. Variation in water depth as the tide changes puts submerged vegetation out of the reach of Whoopers, thus feeding at high tide is difficult or impossible. Not surprisingly, the birds chose to roost then, while at low tide they took advantage of easier access to food (Brazil 1984b).

FEEDING ROUTINES IN TERRESTRIAL AND AQUATIC HABITATS

As a result of their increased use of agricultural habitats in parts of their range, changes in other aspects of behaviour are to be expected, but what form do they take? In particular, swans feeding away from water must establish a pattern of commuting between safe night-time roosts and daytime foraging sites. The daily activity patterns of such birds feeding will differ from those feeding at aquatic habitats.

DAILY MOVEMENTS: ARRIVAL AND DEPARTURE TIMES

The regular movements of animals from a night-time roosting or sleeping site to a daytime foraging area is a common phenomenon, especially among birds. A relationship between roost flights and light intensity has been found for a wide range of bird species (Nice 1935; Hein & Haugen 1966; Siegfried 1971). Where light intensity is the controlling factor, other factors such as cloud and mist are expected to delay morning flight and hasten evening flight (Hein & Haugen 1966). Seibert (1951) found that herons left the roost at a lower light intensity than that at which they arrived, while Siegfried (1971) suggested that a specific light intensity threshold for roosting is modified by environmental constraints and physiological needs. Longer periods of darkness in midwinter lead to a greater hunger in visual feeders, which, in turn, is likely to cause earlier departure in relation to light intensity. Hein & Haugen (1966) found least change in the start of evening flight by Wood Ducks, while the end of the morning flight and duration of evening flight changed most. They also found that on dark or foggy days, Wood Ducks delayed their morning flights and evening flights were prolonged. In general, it appears that conditions of poor visibility lead to adjustments in flight times such that illumination at flight times resembles that on clear days.

Whoopers foraging on land some distance from water also make regular daily movements between roosting and foraging areas. Three factors might be assumed to be affecting them: day length, temperature and safety. Each of these might have some effect on the timing of their commuting activity and their effects might be interactive. During midwinter when days are very short, birds endure long periods of darkness often combined with low temperatures, and hence they are likely to have lower morning energy stores than at other seasons. Compounding the impact of low temperatures and prolonged darkness is that shorter midwinter days also permit less time for feeding among visual feeders. Assuming that the time swans spend at their foraging grounds is positively correlated with their energy requirements, then in midwinter birds should arrive earlier at, and depart later from, their foraging areas. Furthermore, the actual time spent feeding during the day should increase relative to day length.

While hunger might be pushing Whooper Swans to spend as much of the short winter days as possible in foraging, it should be remembered that terrestrial-feeding birds are essentially visually oriented. Not only do they require sufficient light to search for food but also for safe low-altitude flights. During their typically very low-altitude flights between roosting and foraging sites they are, particularly in poor light, very much at risk from objects such as telephone and powerlines and trees. Collisions are a particularly significant cause of swan mortality, accounting for 33–44% of casualties (Ogilvie 1967; Owen & Cadbury 1975; see Chapter Eleven). Thus it becomes increasingly risky for a diurnal species to fly at low altitude as it becomes dark. Foraging on land after dark increases the risk of attack by nocturnal ground predators, such as Red Foxes. Because of the increased risk of mortality resulting from foraging or flying during darkness, Whoopers are also under conflicting pressure to arrive later on their feeding grounds and depart earlier, at a specific light intensity, that permits them to fly safely between roosting and foraging sites.

Study of the morning arrival times of Whoopers at foraging sites, and evening departure times from the same sites, in central Scotland over day lengths of 7–13 hours, revealed a distinct difference in the overall pattern. Morning arrivals were evenly distributed from 60 minutes before local sunrise to ten minutes afterwards, whereas evening departures were normally distributed from sunset until as late as 90 minutes afterwards, although the peak period occurred 31–40 minutes after sunset. Although the swans did tend to arrive earlier at their feeding grounds on short days and later on long days, this was not significant for the small dataset concerned. Their departures from feeding grounds, however, occurred significantly later after sunset on short days and earlier on long days (Brazil 1981c).

Broadly speaking, light levels directly correlate with time of day, such that birds flying earlier in the evening do so at higher light intensities than those leaving later the same evening. Daily variation in weather conditions, however, can lead to marked variability in light intensity; this is largely due to cloud cover, which varies unpredictably. Consequently, Whoopers leaving at the same time relative to sunset on different days departed at different light levels. Light intensity, in particular, has a strong bearing on departure time, with the frequency of departures increasing markedly as light intensity falls. Few departures occurred at light readings above 0.1–0.3 foot candles (fc), and the majority occurred below 0.1 fc (Brazil 1981c). Although light intensity at a given time is correlated with day length, great variation between days can occur owing to the presence or absence of cloud. It transpired that for the period of winter when day length was less than 12 hours, there was a significant negative correlation between departure time and cloud cover. Early departures were associated with heavy cloud cover, as expected from the association of departure times with light intensity. Both day length and hence light intensity under standard conditions are predictable cyclical variables. Cloud cover, on the other hand, is unpredictable. If swans responded only to day length a smaller scatter would be expected in the data. Unpredictable variables such as cloud cover clearly contribute to this scatter (Brazil 1981c). Morning arrivals, it transpired, were spread over a wide range of light levels. In contrast, departures tended to be concentrated at lower light levels. Since duller days tend to be associated with short winter days, in midwinter Whoopers might be expected to leave earlier relative to sunset if they rely on a constant light level to trigger their departure; in other words they might show a simple seasonal pattern. When light readings at the 50% departure time (by which half of the flock had arrived or left the foraging grounds, an endpoint least influenced by chance variation; Davis 1956; Henty 1977) were compared with day length, however, it became clear that they were departing at lower light levels in midwinter and at higher levels in late winter/spring. Thus, relative to sunset, Whoopers chose not to leave earlier but to leave their feeding grounds later, in midwinter than in spring. They also tended to arrive on their feeding grounds at lower light levels, that is earlier with respect to sunrise, in midwinter

(Brazil 1981c). In brief, during the shorter days of winter Whoopers attempt to maximise available time for feeding, whereas during the longer days of spring they are less hard-pressed.

It seems logical to assume that, in addition to evening departure times being correlated with day length and light intensity at departure, low temperatures might also cause Whoopers to feed for longer periods to compensate for greater energy losses. However, neither Henty (1977) nor Brazil (1981c) was able to support this with field data, although a more general seasonal pattern of foraging activity was apparent (Brazil 1981a,c). If low temperatures cause swans to lose more body heat, then their rates of energy consumption will be higher, and if the length of time that they spend feeding is directly related to their energy requirements, both assumptions that seem reasonable, then a simple prediction might follow: on days following nights of low temperature birds should attempt to extend their feeding time, by arriving earlier and/or departing later in order potentially to increase energy intake (Brazil 1981c).

Surprisingly, however, although there was no correlation between departure time and maximum or minimum temperatures in central Scotland, there was a significant positive correlation between arrival and maximum and minimum temperatures. Contrary to what one would logically expect, based simply on energetic grounds, Whooper Swans arrived later, not earlier, on mornings following cold nights! No explanation for this is known, but cold, freezing weather clearly affects their behaviour at the macro level, causing birds to change feeding and roosting areas, extend their migratory movements, and under extreme circumstances to even appear beyond their normal range.

TIME BUDGETS IN TERRESTRIAL AND AQUATIC HABITATS

Time or activity budgets have provided much basic information on the Whooper Swan, as they have for many other species of birds, mammals and insects. Four mutually exclusive behaviours (feeding, preening, roosting and 'head-up') are easily observed among flocks, and a simple method of scans made at 15-minute intervals readily reveals the proportion of a flock engaged in each behaviour (Brazil 1981c). By averaging several scans, during which behaviour is assigned to one or other of the mutually exclusive behavioural categories, the mean proportion engaged in each behaviour and hence the proportion of time spent in each behaviour is obtained.

When considering time budgets, it is important to remember that these are four broad behavioural categories, thus the category 'feeding' (on land) includes not only those obviously pecking, but also those between pecks with the neck below the horizontal line of the body and the head close to the ground. The category 'preening' covers all comfort movements, including wing flapping, while 'roosting' includes birds sleeping or resting with the eyes open or closed, during which the head is usually rested on the back, with the bill at least tucked in among the back feathers. Sometimes, especially when facing into a strong wind, Whoopers will rest the neck curved so that the head lies almost horizontally on the lower neck, pointing forwards. 'Head-up' includes both standing and sitting birds with their heads and necks raised above the level of their bodies. 'Head-up' consists of a continuum of positions less readily separated into 'head-up' and 'extreme head-up' as they are in various geese (Lazarus & Inglis 1977). Subjectively, at least, Whoopers adopt a general 'head-up' when being observant of their environment, whereas once they have perceived a specific stimulus they adopt a position that appears somewhat more tense and directed—'extreme head-up'. Their behaviour then is directed towards the specific stimulus, e.g. as a result of disturbance or birds joining or leaving the flock. 'Extreme head-up' can be reliably recognised when watching an individual continuously, but not when making spot-sample scans. The major factors disturbing Whoopers in Britain, and presumably elsewhere, tend to be people, dogs and shooting (Airey 1955; Brazil 1981c).

The amount of time that Whoopers apportion to each of these behaviours, and how that time varies seasonally is of particular interest. Thus flocks on foraging grounds in Scotland spent 62–75% of their time feeding 14–16% in 'head-up', 4–19% roosting and just 2–4% preening (ranges are for winter, calculated per hour of daylight). Feeding is clearly the most important behaviour, and the time spent feeding each day by birds on land increased during the winter, from November to March. Whereas in November the cumulative time spent feeding on average over an 8–9-hour day was 6.4 hours, by March it was 9.8 hours over a 12–13-hour day (Brazil 1981a,c). In November–February, they followed a marked bimodal feeding rhythm with morning and afternoon peaks. By March this pattern had weakened, with birds feeding at a higher rate for a longer time each day; consequently the midday lull in feeding, marked in earlier winter, had almost disappeared by March, as if birds were attempting to increase their food intake.

Given the prolonged period available for sleep at night, it is unsurprising that roosting occupied only a small period on the foraging grounds. Roosting during the day and feeding are strongly linked in a negative way, with roosting peaking during the feeding lull and otherwise occurring at a very low level. 'Head-up' occurs throughout the day, but at a noticeably higher rate usually at the beginning and towards the end'. 'Preening', which occupies least time, can occur throughout the day.

In contrast to those Whooper Swans feeding on land, those feeding naturally (i.e. without being provisioned) at a freshwater site in central Scotland adopted a completely different pattern of behaviour, steadily increasing the amount of time they spent feeding until *c.*14:00, i.e. 7–8 hours after sunrise, a pattern similar to that among swans on the English Fens (Owen & Cadbury 1975; Brazil 1981a,c). As part of this pattern, roosting was the dominant activity during the morning, preening reached its maximum (18%) at 11:00–12:00, while head-up occurred throughout the daylight period at 18–47%.

Ireland represents the most important wintering area for Icelandic-breeding Whoopers, yet ecology and behaviour there had been relatively little studied. Thankfully, O'Donoghue and O'Halloran (1994) set out to redress that imbalance, and in an aquatic environment. Though I attempted to study behaviour in such a setting, the majority of birds spent significant amounts of time grazing on land, so that I was observing a mixed situation at best. At other sites in Britain where birds spend most of their time on water, they also receive considerable supplementary food. O'Donoghue & O'Halloran (1994), however, were able to study a flock of up to 24 unprovisioned Whoopers on the 50-ha Rostellan Lake, a shallow brackish waterbody in Cork Harbour on the south coast. These birds neither received supplementary food nor did they (with just one exception) graze on land. They can, therefore, be regarded as representing a naturally, aquatic-feeding flock. The depth of Rostellan Lake is such that most areas are within the species' upending range, and submerged aquatic vegetation consists almost entirely of pondweeds.

O'Donoghue & O'Halloran (1994) recorded eight major activities: feeding, comfort (preening, washing, resting and loafing), swimming, standing, walking, vigilance, flying and social interactions (aggression and pre-flight signalling behaviour). Feeding was further subdivided into grazing, stirring of submerged vegetation by foot paddling, upending, 'head and neck submerged feeding' (i.e. dipping), head submerged (also dipping) and surface feeding (dabbling).

They found that the main activities of their birds on water were feeding and comfort activities, followed by swimming and vigilance, while walking, standing and flying represented only a very minor component (<1%). Dominant feeding strategies were upending and dipping. In contrast to my studies (Brazil 1981a,c), O'Donoghue & O'Halloran (1994) recorded grazing on agricultural land only once, so their flock can be effectively regarded as entirely aquatic. They found two very different feeding patterns at different periods: a bimodal pattern, akin to that elsewhere in terrestrial habitats, and of increased feeding during the day, with a peak 5–7 hours after sunrise like that observed in Scotland (Brazil 1981c). Comfort activities mirrored feeding activity, with higher levels when feeding decreased, particularly later in the day. Vigilance varied significantly, but was not correlated with time of day. The two commonest styles of feeding, upending, and dipping, were accompanied by frequent stirring activity. Of particular interest was that upending declined rapidly from eight hours after sunrise onwards, from 55–69% to just 14% ten hours after sunrise, and was replaced by rapidly increasing levels of dipping 17–34% during most of the day but c.65% 9–10 hours after sunrise. Perhaps upending is more energy consuming than dipping, and the birds tire.

The amount of time spent feeding depends largely on the types of food available, but it is generally constrained by the Whooper's poor digestive efficiency. Thus, where birds receive supplementary feeding they spend least time feeding. At

Caerlaverock, where food is provided, they spend just 11% of their time grazing and only 5% feeding in the water, whereas those grazing outside the refuge spend up to 45% of their time feeding (Black & Rees 1984). In contrast, Whoopers spending the majority of their time on fields spend 60–75% of time feeding (Brazil 1981a,c), and where they feed exclusively on water but are not provisioned, they feed for 40% of the time (O'Donoghue & O'Halloran 1994). The lower level of feeding that they recorded, compared with that of birds foraging on agricultural crops, they attributed to the lower crude fibre content of submerged vegetation, which thus provides more nutrition for the same weight of food, and presumably requires less energy to process. One further factor contributing to the large difference is the need for those birds that habitually graze on agricultural land to make daily flights between their grazing and roosting sites. This is likely to be their largest energy demand each day.

O'Donoghue & O'Halloran (1994) found that feeding was more intense among those birds that had just arrived, and presumed that this was necessary to replace lost energy expended during the previous breeding season and migration. Thereafter, feeding levels declined, particularly in February, a further indication that perhaps, despite the short midwinter days, this is not the period when Whoopers are most energetically stressed.

Changes in feeding behaviour among birds feeding in water were also associated with food preferences. For example, for most of the time, upending birds were feeding on pondweeds whereas dipping birds were taking reed stolons. None was recorded as being provisioned (O'Donoghue & O'Halloran 1994).

Other than feeding, the single most important category at Rostellan was comfort behaviour, which included feather maintenance, amounting to 44% of their time (range *c.*30–70%). They spent little time swimming, whereas a flock of Mute Swans spent up to 45% of their time swimming due to the provision of bread at a site with little natural food (Keane & O'Halloran 1992). Another crucial maintenance behaviour is vigilance, during which swans have their heads raised and are looking around. It is by this means that birds are forewarned of potential danger or shifts in the behaviour of other flock members. Vigilance and other behaviours are mutually exclusive, and thus time apportioned to each must be seen as a balance between conflicting behaviours. As such, various factors may affect the balance between vigilance and other behaviours, such as seasonal energetic state, the individual's status, whether it is adult or immature, and whether it is an adult with or without cygnets. Vigilance can be increased in two ways, either by increasing the length of each bout of vigilance or by shortening the interval between periods. It seems that it is the length of the interval between periods of vigilance that is more important for detecting a threat (Hart *et al.* 1984) and this variation has already been detected among Whoopers (Brazil 1981c). O'Donoghue & O'Halloran (1994) also recorded the average interval between periods of vigilance and like Brazil (1981c) found that parents had the shortest interval at just *c.*40 seconds. In contrast to Brazil (1981c), however, they found that non-parental adults had longer intervals (112 seconds) than first-winters (83 seconds), though this may have been a result of site differences and small sample sizes.

Because of the difficulty in distinguishing between a bird which is actively vigilant and one merely raising its head between feeding bouts to breathe or swallow, Brazil (1981a,c) used a loose definition of vigilance to include all swans with raised heads and found levels of 14.5–16.5%. O'Donoghue & O'Halloran (1994), in contrast, found that based on a stricter definition of vigilance their birds spent only *c.*5% of their time vigilant, although this did rise to 20% when the looser definition was applied. The short intervals between feeding bouts, when birds have their head raised, surely account for an important part of each individual's vigilance behaviour. Detailed studies of Barnacle and White-fronted Geese flocks have revealed that vigilance varies with flock size, with a decreasing percentage of birds being vigilant in larger flocks, and in relation to their position in the flock (those at the periphery are more vigilant than those nearer the centre; Drent & Swierstra 1977; Lazarus 1978). Because swan flocks generally are considerably smaller than goose flocks it is very much more difficult, if not impossible, to determine which flock members are peripheral (Brazil 1981c). The flock at Rostellan may have been just too small to produce any detectable relationship between flock size and vigilance (O'Donoghue & O'Halloran 1994).

In Scotland the Whooper Swan has, in adapting to foraging on agricultural land, adopted a feeding pattern similar to that of many geese, and shows a bimodal feeding pattern typical of many diurnal species (Brazil 1981a; Owen & Black 1990). In southern Ireland, O'Donoghue & O'Halloran (1994) found an overall slight bimodality (peaking one and four hours after sunrise) and declining dramatically later. When they examined their data over narrower time frames, however, they found two different patterns being obscured, one a bimodal pattern similar to that of Whoopers on land, and the other of increased feeding throughout the day as in those feeding elsewhere on water, although their peak was *c.*6 hours after sunrise.

In contrast to O'Donoghue & O'Halloran's (1994) purely aquatic-feeding Whoopers in Ireland, and my largely terrestrially birds in Scotland (Brazil 1981a,c), elsewhere in Scotland studies have been made at sites where mixed feeding was more typical. Thus, Black & Rees (1984), who studied a flock around Caerlaverock, found that the main activities while on the refuge were feeding and resting on water, both of which occupied *c.*30% of an average day. Much of the rest of the day was spent in comfort behaviour (22%; preening, sitting, sleeping or bathing), while grazing or digging on land was 8% and other behaviours, including social activity, occupied 10%. Caerlaverock is a provisioned site, so unsurprisingly the percentage feeding peaked at times when food was provided. After their morning feed, the swans typically engaged first in comfort behaviours, then in resting on the water until the afternoon feed. The swans are so familiar with the feed that they commonly assemble more than an hour before the scheduled time. I have observed similar anticipatory behaviour among those wintering at Lake Kussharo in east Hokkaido, Japan. At Caerlaverock, Black & Rees (1984) found that although the time spent feeding peaked in February (the coldest month in

1983–84), grazing increased in March, coincident with the increasing growth rate of the grass. They found that comfort behaviour peaked in October, when the swans first arrived, and in April, when day lengths were longest, but that the total time spent resting on water decreased during winter.

Three species of swans (Whooper, Bewick's and Mute) cohabit at Caerlaverock and, despite sharing habitat and scheduled feeding periods, they behave surprisingly differently. For Whooper, feeding is the most important daytime activity, and whereas comfort behaviour ranked only third, it ranked first for wintering Bewick's. Mute Swans differed in that their primary activity involved resting or swimming (Black & Rees 1984). The extent of these differences has not been explained, but perhaps in a wider context their behavioural differences serve to reduce competitive interactions.

As with my studies in Scotland and those in Denmark (Laubek 1995c, 1998), Black & Rees (1984) found that Whoopers in southwest Scotland also make seasonal changes in foods and feeding patterns. Thus in October, they feed on any barley left after the harvest. During November, however, as potatoes and turnips are harvested, leftovers were eaten in addition, and they began dipping for aquatic plants in the River Lochar, which was bordered by several fields of sprouting winter grass that they also consumed. In December, they spent similar amounts of time feeding on river plants and grasses, and in January–February they spent 40% of their time on old grass and increased the time feeding on new grass to *c.*50%. In March–April the new grass began spring growth and the majority of Whoopers spent the majority of their time there. For those feeding away from the refuge, Black & Rees (1984) found that the new grasses planted in autumn were the most important food overall. Their work also showed that at Caerlaverock, Whoopers have formed a clear attachment to the wintering site, with as many as 78% of ringed swans returning for a second winter commonly staying from October to March, whereas both the size and composition of wintering flocks in central Scotland changed on a near-daily basis (Brazil 1981a,c). But, perhaps the most significant factor distinguishing these two study areas was that the provision of food at Caerlaverock was a consistent and persistent attraction.

Despite it being a major and traditional habitat for them, very few observations of Whoopers wintering in a tidal environment have been published. In Iceland, Boswall (1975) had observed them foraging on a brackish lagoon at Lónsfjördur, and had noted that, whereas the swans dipped increasingly for eelgrass as the tide fell, only a few fed (and those only by upending) when the tide was higher. At high tide he noted that bathing was a more frequent behaviour. Furthermore, the behavioural patterns of Whoopers wintering in a tidal area differ markedly from those using either terrestrial or freshwater habitats. At a tidal area in northern Honshu, where they frequented the shallow waters of a large bay and a small peninsula, behaviour was related to both location and tide. Thus, at the peninsula, 'roosting' was dominant and 'feeding' second, whereas 'head-up' and 'preening' were much commoner at the north shore, feeding occupied less than 20% of time and 'roosting' rarely occurred. The number of swans at the two subsites was linked, with a significant negative correlation between them. Roosting at the peninsula increased as numbers increased and decreased as numbers increased at the north shore. Conversely, feeding decreased as numbers increased at the peninsula and increased as numbers increased at the north shore. These data thus supported the subjective impression that swans foraging along the bay shore moved to the point to roost. Since feeding and roosting are mutually exclusive, these complementary results were expected (Brazil 1981c, 1984b).

At the peninsula, feeding tended to peak during midday (if the data are grouped according to time) or towards low tide if grouped according to tide. Conversely, roosting occurred more in the morning and afternoon, and tended to increase over high tide. At the north shore, feeding occurred throughout the day at the same level, regardless of tide. Roosting, however, decreased from dawn, and increased after high tide.

The available data concerning feeding patterns in a tidal environment provide little that is conclusive, but suggest that is not at all like those shown by swans on non-tidal waters or on land (Cadbury 1975; Brazil 1981a,c). It shows neither an increase in feeding during the day to an afternoon plateau, as on freshwater, nor a clear-cut pattern of morning and afternoon peaks and a midday lull as on land; in fact, at one locality, feeding tended to peak during midday. Unfortunately, too few data were available over sufficient tidal states to make conclusive statements concerning feeding rhythm. Certain results, however, were apparent. On a rising tide, the number of swans feeding at the peninsula and preening at the north shore decreased significantly with time, while 'head-up' at the north shore increased significantly. Conversely, on a falling tide, 'head-up', roosting and preening at the north shore were all correlated with time and tide. Further analysis suggested that time and tide are counteractive on behaviour (Brazil 1981c). The relationship between tidal state and activity among Whoopers raises many interesting questions. For instance, if tidal state is important, then are they able to feed at night to take advantage of it? Feeding on submerged aquatic vegetation such as eelgrass may be as easily accomplished by touch as by sight, which would make nocturnal feeding quite possible, though I am unaware of any such observations.

AN OVERVIEW OF FEEDING ROUTINES

Various factors may affect the daily movements of foraging Whoopers, and may lead to different assumptions concerning their behaviour. For example, assuming that short days and low temperatures lead to higher energy demands then they might respond by spending more time feeding. Consequently, they should arrive earlier at their foraging grounds and/or depart later for their roost site to facilitate this (hypothesis one). Alternatively, owing to the potential dangers of predation associated with being on land during darkness or of injury while flying at night, Whoopers should always arrive at their feeding grounds after first light and leave before darkness falls. In consequence of this (hypothesis two), the time spent feeding would be limited simply by day length and therefore any adjustment in relation to energy demands would have to be made in the rate of feeding, rather than in time spent feeding. Whoopers are highly visual animals and do not usually feed or fly after dark. The most pertinent stimulus, therefore, is probably light intensity.

It is important to remember, however, that circumstances at arrival and departure points are different. Measurements of light intensity at departure are easily made at the sites where the birds fed and where they would actually perceive and respond to such a stimulus. However, measurements at arrival differ because they can only be made when the birds reach the foraging area, not when prompted to leave their night-time roosts. Assuming the same mechanism to be at work, Whoopers respond to a stimulus perceived while still at the roost, which causes them to leave for the foraging area, so there is a delay, equal to the flight time, between departure from the roost and measurement of arrival at the feeding grounds. Even though this flight time was short, *c*.3–5 minutes in my Scottish study area, and varied between days and subgroups, it often occurred when light levels were changing most rapidly. Since observation on the foraging grounds was the prime objective, light intensity was measured on arrival at these areas, rather than at departure times from the roost site.

Arrivals were evenly distributed with respect to sunrise and without any tendency to occur at a fixed time, whereas departures suggested a tendency for movements to occur at a fixed time in relation to sunset. The latter agrees with hypothesis two and suggests that the change in roost flights is seasonal whereas the former does not. Departures also occurred relatively later on short days than on longer days, and arrivals tended to occur earlier on short days. Both factors again supporting a seasonal pattern in roosting flights. Early departures were also associated with heavy cloud cover, which tends to reduce light intensity, thus replicating shorter days (Brazil 1981c). Seasonal variation in roosting behaviour has been reported in many bird studies, including those of various woodland passerines, as well as among various waterfowl, showing that flight or roosting times may be related to times of sunset and sunrise. Furthermore, European Starlings have been found to roost later, relative to sunset, as midwinter approaches with roosting occurring earlier on days of poor visibility, in much the same way that I found among Whooper Swans (Hinde 1952; Hein 1961; Siegfried *et al.* 1977; Brodie 1980; Brazil 1981c).

It seems reasonable and likely that for a temperate-zone species, experiencing a steadily and seasonally shifting photoperiod, that a seasonal response would be the most basic. Seasonal photoperiods have long been known to be important in controlling cyclical changes, e.g. song, gonad growth and hence breeding, migration, hibernation and even thermal tolerance (Hoar 1956; Roberts 1964). That departures occurred later on short days, and more often at lower light intensities, are two ways of stating that on short days of midwinter Whoopers stayed longer on the fields relative to sunset than in spring. This clearly contradicts the second hypothesis and agrees with the first. It suggests that in midwinter Whoopers experience some energetic stress and need to feed longer relative to the available daylight/day length. The evidence that departure times occurred earlier on nights with heavy cloud (Davis 1956; Brodie 1980) supports the earlier suggestion that light intensity was the major stimulus affecting their diurnal movements.

Delaying departure from the foraging grounds because extra feeding time is required to meet energetic needs seems a reasonable strategy, yet on most evenings the last few minutes, often ten or more, are not actually spent feeding. Waves of alertness ('head-up') tend to spread through the flock, while wing flapping (a flight-intention movement) becomes more frequent. In addition, some begin head bobbing and calling and frequently small groups walk about through the flock as if to stimulate others to join them. They sometimes feed again briefly, but eventually all flock members stop feeding and sufficient numbers become alert to trigger the final departure. Where several subgroups departed separately, it was usually those that had started the wave of head bobbing that left first. During this time that this was occurring, visibility typically declined until it became poor. The behaviour involved during this pre-flight period has since been studied in considerable detail by Black (1988; see Chapter Seven). A further possibility remains: contrary to the assumption of hypothesis one, it might be that for much of the winter Whoopers are not actually under extreme energetic stress and therefore can afford to spend some time on the foraging grounds not feeding.

Temperature is a factor that one would expect to play a major role in determining food requirements, foraging time, and hence arrival and departure times at the feeding grounds, yet in my Scottish study it did not affect departure times from the foraging grounds. It did, however, affect arrivals in a way contrary to expectations. Low temperatures and frosts are typically associated with clear skies and higher light levels, conditions that should surely stimulate birds to depart earlier from the roost. However, I found that arrivals at the feeding grounds occurred later at low temperatures. Temperature has been shown to be a fundamental influence in the timing and magnitude of morning flight in some geese (Raveling *et al.* 1972; Ogilvie 1978), and my results were consistent with theirs, whereas Owen (1980) had considered that morning flights by geese were related to light intensity and the conditions of the feeding grounds. Low temperatures appear to override a commuting system linked primarily to light intensity, such that although sufficient light is available to trigger commuting, the low temperature stalls it. It could be that on cold mornings, especially when frosts are involved, the energetic cost of warming up food for processing, is too great to be worth the cost of flying to the fields while it is still so cold. European Wood Pigeons, which commute between woodland roosts and foraging grounds on agricultural fields, fed first on tall *Brassica* plants, which were, because of their exposure to sunlight, warmer than much lower clover plants, but then moved to feeding mainly on the latter as the temperature rose (Kenward & Sibly 1977). For species that have no readily available alternative food supplies (alternative to primary foods that may still be cold or even frozen), inactivity might be the most adaptive response to severe cold so as to conserve energy (e.g.

Markgren 1963; Raveling *et al.* 1972). In less severe conditions activity might merely be postponed until air temperatures rise.

As low temperatures are commonest in midwinter, while days are still very short, any negative effect of low temperature should magnify any negative effect of short day length, yet among Whoopers in central Scotland this was not the case. It seems that any tendency for them to arrive earlier on short days was counterbalanced by their tendency to arrive later on colder days, many of which would also be short. The departure times for their roosts appeared to be independent of temperature. Neither hypothesis adequately explains such results. Light intensity is clearly an important proximal stimulus to commuting flights, but its effect varies seasonally and with day length. The obvious seasonal change is in the available feeding time, which is reduced as days shorten and extended as days lengthen. Low temperatures act contrary to this, as they tend to cause later arrivals on short days where available feeding time is already limited. It appears that any increased cost of thermoregulation is outweighed by some effect on feeding effectiveness. Consequently, the impact of temperature on field foraging Whoopers is more complex than previously envisaged, leaving much unexplained.

Wind might also affect daily movements as it, coupled with low temperature, might increase energy losses or even exacerbate risks of commuting. The microclimatic advantages of roosting at riverine sites where banks and bends provide protection appear obvious, but birds could still be affected by winds parallel to the river. Wind velocity has been widely shown to directly affect heat loss, and thus adaptations such as roosting in flocks are known to aid thermoregulation by reducing the effects of wind speed (e.g. in waders; Whitlock 1979). Presumably, waterfowl could easily shelter by moving round a bend in the river. It seems unlikely therefore that wind speed can help explain the daily variation in arrival or departure times.

Henty (1977), who also studied Whoopers in central Scotland, found 'no noticeable effect of frost during the preceding 24 hours although continued hard weather disrupted the whole commuting system'. If swans were already under energetic stress, then cold weather would obviously be expected to exacerbate this. The effects of low (subzero) temperatures during night-time, which is a critical period at high latitudes for many species (Kendeigh 1961) can be ameliorated behaviourally in several ways, including by reducing the amount they expose themselves to the elements. Some species seek dense vegetation (Frazier & Nolan 1959; Swann 1975) or roost communally, like Winter Wren (Armstrong 1955), Eurasian Treecreeper (Löhrl 1955) and swans or, as in the case of Common Dippers (Shaw 1979), in groups under bridges. Physiological studies of heat loss from roosting swans under different conditions might prove very interesting.

Despite the apparent lack of any effect of temperature or wind speed on the daily commuting pattern, there was a marked effect on the commuting system in general caused by prolonged cold. Birds actually left the study area entirely, if snow cover was complete, during long periods of low temperatures, severe frosts or after severe flooding (Brazil 1981c). Rather than being merely energetically constrained,

Whoopers may be prone to catastrophic changes in food availability as when it is frozen or covered by frost or snow. Such cold-weather movements have been widely reported among Whooper Swans.

Whooper Swans, in Scotland at least, tended to arrive earlier at their feeding fields and depart later on short midwinter days than on the longer days of autumn, late winter and spring. On the very long days of spring (longer than 12 hours), however, they also arrived very early and left very late, although day length was increasing such that their feeding day was almost doubled at this season. Time budgets suggest that their greatest nutritional requirements may be in spring, immediately prior to migration, not during midwinter as I had earlier expected (Brazil 1981c). It is likely that staying longer on the feeding grounds does enable the birds to acquire extra food, but the factors affecting their departure times operate differently at different times of year. In midwinter they are probably trying to maintain weight whereas in spring they are trying to increase it. Although as yet unstudied in either the Whooper or the Bewick's Swan, it is likely that the crucial fat deposition necessary to make the spring migration and achieve breeding condition occurs in the final month or two of the winter, as it does in geese and shorebirds (Minton 1974; Evans & Kear 1978).

It appears that Whoopers do compensate a little for short winter days by departing at lower light levels, but that energy requirements are probably much higher in spring than midwinter. Neither of the two hypotheses proposed accurately predicted my field observations. The most reasonable explanation is that Whoopers are responsive primarily to light intensity, and in midwinter may have to extend their feeding time a little, but in spring they must increase feeding time enormously. That they are not nocturnal suggests that, while for energetic reasons they might tend to increase the time spent on land after dark, the risk of mortality from collision or predation acts negatively. Collision is probably, by far, the greater risk (Owen & Cadbury 1975) and the encumbrance of the added weight of a full gut makes flights from the foraging grounds to the roost more dangerous than vice versa (see Chapter Ten).

There is no clear ecological explanation for the observed differences in activity patterns on land, on freshwater, and on tidal areas, which deserve further detailed study.

FEEDING AND VIGILANCE IN WINTER

Prior to 1980, feeding ecology and behaviour had been studied in detail only on the Finnish breeding grounds, where they feed aquatically during summer (Haapanen *et al.* 1977). What little was known of those topics in winter was drawn together by Scott & The Wildfowl Trust (1972) and Cramp & Simmons (1977), then expanded by Brazil (1981a,c).

The two behaviours occupying most time in winter are feeding and 'head-up'.

Although completely different, because they are completely interdependent, they are best discussed together. The rate of foraging in a large bird such as a Whooper Swan is easily measured in terms of its peck rate, though that is only possible when they are feeding on grass, which is most common in January–May (Brazil 1981a,c). Ideally it would be interesting to compare foraging rates of parents, cygnets and adults without young. However, because of difficulties in identifying parents when grazing, except where these are individually marked, relatively few peck-rate data has been available for them. Consequently, I will refer chiefly to just two classes, adults and cygnets. In order to eliminate any effect of time of day on feeding rate in the following analyses, dates were selected where similar quantities of data existed for both adults and cygnets from similar times of day. With the limited data available it was not possible to also eliminate the effects of flock size, although this was deemed negligble (Brazil 1981c).

Ignoring any effect of flock size, peck rate per minute (p/m) was found to increase in an essentially linear fashion throughout the day in January, reaching a maximum of over 65 p/m at the end of the day (Brazil 1981a). Owen (1972) had previously proven the same relationship for the White-fronted Goose (although its peck rate is twice as fast). Peck rates correlated with time of day in three consecutive winters, such that they tended to be higher during late afternoon than at any other time. The increase in peck rate during the day was thought to be related to the approach of a long period when foraging was not possible, by increasing peck rate it would be possible to overload the digestive system, thereby maximising the quantity of undigested food before flying to roost. The long elastic oesophagus acts as a temporary food store until the processing capacity of the gizzard catches up. If increasing the peck rate can be used as a short-term strategy, i.e. on a daily basis, then perhaps it might also be used as a long-term strategy, i.e. as a seasonal means of increasing food intake. Although peck-rate data from adults and cygnets in late winter/spring gave no indication of any systematic change with respect to season, time budget data did show that time spent feeding increased in spring. As a consequence, food intake could be increased even while feeding rate remained essentially the same.

Peck rates per minute are a reasonable method of comparing foraging rates between age classes, and can be presumed to be related to relative energetic needs. Unfortunately, in my own study, data were insufficient to elucidate details, but the pattern was clear: the same linear increase was observed with adults commencing in the morning at c.45 p/m increasing to 55 p/m, while cygnets began at c.55 p/m and increased to 65 p/m (Brazil 1981c). The subjective impression from the field was that cygnets fed faster than either adults or parents. When peck-rate data from several years were combined, it became clear that the modal peck-rate class differed among the various classes. In increasing order they were: parents 31–40 p/m, adults 41–50 p/m and cygnets 51–60 p/m. Only with more data from a wider range of winters might it be possible to tease out the details of the differences in feeding rates between different classes.

Behaviour associated with breeding commences even before spring migration, with increased feeding and fat deposition (see Dorst 1971; Owen 1980). Whereas most small migrants are able to replenish their energy supplies after migration, for many arctic breeders, including swans and geese, this is not usually the case. They must store sufficient fat to provide enough energy to last through display, nest building, sperm or egg production, and incubation, before they will have the opportunity of good feeding again. Many such species begin nesting almost as soon as they arrive on the summering grounds (Ogilvie 1978).

DROPPING RATE

Whoopers grazing on land are essentially goose-like in many aspects. Birds arriving at their feeding grounds from the roost are assumed to have an empty gut, as they do not feed at night. The first period of feeding steadily fills the gut and until this is achieved no droppings are produced. Once full, the gut works like a conveyor belt, with droppings excreted at regular intervals. Dropping rates have been measured for many grazing species, especially geese (Owen 1980) to enable a calculation of daily food requirements. This requires knowledge of dropping weight, length of feeding day and total number of droppings per day. Some data is also available for Whoopers, making it possible to draw some conclusions concerning their needs (Brazil 1981c).

A simple technique involves selecting at random a newly arrived bird from a flock and watching until it first defecates, then further measuring the interval from the first defecation until the next. Unfortunately, this method is less than satisfactory for Whoopers, which have a very long inter-dropping interval, making it very likely that they may be lost among the flock before defecation occurs. In fact this proved so difficult, because the time was so long, that I was only able to successfully observe five birds from their arrival until production of the first dropping (in January and November 1980). The mean time to first dropping was 82.6±23.9 minutes, a period significantly longer than the inter-dropping interval measured later in the day using either the time to dropping (ttd) or complete inter-dropping interval methods (Brazil 1981c), but only a third as long as the Trumpeter Swan, which takes 239 minutes from feeding to first defaecation (Squires 1991).

Instead, therefore, Owen's (1972) method used for measuring the interval between bouts of vigilance proved more suitable, in which individuals are randomly selected and the time until first vigilance (or in this case defecation) is recorded then doubled. Even using this method it was difficult to accumulate a large sample size, but both methods produced rather similar results despite the small samples. After the first dropping, the mean interval between droppings (complete measure) was 14.5±3.6 minutes (N=7), whereas the mean ttd was 7.48±5.3 minutes, giving a mean interval by the ttd method of 14.9 minutes (N=26). This inter-dropping interval of 14–15 minutes was much longer than that typically found for White-fronted and Barnacle Geese, which average only 3.5 minutes between droppings (Ebbinge *et al.* 1975; Owen 1980). This difference is, presumably, a function of the great difference in body size between the Whooper Swan and geese.

FOOD INTAKE

The little information available on the daily food intake of swans is from summer (Mathiasson 1973; Fridriksson *et al.* 1977). Mathiasson (1973) found that the amount of food eaten by moulting Mute Swans depended on the food species. For example, when feeding on eelgrass they consumed 0.077 kg (dry wt)/kg body wt/24 h and when feeding on *Ulva lactuca* they consumed 0.018 kg/kg body wt/24 h. The large difference between these is presumably due to the greater digestibility of the algae.

The only previous work specifically on Whoopers was that by Fridriksson *et al.* (1977), who estimated their daily requirements to be 400 g dry wt (2,000 g wet wt). Although this is an apparently reasonable figure (given consumption by geese of *c.*700 g wet wt; Owen 1972; Ebbinge *et al.* 1975), it cannot be relied upon as the experimenters failed to exclude sheep from the study area and therefore were uncertain as to exactly how much of the biomass was removed by swans. In North America, Trumpeter Swans were found to consume considerably more— 4.5–5.5 kg (wet wt) each day (McKelvey 1981).

Subsequently, data collected by Brazil (1981c) facilitated a calculation of the daily intake, for comparison with geese. It is assumed that the increase in feeding rate, which occurs during the day (see above), leads to a progressive overloading of the gizzard, so that on departure birds have an oesophagus full of undigested food (Owen 1972). It is also assumed that this does not lead to a corresponding increase in dropping rate. In a midwinter day of approximately 8.5 hours, 34 droppings would be produced by a Whooper Swan, compared with 150 by a White-fronted Goose (Owen 1971). When swans fly to roost they may also carry a full oesophagus, thereby effectively extending the feeding day, because they still have that food to digest at the roost. The only data available, however, was from one, which struck wires as it flew from its feeding grounds to its roost site; its oesophagus held 7.8 g dry wt. In the absence of specific information, it was assumed that the digestive efficiency of the Whooper Swan was the same as that of geese; at its lowest *c.*25% (Owen 1980), hence daily intake is equal to 1.33 times the total dropping weight, plus the weight of food in the oesophagus. The mean dry wt of droppings was found to be 4±1.2 g (N =103), and the total dry wt of the mean 34 droppings produced in a day was 137.4 g. The amount carried in the oesophagus was potentially 8 g dry wt, roughly 75% of which would be egested. Thus, the total dry wt egested per day was 143.4 g. The total daily intake would, therefore, be 191g dry wt or approximately 955 g wet wt. Considering the number of assumptions involved and the possibility of error in each of the measurements, the resulting figure for intake seems reasonable. It compares favourably with the 700 g obtained by Owen (1972) and Ebbinge *et al.* (1975) for White-fronted and Barnacle Geese and, as was expected for a heavier bird, it was larger. The digestibility of grass increases as growth increases in late winter/spring and the assimilation efficiency of geese can reach nearly 40% (Owen 1980). If this is also assumed to be the case for the Whooper, then maximum daily intake would be 238 g dry wt or approximately 1,190 g wet wt (Brazil 1981c).

VIGILANCE BEHAVIOUR

All other behaviours are interrupted at frequent intervals by bouts of 'head-up', during which Whoopers appear to be scanning their environment. Such attention to unspecified stimuli is generally called 'vigilance' (Dimond & Lazarus 1974), the specific functions of which are not always readily identifiable in the field, though the habit is assumed to be adaptive. The amount of time allocated to vigilance does, of course, limit the time available for other activities, such as feeding (Fig. 6.11). The motivation for such allocation varies between individuals but, more particularly, it varies between different ages or reproductive classes, that is: between adults unaccompanied by cygnets, parents accompanied by their cygnets, and cygnets themselves.

Cygnets are initially smaller-bodied than adults and so may be assumed to be gaining weight into their first winter. They may be expected, therefore, to spend as much time feeding as possible, perhaps at the expense of vigilant behaviour, particularly as their parents usually accompany them. Cygnets might also be unskilled in vigilance behaviour; their parents might, therefore, be more vigilant in order to protect their offspring, and to provide them with the extra time they require for feeding.

However, all three (adults, parents and cygnets) spend about the same time with their heads up at each bout. What differs significantly between them is not the length of each 'head-up', but the interval between bouts. As might be expected, parents most often adopted 'head-up', and both parents and adults used 'head-up' more often than cygnets. As a result, parents spent most time head-up, cygnets least

Figure 6.11.   Providing food at wintering sites affects behaviour, allowing more time for social interactions, vigilance behaviour and loafing.

and adults an intermediate amount of time, although the pattern was not uniform throughout the winter (Brazil 1981c). In November 1980, for example, when many families were observed, the mean periods between head-up bouts were: 22.4 ± 30.7s for parents, 49.6 ± 35.5s for adults and 71.3 ± 45.0s for cygnets. The differences between parents and cygnets, and parents and adults were highly significant. It is assumed that the greater level of vigilance on behalf of parents represents an important aspect of parental care. Cygnets spend less time 'head-up' than adults or their parents and, as a consequence, are able to spend far more time feeding. As winter progresses, cygnets grow in size and mature behaviourally. It is unsurprising, therefore, that by March there is no longer a difference between cygnet, adult or parental levels of vigilance, as by then cygnets have virtually reached adult size.

Kakizawa (1981) showed that in winter, Whooper parents exhibited more aggression than adults, but that this level of aggression declined as the food requirements of cygnets declined, presumably parents thus 'defend' their offspring's freedom to feed. The amount of time spent 'head-up' by adults was intermediate between that of cygnets and parents. It is unknown whether this is typical or whether adults gain an advantage from joining flocks with families, 'cashing in' on the presence of parents and their high levels of vigilance.

THE RELATIONSHIP BETWEEN FLOCK SIZE AND BEHAVIOUR

Flock sizes vary considerably, but in general they are small, mostly of fewer than 50. The size of a Whooper flock might be considered to affect the behaviour of flock members and to examine that I studied flocks ranging in size from one to 150 birds in Scotland, and examined the mean proportion of each flock engaged in each behaviour (Brazil 1981c).

The most obvious feature was that the number vigilant significantly correlated with flock size. The significantly (negative) correlation between the percentage of birds 'head-up' and flock size indicates that 'head-up' does decline with increasing flock size, despite its variability within smaller flocks. The actual relationship appears to be curvilinear. There is an initial steep decline in 'head-up' in flocks numbering 1–20, after that the incidence of 'head-up' levels out (Brazil 1981c). The same relationship has also been shown for White-fronted (Lazarus 1972, 1978), Barnacle (Drent & Swierstra 1977), Pink-footed (Lazarus & Inglis 1977) and Brent Geese (Inglis & Isaacson 1978), all of which are grazers, but not apparently for any other swans, though one might expect the same to apply. In geese, which occur in flocks of up to several thousand, an effect occurs over a range of flock sizes 1–100, whereas in swans, which typically occur in much smaller flocks, an effect is apparent over a range of flock sizes 1–20. The main advantage is probably gained by individuals or families that band together, rather than by larger flocks of adults that coalesce. Parents may gain extra time for feeding, by joining a small flock, as they need not spend as much time in vigilant behaviour as when in a single family unit (Brazil 1981c).

If Whoopers in larger flocks are able to reduce the amount of time they must spend vigilant, then they must have more time available for other behaviours, but which? Extra time spent feeding, it seems, comes as one obvious benefit of reduced vigilance. Both the number and the percentage of birds feeding were significantly positively correlated with flock size (see Figs 6.12 and 6.13). As Whoopers are capable of altering their feeding rate quite rapidly, it might be of interest to see whether birds in larger flocks feed at a different rate from those in smaller flocks. That is, do they gain in total food input from being in a larger flock, by spending less time vigilant and hence having more time available for feeding?

When looking at the peck rates of adult and immature Whoopers belonging to a range of flock sizes, it appears that flock size has little effect. Over a range of flock sizes of adults from 2–59, and of cygnets from 1–59, peck rate showed much variability, but no marked relationship, to flock size. This suggests that birds joining larger flocks might actually gain in total food input, as while their peck rate remains unaffected, the total amount of time they spend feeding increases.

Peck rate among Whoopers shows the same general tendency to increase throughout the day as it does in geese (Owen 1972; Brazil 1981a,c). It is also clear that the proportion of time spent feeding peaks in the morning and afternoon and shows a midday lull. If peck rate is roughly correlated with ingestion rate then in the morning they spend a high proportion of time feeding, but do so slowly. Bite size or selectivity might also vary, but no data are available. During the midday period slightly less time is actually spent feeding, but it is at a faster rate, thus intake probably does not differ greatly from that in the morning. In late afternoon, both the proportion of time spent feeding and feeding rate are high. Consequently, during the latter part of the day food intake probably rises rapidly.

A daily intake of 955 g (wet wt) was calculated for Whoopers in midwinter; this was expected to increase up to c.1,200 g per day in spring. Whether they maintain body weight during midwinter or endure a short-term loss as do geese (Owen 1980) is unknown. Peck rates of adult Whoopers were roughly constant in February–May, and of cygnets in February–April. During this period total time spent feeding each day increased. As grass growth increases in spring its digestibility also increases, hence more food is retained (Owen 1980). The increase in time spent feeding and in the digestibility of food in spring probably leads to a rapid increase in weight by Whooper Swans, facilitating fat deposition for spring migration and the approaching breeding season.

It is uncertain what precise function vigilance, by Whooper Swans, serves. It may be that different functions are served for different classes. The most pertinent functions appear to be: detection of danger or disturbance, either from predators, people or intraspecific competition; location of environmental resources, such as food or roost sites; and intraspecific communication, especially the location and movement of conspecifics, parents or offspring (Dimond & Lazarus 1974). Parents, adults and cygnets spent the same length of time 'head-up' at each bout, but parents were 'head-up' more often than adults and much more frequently than cygnets. Consequently, parents had least time available in which to feed. It was

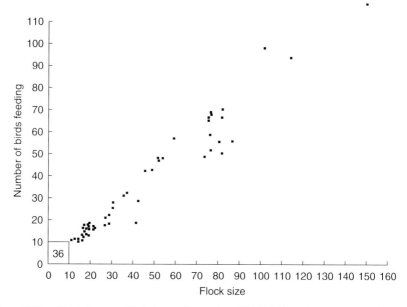

Figure 6.12.   The influence of flock size on the number of birds feeding. (n= 87, r= 0.984, p<0.01).

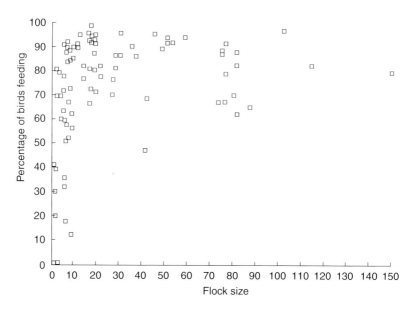

Figure 6.13.   The influence of flock size on the percentage of birds feeding. (n= 87, r= 0.307, p<0.01).

found that as flock size increased, the proportion of the flock that was vigilant declined and the proportion feeding increased. Parents may, therefore, without additional risk, be more able to reduce the time they spend vigilant by joining a flock than is possible if they remain as an isolated family unit. Since their time spent feeding increases with increasing flock size, and as peck rate remains the same, then their food intake should also increase as flock size increases.

The particular advantages of flock sizes of more than $c.30$ are not obvious. It should be noted, however, that in central Scotland after 1 January, i.e. the period when they were mainly feeding on grass, flocks containing fewer than 30 were particularly common (Brazil 1981c), and more general studies in Ireland and the UK have shown that the majority of Whooper flocks are small.

It may be that the advantages and disadvantages of flocking are not consistent throughout the winter. When birds are feeding on a potentially very patchy food supply, such as waste grain in autumn, there may be greater advantages from being in large flocks (Thompson *et al.* 1974; Ward & Zahavi 1974); whereas when grazing on grass, the main advantages may be in terms of detecting danger or intraspecific competition. This, coupled perhaps with the early movement of family groups north, may contribute to the higher proportion of small flocks.

Drent & Swierstra (1977) and Inglis & Isaacson (1978) showed that the behaviour of geese in a flock even affects the attractiveness of that flock to other birds. Using models, they showed that birds in the feeding posture were more attractive than vigilant birds. This implies that, rather than simply joining any flock in order to increase the size of the flock they are in, birds are capable of deciding whether to join a flock and this decision is dependent on the relative merits of that flock. Whooper Swans are very goose-like, and the same might also be true of them.

CHAPTER 7

Social behaviour

SOCIAL BEHAVIOUR IN WINTER

The Whooper Swan shares many similarities in its general behaviour with other 'northern swans'. They are all highly social, commonly occur in flocks, and are generally rather goose-like (the Whooper especially so). The long-lived Whooper has strong pair-bonds, and a strong family bond that lasts not only throughout their offspring's first winter but also through the subsequent spring migration until, presumably, immediately prior to their next nesting attempt, although family break-up is poorly documented. While the family might be presumed the basic social unit among Whoopers, the proportion of the population breeding is very low, and some fail to produce young, thus most birds lack families, and occur as pairs or in flocks, which are a loose, variable association of lone adults, mated adult pairs without offspring, families, and occasionally lost or temporarily unattached cygnets. There is no evidence to suggest any particular social bonds between Whoopers other than pair members, and between parents and their offspring of the year. Whoopers have a range of vocal signals that serve to keep social units, pairs, families and flocks together. They call loudly in flight, and this feature draws attention to them when they are on migration. Gatherings at staging areas also call loudly at night, and these calls are presumed to serve as a sound beacon enabling subsequent arrivals to find the roosting site. This has been taken advantage of at Onuma, northern Hokkaido, Japan, where calls broadcast from loudspeakers were successfully used to attract swans to use a hitherto unused lagoon.

Social behaviour may be classified into five broad categories Brazil (1981c):

1. Pre-flight behaviour
2. Aggressive behaviour
3. Territorial displays between males
4. Threat display at the nest (to people, dogs, sheep etc.)
5. Greeting or triumph ceremonies

To which may be added, general family behaviour.

Pre-flight and aggressive behaviour, and greeting or triumph ceremonies may be observed year-round, and are particularly noticeable in winter flocks, whether mainly aquatic or terrestrial, and are described here, while categories 3–4 may only be observed on the breeding grounds and are thus described in detail later.

In this long-lived species the strong pair-bond, and the strong family bond during the first winter, produce a number of interesting social behaviours. Among Bewick's Swans, it is not unusual for offspring from a previous year to re-find their parents in a second or even subsequent winter, even though the parents are now accompanied by a new family (Rees *et al.* 1997c), whereas among Whoopers this is generally thought not to occur. Mathiasson (1991) however, studying Whoopers repeatedly returning to the same winter quarters, where they spent their first winter, in Sweden, found that in some cases cygnets from previous years were still accepted by their parents at ages of 2–4 years (even if they were paired), although their parents simultaneously vigorously rejected other swans in their immediate surroundings.

Despite the strength of the bond between parents and cygnets, some parents do lose cygnets and occasionally parents depart on spring migration leaving their brood. Very rarely, it seems, 'lost' cygnets may become 'adopted', amalgamating with other families, something that I have observed among Bewick's in Japan (Brazil 2002a), and which Black & Rees (1984) and Rees *et al.* (1990a) have observed once each among Whoopers in Scotland and England. In the former case, a cygnet became separated from its original family (two parents and a brood of three) and joined a second family (a brood of three with a single female) in January 1984. Although the cygnet joined its new family in triumph ceremonies and aggressive activities it was not permitted normal close proximity to family members. In this unusual case, the cygnet's original family was still present yet the cygnet followed its 'adopted' family and eventually they left together (Black & Rees 1984), and is the only unambiguous record of adoption for this species. However, the possibility of brood amalgamation remains, and may explain some of the unusually large broods previously described (e.g. Brazil & Spray 1983).

For swans wintering in classically 'natural' habitats the majority of each short winter day is spent obtaining food. Not surprisingly, provisioning swans at winter gatherings or the ready availability of monoculture agricultural crops, affects their behaviour. As Black & Rees (1984) showed at Caerlaverock, Scotland, which must be true for other provisioned sites, a unique environment is created in which Whoopers are able to fulfil their daily and seasonal life-cycle requirements in less than half the time it takes swans that do not use such refuges. So, due to the provision of food at their wintering sites, Whoopers may be able to reach optimum breeding condition more swiftly or more easily, and also have more time for other behaviours.

Scott (1978) showed that, for Bewick's, much pair-bond activity and reinforcement, including triumph ceremonies, occurred while loafing. If the same is true of Whoopers, and if provisioning permits them to spend more time loafing, then birds at such sites as Caerlaverock, Martin Mere and Welney may have time to develop stronger bonds, thus facilitating enhanced breeding performance. Only long-term comparative studies will confirm this. As a result, the most significant effect of provisioning may be the increase in time available for loafing on water and thus engaging in social behaviour, compared with swans at non-provisioned sites.

DISPLAYS IN WINTER

PRE-FLIGHT SIGNALLING

The 'northern swans' all use conspicuous ritualised pre-flight signals that seem to indicate to other family/flock members their readiness to depart a feeding or roosting site. These involve various movements at varying intensities, but generally include a pronounced pumping of the raised head and neck while calling and, sometimes, a lateral shaking of the head (Fig. 7.1). This behaviour was described by Brazil (1981c) as typically consisting of three parts: 1) 'head pumping' (because of

Figure 7.1.   Pre-flight behaviour.

its often vigorous nature; among geese it is also called head tossing (Raveling 1969), during which the neck is stretched vertically and the head is rapidly raised and lowered by several cm in rather energetic manner; 2) increased calling, and 3) wing flapping, during which the whole wing is spread and flapped vigorously once or twice.

It has long been clear that flocks or units within them, whether pairs, families or subgroups are, to some extent, able to synchronise their take-off and departure. Brazil (1981c) noted that even when feeding appeared a priority, usually the last ten minutes or more that flocks were at feeding fields was spent in social, non-feeding behaviour. Black (1988), who studied this particular period intensively, revealed the mechanism that makes that synchronisation possible. Interested in the adaptive significance of the pre-flight ritual, he found that the rate of signalling increased significantly near take-off, providing the first quantitative evidence that a threshold of excitability is responsible for triggering synchronised flight.

The pre-flight sequence includes: head bobbing or head pumping, during which the head and neck are moved vertically, head shaking, when the head is moved laterally, and wing flapping (Johnsgard 1965; Raveling 1969; Scott 1978; Brazil 1981c). Black (1988) confirmed that head bobbing was by far the commonest movement, contributing up to 91% of all movements, and occurring in 100% of all pre-flight sequences. Head shakes occurred in 81% of all pre-flight sequences, but made up only a further 8% of movements. Head shakes were, however, far less common among field-feeding swans in central Scotland (Brazil 1981c) indicating some variability in pre-flight behaviour. Wing flapping was the least common such behaviour, contributing just 0.6% of movements and occurring in only 23% of pre-flight sequences (Black 1988).

Head bobbing, the commonest pre-flight signal, involves both lowering and raising the head. It may be performed at varying intensities, some involving just shallow downward dips of the neck and others involving deep dips. The head raise at the opposite end of the movement also varies and may either be moderate or extreme. As in aggressive encounters and in bow-spritting behaviour (see Chapter Eight), the head-bobbing sequence ranges from gentle to an almost violent motion. A single upward nodding to the head's highest horizontal position signals the beginning of the pre-flight sequence, which can be deemed to have ended when the swan hunches forwards and opens its wings in readiness for take-off.

It is the distinct four-syllabic call that accompanies each raise of the head that distinguishes the pre-flight head bob from other head bobs that might also serve as simpler contact signals. The frequency of movements and volume of calls both increase towards the end of the sequence. Then, as soon as the calling birds take off, their calls change to a higher pitch.

Take-off, it has been assumed, results when pairs or parents perform ritualised pre-flight movements in rapid succession, during which varying degrees of leadership may be exercised (Raveling 1969; Black 1988). Interestingly, Black (1988) found that flight is not triggered either by a specific number of signals or after a fixed period of time. The rate of signalling becomes significantly faster, however, in the final third of the pre-flight period, suggesting that take-off is triggered once a certain threshold of excitability is reached, apparently dependent more on the rate of escalation of the movements than their number.

Not all pre-flight sequences result in flight. Some degenerate, as if insufficient momentum has been gained, and the intention to fly may be subsumed by feeding, preening or other activities. Those sequences that do lead to flight may involve just a few signals followed by rapid take-off or several hundred signals that may last tens of minutes. This may vary depending on the circumstances of the flocks in question. For example, in central Scotland, where failing light at the end of the day was a clear stimulus for departure to roost, a short period of pre-flight signalling was often quite rapidly followed by flight, presumably because delayed departure was physically dangerous. One individual was observed to die when it hit overhead power lines, directly as a result of delaying its departure (Brazil 1981c).

Black (1988) was also able to demonstrate that the extent of pre-flight signalling depended on the status of the birds. Thus lone birds differed from paired individuals, performing fewer pre-flight signals and leaving more quickly. It transpired that 48% of first leavers were singles.

Pre-flight signalling is both a time and energy consuming behaviour that draws attention to the birds performing the behaviour, and which is more heavily invested in by birds with mates than by singles. What then is its function, and what are the advantages accrued to its performance? Pre-flight signalling might function to facilitate the synchronisation of take-off and thus reduce the risks of flock, family or pair separation, but which is more important?

The advantages of being in a flock are widely described. Similarly, the advantages of a family or a pair remaining together can be considerable. Owen *et al.* (1986) showed that among geese and swans, re-pairing results in a substantial reduction in lifetime reproductive success, and that it is to a pair's reproductive advantage for them to remain together in long-term association. Furthermore, large waterfowl in general experience numerous energetic and social benefits in maintaining family associations (Scott 1980a; Black & Owen 1999a,b). It is probable, therefore, that a crucial function of pre-flight signalling is to prevent mates and family members becoming separated. In this respect it seems that mates could be more important than families, as Black (1988) found no significant differences in the duration or number of pre-flight signals between pairs and families, suggesting that synchronisation was between mates rather than parents and offspring.

Flight departure is dependent not only on the activities of the signaller, but also the attentiveness of the other parties and when a parent initiates a pre-flight sequence its take-off is determined more by its mate's attentiveness than by that of its offspring. Interestingly, cygnets are willing to invest in pre-flight signalling, yet rarely initiate flight. Instead they usually continue signalling until their parents join in (Black 1988).

Under artificial feeding regimes, a dominance rank order exists such that families rank highest, then pairs and then singles. Dominant birds feed for longer periods. Less dominant individuals may be more at risk because of their social position and energetic state, and might benefit more from joining and remaining within a flock. There, an individual experiences a lower risk of predation disturbance, while at the same time its efficiency at finding food may be enhanced by needing to spend less time vigilant. It follows that flock members are likely to benefit if they watch for pre-flight movements by other birds and decide whether or not to follow. Conversely, signallers may use visual signals to manipulate other flock members, in addition to manipulating the behaviour of their mate (Scott 1980b; Black & Rees 1984; Krebs & Dawkins 1984; Black 1988).

Considerable advantage exists to mates remaining in close visual contact, particularly close to departure, as individuals that began the pre-flight sequence without their mate's participation gave more than twice as many signals as those which signalled with their mate's cooperation. Of even greater significance is that when both mates are attentive from the outset, and commence pre-flight sequence together, they are able to leave more than four times faster than other birds (Black 1988).

Whoopers are monogamous and maintain continuous proximity to their mates. Each partner benefits from mate protection, increased feeding success and offspring care. They maintain proximity by calling and signalling, and the importance of this is indicated by the fact that the initiator of a pre-flight sequence invests substantially more in signals and waits until their mate (and family) join them before flying. Other monogamous birds, which do not maintain continuous proximity to their mates, have not evolved such sophisticated signals to advertise their pre-flight intentions (Black 1988).

Significantly, pre-flight signalling also depends on situational clues—both current location and the site to which birds intend to travel—as well as on the status of the individuals concerned, and these may strongly affect the length of time engaged in signalling and the manner in which they depart. Whereas I had studied field-feeding birds, and departures from terrestrial feeding sites to aquatic roosts, as light levels fell towards evening (Brazil 1981a,c), Black (1988) studied birds frequenting pools at a provisioned refuge, where departure was typically of roosting or resting birds leaving to feed, thus the motivation for departure was quite different. Whereas Black (1988) found that pre-flight signalling was often lengthy, averaging three minutes longer when the birds were heading for non-refuge sites than when departing to feed at the refuge, I found that pre-flight signalling typically occupied a quite brief and intense period at the end of a long day of feeding, and frequently resulted in entire flocks departing together.

Black (1988) observed a tendency for families to fly to feeding areas alone. They achieved this by performing pre-flight signals away from other flock members, and by threatening would-be followers, so as not to attract others. They are presumed to adopt these countermeasures because, as a family, they have less to gain in terms of decreased vigilance and increased feeding inherent in a flock situation. Some support for this comes from the evidence of Hewson (1964), Brazil (1981c), and Salmon & Black (1986), all of who found a tendency for families to occur in small flocks, often as discrete units, whereas singles and pairs tend to occur in larger flocks.

Status, arrival time at wintering sites, feeding rate and daily departure times are closely interrelated. Birds arriving later at a wintering site are less likely to achieve dominance, which has serious implications for their opportunity to feed, as they suffer more displacements, are less successful feeders and tend to be unpaired individuals with less experience of the site. They are also, usually, the first to fly from a feeding situation. Lone birds are displaced significantly more frequently than pairs or family adults, and as a result are less able to feed, because the time birds can feed is inversely proportional to the number of displacements they suffer and number of threats they perform. As each threat results in a displacement, birds performing most threats and which are least frequently displaced are able to monopolise a given area—particularly areas of high food availability. Conversely, subordinate individuals receive most threats, are displaced more often, and as a result are pushed to areas of lower food availability (Black 1988).

Whereas families more often occur in small flocks, individuals and pairs without young, more frequently occur in larger flocks, and are presumed to join such flocks because of the benefits gained while group foraging (see Brazil 1981c). Strategies such as pre-flight signalling, extended signalling and 'neighbour pulling' are important in successfully attracting others to join, and serve to establish or maintain such flocks.

For birds leaving an aquatic site, such as that studied by Black (1988), a 'clear' runway is essential. When birds already engaged in pre-flight signalling swim among groups of loafing neighbours, they are clearly postponing their departure as flight from such a position is not usually possible. That such birds have been observed to continue signalling suggests that they are attempting to induce others to follow, a process that Black (1988) defined as 'neighbour pulling' and demonstrated to be successful in attracting followers in 61% of attempts, compared to only 35% when the behaviour was not employed. Pre-flight signalling does, it seem, significantly affect the behaviour of flock members in such situations. Among field-feeding Whoopers, take-off is easier as 'runway space' is usually available as soon as birds turn away from the flock (assuming there is not a strong wind), thus it should be easy for small parties to leave independently, nevertheless there is a strong tendency for all members to depart together.

In conclusion, pre-flight signalling between pair members increases simultaneously, having reached a certain threshold of excitability, in the final 30 seconds before take-off. Four factors appear to influence this threshold in pre-flight behaviour. 1) Mate attentiveness and proximity. Birds with non-attentive mates were delayed and their signalling averaged four times longer. 2) Maintenance of flock cohesiveness. Birds that signalled for longer, while swimming among uninterested birds, were able to manipulate flock members successfully, attracting followers in 61% of instances. 3) Energetic and dominance status. Subordinate and hence less successful feeders, and potentially hungry birds, flew quickly after few signals. 4) The eventual destination and type of feeding opportunity at the eventual destination, of the signaller. Birds that flew to nutritious barley provisions spent less time engaged in pre-flight signals than when flying to forage on grass (Black 1988). Behaviour observed under semi-artificial conditions at wildfowl refuges differs from that in non-provisioned situations, perhaps mainly because, in the former situation, birds have more time to engage in social behaviour.

AGGRESSIVE BEHAVIOUR

In Europe, the aggressive Whooper Swan has been reported as dominating both Bewick's and Mute (Black 1988), and in North America feral Whoopers have proven not only extremely aggressive towards Mutes and capable of displacing them territorially, but in one instance, on 7 April 1995 at Clark Pond, Massachusetts, even attacked and killed a Mute (Berry 1997), although conversely during the 1980s at the Utonai reserve, in southwest Hokkaido, concern was expressed locally that the increasing and spreading Mute Swan population may actually dominate the Whooper. There may indeed be behavioural differences between the populations, or possibly seasonal differences, that deserve further study. Aggressive behaviour is usually reserved for conspecifics, for other swan species, and occasionally other waterfowl. Aggression usually occurs in the form of simple displacement, particularly of smaller species, although some ducks have been killed by Whoopers. Aggression may be directed towards other animals and people, though on the wintering grounds the response of most birds is simply to move away and avoid such intrusion.

The usual components of aggressive or threat displays are as follows and the whole may appear as an escalating sequence as the individual becomes more aggressive:

1. **Ground Staring** (or head submerging, if they are on water), during which the neck is strongly and rigidly arched, the neck feathers ruffled, and the bill pointed at the ground and held a few cm above it if on land or the head is submerged if on water. The wings may be closed or slightly spread and the bird may hiss (see Johnsgard 1965; Kear 1972; Cramp & Simmons 1977; Fig. 7.2).

Figure 7.2.   Ground staring.

2. **Bow-spritting** and head throwing, during which the neck is held stiffly forward at *c*.45° from the horizontal, like a ship's bowsprit. This position is often held only briefly; the neck relaxes, then both the head and neck are jerked or thrown forwards again, typically accompanied by calling (see Fig. 7.3).
3. **Carpal Flapping**, during which the carpals are held away from the body without spreading the wings, and just the 'hand' is flapped or quivered vigorously (this differs markedly from the full wing flap, which is often observed after preening. Carpal flapping is often accompanied by bow-spritting, with the neck being alternately extended and bent repeatedly, often while calling (Johnsgard 1965; Kear 1972; Cooper 1979; see Fig. 7.3).

Carpal flapping and bow-spritting displays are also used as a threat, a greeting, and as a triumph ceremony (Fig. 7.4), and bow-spritting sometimes occurs while in flight (particularly when gliding to land (Fig. 7.5)), when the waving movements

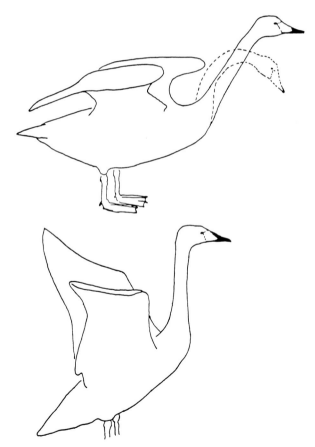

Figure 7.3.   Sketches of 'bowspritting' and 'carpal flapping' (upper), and of 'full wing flap' (lower) (Mark Brazil).

Figure 7.4.   Pair engaged in threatening behaviour, bowspritting, calling and wing flapping.

Figure 7.5.   Greeting ceremonies may commence even before landing.

of the wings differ noticeably from those during normal flight, and the birds gradually lose altitude. When an aggressive display escalates to attack, the wings are usually fully spread (Johnsgard 1965; see Figs 7.3, 7.6). Other forms of aggressive behaviour have also been described, mostly from situations where birds occur in dense concentrations.

WATER BOILING

Where birds are drawn by the availability of abundant food they occur in unnaturally dense concentrations. At such sites, behaviours may be observed not seen at other localities or at other seasons. Kemp & Revett (1992) described the

Figure 7.6.   Fully spread wings during a face-off may precede physical attack.

'water-boiling display', which is regularly used by both Whooper and Bewick's, where provisioning with waste potatoes has led to swans feeding at very high densities. They noticed this behaviour at Welney, Norfolk, from the mid 1980s onwards, and found it to be most impressive and loudest in Whoopers.

'Water boiling' typically though not always, follows an aggressive encounter with a conspecific. The behavioural sequence involves creating a pool of 'boiling water' as a result of the swan raising itself clear of the water while paddling vigorously, and producing a loud churning noise.

'Water boiling' is very variable in intensity, ranging from light paddling which hardly lifts the swan at all, to an impressive and violent two- or three-second burst that lifts it well clear of the water, and disturbs the water so violently that it may take up to ten seconds to subside. The wings are not usually opened during 'water boiling', but the latter sometimes occurs in mild form as part of the more usual aggressive wing-flapping encounters (Kemp & Revett 1992). 'Water boiling' appears to be an extension of aggressive behaviour under particular circumstances on water and is perhaps derived from the feeding-related foot-paddling behaviour. Its exact significance is unclear.

AGGRESSION AND COPROPHAGY

At Caerlaverock, where Whoopers and other waterfowl are attracted in large numbers to the WWT reserve by the regular provision of food, particularly barley, unusually high levels of aggression have been reported, more so than is normal when birds feed on grass (Black 1988). Aggression appears to increase in late February–March as a result of the unusual development of seasonal coprophagy, first documented in wildfowl by Black & Rees (1984). They observed Whoopers engaged in aggressive confrontations over faeces, both their own and those of others. Coprophagy at Caerlaverock increases and becomes quite common by late March. The cause seems to lie in the fact that Whoopers do not fully digest barley grain when it is eaten alone in large quantities. At such times they produce rather liquid faeces containing many undigested grains. During late winter and early spring, when they must lay down fat reserves, such faeces actually represent valuable and easily accessed food. Conflict and aggression occur upon defecation. Defecating birds tend to turn their heads and necks to examine their own droppings. When neighbouring birds approach, also showing interest in the droppings, conflict ensues (Black & Rees 1984).

GREETING OR TRIUMPH CEREMONY

It is common, at gatherings of Whoopers, to witness bouts of vociferous displaying triggered by new arrivals, who engage in greeting ceremonies with those already present, and may trigger pairs to continue displaying to each other, which can continue for many minutes (Fig. 7.7). I have observed this frequently in Japan,

Figure 7.7.   Triumph display.

particularly at Kussharo-ko in Akan National Park, where birds roost on the snow-covered lake ice and fly in the morning to a lake-edge feeding area. Each successive group arriving seems to trigger a cacophony of calls and greeting ceremonies. As Höhn (1948) noted, greeting ceremonies can easily slide into sexual or pair-bonding displays in which couples stand 'breast to breast with wings still held away from the body and continuously waving the carpal portion.' In such displays the bill and head are typically steeply raised and the neck alternately stretched upwards and bobbed low, with the couple alternating movements. Occasionally, if a couple is too close to another this may trigger outright aggression—usually a wing or a tail is grabbed, and pursuit may continue for tens of metres across the ice (Brazil pers. obs.).

Greeting or triumph ceremonies (see Fig. 7.7) are commonly performed when mates rejoin after an absence, after disturbance or after other displays (see Kear 1972). They usually comprise: 1) 'head-up' and/or bow-spritting; 2) calling; which may be synchronised or antiphonal (Hall-Craggs 1974), and 3) wing flapping, and often accompany or follow the rushing together of the pair. Families, including cygnets, also perform triumph ceremonies, e.g. following an interaction between families at a feeding site (Black & Rees 1984).

Kakizawa (1981), who studied Whoopers at Hyoko, an artificially provisioned site in Japan, where they occur at considerable densities, observed social hierarchies in which: larger families were dominant over smaller ones; earlier arrivals were dominant among families of equal size; and families with particularly aggressive mates were dominant. Exceptions and reversals were also noted, however, illustrating how complex, and as yet poorly studied, this area of behaviour is.

TERRITORY AND TERRITORIAL BEHAVIOUR IN SUMMER

Long-lived, monogamous species such as arctic-nesting swans and geese exhibit a high degree of behavioural cooperation; both individuals of a pair gain by investing time and effort in their offspring, not just during incubation and fledging, but also throughout the year.

All white swans are strongly territorial, defending a nest and an area around it suitable for the young to feed in (Kear 1972), although how best to define territory and assess its extent in Whoopers is difficult. Haapanen *et al.* (1977) showed that some pairs in Finland remain in the nesting territory only until the eggs hatch, before moving to feeding sites some distance from the natal area. In these cases, should these areas be considered part of a pair's very much larger breeding territory? Many other pairs remain in their nesting territory until their young fledge. During the Finnish study, the density was so low that Haapanen *et al.* (1977) recorded only one intrusion into a territory by an 'outsider'; no interactions were observed which involved other breeding pairs. During the time that I studied breeding behaviour in detail in Iceland, I was able to add some new facets to our knowledge of territoriality (Brazil 1981c).

For most passerines, and even for many non-passerines, territories are contiguous occupied/defended areas, the available habitat being relatively uniform. Territorial behaviour is thus regular, typically performed on a daily basis, during the breeding season between pairs or males from adjacent territories. With a bird such as the Whooper, dependent on particular nesting areas that may not be available in some regions other than at low densities, territories are unlikely to be contiguous, and pairs, once they have selected and settled on their own territories, may never interact with their nearest neighbours. Instead, they may interact only with intruding non-breeders, which occasionally overfly or attempt to visit already occupied habitat. In some areas, pairs do occur at densities that mean adjacent pairs are in visual contact and this leads to differences in behaviour.

To distinguish between the circumstances leading to the different behaviours, it is of value to identify inter-pair territoriality or displays, as those interactions occurring only between breeding pairs, as different from 'interactions with intruders'. The latter typically involves only one of the interacting birds having a territory. Territorial birds often overfly the boundaries of their territory (as defined during interactions with adjacent breeding pairs) while chasing intruders. Aerial chases have not been observed between members of adjacent pairs (Brazil 1981c).

Prior to 1981, there were no published records of inter-pair territoriality. This was no doubt owing to the understudied status of the species and its low density over the vast majority of its breeding range. The factors limiting the size of the breeding population are unknown, although the proportion of known age birds surviving until breeding age is only *c*.12% (Brazil & Petersen in press). Large areas of their range are apparently suitable, but unoccupied. Although Scott *et al.* (1953) recorded only 20 pairs within 11,400 ha in Iceland, Kear (1972) suggested that the true territory size in both Whooper and Bewick's has been exaggerated. Especially in upland areas of Iceland, large areas that do not hold swans are probably unsuitable because of poor food availability. It seems that Whoopers in Iceland defend small areas and territory size is likely to be even smaller than the 570 ha per pair calculated by Kear (1972). More recently, Einarsson (1996) found territory size in Iceland to be variable, depending on habitat, but up to 0.7 km^2 on large lakes.

Before 1981, there were no records of more than one pair of Whoopers nesting at the same pond or lake. Consequently, observations of three pairs nesting simultaneously at Arnarvatn, a lake just southwest of Lake Myvatn, were of particular interest. Direct inter-pair territorial behaviour was observed for the first time, and frequent observations of conflict between pairs and intruders. Furthermore, a previously undescribed sequence of behaviour was recorded. As in Haapanen *et al.*'s (1977) observations, pair/intruder conflict almost invariably resulted in the intruder being expelled from the nesting area. In this instance, the Trumpeter appears to be very similar to the Whooper Swan, in that Hansen *et al.* (1971) reported only four known cases of more than one pair of Trumpeters nesting at the same waterbody. All four involved large lakes where nests were well separated and, perhaps for that reason, no intra-specific defence was observed. Hansen *et al.* (1971) also noted that aggressive behaviour towards other swans was lower after hatching and during wing moult. This could be due either to seasonal changes in numbers of other swans or to a real decline in aggression.

During summer studies in Iceland in 1978–79, I observed 96 display sequences: 27 territorial displays between breeding pairs, 27 greeting or triumph ceremonies and 42 directed at intruders. More displays at intruders and fewer territorial displays were recorded during the incubation period than during fledging. Unfortunately, the difference between parts of the breeding season was confounded by the difference in years during which observations were made. The incubation period was studied in 1979; the fledging period in 1978. The difference in the numbers of different displays between years, however, was significant suggesting that so too was the difference between the incubation and fledging periods (Brazil 1981c).

Summer 1979 was cold and followed an exceptionally late spring. A high proportion of Whoopers breeding in the highlands were thought to have failed and cygnet production was exceptionally low (Brazil & Kirk 1979; Brazil 1981c). The intruders visiting Arnarvatn were probably unsuccessful breeders moving between their own territories and the Lake Myvatn moulting flocks. They were often noticed only because of the violent response to them made by breeding pairs.

During incubation, females rarely left the nest and eggs, and males usually remained in the immediate vicinity; during this phase there was little opportunity for inter-pair aggression. Unsuccessful breeders en route to moulting flocks, however, disturbed breeders whenever they attempted to overfly or land in their territories. Very soon after hatching, the strong attachment to the nest is lost and throughout the fledging period families move about their territories to feed. During fledging therefore there is a greater opportunity for inter-pair aggression where pairs have bred in close proximity. During the same period, most other non-breeding adults are in moult, consequently fewer were capable of flight and thus unlikely to pass over occupied territories. It seems more reasonable, therefore, to suppose that the differences in display frequencies were due to differences in intruder frequency and parental behaviour at different stages of the breeding cycle, than differences between years.

DISPLAYS USED IN TERRITORIAL AND NEST DEFENCE

Interesting general and specific differences exist between the displays used by the two major groups of swans, the 'northern swans' and Mute, Black and Black-necked. For example, although both use wing flapping as a highly visible expression of aggression, the latter also tends to raise their secondary feathers, while the former ground stare, bow-sprit, head throw, and carpal flap.

The general nature of the sequences in the displays of the 'northern swans' was well documented by Banko (1960), Johnsgard (1965), Hansen *et al.* (1971) and Haapanen *et al.* (1977). The details of those in the Whooper Swan's displays, however, were not well described, and previous references to intra-specific aggression invariably applied only to interactions between breeders and non-breeding intruders, not between adjacent territory holding breeders (see Cramp & Simmons 1977).

Neither Banko (1960), Hansen *et al.* (1971), Haapanen *et al.* (1977) or Cramp and Simmons (1977) recorded intraspecific territory defence by Trumpeter or Whooper Swans, which seems to have been first described, along with a unique display behaviour, by Brazil (1981c). The usual components, in sequence, of such displays are six-fold. 1) 'head up', i.e. a general vigilant or alarmed posture. 2) Approach, when one bird begins to swim towards a second. 3) Run/patter/fly, during which one bird begins normal take-off sequence by running along the surface beating its wings with most of the weight appearing to be borne on the feet (see Fig. 7.8). This may develop into pattering; almost airborne; the weight is taken by the wings, the feet paddling loosely along the water surface. The bird may actually become airborne in extreme cases. These three are sequential and depend on the distance travelled. 'Flying', in these circumstances, lasts only a few wingbeats. 4) Turn; on ceasing the run/patter/fly sequence, the bird turns quickly in a very tight circle to face the way it has just come. 5) Head dip or head plunge; this may occur as the bird turns or immediately afterwards. Head dipping appears to be the same as ground staring, during which just the head is submerged. Head plunging is more vigorous, and both the head and neck are submerged; followed by 6) either swimming away or the 'butterfly' display. Following the turn, and the head-dip or head plunge components, the displaying bird either swims away normally or it may use what is most simply described as a butterfly stroke. Both wings are spread on the surface of the water, raised and thrust forward simultaneously, and thereafter pressed into the water and pulled backwards, the whole sequence being repeated in rapid succession, like the flailing motion of the swimmer's butterfly stroke (Fig.

Figure 7.8.   Run/Patter/Fly aggressive pursuit.

7.9). One instance, which I interpreted as a case of extreme aggression between adjacent territory holders, was observed when, instead of the normal alternation between antagonists run-pattering towards each other, the aggressor flew directly up to the second bird and grasped hold of it with its bill. A chase ensued along the water while the aggressor maintained its grip. It finally released its hold and turned away using the butterfly stroke, whereupon the second bird began calling and bow-spritting.

Such interactions were regarded as having begun when the aggressor, previously in the 'head-up' posture, suddenly began to swim or fly towards a member of the adjacent territorial pair. Initial flights were usually terminated some distance from the second bird, then the aggressor either took a separate short flight or ran/pattered noisily along the water towards the other bird. Then stopping or landing, the aggressor usually turned on the spot, to face away from the second bird which usually reacted by flying or run/pattering towards its antagonist. From here the description applies to both individuals, with the second usually repeating the aggressor's behaviour, but one step behind. The aggressor, having turned its back on its opponent, usually head dipped or head plunged, the wings frequently spread flat on the water at this stage (Fig. 7.9). The aggressor then usually swam away, either raising its wings, with its head and neck low and retracted along the body (Fig. 7.10) or adopting the 'butterfly' display (Fig. 7.9). Swimming away was either followed by a repetition of the entire intra-territorial display sequence, beginning with run/patter/fly or by a greeting or triumph ceremony, or by feeding.

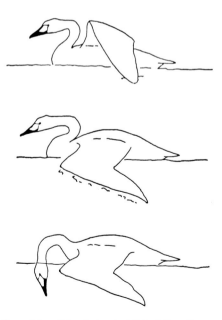

Figure 7.9.   The 'Butterfly stroke' and 'head plunging' (Mark Brazil).

Figure 7.10. Raised wings with head back (Mark Brazil).

THREAT DISPLAYS AT THE NEST

Threat displays at the nest may be elicited by people, dogs, and sheep (and presumably by any other potential animal predators). Both sexes may be involved, either separately or together, depending on when the nest is approached. The display usually comprises a sequence or range of several behaviours including: 1) 'head-up' or 'head-low' by the female on the nest, 2) ground staring, 3) wings spread, 4) wingbeating. The display continues until the intruder is driven away (Plate 9), ceases to be a threat or forces the swans from the nest.

VOICE

The Whooper Swan is very vocal, with a haunting, far-carrying call evocative of wary pairs on their wild northern breeding grounds. Its tumultuous clamouring elicits images of close-packed gatherings at staging areas or crowded wintering sites. Its nocturnal contact calls are typically softer and less intensive than those during the day, but when disturbed it gives persistent alarm calls, day or night.

The upward loop of the windpipe inside the sternum of the Whooper Swan, though shorter than in Trumpeter, does nevertheless lend the Whooper a deep bugle-like quality to its voice. Though not renowned for the scope and variety of its vocalisations, Whoopers do use voice in a range of situations. A rhythmic, musical clarion of loud bugle-like double notes accompanies their wingbeats and makes it difficult to miss the dramatic passage of flocks. This typical flight call has been variously rendered as *ahng-hä, gang-go, whoop-a* or *whoop-eh*, on a rising cadence (Witherby *et al.* 1940; Dement'ev & Gladkov 1967; Wilmore 1974), personally I find it sounds more like *who-oop*. Thus, the departure of Whoopers, whether on long migrations or merely from their daytime foraging grounds to their roost is an inspiring sight, and one that may well have contributed to the prevalent link in the mythology surrounding swans, between melancholy and change or departure (Chapter Two). The harmonious, monosyllabic calls given while swimming have been described as a loudly trumpeted *gang* (Dement'ev & Gladkov 1967). When excited, however, they produce a wide range of calls including hissing.

Other calls described by Witherby *et al.* (1940) included short, high-pitched conversational notes (*kloong*) given either singly or in quick sequences of 2–3, rarely 3–4 calls while in flight or on the water; also a longer connected series of softer, feebler notes with a rolling, twittering/trilling effect (*kilkliih*), given when excited, alarmed or angry, and sequences of lower pitched notes to express impatience, with harder variants of the typical call in alarm. Young Whoopers have a gruffer, lower pitched voice and may also utter soft contact calls, and higher pitched squeaky calls if distressed (Rees *et al.* 1997a).

Whooper dialect is certainly difficult to transcribe, but Brendan Lehane (in Wilmore 1974) made a passable, almost poetic attempt, describing their various calls as: *ahng-ha oglank oklang ONK ONK, Loo-ang A Klooang ok ok ok Whoop-aaa.* To my ear, the commonest calls are the oft-repeated *whoo-whoo* or *hu-hu* notes. Pairs will duet, but more commonly heard are the haunting sounds of the Whooper, weaker and softer conversational contact notes uttered regularly on water, softer still through the night. In display, when pairs or families engage in fracas, they become more animated, combining bugling and whooping with head pumping and bow-spritting, the powerful sounds swelling to a climax as emotion or alarm spreads through a flock; at such times they produce an escalating series of loud *kloo-kloo-kloo* notes. Perhaps the vigorous head pumping, which resembles the astonishing exertions of a Japanese fish auctioneer selling tuna, serves not only to force air more powerfully and thus more loudly through the long windpipe, but also to transform its distinctive call into a visual signal. Unlike Trumpeters, Whooper Swans often terminate a bout of calling by rigidly outstretching their necks in a bowsprit-like pose (Johnsgard 1965; see Fig. 7.3). During aggressive or excited calling, the neck feathers are typically erected making them appear thicker necked, while when alarmed these feathers are depressed and the sleekness of the long neck and junction of the larger head are more conspicuous.

The Whooper's calls are so closely linked with its displays that it makes sense to define its calls in conjunction with specific situations, thus calling may be heard under several circumstances, including: a) pre-flight signalling; b) flight, presumably as a contact call; c) as a contact call among feeding flocks during the day; d) as a contact call among roosting birds; e) aggressive behaviour; f) triumph and greeting ceremonies; g) territorial displays; h) driving away intruders from the nesting territory; and i) nest defence.

Male and female voices are essentially very similar, though debatably at different pitches. I find their calls powerfully attractive, reminding me of many exciting encounters at ice-clad caldera lakes and remote coastal bays and marshes, but the pragmatic Russians Dement'ev & Gladkov (1967) cut through such romantic attachment and described them simply as a 'loud grunting honk'. However they are interpreted, the calling of one member of a pair is usually sufficient to stimulate a duet (Thorpe 1972). At first this may lead to them calling alternately, but after a while they may synchronise. There is some confusion as to which member gives the higher calls. According to Johnsgard (1965), the pen gives the slightly higher pitched call, but Armstrong (1947, 1963) suggested that the pen calls a semitone lower than the cob, whereas Scott (1950) placed cob calls five tones higher than the pen's. To my knowledge, the issue has still not been resolved. Nevertheless, not only does it duet, but the Whooper is capable of maintaining a precisely timed antiphonal duet. A sound recording from southern Iceland revealed that a pair engaged in controlled antiphonal calling, with gradual and matched increases in the durations of the constituent calls. Apart from the regular alternation in calling by birds A and B, the most interesting aspect of the series is that the calls gradually increased in duration from the beginning of the series (0.1 and 0.09 seconds) to the end (0.48 and 0.49 seconds; Hall-Craggs 1974), indicating a very high degree of precise control.

According to H. W. Robinson, cited by Witherby *et al.* (1940), it is the Whooper Swan that provides the basis for the legendary song of the dying swan that I referred to in Chapter Two. Robinson allegedly heard that rarest of all the Whooper's calls, its very last 'the final expiration of air from its long convoluted wind-pipe producing a wailing, flute-like sound given out quite slowly.'

CHAPTER 8

Breeding biology and behaviour

INTRODUCTION

The Whooper's breeding season is an all-or-nothing, one-hit affair. Just one attempt is possible each year because of two factors, their long reproductive cycle and the shortness of the northern summer. Habitat quality is very variable across the range, thus they must establish a suitable territory in reasonable habitat providing sufficient food, sufficiently early in the season to breed successfully, which is still only possible if other factors, such as the weather, do not interfere. Here I will provide an overview of the breeding season timetable and the behaviour of Whooper Swans at each stage, relying on a combination of my own studies, and those undertaken more recently in Iceland, Finland, Russia and China.

The Whooper Swan, like high-latitude nesting geese, is long-lived, monogamous and experiences delayed maturity. The sex ratio is virtually equal. On the basis of cloacal examination of 129 adults in Iceland, 49.6% were female and 50.4% male; a slightly higher proportion of females (56.4%) among 55 cygnets probably just reflected the greater difficulty in sexing young (Black & Rees 1984). More recently, a sample of over 4,000 from Britain and Iceland proved to be 47% male (Rees *et al.* 2002). As with other swans, juvenile plumage is retained for one year or longer. The factors limiting the size of the breeding population are still unclear, although probably fewer than 20% survive until breeding age (Brazil & Petersen in press). Of these, not all attempt to breed, and among those doing so, a proportion fail, thus only a small percentage breeds in any year, *c.*20% in Finland for example (Haapanen *et al.* 1973b; Haapanen 1991) and reproductive output can vary enormously due mainly to weather conditions (Brazil & Kirk 1979). A number of studies from various regions, including that of 184 Whoopers in a 500-km^2 area of the Magadan State Reserve, have found that only *c.*30% of the population consists of breeding pairs. This may be best seen as a maximum; in the Anadyr of eastern Siberia, it may even be as low as 15–20% and appears to be declining through unfavourable weather conditions over consecutive summers (Krechmar 1990; Kondratiev 1991).

The ice-free period is the critical factor affecting summer distribution and breeding success of both the Trumpeter Swan in Alaska (Hansen *et al.* 1971), and the Whooper in Eurasia. With an incubation period of 31–35 days and fledging taking a further 90–110 days (e.g. Ma 1996), the breeding period typically takes a minimum of 130 days in Whooper Swan, and may be as long as 150. Thus, in regions with fewer than 140 ice-free days in summer, breeding is unlikely to be successful (Haapanen *et al.* 1973a, 1973b). The ice-free summer period presumably affects Whoopers breeding in the highlands of Iceland, but those nesting in the lowlands are likely to be affected only by the time of ice break-up in spring, not by freezing in autumn (Brazil 1981c). In China too, timing of their arrival on the summering grounds is critical. Birds may spend some time at small ice-free areas but generally pairs move quickly to their territories over which there is much aggressive interaction (Ma & Cai 2000).

Spring thaw, and autumn freeze (along with food availability) are critical factors delimiting the northern extent of the range, and influencing breeding success. Selection pressures would operate against late nesters and re-nesters as the likelihood would be high that cygnets would still be flightless at the time of the autumn freeze. Young still incapable of flight are not infrequently found as late as the middle third of September (e.g. Vengerov 1990b) and at high latitudes that is late indeed. The very critical timing of fledging is best understood through an example from Chukotka. On 15 September 1979, Krechmar (1990) found that none of the five broods he was studying were capable of flight, and the first bird able to fly was not seen until 25 September. Meanwhile, freezing of most of the lakes in the region occurred from 20–21 September, and flightless swans were only able to survive in small lacustrine or riverine polynias still rich in food.

Those factors delimiting the southern extent of the range remain unknown. The large size, and length of the breeding season, among all swans apparently precludes them from making more than one attempt per season. A single example of double-brooded Mute Swans in temperate Britain (Absolom & Perrins 1999), the only example from any swan that I have found, serves only to confirm how rare this situation is.

Like all white swans, Whoopers are strongly territorial. They defend a nest site and an area around it suitable for them and their young to feed in (Kear 1972). They defend it not only against conspecifics, but also show aggression towards other species, e.g. Greylag Geese, Common, and Demoiselle Cranes (Ma & Cai 2000). Responsibility for territorial defence lies primarily with the male, while the female incubates. Once the clutch has hatched, both adults tend the brood, though in what proportion will be seen later.

Haapanen *et al.* (1977) showed that some pairs hold a nesting territory only until their eggs hatch, and then move to feeding sites some distance from the natal area, while in other areas territories are maintained throughout the summer (Brazil 1981c). When a pair occupies a new site they may, in many cases, defend the territory for one or even two summers but not breed until the third or fourth year (Haapanen & Hautala 1991). Their first efforts may involve territorial occupation and defence, without nest building. In subsequent years nest building may follow, but without egg laying, and perhaps only in later years do some pairs successfully hatch eggs and rear young.

That large areas of the range were apparently suitable, but unoccupied, may have been generally true 40 years ago, but over the intervening period populations are thought to have grown considerably in some regions, notably Scandinavia. Scott *et al.*'s (1953) observation of only 20 pairs in 11,400 ha in Iceland may no longer be typical, and would certainly have changed considerably in Finland, where the population has multiplied many times in the five decades since hunting ceased. Although breeding territories are generally assumed to be large, early estimates of size were based simply on overall habitat available divided by the number of pairs present, but Kear (1972) was first to suggest that true territory size had probably been exaggerated. Especially in upland areas of Iceland, large areas without swans are probably unsuitable because of poor food availability. Brazil (1981c) discussed the situation in parts of Iceland in detail, concluding that Whoopers there defended small areas, and that territories were likely to be even smaller than the 570 ha/pair calculated by Kear (1972). Martin (1993) considered that, in the taiga scrub zone, they averaged only *c.*100 ha/territory, while in China, Ma (1996) found pairs to be territorial when nesting with each pair defending no more than 50–100 ha of swamp/open water. In Omsk, Whoopers prefer small lakes (<10 ha) surrounded by deciduous forest or dense thickets and swamps (Yakimenko 1997). Where their nesting territories are tiny, of necessity they visit large waterbodies to feed, sometimes several km distant, where, presumably, they cannot maintain anything more than a very temporary feeding territory. While the norm across the range may be for pairs to each occupy their own pond, lake or swamp, in some cases rich habitats support several territorial pairs, as I observed in Iceland, Ma (1996) in Bayinbuluke, and Karyakin & Kozlov (1999) in the forest-steppe of the Ural region. In the Ob Valley, Vartapetov (1984) found 3–4 broods no more than 100–150 m apart, the highest concentration reported so far.

As mentioned, Whooper Swan delays sexual maturity. It may not breed until five years old in Russia (Dement'ev & Gladkov 1967) or even seven in Iceland

(Gardarsson and Skarphedinsson 1984), although a tame female in Finland was sexually mature and bred when just four, which Haapanen *et al.* (1973b) and Haapanen (1982) considered normal. This variation, 4–7 years old at first breeding, led Haapanen (1991) to adopt an average age of six when calculating mortality rates. Haapanen (1982) also referred to another aspect of breeding considered normal, that is to produce fewer eggs and rear fewer cygnets in the early years of breeding than later, presumably through inexperience or physical immaturity. The maximum age reported for a Finnish Whooper is in excess of nine years (Haapanen 1982) giving each individual a minimum of five or more possible breeding opportunities, although an Icelandic bird has reached over 20 in the wild making as many as 15 breeding opportunities possible. Turnover of pairs on territories is high, as many as 35% were only present for one summer and only 27% were present for five (Einarsson & Rees 2002). In reality of course, with so few pairs breeding and so many factors influencing breeding success, relatively few potential breeding opportunities successfully produce offspring.

Long-lived, monogamous swans and geese exhibit a high degree of behavioural co-operation; both individuals of each pair investing heavily in time and effort on behalf of their offspring for almost an entire year.

HABITAT SELECTION

Whoopers prefer to nest and summer at shallow lakes and marshes within forest-tundra, taiga and steppe, including northern forest-steppe, southern forest-steppe and true steppe. They are generally solitary breeders with typically one pair nesting at a waterbody (Brazil 1981c; Koskimies 1989; Batten *et al.* 1990), the minimum size of which is limited by the need for take-off room (*c.*100 m). They tend, therefore, to nest at very low densities, with perhaps just 5–20 pairs/100 km^2, although suitable habitat within that area may be considerably smaller (Koskimies 1989; Gibbons *et al.* 1993). Though they do occur at large and medium-sized lakes, they are more common on relatively small waterbodies. Pairs usually frequent lakes with a surface area of less than 1 km^2, which they occupy exclusively. In northern Norway, Myrberget (1981) found that Whoopers typically nested at rather small lakes, 0.5–2.0 km long and 0.5–1.0 km wide with successful pairs no closer than 3 km. There are notable exceptions, however, when several pairs nest at the same body of water, as has been observed in Iceland, Russia and China.

Those breeding in Yakutia, for example, have a distinct preference for small wetlands, with 68.6–77.4% of pairs at lakes with an area of just 0.01–1.0 km^2. Lakes of 1.1–6.0 km^2 were considerably less attractive (only 17.0–20.1%), and larger wetlands were only seldom used (4.7–11.4 %) (N=172). A closer examination revealed that they preferred round grassy lakes with sloping, indented banks (43.8%) with developed coastal-water vegetation, more seldom slightly overgrown lakes with boggy banks (16.1 %) and lake-like extensions of grassy rivers (1.8%).

Moulting Whoopers were on large (6–10 km^2) lakes with indented coastlines. In 8–9%, birds were observed on open ground near wetlands, and they fled into forest at the approach of a low-flying aircraft (Degtyarev 1990).

Experience in Iceland and Scandinavia and the European literature indicate that it is usual for Whoopers to nest in isolated territories, at wetlands separated by considerable distance, yet this is not always the case. On the Kola Peninsula, Semenov-Tyan-Shanskiy & Gilyazov (1990) found the closest nests 3 km apart. A little further east, Mineev (1995) found that pairs nested on seldom-visited lakes 3–5 km apart, whereas in undisturbed areas in eastern Russia they nest at sites 1.5–6 km apart. In the Urals, whereas Whoopers breed solitarily at lakes along the eastern slopes, in the forest-steppe they occur in a more concentrated fashion, much as I observed at one locality in Iceland[1], with 2–3 pairs per waterbody. Concentrated sites have also been found in the southern lowlands of Iceland, where as many as 5–6 pairs may occur in eutrophic wetlands at sea level (Gardarsson *in litt.*). At such sites in Russia, the minimum distances between neighbouring nests were just 500–800 m. On Kurlady Lake, where there are numerous islands of floating vegetation, 8–20 pairs breed, with a further 5–17 summering (Karyakin & Kozlov 1999). In the central portion of Bayinbuluke, China, the density reaches an astonishing 2–4 pairs/km^2, with nests within 150 m, although in most areas it is <1 pair/km^2 and nests at least 500 m apart (Ma & Cai 2000).

Although most commonly found at freshwater habitats, in some regions they also breed at brackish and mineralised lakes (Yakimenko 1997). It should be noted that in Iceland, where almost all woodland cover was removed historically, and kept clear by sheep-grazing, the majority of nesting habitats are exposed, though when considering the entire range this is not typical. In northern Finland, most breed at small pools amid huge aapa fens, while in central Finland they prefer undisturbed eutrophic lakes within forests (Koskimies 1989).

Whoopers generally nest at relatively low altitudes, many near sea level. Myrberget (1981) found that all successful nests in northern Norway were below 100 m. In Sweden, although pairs are occasionally found above 500 m, the majority were at 100–400 m (Nilsson *et al.* 1998). In Iceland they breed from near sea level to 700 m (Gardarsson & Skarphedinsson 1984). Despite this general pattern, in China Whoopers breed at Bayinbuluke, at an exceptional 2,400 m in the Tien Shan, which is apparently the highest altitude breeding area in the world.

Although commonly nesting in areas rich in nutrients (eutrophic) or with average levels (mesotrophic), they sometimes use lakes or peat lands with ponds that are low (oligotrophic) or even very low in nutrients and even breed at brackish mineralised lakes (Yakimenko 1997). As Ohtonen & Huhtala (1991) in Finland, and Rees *et al.* (1991a) in Iceland have demonstrated, the habitat chosen for breeding has a significant effect on reproductive output, and begs the question as to why Whoopers would choose to nest at oligotrophic sites. Is it because eutrophic sites are occupied first? Do pairs attempt to move from oligotrophic to eutrophic sites during their reproductive lifetime, or, having once occupied a site, is familiarity advantageous? These questions have no answers, as yet, but they remain intriguing topics for swan biologists to tackle.

Haapanen & Hautala (1991) found that at Kuusamo, northern Finland, the first Whoopers arrive in early April, some even in late March, many weeks prior to breeding. Pairs defend pre-breeding habitat, slow-flowing shallow rivers. Most stay, however, in small flocks of 11–40. Elsewhere, as in Russia, breeding pairs commonly arrive within their nesting region well before nesting is possible. However, in the Anadyr Basin in northeastern Russia, even though it appears earlier in spring in the middle reaches than any other waterfowl, nevertheless, they are unable to nest until the last third of May (Krechmar 1982a).

THE NEST

NEST SITE SELECTION

In general, all swans nest close to water, preferring sites with some natural protection from terrestrial access by predators, e.g. islands or promontories, or affording a view across their territory, such as on raised hummocks, or in many cases, both. Whoopers are no exception. Their nests are usually built in the open, near water (typically within 5 m), on a hummock in a marsh, on a river bank, lakeshore, islet or promontory, such sites become clear of snow early, before the main thaw, though they are surrounded by shallow water later (Cramp & Simmons 1977; Brazil 1981c; Gibbons *et al.* 1993). Though frequently in exposed locations or at most surrounded by sedges, rushes and grasses, they are occasionally built among, near or even adjacent to small birches in Iceland and Russia (Pearson 1904; Plates 5 & 6).

Water obviously provides extra protection from predation or disturbance, and many waterfowl have markedly greater nesting success on islands than on the mainland (e.g. Hammond & Mann 1956). Brazil (1981c) found that 74% of nests in a northeast Iceland study area were partially or completely surrounded by water (Fig. 8.1). Three nests could not be pinpointed, because of the impenetrability of the area, but these were the only nests that could have been sited in emergent vegetation. This contrasts with Trumpeter Swan, which commonly nests in emergent vegetation, often bounded by a moat of water, or on a semi-floating sedge mat (Banko 1960; Hansen *et al.* 1971), a habit unknown among European Whoopers but widely reported in Russia. Tundra Swan, is also attracted to islands and promontories, choosing elevated, dry sites that are already snow-free before the general thaw, and sometimes surrounded by water after the thaw. Mute and Black Swans also select sites on islands or headlands in or near running or standing water (e.g. Campbell 1960). Interestingly, in Myrberget's (1981) northern Norway study, most Whooper nests were within 1–2 km of the sea, and one pair nested on a small island in the sea. This appears to be the only documented case of this kind. In

Figure 8.1.   The massive nest is usually partially or completely surrounded by water.

Xinjiang, 80–90% are in marshes or open, non-flowing water, with a massive nest mound rising up to 1 m above water being constructed from local vegetation. A small number are built on small islets or hillocks. As the ground thaws, the nests sink and pairs must continually add material to accommodate this. Nests are like large truncated cones rising on average 41 cm above the water and with a diameter at the surface of 1.1–2.8 m (Ma & Cai 2000).

Site fidelity in large, long-lived birds that lay only one clutch per year has clear advantages, not least facilitating an earlier start to nesting than would otherwise be possible. Though long assumed in Whoopers, and circumstantial evidence has suggested that the same pairs may be involved in using territories and even nest sites between years, that evidence was slim. Bowler *et al.* (1993), however, were able to confirm, that not only did adults show a high degree of fidelity to the same nesting territory, but also that that young may do so too. They found a bird ringed as a cygnet in 1988 nesting in a territory adjacent to its parents, and in the territory where it had been reared. Einarsson (1996) found a high level of nest-site fidelity in Iceland with 57% of sites occupied by pairs both members of which had nested in the same territory the previous year. The prevalence of nesting on a previous site may be related to the accumulated pile of vegetation warming and hence thawing early. As Elkins (1983) has suggested, such sites have the added advantage of usually being above the level of average spring thaw floodwaters. In Kuusamo, Whoopers begin to visit their breeding sites while their nest mounds are still under snow, indicating that not only are they able to locate the sites under such conditions, and remember them exactly from the previous year, but that habitat selection must have taken place prior to arrival, i.e. during a previous season (Haapanen & Hautala 1991).

NEST CONSTRUCTION

Having arrived in the vicinity of their breeding grounds perhaps as early as late March or early April, Whoopers begin to visit their old nesting sites in late April–early May depending on latitude and altitude. At first, these visits may be only occasional, but later they become daily and may involve flights of several km each way from a staging area. In Kuusamo, the average daily temperature usually reaches $+5°C$ by 15–20 May, by which time there is usually sufficient free water nearby for breeding birds to commence breeding (Haapanen & Hautala 1991).

In Iceland, nests are normally built or refurbished during May, although in late springs this may occur in June. One pair was still nest building at Arnarvatn (*c.*250 m) on 3 June 1979, while at nearby Helluvadstjorn (*c.*300 m) although a pair had built their nest by 12 June, much of the lake was still frozen (and the nest was subsequently deserted, without eggs being laid by 1 July), though most would normally have laid eggs long before (Brazil 1981c). Typically, once the breeding sites are ice-free, nest building takes just a few days and then egg laying begins.

Although commonly portrayed as sharing the nest building (e.g. Wilmore 1974; Martin 1993) I feel that this is an over-simplification. In northeast Iceland, construction was performed almost entirely by females, although males helped by pulling up vegetation and passing it closer to the nest. At nest sites I observed, the male's contribution was not, however, always productive, as it was not always directed towards the nest! The success depended on the male's orientation to the nest, as any material the male extracted was *always* passed backwards and deposited by his flank. On occasions, when the male was facing the nest mound, this actually led to material being moved further away, not closer (Brazil 1981c). The female's behaviour is far more effectively directed. Neither sex has been observed transporting nest materials. It seems that nests are built from the available materials immediately surrounding the site, as in Trumpeter Swan (Cooper 1979). Once construction has been completed, material is still added by the female during incubation, and occasionally by the male during the female's breaks.

The substantial mound invariably consists of the common plant species of the immediate vicinity, usually grasses, sedges and mosses. One nest, in Iceland, also contained dried stems of *Angelica*, which grew commonly nearby (Brazil 1981c). Elsewhere, as in the UK, nests have been described as being of reeds, as well as those materials mentioned above (Batten *et al.* 1990; Gibbons *et al.* 1993). All those nests that I examined contained swan down throughout the upper part of the mound, with no apparent greater tendency for it to be used as a lining, and I found none supporting Martin's (1993) statement that nests are 'lined with fine grasses and a little down'. Down appears to be added incidentally throughout incubation. Wilmore (1974) considered that down was added to the nest as a result of nibble preening, perhaps in response to skin changes, and to the presence of loose down around the brood patch, and this remains the best explanation. There seems little or no evidence of a cup lining, and down and other materials are readily dispersed throughout the nest by the female covering and uncovering the eggs before and after incubation breaks (Brazil 1981c).

The dimensions of four nests I measured in Iceland were roughly 1 m in diameter, 30 cm high, with a cup of 30 cm and 5–10 cm deep formed by the weight of the female and by her pressing with the feet, breast and bill. These were evidently similar to, though slightly smaller than nests measured by Bulstrode *et al.* (1973), which also measured 1 m across, but were up to 1 m high with a cup diameter of up to 50 cm, and up to 20 cm deep. Martin (1993) refers to nests being 1–3 m at the base and 50–70 cm high, though all the larger dimensions are exceptional. Continued use during incubation steadily flattens nests so that initial dimensions are typically the larger.

The proximity of most nests to water, combined with the spring thawing of ice and rain that may bring floods, means that while water may provide some protection it also brings the risk of nests being flooded or swamped by wind-driven waves, a significant factor affecting breeding success with up to 22% of nests lost in this way (e.g. Vengerov 1988b). In Magadan (eastern Russia), one nest, considered unusual by the discoverers being situated quite high on a bank beneath dwarf cedars and alders, stood 85 cm tall and measured 2 m at the base. Its location and construction were considered to be connected with the real possibility of the area flooding in spring (Krechmar 1982a; Krechmar & Krechmar 1997). The nest mound becomes somewhat flattened during incubation due to the weight of the sitting female, and of the male which occasionally stands on the nest rim during female incubation breaks (Brazil 1981c). Nest sites are likely to be traditional, used by generations of swans, as the most suitable locations are likely to be within a few metres of each other.

New nests are often built using the previous year's construction as foundation, although in contrast those nesting in the Anadyr of eastern Russia usually don't. In this region, where winter snow cover is considerable, spring melt unpredictable, and the breeding season critically short, nesting pairs cannot afford to wait for the snow to melt on their old nest, but instead build a new one in the vicinity, but in a more favourable site for the prevailing weather that season (Krechmar 1990).

In Omsk and on the Ural-Tobol plateau, and doubtless elsewhere, Whoopers nest in dense reedbeds, and in dense thickets of Bulrush and Marsh Horsetail (Nechaev 1991; Samigullin & Parasich 1995; Yakimenko 1997). Nests may also be commonly built on top of Muskrat houses flattened by the swans; a habit reported from widely separated areas. In Tyumen region (Ob River), Vengerov (1990b) found four of 18 nests built on muskrat houses. More rarely, though reasonably widespread, it seems Whoopers may build their nests on old floating vegetation (Samigullin & Parasich 1995; Golovan & Kondratiev 1999). In North America, Trumpeter Swans will use floating platforms, muskrat houses or even American Beaver lodges (Mitchell 1994).

Although nest construction is essentially the same throughout the range, it varies in the details. In Heilongjiang, China, as much as 80% of nest material consists of the sweet-grass *Glyceria spiculosa*, with *Typha* spp., wild rice *Zinania caducifolia* and *Urticutaria intermedio* also used (Wang 1987; Li 1996). Nests vary considerably in size but commonly range in diameter near the base from 1.0–2.7 m and

70–160 cm at the highest point; they range in height (from ground or water level) from 30–70 cm, with as much as 50–120 cm below water level; they commonly have a cup diameter of 30–70 cm and a cup depth of 11–20 cm. The cup may be lined with dry sedge or sphagnum or other thinner, dry material (Bianki 1988; Semenov-Tyan-Shanskiy & Gilyazov 1990; Vengerov 1990b; Mineev 1995; Samigullin & Parasich 1995; Li 1996; Golovan & Kondratiev 1999; Yumov 2001). Trumpeter Swan takes 11–35 days over nest building (Hansen *et al.* 1971; Cooper 1979) and the Whooper presumably takes a similar period, although Krechmar (1990) showed, using automatic photography, that nests are constantly built up during incubation, activity that is especially intensive where nests are surrounded by water.

PAIR BONDING

Whoopers are essentially monogamous birds for whom choosing and bonding with a mate must be highly significant in determining lifetime reproductive potential. As with several geese and swans, there is a tendency for males and females to pair early and remain together until they either lose each other or one dies, although 5–6% of pairs end in divorce, i.e. re-pairing while a previous mate is still alive (Rees *et al.* 1997a; Rees & Bowler 1997). Males assist their mates to differing extents during breeding, but typically defend an area around the nest, and assist by nest-guarding, escorting and protecting the young once out of the nest, and by pulling up food for the cygnets while their necks are still too short to reach deeply submerged plants. Male incubation, occasional in the Trumpeter, and regular in the Tundra Swan during female absences, does not occur in the Whooper. Cygnets further benefit from a prolonged pair-bond, travelling with their parents to and from traditional wintering grounds, thus learning both about suitable feeding, resting and wintering areas, and migration routes.

Lack (1974) highlighted a further advantage of prolonged pairing, as there is a tendency among birds to commence breeding earlier when pair members already know each other. Among arctic and subarctic nesters, where the breeding season is short and where time available for the young to complete their development is critical, early breeding confers a considerable advantage.

Given so few long-term studies of known-individual wild Whoopers, it has not been possible to establish precise details of the age at which they pair, the age at which they first breed (though 4–7 is considered common) or for how long they pair. Not, that is, until important studies by Black & Rees (1984), who were able to recognise individuals wintering at Caerlaverock, and by following cygnets and yearlings until they had paired were able to establish when pair-bonds were initiated. The earliest pair-bond was at a very young age, single males and females that became firmly paired by their second winter and were with the same mates two years later. By two years of age, or their third winter, pairing was much commoner, though mainly among females. Whereas seven of 15 females were already paired, only one of seven males was, indicating that females mature more rapidly. By three years old, or their fourth winter 14 (87.5%) of 16 known-age birds were paired (Black & Rees 1984), and perhaps some attempted to first breed the following summer. Normal first pairing occurs at 2–4 years and first breeding at 4–6 years (Einarsson 1996; Rees *et al.* 1996). In captivity, sexual maturity is reached at 2–3 years although copulation has been observed among one-year-olds (Zubko 1988) and copulation and breeding has also been reported at one year old (Rees *et al.* 1997a), while pair-bonds have lasted 4–13 years (Zubko 1988).

Black & Rees's (1984) observations, that of five pairs ringed at Caerlaverock in 1979–80 four were still together in 1983–84, go some way towards confirming long-held suspicions of the long duration of the pair-bond. The longest known wild pairing is 11 years for a pair that returned each winter to Caerlaverock between 1981–82 and 1991–92 (Whooper Facts 2001).

Winterers at the same site appear to be an important group from among which mate selection occurs, but is by no means the only group. Black & Rees (1984) found that of 33 ringed birds that returned with new mates, 24% had mated with birds also at Caerlaverock the previous winter. That does, however, beg the question as to where the other 76% found mates! Perhaps staging areas, or for young birds, moulting areas, are important, and even some of those 24% may have already met at staging areas prior to the wintering grounds. The exact timing and location of pairing is poorly defined.

Pair formation leads to conflicts of allegiance, as after exploring several sites in their first few winters, adults often settle into a pattern of repeatedly visiting the same one. Individuals may then be led to different sites by their partners. That two returnees re-paired after having been seen in a previous winter each with a different mate, provides evidence that Whoopers are not unwilling or unable to find a new mate having lost one, or even to change while their first is still alive.

At provisioned wintering sites at least, it seems that Whoopers engage in higher levels of courtship display in spring than do Bewick's (Scott 1978; Black & Rees 1984). The latter do, of course, tend to leave much earlier than Whoopers and may engage in such behaviour at staging areas en route north, while the later-departing Whoopers have more time to engage in such behaviours while on the wintering grounds.

Abnormal pair bonds have been reported, with Tundra Swan in Alberta and Mute Swan in Europe, and even hybridising. Whoopers have even formed *menage à trois* bonds with Mute Swan pairs in Massachusetts where they became accepted and very protective 'godfathers' of the Mute Swan broods.

COPULATION

Despite spending four winters observing Whooper behaviour in central Scotland (1977–80), and almost every winter since in northern Japan, I have observed relatively little clearly socio-sexual behaviour, except where birds have been concentrated by food, where social interactions are commonplace. Even so, I have observed no behaviour indicative of pairing or mating on land or on water, although apparent pair-bond reinforcement is not uncommon. Furthermore, I have come across no published records of mating in the wintering areas, though Wilmore (1974) refers to them mating 'spasmodically' after their first arrival in Iceland, with mated pairs eventually flying into the interior to nest. In China, Ma & Cai (2000) observed mating during nest building, egg laying and even incubation, suggesting that it can occur over a rather prolonged timeframe, but only on the breeding territory.

According to Martin (1993), although some courtship occurs in winter, most occurs among non-breeding flocks in summer, the 'engagement' lasting until possible first breeding the following year and probably beyond. He describes males and females greeting each other with head bobbing, or vociferously using the 'triumph display', and facing each other during courtship, head turning with their neck feathers erect, while Wilmore (1974) describes the culmination more poetically, as the couple beat their open wings 'close together in a whirl of reciprocatory pleasure'. Wilmore (1974) describes Whoopers as 'monogamous, but not above flirting' and as changing partners more frequently than other 'northern swans'. Rivals will, apparently, fight savagely, facing off and beating with their wings and bow-spritting, ground staring and head-plunging (if on water). Nevertheless, despite an extensive literature search, I have found no published descriptions of copulation. During summer 1979, however, I was fortunate to observe by chance copulation twice in Iceland.

On the occasion of the first, near Reykjavik on 24 May 1979, the female was already attending a nest on a small island, though it was unknown whether eggs had already been laid. At 10:31, the female aerial-chased 12 other intruding adults onto another part of the lake, after which she landed and began to head dip, with her wings half spread across the water's surface. She regularly raised her wings and beat the water. Her mate approached, and at 10:34 both displayed, *c.*30 m apart. They both bow-spritted and called briefly before swimming towards each other. At 10:36 they repeated the sequence, but this time only 1 m apart. At 10:38 both were head dipping and splashing, lunging forward vigorously so that the lower neck and breast became submerged and water ran across their necks. Their movements had become synchronised. At 10:39, the male mounted from the side, grasped the female's neck feathers in his bill, held her neck down in the water, and mated with her. Approximately 30 seconds later both called, the male slid from the female's back and they turned to face each other with their necks arched, still calling. The female then flapped her wings, as is often the case after bathing. Both then commenced preening, standing in shallow water near the nest island.

The second occasion was on 30 May 1979 at Myvatn. There too, a nest was already completed, but it was unknown whether eggs had been laid. At 15:05 the male was feeding *c.*5 m from the nest, while the female was roosting on the nest.

Both assumed the 'head-up' position and at 15:07 the female left the nest and swam off followed by the male. After a minute the male was 10 m and the female 20 m from the nest island. At 15:09 they swam towards each other and began to bathe, alternately lunging into the water (as described above). The male then quickly mounted from the side, while holding her nape feathers in the bill. The male held the female's head underwater during copulation. The female shuffled sideways and at 15:10 the male slid from her back and rapidly flicked his tail sideways, as if defecating. The female, meanwhile, preened and wing flapped. The male proceeded to preen in the shallows by the nest island while the female tail wagged. At 15:12–15:16 the male stood 'head-up' in the shallows while the female preened before gradually moving towards the nest rim, pausing to preen briefly. She finally resettled at 15:19, at which time the male was still preening.

Johnsgard (1965), oft-cited, e.g. by Scott and The Wildfowl Trust (1972) and Ogilvie & Pearson (1994), described the copulation behaviour of Trumpeter Swan and, though not having seen that of the Whooper, he considered it likely to be almost identical. The two copulations described from Iceland differ from his description of copulation in Trumpeters, in that the pre-copulation displays involved not only head dipping, but also thrusting the lower neck and breast forwards into the water so that water ran across the back of the neck or onto the back, behaviour obviously derived from, and closely resembling, typical bathing behaviour. The pre-copulation display was short, less than a minute (though considerably longer than Johnsgard's (1965) 10–15 seconds!), as was copulation. In the first instance the pre-copulation displays were synchronised, in the other they were alternate. In both cases, mounting was from the side, the male grasping the female's nape and holding her neck close to the water or submerged. In neither case did the male spread his wings as copulation ended, unlike Trumpeter Swan (Johnsgard 1965). After the 24 May copulation, the pair faced each other with their necks arched as they called (see Johnsgard 1965), the female wing flapped then both began to preen. The second male flicked his tail sideways, while the female preened and wing flapped. In both cases preening followed, or was even part of, the post-copulation display.

In the light of these and Ma & Cai's (2000) observations, it may be that mating behaviour is rarely reported because it occurs after migratory and family flocks have disbanded, and after pairs have arrived at their territories. Copulation is very brief, thus reducing the likelihood of it being observed. Copulation by Whooper and Trumpeter Swans is very similar, but it appears that the pre-copulatory display is prolonged and the post-copulation display, like the territorial displays, is more extreme in the Whooper than in Trumpeter.

Although extremely rare, polygamy has been recorded on three occasions among Trumpeter Swans (once in the wild, twice in captivity) with two females sharing a nest in two cases (Mitchell 1994). At sites where Whoopers nest at high densities, as in certain parts of Russia and China, polygamy may be a possibility.

EGGS

Whooper eggs are a large elliptical shape, broader at one end and tapering to a bluntly rounded point at the other; however, in size, shape, and even texture there is much variation, with some being smooth, others ridged or grooved, some elongated but narrower, some broad and fat (Figs 8.2, 8.3). Colour tends to be a creamy off-white or ochraceous, but while fresh eggs are clean, older eggs acquire staining from the nest vegetation and, in iron-rich waters, from the female's feathers, becoming brownish after several days. They may also be scratched by the female's claws.

Figure 8.2. Whooper Swan nest and clutch.

Eggs are very similar in size across the range, averaging 113 × 73 mm (Martin 1993), but Rees *et al.* (1991a) found that dimensions varied significantly less within clutches than between clutches, thus females tend to be relatively consistent in the size of eggs they produce. According to Rees *et al.* (1991a) 152 eggs averaged 112.8 mm by 71.5 mm, with an average volume of 295.57 ml (Table 8.1). There is evidence from Iceland that dimensions vary with habitat (or altitude) and there may be a similar pattern from south to north at increasing latitudes in its continental range. In the Icelandic lowlands, Einarsson (1996) found eggs to average 113 × 71.5 mm (range 93–126 × 61–79; N=643), whereas in the less productive highlands they averaged 112.7 × 70.3 mm (99–121 × 64–75; N=169). Eggs in China were extremely similar to those elsewhere (112 mm (98–120) by 71 mm (67–75) (Li 1996), although extremely large eggs up to 122 × 78 have also been recorded (Ma 1996).

Figure 8.3.   Eggs vary in size and shape.

In northeastern Russia, Krechmar's (1990) measurements of 87 eggs revealed that there too they were remarkably similar, ranging in length from 98–120 mm, in breadth 65–75 mm, and averaging 112 × 71. Here, weights of fresh-laid eggs were 298–368 g, averaging 336 (N=17). Mineev (1988, 1995) working further west found that eggs were rather shorter and fatter than those in the northeast, averaging 109 × 75 mm (N=21, range 106–119 × 70–78), although their weights were very similar, averaging 341 g (N=6; 303–368). Those from another widely separated region, Kazakhstan, were 106–115 mm by 69–74 mm (Kovshar & Levin 1982).

Fresh eggs in Iceland are lighter, averaging just 304 g (N=24), while those in China average slightly heavier at 310 g (range 280–365), however the largest in China weighed in at a monstrous 427 g (Rees *et al.* 1991a; Li 1996; Ma 1996). Weights decline during incubation by an average of 49 g, losing a mean 1–2 g/day.

Eggs are assumed to be laid at approximately 36-hour intervals (although Martin (1993) considered the interval to be 48 hours, as it is in the Trumpeter Swan), although with that species there is some variation, with some noting 39–48 hours

Table 8.1.   Egg dimensions in Iceland (after Rees et al. *1991a).*

Eggs	Length (mm)	Breadth (mm)	Volume (ml)
N	152	152	152
Mean	112.8	71.5	295.57
Range	101.6–121.5	65.1–77.3	231.86–351.83
SD	3.9	2.3	22.17

(de Vos 1964; Cooper 1979) and others 50 hours per egg (Tillery 1969). Rees *et al.* (1991a) found no correlation between the onset of laying and final clutch size.

Laying begins as early as late April in Iceland and China, with the peak in early May in China (Brazil 1981c; Li 1996; Ma 1996), but in Britain the few clutches have commonly been laid during mid May (Batten *et al.* 1990). In contrast, in northern Norway most are laid during the second half of May (Myrberget 1981). On the Ural-Tobol plateau, clutches are completed between mid May and mid June (Samigullin and Parasich 1995) and in the far northeast of its range and on Sakhalin laying is during a rather concentrated period during the last third of May finishing by early June (Krechmar 1982a 1990; Nechaev 1991). Across the species' range, however, egg laying typically in late May or early June may be delayed in cold springs to mid or even late June.

At Bayinbuluke, Ma & Cai (2000) found clutches were laid in late April to early June, at intervals of 1–3 days. Typically early eggs are larger than later eggs in a clutch. In one particular five egg clutch, they recorded three eggs weighing more than 400 g. The largest of all weighed 427 g, measured 122 × 78 mm and had a volume of 383 ml, making it the largest swan egg ever recorded. All three of these failed to hatch despite the embryos being fully developed, probably because of the extreme thickness of the shells (0.94 mm). Other Whooper eggs found to have broken during incubation were subsequently noted to have very thin shells (only 0.54 mm; Ma & Cai 2000).

CLUTCH SIZE

Whooper Swans lay a single clutch of 2–8 eggs (typically 4–5). Although the possibility of clutch replacement is widely disregarded, it has occurred at the Askania-Nova Zoo, Ukraine (Zubko 1988) and in Tyumen (Ob River), where Vengerov (1988b) refers to incomplete clutches in early June being repeat clutches following earlier nest predation or flooding. Variation in clutch size seems not to be regionally or latitudinally driven, but is related perhaps to the age of the pair and productivity of the territory. Younger birds, or perhaps young pairs, may lay smaller clutches.

Clutch sizes of 4–7 eggs were reported for wild Whoopers in the former Soviet Union (Dement'ev & Gladkov 1967) and of 3–5 in Iceland (Scott & The Wildfowl Trust 1972), and on the basis of this limited data Cadbury (1975) deemed it plausible to assign large broods noted in the UK in winter to the continental population. During my studies in Iceland I was only able to examine five full clutches, one of three, three of four and one of five eggs, i.e. exactly within the range quoted by Scott & The Wildfowl Trust, but again the sample was extremely limited (Brazil 1981c). Myrberget (1981) found 12 clutches of 3–6 (mean 4.8) in northern Norway. Observations of broods of six and eight in Scotland and of seven in Iceland (Brazil & Spray 1983), and of clutches of 3–8 (Batten *et al.* 1990) however, indicated that larger clutches may also occur in the Icelandic population, while Haapanen *et al.* (1973b) had previously recorded a much broader range, of 2–7 eggs, in Finland, suggesting that smaller clutches might occur there too. Brazil & Spray (1983) suggested, therefore, that the ranges of sizes for both the Icelandic and Fennoscandian/Russian populations more likely completely overlap (Haapanen *et al.* 1973b; Baumanis 1975), and that previous samples had simply been far too small to be representative.

Considerably larger datasets have since been collected in Iceland and Russia confirming the extent of overlap in clutch size and indicating that a clutch size of at least 3–7 should be considered normal across Russia (Kornev & Korshikov 1995; Vysotskiy 1998). In Mongolia and China, clutch size has been reported as 3–8 eggs, most commonly 4–6 and averaging five (Bold & Fomin 1996; Li 1996; Ma 1996). Given records of nine-egg clutches in Finland (Haapanen 1982), Germany (Deutschmann 1997) and Russia (Vengerov 1990b) and one of ten from Latvia (Baumanis 1975) along with complete two-egg clutches from Iceland and Finland (Ohtonen & Huhtala 1991; Einarsson 1996) that range should now be regarded as 2–10, with 3–7 being most common. In captivity, at Askania-Nova Zoo, large clutches of 9–11 eggs have been laid several times, and the largest clutch ever recorded was also laid there—an astonishing 14 (Zubko 1988) (although a captive Trumpeter once laid 17 eggs (Lumsden 1988)). Experimental egg removal has shown that Whoopers can, under certain circumstances lay a second clutch 19–23 days later, with repeat clutches containing 1–2 eggs fewer (Zubko 1988). In the wild, second clutches have only been recorded after loss of the first clutch; there too they are smaller than first clutches (Bold & Fomin 1996).

Clutch size among the Anatidae is known to decline during the breeding season for a variety of reasons (Sowls 1955; Lemieux 1959; Hilden 1964; Dane 1966; Klomp 1970; Johnsgard 1973; Eisenhauer & Kirkpatrick 1977) and varies between years. Contrary to Lack's (1967) suggestion that food supply only affects the laying date, Bengtson (1971) found that clutch size was lower in poor food years. Arctic breeders are particularly good examples of species whose clutch sizes are affected by weather conditions (Johnsgard 1973; Eisenhauer & Kirkpatrick 1977), being lower in late seasons (e.g. Milne 1976).

Haapanen *et al.* (1973b) noted some important trends in the Finnish Whooper population, firstly that mean clutches were smaller at higher latitudes (mean = 3.4 eggs in north Finland and 4.8 in south Finland) and that first clutches were smaller than those of more mature breeders. Overall mean clutch was 4.4 (range 2–7), rather similar to that confirmed from Iceland, namely 3.8–4.7 (Bart *et al.* 1991), and, interestingly, smaller than that reported for Trumpeter Swan (mean = 5.1; Banko 1960). In northeastern Russia, Krechmar (1990) also found a relationship between spring conditions and clutch size, averaging 5.3 eggs (13 clutches) in 'normal' springs, but only 3.9 (11) in delayed springs with heavy snow.

It seemed likely, from the habitats available to Whoopers and of the varied thawing dates at differing altitudes, that birds nesting in different regions might experience quite different breeding success, contribute disproportionately to the percentage of juveniles raised, and that conditions might even influence birds sufficiently early to affect egg production. By the late 1970s, this subject already interested Gardarsson, but it was to be some years before sufficient resources were available to study such potentially interesting variability. Then, in the late 1980s, studies began almost simultaneously in Iceland, by Rees *et al.* (1991a), and in Finland, by Ohtonen & Huhtala (1991).

In Iceland, Rees *et al.* (1991a) compared lowland nesting with highland Whoopers, finding that females in the former region tended to lay more eggs (mean clutch size 4.7) than in the highlands (3.8). In the lowlands, the modal clutch size was five, and in the highlands four, with nests containing more than five eggs only in the lowlands. In summer 1993, those in the lowlands seemed to enjoy a good season, with a mean clutch size of 4.5 eggs and mean brood size of 3.8, whereas those nesting in the highlands were less successful, with a mean brood size of only 2.7 (Bowler *et al.* 1994). This comparative study of lowland and highland breeding success has continued to reveal the basic pattern that where conditions are more severe, the summer shorter for example, productivity is lower.

Meanwhile, in Finland, Ohtonen & Huhtala (1991) examined a total of 112 nests with eggs, 40 on fens and 72 on ponds or lakes. The fens were lower in plant productivity and hence pairs there were mobile, travelling in search of food with their broods (mobile pairs), while ponds and lakes were richer in vegetation enabling families to stay put (sedentary pairs). The latter produced significantly larger clutches (mean 5.1) than mobile pairs (mean 4.4) (see Fig 8.4; Table 8.2).

Figure 8.4.   Clutch size varies in relation to habitat.

Table 8.2.   Clutch size in Finland in relation to habitat (after Ohtonen & Huhtala 1991).

Habitat	2	3	4	5	6	7	N	Mean ± SE
Fens	2	7	9	18	3	1	40	4.40 ± 0.17
Ponds	1	4	13	31	16	7	72	5.08 ± 0.13

While Rees *et al.* (1991a) were able to link clutch size to altitude, Ohtonen & Huhtala (1991) demonstrated a more tangible link between clutch size and productivity of the nesting territory. They suggested that, as horsetails *Equisetum* sp. are typical vegetation of Finnish lakes and ponds, and an important food resource for Whoopers, one reason why sedentary pairs may have been able to lay larger clutches is because of the quality of the food females ate during laying. Clutch size has also been shown to vary with habitat/altitude in Iceland, with fewer large clutches and more small clutches produced in the highlands (Einarsson 1996). Further evidence of the potential significance of food supply on output is also available from Finland, where Haapanen (1982) described the breeding success of a tame Whooper that regularly received extra food from visitors to its nesting territory. Whereas there had been no previous records of clutches larger than seven in Finland, this female's average clutch size was exceptionally high at 7.4, 2.6 more than usual in southern Finland (see Table 8.3).

Table 8.3.   Clutch size of a tame Whooper Swan in Finland (after Haapanen 1982).

Year	Number of eggs	
1974	•••••	5
1975	•••••••••	9
1976	•••••••••	9
1977	••••••••	8
1978	••••••	6
Sum	37	37
Average	7.4 ± 1.6	

The high average production of the Finnish female is especially noteworthy. Haapanen (1982) considered that the extra energy available from being provisioned might have affected clutch size. If so, then these are interesting implications for other populations. Are swans wintering in parts of Japan where food supplied is plentiful throughout the season, and those provisioned in winter in the UK, producing larger clutches and perhaps broods than their non-provisioned relatives? Is human interest contributing to regional population increase?

Clutch or egg volumes may also be dependent on habitat type or food quality, as suggested by Ohtonen & Huhtala (1991), who aged eggs (using the floating test) and found no difference in laying date between their two different habitat types. It has been suggested by Hansen *et al.* (1971), Haapanen *et al.* (1973a 1973b) and Nilsson (1979) that the most significant factor affecting nesting is the weather during the winter and spring preceding nesting, not that at the time of nesting. Milder winters/earlier springs are often followed by more successful breeding seasons with higher fledging success, but this is not the only factor affecting output, as Ohtonen & Huhtala (1991), and Rees *et al.* (1991a) demonstrated. Egg production is also dependent on habitat quality. At the highly productive Elizarovskiy reserve in Tyumen, Vengerov (1988b) found a high average clutch size, of 5.8 (N=23; range 5.5–6.4) in 1984–87.

Habitat quality, it seems, dictates whether families are sedentary or mobile. The pressures and energy requirements placed on different broods under these circumstances differ. Although, according to Lack (1967) clutch size in nidifugous birds, including waterfowl evolved primarily in relation to the average availability of food for the female, habitat may also play its part. As egg and clutch size among waterfowl are inversely related (Lack 1967), producing larger eggs may give the chicks a greater chance of survival, although possible only at the expense of producing fewer eggs. Ohtonen & Huhtala (1991) concluded that, for mobile pairs, it would be an advantage to invest in egg quality, and to produce fewer, larger eggs, to increase the survival chances of their young during the critical post-hatching period. Conversely, sedentary pairs on ponds and lakes rich in vegetation where post-hatching mortality might be expected to be lower might produce more surviving offspring by investing in more eggs.

Indeed, Whoopers nesting on food-poor fens do appear to invest in egg quality, in accordance with predictions. At food-rich ponds, however, not only did clutch size appear to increase as expected, but egg size also increased (Ohtonen & Huhtala 1991). Similarly, Rees *et al.* (1991a), in Iceland, found that females laying larger clutches also tended to produce larger eggs. As chicks from larger eggs are more likely to survive, this suggests that Whoopers might be able to increase their numbers rapidly when food availability is high, or that pairs may be able to increase their lifetime reproductive success significantly if they can occupy a food-rich territory. Haapanen (1982) found that where Whoopers are regularly fed by people, they produce exceptionally large clutches, reinforcing the view that food may directly affect laying capacity. Furthermore, in addition to habitat quality affecting clutch size, laying date also affects output, with later nests holding fewer eggs (Ohtonen & Huhtala 1991).

Altricial birds show a trend of decreasing clutch size towards the equator, and on more remote islands. Interestingly, precocial species including geese and swans buck this trend, with a tendency for smaller clutches at higher latitudes (e.g. Gill 1994) probably because of the generally harsher conditions further north. For each egg laid, an extra 36–48 hours is required to complete the breeding cycle. The exact inter-egg laying interval in Whoopers is unknown, but an eight-egg clutch might require up to ten days longer than a three-egg clutch to hatch and raise, and where the summer ice-free period is already restrictive, laying a large clutch may preclude successful breeding. That harsher conditions are implicated, whether through temperature or through food supply, is further supported by the fact that at poorer habitats clutches tend to be smaller (Bart *et al.* 1991).

INCUBATION

Once the first egg is laid, the behaviour of the pair changes. The primary focus now for the female is incubation, while for the male it is nest-, female- and territory-guarding. Among the Anatidae, the incubation period of swans is longer than for other species, while among them the 'northern swans' have the shortest incubation periods. Incubation, which begins in earnest with production of the last egg, lasts *c.*35 (30–42) days in Whoopers (Bianki 1988; Batten *et al.* 1990; Mathiasson 1991; Martin 1993; Ma 1996). Throughout that period, for both Whooper and Trumpeter Swans, males do not usually incubate (Cooper 1979; Brazil 1981c; Krivtsov & Mineyev 1991; Ma 1996), although Henson & Cooper (1992) did record one male Trumpeter nest sitting, and Mitchell (1993) considered that males incubated, but only 1.0–1.75% of time.

Though highly attentive to their nests, it seems that Whoopers may not be particularly discriminating about the contents. Ma & Cai (2000) reported several interesting and unusual cases of nest parasitism involving Greylag Geese and Red-crested Pochard laying eggs in Whooper nests. Although some eggs were laid after swans left, others appeared during female incubation breaks.

Despite the necessity for females to absent themselves for varying periods during incubation to feed, it is obviously crucial that the temperature of the eggs be maintained. Krechmar (1982b) found that despite often quite prolonged absences by the female, the temperature remained rather constant, fluctuating in the day by no more than 2.7°C regardless of changes in ambient temperature. Their large mass, the insulation capacity of the nest, combined with a complex of behavioural reactions, all appear to ensure a constancy, and allow what is for such a large bird, a relatively short incubation period. Nest temperatures in the vicinity of the clutch are 28–34°C, slightly cooler than the Trumpeter (29.7–38.2°C; Hampton 1981) and even during incubation breaks the nest and eggs retain heat such that temperatures does not fall below 28°C (Ma & Cai 2000).

Cramp & Simmons (1977) and Martin (1993) refer to male Whoopers occasionally relieving their mates during incubation, but I have traced no published records. Although males in Iceland were not confined to the nest, as were females, they did restrict themselves to its vicinity spending most time within 100–150 m, usually 'head-up' or feeding (Brazil 1981c). Rather than incubate, males remain close by ready to defend the area from intruders. In preparing for an incubation break, the female covers the eggs with vegetation using her bill. In her absence, her mate may approach the nest very closely, even standing beside it. On her return, he moves away and she uncovers the eggs by shuffling onto them (Gordon 1922; Banko 1960; Haapanen *et al.* 1977; Brazil 1981c).

The long-term stability of the pair-bond in swans probably serves to reduce time and energy expended in pair formation prior to breeding, as was suggested for geese by Owen (1980), and leads to sharing of investment and marked sexual differences in behaviour. Not only are there marked behavioural differences, but also these differences vary between species, being apparently directly related to ecology, and though male incubation is very unusual among Anatidae there are interesting exceptions among swans.

For both the Whooper and Trumpeter swans incubation is usually, if not always, by the female alone (de Vos 1964, Haapanen *et al.* 1977; Cooper 1979; Brazil 1981c; Krechmar 1982b; Batten *et al.* 1990; Mitchell 1994; Ma 1996), whereas in the Tundra Swan the male normally takes part (Johnsgard 1965; Evans 1975, 1977; Scott 1977). As a consequence of this difference in parental behaviour, the length of time that a Whooper Swan's clutch may be left unattended is up to three hours a day, whereas in Bewick's it may only be left for 5–7 minutes (Krivtsov & Mineev 1990), which contributes significantly to its capacity to nest at higher latitudes. Increased incubation constancy is particularly important considering the short duration of the arctic summer (Lack 1968; Kear 1970b; Scott 1977; Cooper 1979) and may be one factor enabling the two smaller 'northern swans' to breed successfully at such high latitudes.

Hatching was described as synchronous by Martin (1993), but this is not so, it takes place over more than one day, as in Trumpeter Swan (Brazil pers. obs.; Cooper 1979). In Whoopers, it occurs over a wide range of dates from June in more temperate regions to late June–early July in northeastern Russia (Krechmar 1982a), and sometimes even August (Mathiasson 1991).

Krivtsov & Mineyev (1991), who studied the daily time/energy budgets of Whooper and Bewick's Swans in Russia, found that they require relatively little energy during incubation relative to birds in temperate latitudes (1.9 and 1.8 BM, opposed to 2.5 BM). Their behaviour assists in maintaining low energy levels appropriate to the short cold summer and low calorific foods in their subarctic breeding areas. Whooper Swans spent considerably more time sleeping and flying during breeding than Bewick's, and less time feeding and incubating. Both spent approximately equal amounts of time in resting, swimming and comfort behaviour. What they clarified was that for Whooper Swans, flying is very costly in energetic terms, more so than for the lighter bodied Bewick's. In consequence, there is apparently no energetic advantage for Whoopers in feeding on diffusely distributed foods over large areas, as they would consume too much energy in flying to obtain it. Instead, they need a large concentration of food per unit area, and as a result, in contrast to Bewick's, Whoopers feed almost exclusively from shallow waters usually within 2.0–2.5 km or less of the nest. Thus, one of the reasons why Whoopers are unable to nest in tundra is that food sources are too widely dispersed, requiring large energy expenditure for them to recoup it. Whoopers at the northern (or altitudinal) limits of their range are likely, therefore, to be energetically stressed, and dependent on food availability.

Although territorial conflict typically occurs between males, female Whoopers in

Iceland, and female Trumpeters in Alaska, will also defend the nest area from intruders. Whereas defence of the nest itself is essential to ensure the short-term survival of the eggs and young, and thus may reasonably involve both sexes, it is thought that the importance of the surrounding territory is to ensure a food supply for the adults and the young, and that defence of this area is undertaken primarily by the male (Cooper 1979; Brazil 1981c).

Studies of breeding Whoopers in the northwestern range have commonly indicated that individual pairs occupy separate waterbodies, such that a single pair has exclusive use of the available food (e.g. Bulstrode *et al.* 1973; Haapanen *et al.* 1977; Rees *et al.* 1991a). At one site in northeast Iceland, I found two and even three pairs breeding *c.*500 m and 600 m apart in 1978–80, and engaged in active intraspecific territorial behaviour presumably over food resources (Brazil 1981c). Such behaviour over food resources presumably also arises in Russia and China where breeding at close proximity has also been discovered.

FEMALE BEHAVIOUR DURING THE INCUBATION PERIOD

Females spend almost all of their time on the nest, mostly roosting or 'head-up'. Periodically, they stand and turn their eggs using the bill, and occasionally stand, turn and settle down again facing in a different direction. Each and every movement ends with a short period during which the female rocks or shuffles from side to side over the eggs (Brazil 1981c). Cooper (1979) interpreted this behaviour, which I have called settling, as the female repositioning her feet under the eggs, although it may also include wrapping the eggs below the webs, behaviour that has now been observed in the Trumpeter Swan (Lumsden 2002), though not yet confirmed in the Whooper Swan. As swans have only a poorly developed brood patch, and the feet are potentially the best means of transferring heat to the eggs, settling would appear to be an important aspect of incubation behaviour among 'northern swans'. Incubation must, of course, be interrupted periodically for comfort and feeding bouts. In China, Ma (1996) found that females spent 80% of time on the nest. Four different females that I studied in Iceland devoted slightly different amounts of time at the nest, 81–94% (see Table 8.4). In Finland, two females spent 85% of the time incubating (Haapanen *et al.* 1977), while females in the Anadyr spent 87–89% on the nest (Krechmar 1982b, 1990). In China, females left their nests 3–5 times/day, usually during clear warm weather and most often around noon or in the afternoon. Nest absences were fewer and shorter on cloudy or rainy days. Nest absences lasted 20–90 minutes under normal circumstances, but when disturbed by animals or people extended rarely to 2–3 hours and, exceptionally, eight hours (Ma & Cai 2000).

The behaviours recorded in Iceland, in order of time spent, were: roosting, 'head-up', preening, nest fixing, settling, egg care, displaying and covering the eggs. The rank order was very much the same for all four females. Interestingly, all spent less time roosting (43%, 50%, 53% and 60%) than did those studied by Haapanen

Table 8.4. *Nest attentiveness by female Whooper Swans (after Brazil 1981c).*

Whooper Swan	Nest attentiveness
Iceland (four females)	
Arnarvatn (western)	81%
Arnarvatn (southern)	86%
Myvatn	87%
Ellidavatn	94%
Finland (two females)	85% (Haapanen *et al.* 1977)
Northeastern Russia	87% (Krechmar 1982b)
Trumpeter Swan (two females)	94.7 and 95.7% (Cooper 1979)

et al. (1977) in Finland (70%) though the reason for this is not apparent. Haapanen *et al.* (1977) gave no details of how females occupied the rest of their time. Because of the position of my hide overlooking Arnarvatn three of the four were able to move out of view during their incubation breaks. Females differed in the period of their incubation breaks that they were out of view 1.2–16.3%. The other female, however, spent an even greater time out of view (61% of incubation breaks), leaving the nesting bay entirely. A closer examination revealed that the immediate area was almost devoid of food, necessitating her departure to feed.

During incubation breaks, feeding was the most important activity, occupying 59% of the southern female's breaks and 76% of the western female's[2] (Brazil 1981c). These results are similar to those of 64% and 72% for two females in Finland (Haapanen *et al.* 1977), but all four differed considerably from the 38% spent feeding by Trumpeters (Cooper 1979). At Arnarvatn, the southern female fed mainly by grazing on shore and the western female by dipping, and the southern female spent more time 'head-up', preening and pulling nest material.

It is clear from such work that although individual variation occurs between incubating females, the overall way in which they allocate their time is rather similar, suggesting that the behaviour patterns observed were species specific. For example, roosting was the commonest and 'head-up' the second commonest behaviour during incubation, while feeding was the commonest behaviour and 'head-up' the second commonest behaviour during incubation breaks for all of those birds studied.

INCUBATION BREAKS

While overall behaviour patterns may be species specific, there is considerable individual variation. The number of incubation breaks taken each day, for example, varies between females. Overall, the number of daily incubation breaks is 1–12 and these vary in length naturally from 15 minutes to about two hours. Haapanen *et al.*

(1977) found that two Finnish females averaged 6.7 and 7.4 breaks per day, whereas I found that two females in Iceland took only 2.6 and 5.1 per day (Brazil 1981c). In contrast, Krechmar (1982b, 1990) found that females in Anadyr left the nest as little as once per day, but as many as 12 times, although their mean of 5.4 was similar to that in Iceland and Finland. Krechmar (1990) opined that the total duration of incubation breaks in Chukotka depended less on temperature than on rain and wind. It appears, for the two Icelandic females at least, that they tended to compensate for the frequency of their absences by adjusting the length of time spent away. Thus the southern female took more breaks with a mean length of 31 minutes, whereas the western female took fewer but longer breaks averaging 79 minutes. Bulstrode *et al.* (1973) had previously reported that incubating females in Iceland left their nests for periods of 40 minutes every three hours. Both those observed by Bulstrode *et al.* (1973) and those by myself remained away for considerably longer periods than the 21 minutes recorded for Trumpeter Swan (Cooper 1979).

Whereas Haapanen *et al.* (1977) found that the length of breaks increased towards the end of the incubation period in Finland, no such correlation was evident in Iceland, nor was there any relationship between the date and total time spent off the nest for the southern pair. There was, however, a highly significant correlation for the western pair. Total time off the nest increased with date until the western female ceased incubation entirely, only a few days before the eggs were expected to hatch. It seems likely that some of the differences in behaviour between these two pairs, particularly their attentiveness at the nest, had a direct bearing on their differences in reproductive success (Brazil 1981c).

MALE BEHAVIOUR DURING THE INCUBATION PERIOD

Although according to Cramp & Simmons (1977) and Martin (1993), males may sit on the nest during their mate's absences, neither Bulstrode *et al.* (1973), Haapanen *et al.* (1977), Brazil (1981c) or Krechmar (1982b) observed such behaviour nor have I been able to trace any actual records of it. Thus we can assume that male Whoopers do *not normally* participate in incubation or that if they do it is extremely rare.

Because of this, they have more time available for other activities, particularly vigilance, but also feeding. Haapanen *et al.* (1977) found that two Finnish males spent 34–35% of their time feeding, a proportion not dissimilar to Iceland, where two males spent 31% and 44% of their time feeding. A further male was observed feeding only 11% of the time, but spent 28% of time out of sight, and was probably also feeding. A fourth male spent 39% of time feeding (Brazil 1981c). The four commonest behaviours for all four males I studied intensively were feeding, 'head-up', roosting and preening. The southern male spent less time feeding than the western male, but spent longer in all other behaviours except displaying; in which they spent the same amount. The southern male spent more time 'head-up' than his mate (28% and 19%), whereas the western pair spent almost the same amount 'head-up' (male 27%, female 29%). So, although not incubating and with presumably more time available for vigilance, it was not clear that males invested particularly more heavily than their mates in such behaviour, although as we will see they may invest particularly heavily at specific times.

Bulstrode *et al.* (1973) observed males in Iceland joining their mates during their incubation breaks (temporarily abandoning the nest) and that the males were more alert then. Bouts of male alert behaviour lasted up to 1–2 minutes and occurred every 90 seconds while the female was off the nest, as opposed to 10–20 seconds spent alert every 2–3 minutes when the female was on the nest, indicating that not only is vigilance an important male role, but that males adjust their behaviour in relation to that of their mate's. Conversely, in northwest Russia, Mineev (1995) found that males seldom escorted their mates during feeding breaks, and in most instances remained 50–100 m from the nest. The attentiveness of males during the female's absence suggests to me that they are vigilant in order to lessen the likelihood of harassment or disturbance reducing their mate's feeding time. However, Bulstrode *et al.* (1973) and Brazil (1981c) found that intrusion into nesting areas by other swans was extremely unusual, if not entirely absent, during this part of the breeding season.

THE RELATIONSHIP BETWEEN MALE AND FEMALE BEHAVIOUR DURING THE INCUBATION PERIOD

Whoopers pair for long periods, possibly 'life', their long-term association and cooperation presumably facilitating more effective reproduction. It seems reasonable therefore to expect that a degree of cooperation and coordination might exist between them such that, for example, the nest might be guarded or eggs incubated for the maximum time possible. Such coordination, representing highly developed parental care, requires that each monitor the behaviour of the other and modify their own behaviour accordingly. In order to examine this assumption more closely, the frequency with which behaviours were exhibited by both males and females at the same (synchronised) and different times (asynchronous) was tested in Iceland.

Striking differences were apparent between two intensively studied pairs in Iceland making generalisations impossible. One, which hatched and reared three young successfully, showed a much higher frequency of significant associations between male and female than another unsuccessful pair; 47% of all behaviours were significantly associated for the successful pair, whereas only 38% were significantly associated for the unsuccessful pair. The most striking example of co-ordinated behaviour was that occurring when females left the nests for incubation breaks. Their mates then spent more time 'head-up', which was significantly synchronised with female absence by both males during these periods (Brazil 1981c). In general, it appeared that male and female behaviours were coordinated such that one was usually attentive at the nest or its vicinity. There was a tendency for some other activities to occur simultaneously, presumably in response to external stimuli, e.g. roosting. Both males tended to be 'head-up' if the females were roosting, and in view when their females were off the nest, dipping, grazing or 'head-up' on water.

Feeding by dipping was synchronised with periods when females were 'head-up' on the nest, and was asynchronous with 'head-up' on land or on water, preening on land, grazing or displaying. Conversely, feeding by upending in males was synchronised with periods when females were roosting on the nest, suggesting that males indulge in more protracted feeding when they know their mates are undisturbed. Roosting was synchronised in both pairs, and like Bulstrode *et al.*'s (1973) findings, males tended not to roost if their mates were off the nest, nor did males roost when females were 'head-up'. Male display behaviour tended not to occur when females were roosting, presumably because anything likely to stimulate display would also be likely to arouse the female (Brazil 1981c).

Distance from the nest during the incubation period

Proximity to the nest might also be a function of the degree of vigilance among males, and might perhaps be related to food availability for females. In northwest Russia, Mineev (1995) found that whereas at first females fed close to the nest, by the end of the incubation period they were feeding up to 1 km away. This change could be for a number of reasons including increased familiarity with incubation, a reduced perception of danger or the steady depletion of food closer to the nest.

An examination of the pattern of mean distance from the nest in Iceland, over the 24-hour period, revealed that males and females differed. Males spent time furthest from the nest in the early morning (01:00–06:00). For most of the day, however, they remained mainly within 100 m of it. Both sexes tended to be either very close to the nest (<50 m) or well away from it (>150 m), and of course, being sole incubator, females tended to be either on or very close to it almost constantly, making only brief journeys further away to feed (Brazil 1981c) (see Fig. 8.5).

Certain behaviours were directly associated with their distance from the nest. Many behaviours, especially by females, could occur only under specific circumstances or in particular situations; e.g. feeding could only occur when females were off the nest, whereas nest fixing could only occur whilst on it. Males tended to be more out of view in early morning, when they were furthest from the nest and while the female was on the nest. Males fed more by dipping early in the morning and were 'head-up' more from 06:00, i.e. during the period when females were most often recorded off the nest. It seems that roosting by males alternates with 'head-up' and may occur cyclically during high-summer 24-hour daylight, with peaks every 2–7 hours.

Two females studied intensively fed in different ways, and in different areas, due entirely to the location of their nests. The western nest was on a headland backed by heath, hence the female fed offshore mainly by dipping, whereas the southern

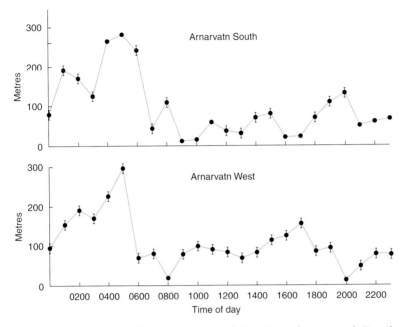

Figure 8.5. *The mean distance of males from the nest during the incubation period. (In order that standard deviations can be shown they have been reduced by a factor of ten).*

nest was situated between two grass meadows bordered by sedges, hence it fed more by grazing and was thus able to feed close to the nest. Both tended to prefer areas well away from the nest when they fed by dipping.

Even though daylight was continuous for much of the incubation period, females roosted more consistently at night. It has been shown for other high-latitude breeding birds too that a prolonged period of inactivity occurs, despite continuous daylight. Birds have a marked diurnal temperature cycle, the amplitude being partially related to body size, which may affect activity patterns regardless of continuous daylight.

Both females mainly left their nests later in the day, after 06:00 for the western female and mainly after 12:00 for the southern female, although the latter also left at other times during early morning. The southern female grazed between 06:00 and 07:00, 09:00 and 10:00, and also more frequently after 12:00 (Fig. 8.6). In China, Ma (1996) found female activity away from the nest peaked in the afternoon when weather was warm, although this varied with season. Nevertheless, the indication is that females are less active at night despite available daylight.

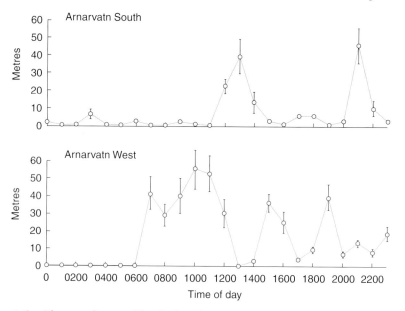

Figure 8.6.   The mean distance of females from the nest throughout the 24 hr. period.

BROOD SIZE

Hatching in western, southern and eastern parts of the Whooper's range occurs in June, while in the cooler central and northern parts it may occur during July, and even sometimes early August (Mathiasson 1991), with local variation depending on altitude. Thus, on the Ural-Tobol plateau hatching is in late June–late July (Samigullin & Parasich 1995). It normally occurs after 30–32 days according to Ma & Cai (2000), but under exceptional circumstances females may continue incubating long after hatching should occur, thus one, whose eggs failed to hatch, incubated for two months, perhaps indicating that they do not have an inherent ability to judge when hatching should occur, and that they depend on clues from within the eggs, perhaps calls. They remain attentive at the nest on the day of hatching.

Not all pairs lay eggs, and of those that do, not all hatch. Hatching losses are thought to be very low among Whoopers, although unfertilised eggs have occasionally been found (Myrberget 1981) as have dead newly-hatched young (Fig. 8.7). In China, only 60% of 78 eggs hatched (Ma & Cai 2000). Hatching takes 36–48 hours, depending on clutch size or 20–40 hours, during which time the chick chirps noisily, calls that may help communicate readiness both to the parents and other eggs. After hatching, the eggshells are left in the nest, which is abandoned within 1–2 days. During the first 2–3 days the cygnets retain some attachment to the nest, sometimes returning to rest or be brooded by the female (Brazil 1981c; Krechmar 1990; Ma 1996). I am unaware of any cases of the adults carrying the

Figure 8.7. Almost fully developed embryo from an abandoned egg.

young on their backs although this has been reported for Trumpeters (Bailey *et al.* 1980), but they are very attentive and remain in close proximity.

In China, during the first 1–3 days, chicks seldom leave the area of the nest. At this stage they weigh a mean 198 g or just 59% of the fresh egg weight, and they continue to lose weight as they live off what remains of their internal energy resources from their egg's yolk, but begin to gain weight from about day five, then grow rapidly to 550–600 g by day 12, and 3,100–3,700 g by 40–45 days. They fledge during September after 90–105 days (Ma & Cai 2000).

Considerable losses occur during the first months. Embryonic mortality is apparently insignificant, but the behaviour of the parents may, along with factors such as weather, influence the likelihood of eggs hatching. Once the clutch has hatched, what factors affect brood size (and when)? Brood size may decline at many different stages, such as in the first few days during the first forays away from the nest; during longer journeys (if the parents choose to move to another location); during the period leading up to fledging, at the early flying stage, and in late summer/early autumn as the weather cools.

Mean brood size just after hatching is broadly similar, even in very different regions, with 3.7 recorded in north Norway, 3.8 in Swedish Lapland, 4.0 in Yamal and 4.5 in Baraka[3], with numbers declining in relation to adverse climatic conditions in September, particularly in the Pechora (Mathiasson 1991). However, average brood size can vary considerably between years (and declines from hatching) (Poyarkov & Johnson 1996), thus Vartapetov (1984, 1998) found that in the Ob Valley area, among 64 broods in 1994 and 71 in 1995, the number of cygnets was 1–8 (two broods of eight were observed there in September 1995), and mean brood size in 1994 was just 3.8, whereas it was 5.0 in 1995. An even higher average brood size of 5.3 (N=34) reported from the Elizarovskiy reserve, also in the Ob Valley, perhaps indicates that this area provides particularly suitable habitat. In strong contrast, in Yakutia not only did few pairs breed, but their output of young was very low, with mean brood size varying from just 2.6 to 3.3 (N=25), and in marginal areas it was even lower at just 1.7 (N=6; Degtyarev 1990).

Myrberget (1981) found that some families in Norway failed quickly, with three of his 37 hatched broods losing all of their cygnets soon after hatching. He also confirmed results from Iceland that serious mortality occurs *before* autumn migration, with the mean brood size in his study area falling from 3.7 in early July to 2.6 by August/September. Reynolds (1965), who calculated mortality rates of Mute Swans in each week of life, found that by far the greatest mortality (22%) occurred in the second week. Whoopers also suffer significant mortality during the critical early period when the small cygnets are very susceptible to poor weather conditions and only just learning to feed (Krechmar 1982a).

The final, or surviving, brood size is also related to the degree of mobility of the parents, although the causal factor here is presumably food. Food quality or availability in the natal territory appears to influence whether families are sedentary or mobile. The moving of families has been reported from Iceland and Finland and is a risky business, with cygnets easily lost en route (Haapanen *et al.* 1973a; Brazil 1981c; Ohtonen & Huhtala 1991; Rees *et al.* 1991a). Rees *et al.* (1991a) reported families moving up to 3 km when the marsh around their nests dried out, and though such distances may sound short, for a brood of small cygnets perhaps just weeks old they are considerable.

When Ohtonen & Huhtala (1991) compared the brood sizes of Whooper Swan pairs in different habitat types, they found that by late July sedentary pairs had significantly larger broods than mobile pairs. While this difference is partially the result of sedentary pairs starting with larger clutches, some of the losses experienced by mobile pairs are considered to be directly associated with their mobility, while the relationship between clutch size/egg volume appears to be different in the two habitats (Ohtonen & Huhtala 1991).

Haapanen *et al.* (1973a), who were first to point out the difference between sedentary and mobile pairs, considered that higher cygnet mortality during the first weeks after hatching among the latter was the most probable reason for their having smaller broods. As further confirmation of the costs of mobility, Myrberget (1981) found that pairs in Norway that moved their young between bog and a more favourable lake reared only one cygnet but, when nesting at lakes, reared three. It seems then that pairs nesting at sites from which they must subsequently move their young are in effect using suboptimal habitat.

Whereas habitat differences or the differing degrees of mobility in different habitats do lead to differences in brood size in Finland and Norway, studies in Iceland of success at different altitudes, by Rees *et al.* (1991a) revealed that by August (1988) mean brood size (3.1 cygnets) did not differ between highland and lowland sites. While mean brood size did not differ, cygnet size did. They found that cygnets reared in the highlands were significantly smaller and lighter. There could be several reasons for this: lowland sites are generally assumed to provide richer feeding grounds, and adults are, for example, significantly heavier than in the highlands, so habitat quality and a difference in age might complicate the matter. Not only did Rees *et al.* (1991a) find highland-breeding Whoopers to be lighter and have smaller cygnets in late summer, but they also found that they lost significantly more cygnets between summer and winter. With highland birds taking significantly fewer cygnets to their wintering grounds, it seems that in addition to the significant annual variation in breeding success, there is also an important regional or habitat-related effect that may mean that birds able to secure territories in rich habitat areas contribute disproportionately to the next generation, and in particular to the percentage of cygnets reported from wintering grounds.

Rees *et al.* (1991a) were able to relocate 26 of their study pairs over the summer and found that pairs with small clutches (three eggs) tended to lose fewer eggs or cygnets than pairs with larger clutches (six). Nevertheless, clutch size and subsequent brood size still proved to be positively linked.

Following those same ringed breeding birds into the following winter (1988–89), Rees *et al.* (1991a) found that although pairs with larger clutches led more cygnets to their wintering grounds, mean brood sizes of lowland families fell from 3.1 to 2.3 cygnets, but those of highland families declined even more dramatically and were significantly smaller, at just 1.4 cygnets. At some point after August, therefore, when brood size still averaged 3.1 cygnets in both regions, but before they were relocated on the wintering grounds, both lost a large proportion of their cygnets, though the correlation between brood size in August and number of cygnets surviving autumn migration was strong for lowland families (Rees *et al.* 1991a).

Gardarsson & Skarphedinsson (1984), in a large-scale study, examined breeding across Iceland. They found that overall mean brood size was 2.6 cygnets. This average, however, masked important regional variability; in the east broods were actually smallest (2.3), while those in the southwest were largest (2.7). In October, these regional differences in brood size were, perhaps unsurprisingly, not statistically significant, for by that time large numbers have already left the vicinity of their breeding areas in readiness for autumn migration or local wintering. The authors concluded, therefore, that the similarity between regions might have resulted from the swans already having mixed in autumn. They were also able to look at the changing sizes of broods by month and found that those in early August, essentially late summer, were significantly larger than those in autumn (September–October) prior to migration (Table 8.5). In Finland, Haapenen (1991) found a north-south difference. In early September mean brood size in the south was 3.2, but in the north it had fallen to 2.5. A similar north-south difference is evident in the Russian Far East too, where in the Penzhina region it was 2.9, but further north in Anadyr it was 2.2 (Kondratiev 1991.

Gardarsson & Skarphedinsson's (1984) broader examination of brood size showed that later differences (after September) were not significant, leading them

Table 8.5.   *Seasonal change in mean brood size*
(after Gardarsson & Skarphedinsson 1984).

Month 1982	Location	Mean brood size
August	Iceland	3.16
September	Iceland	2.64
October	Iceland	2.60
November	Scotland	2.56

to conclude that post-fledging mortality was not of particular importance and that for birds that had survived until then the migration to Britain and Ireland was not especially hazardous. Their results indicate that the most significant period for cygnet mortality is that between hatching and fledging. This corresponds in Finland with the period of most serious losses, particularly of mobile families.

Haapanen (1982) was able to follow the reproductive history of an unusually confiding female in Finland, which was not only exceptional in the large size of her clutches but also in her exceptional brood sizes. She fledged 29 cygnets in just five years averaging 5.8 per year, a number the author calculated would have taken a normal female nine successful breeding seasons to achieve (Table 8.6). The brood of eight fledged in summer 1976 is believed to have been the record for a wild Whooper. The previous maximum recorded from Finland was of six, although apparent broods of seven have been reported from Sweden, and an apparent brood of eight has been reported from Scotland (Brazil & Spray 1983).

The length of the summer is often considered a critical factor limiting breeding success. Regardless of brood size immediately prior to migration, the abrupt ending of the summer can, theoretically, ruin all chances of parents migrating south with their offspring. Until fledging in northeast Russia, Whooper cygnets keep to overgrown shoals of rivers, and in the case of early frosts families can actively keep the water ice free until flight is achieved (Krechmar 1982a). An early freeze or other factors forcing parents to migrate without their brood would presumably usually doom the entire brood. To my knowledge, only Myrberget (1981) has reported

Table 8.6.   *Brood size of a tame*
Whooper Swan in Finland
(after Haapanen 1982).

Year	Cygnets fledged
1974	2
1975	6
1976	8
1977	7
1978	6
N	29
Average	5.8 ± 2.0

observations that relate to this; in three cases, parents departed while their single cygnets were still unable to fly. In one case the cygnet disappeared (its actual fate unknown), the other two both died soon after the departure of their parents. However, in Russia, there is anecdotal evidence that families unable to leave the vicinity of their breeding range may resort to the relative and temporary safety of polynias as breeding territories freeze.

Brood size, which is limited of course by clutch size, declines from hatching and there are clear links to the situation in winter. Brazil (1981c) observed 186 broods from 50 flocks in Scotland, noting that, as parents without cygnets are indistinguishable from adults which have not bred, all failures, that is broods of zero, are missed. In winter, therefore, it is impossible to calculate the proportion of adults that have bred nor is it possible to calculate an absolute mean brood size. However, comparisons of mean brood sizes between years provide an indication of breeding success; the reduction in mean brood size from 2.5 in 1978–79 to just 1.3 in 1979–80 clearly reflected the very poor breeding season of 1979, as did the exceptionally low percentage of cygnets recorded both locally (3.7%) in Scotland and in the national survey that winter (5.12%; Brazil & Kirk 1979).

Boyd & Eltringham (1962) suggested that poor breeding success was as likely to be due to failure to hatch or rear *any* young as to reduced brood size. Brazil's (1981c) results suggest, however, that the reduction in the percentage of cygnets, in 1979–80 at least, was due to higher mortality of cygnets and/or a general reduction in clutch size, and was not due to failure by one population and normal breeding success by another, as the latter would result in a reduced proportion of cygnets but normal brood sizes.

Frequency of different brood sizes can differ significantly between winters, dependent on the conditions of the previous spring. In 1979, poor breeding success was as a consequence of a very late spring in Iceland, leaving highland breeding areas frozen too late for many pairs to breed at all, and even many lowland pairs lost eggs or whole clutches. Those young that did hatch were about three weeks later than usual (Brazil 1981c). Smaller broods may have been a result of smaller clutches due to re-absorption of eggs by females forced to delay laying. It is a frequent phenomenon in waterfowl that clutch sizes are smaller in late seasons and the most acceptable view seems to be that this is related to the amount of fat reserves and their rate of depletion (Owen 1980), factors that surely also relate to migratory swans.

Boyd & Eltringham (1962) assumed that mean brood size declined slowly during winter, and that this explained the decline in the percentage of cygnets present in early winter after initial autumn arrivals. This assumption, however, fails to take into account the arrival of further adults (without young) in early winter, which, by dilution, also leads to reduced percentage of cygnets. Several hundred Whoopers remain to winter in Iceland, perhaps including some non-breeders, some failed breeders or some late breeders. If severe winter weather strikes Iceland, some of these may migrate later, after the peak arrivals into Ireland and the UK in October– November. Earlier, I considered it likely that the long autumn/spring migrations were times of greatest risk for cygnets, not the intervening period, with, of the two, autumn representing the greater risk, given that the cygnets are smaller (Brazil 1981c). On reflection, the dangers of impacts with wires and of lead poisoning might make the winter more dangerous. Nevertheless, Gardarsson & Skarphedinsson (1984) have shown that the most significant period of cygnet loss is actually within the first month or two of life.

Post-hatch brood amalgamation is common among waterfowl in general. It has been reported in Trumpeter Swan occuring where several pairs share breeding marshes or lakes and while cygnets are 1–14 weeks old (Banko 1960; Mitchell 1994; Mitchell & Rotella 1995, 1997), and once in Bewick's Swan in Japan (a pair with 12 cygnets; Brazil 2002a). Although it is also likely to occur among Whoopers, the only documented cases that I am aware of involved a single cygnet being adopted by another family at Caerlaverock (Black & Rees 1984), and once at Welney (Rees *et al.* 1980). It is important to know the extent to which brood amalgamation occurs when considering large broods, as their origins may not be simple, and much clearly remains to be learned.

CYGNET GROWTH

Cygnets are precocial, nidifugous and only retain an attachment to the nest site for the first 2–3 days, where they are brooded by the female at night and to which they return if the weather becomes inclement. The overall pattern of growth is prolonged, fledging taking from 87 (78–96) days in Europe (Haapanen *et al.* 1973a, though this was earlier 80 (77–84) days among captive-bred young suggesting that perhaps food availability/habitat quality may affect the speed of growth and hence date of fledging. This period appears to be longer elsewhere, taking 90–105 in Russia and China (Bianki 1988; Vengerov 1990b; Degtyarev 1987; Ma 1996). The similar Trumpeter takes 90–122 days (usually 90–102) (Mitchell 1994). There have been few opportunities for researchers to study in detail the growth and development of cygnets, either in the wild or in captivity, although in China, Ma (1996) found that wild cygnets 1–3 days old weighed an average of 198.2 g or just 58.7% of their original egg weight, while three downy chicks trapped on 1 July as they left a nest weighed 190, 196 and 200 g, and four-day-old chicks away from the nest on 3 July weighed 251–295 g (Krechmar 1990). At Moscow Zoo, Ostapenko *et al.* (1990) studied cygnet growth and found them to lose weight slightly over the first two days after hatching before increasing in weight more than ten-fold in the first 30 days (Table 8.7).

One who has made a detailed study of cygnet growth is J. M. Bowler, who studied the growth of ten captive-bred birds, in Wales in 1990, from hatching during the second week of May to their 23rd week, which equates to the age at which most wild cygnets would commence migration. They consisted of three males and seven females, and were given a 24-hour source of light during their first 25 days in a pen, thereafter they were kept in a larger pen with only natural daylight, were fed on a high-protein pellet diet, with grass available in the pen (Bowler 1992).

Table 8.7. Cygnet growth (after Ostapenko et al. *1990).*

Whooper Swan cygnets	Days					
	1	2	10	14	21	30
Weight (g)	191.35	190.98	372.0	485	864	2,150

As was predictable, the cygnets grew rapidly, making more substantial weight gains and showing steeper growth curves than typical of other waterfowl (Kear 1972), as would be consistent with their need to concentrate their growth at high latitude, to be sufficiently mature to make their first autumn migration. Although growth curves are typically sigmoidal, the weight of these cygnets increased linearly over the first seven weeks, from a hatching weight of 210 g (Kear 1972), with daily gains of 127 g, but then reached a plateau as they approached adult weights of 6–9 kg, with only a slight increase between weeks eight and 23 (Bowler 1992). This represents an approximately thirty-fold increase in weight, which is greater than previously recorded for any swans. Although the males were consistently up to 10.4% heavier until week ten, the difference was not significant, whereas among Mute Swans over the same period the difference is marked, and males average 28% heavier than females at fledging, confirming Kear's (1972) statement that the 'northern swans' show a less marked difference between the sexes (Mathiasson 1980; Bowler 1992).

Skull, cranium and tarsus measurements followed similar patterns, with very rapid enlargement and elongation over the first seven weeks of 2.09 mm/day for the skull, 0.61 mm for the cranium, and 1.80 mm for the tarsus. Growth levelled off quickly, with no further growth in tarsus length after week 13, very slow growth in the skull continued until week 18 and cranium height increased slowly until week 23. Tarsus length had reached 99.8% of mean adult length of 118 mm (Cramp & Simmons 1977) by week 18. Development of the tarsus, relative to other parts of the body, was even faster among wild cygnets than the captive birds (Bowler 1992) and is presumably related to their need to keep up with their parents. Male cygnets of both Whooper and Whistling Swans have significantly longer tarsi than females (Limpert *et al.* 1987; Bolwer 1992). In contrast to the rapid linear growth, followed by a plateau, of skull, cranium and tarsus, and also in weight, total wing length follows a distinctly sigmoidal growth curve, because during the first six weeks there is relatively little growth until the flight feathers first appear in week six. Subsequent growth is rapid, combining both primary growth and development of the forewing. The primaries reach their maximum length by week 18, and then decline as a result of abrasion. Primaries and secondaries develop evenly and with little individual variation. Geese fledge when their wing feathers have reached *c.*85% of final length (Owen 1980), and the Whooper cygnets reached this stage at around day 80 at a mean weight of 7.5 kg, although first flight was not recorded because they were pinioned

(Bowler 1992). Assessing the exact time of fledging is very difficult in the wild and though dates are often unreliable, it appears that fledging in this study was faster than among most other comparable studies. Among wild birds, Ma (1996) found cygnets weighed 6.5–7.0 kg at fledging in September.

Bowler (1992) defined six stages of development for young Whoopers. During stage 1, from hatching to two weeks of age, chicks are covered with down, the body appears rounded, and the neck and legs short. During stage 2, at 3–4 weeks old, the still-downy chicks have a longer body, and the neck and legs look long. Stage 3 occupies weeks 5–7; during this period, the contour feathers appear, contrasting with the remaining down and giving them an untidy look. The neck and tail become obvious and the legs are still pinkish. During stage 4, weeks 8–11, the contour feathers become increasingly uniform and grey, but some down remains. Although the primaries are by now well developed, they are still flightless. Body shape approaches that of adults, and the legs are becoming blackish. In stage 5, weeks 12–14, fledging occurs, no down remains, the contour feathers are uniformly grey, and they are now capable of flight. In the final stage, from week 15, white patches of contour feathers appear on the mantle, flanks and belly and increase in extent as the bill pattern becomes better defined (Fig. 8.8).

In terms of plumage development, the first contour feathers to appear are the scapulars and underwing-coverts, at 4.3 weeks, followed by the head and greater wing-coverts, a timeframe very similar to the four weeks required by Alaskan Trumpeter Swans. Contour feathers on the neck, breast and flanks, wings and tail become evident in week six, but distinctive clumps of down remain on the

Figure 8.8.   Six stages in the development of Whooper Swans (after Bowler 1992).

forewing where they protect the developing flight feathers. By week eight, rapid contour feather development gives cygnets a uniform appearance and down remains only on the wings, lower back and head. By weeks 10–11 the only down is on the underwing. By week 15 they appear slightly less grey with white areas developing on the mantle, flanks and belly; these continue to spread during the remaining weeks until week 23 (Bowler 1992).

Whitening continues into the first winter, with most attaining a pale whitish-grey plumage early after the turn of the year. Over the same timeframe, the bill, which is initially pinkish with a black tip, becomes white with a black tip. The black area darkens and the white area begins to develop a faint hint of yellow early in the winter, and by departure from the wintering grounds, the bill is at least pale yellow and black, and much of the plumage is white (Brazil pers. obs.). Leg colour also changes markedly, from greyish pink during the first seven weeks, to greyish black by week eight (Bowler 1992) and thereafter is fully black. Bowler (1992) also found consistent differences between males and females, with the former heavier and longer in measurements, although the differences were not always significant. Significant sexual differences in skull and tarsus length were noted by week 10, as were near-significant differences in cranium height and body length. In comparing wild with captive birds, Bowler (1992) found that in weeks 8–10, wild birds were significantly smaller than captives in all measurements except one, tarsus length. It seems that primary feather development is also considerably slower, by two or more weeks, among wild birds. Such differences may be attributed to the consistent diet and limited exercise of the captive birds.

That cygnet growth rates are related to habitat quality and probably the availability/nutritional quality of food has recently been proven by Knudsen *et al.* (2002), who found that cygnets raised in peatlands and oligotrophic lakes were significantly lighter (35% and 19%) and had shorter skulls (12% and 6%) on average than those reared at eutrophic lakes. Furthermore, fledging was estimated to be delayed by 20 days in peatlands and by ten days at oligotrophic lakes compared with eutrophic lakes. The differences continued into the survival ratios, with 76.4% of cygnets from eutrophic lakes being recorded on the wintering grounds compared with 71.4% of those from oligotrophic lakes and just 51.9% from peatlands.

FAMILY LIFE

The few detailed behavioural studies that have been made during the breeding season are mostly from Iceland and Fennoscandia, but even among these very little work has been done on the behaviour of individual family members.

FAMILY LIFE FROM HATCHING TO FLEDGING

Behaviour between hatching and fledging, a period of *c*.12 weeks (Batten *et al.* 1990), has been little studied in Whoopers and I have been unable to find modern material to supplement my own limited observations made in Iceland 25 years ago. Because of that, I am heavily reliant on those observations of just a few birds in this description. I hope, however, that these will prove representative or perhaps sufficiently unrepresentative to stimulate others to study the same period.

Cygnets are nidifugous, hatching already covered with down and with their eyes open. They are able to move immediately, so the nest remains a focus for their ever-increasing activity only for a very brief period (2–3 days) (Fig. 8.9). Damp and grey at first, they soon become dry and fluffy. Cygnets discover their own oil glands when one day old and thereafter spend a great deal of time nibbling and preening, receiving vitamins D in this way (Wilmore 1974). Newly hatched cygnets may be assumed to have advantages over other, smaller birds, because being large, they have a relatively small surface area to volume ratio, and thus lose heat slowly. Furthermore, swan eggs are very large, providing substantial yolk on which the cygnets rely during the first day or even week. Mortality is most likely, therefore, in the second and subsequent weeks (Wilmore 1976; Brazil 1981c). Cygnets found dead within days of hatching may have had too little yolk reserve, and birds reported as supposedly feeding on insects in their first week may just have been indulging in exploratory pecking at unfamiliar moving objects. The cygnets soon begin exploring, but the family bond is strong, and they rarely stray far from each

Figure 8.9.   The attentive parents soon lead the cygnets away from the nest.

other or their parents. Once they start feeding they are at first dependent on their parents, who, in pulling up aquatic vegetation from deeper water make food available, and also escort them, and brood them while roosting (Brazil pers. obs.).

During a study of this early period, I examined the time budgets of three different families in Iceland. Although there was variation both within and between pairs regarding the time allocated to specific behaviours, nevertheless certain generalisations can be made. Only 5–6 behaviours were regularly recorded: 'head-up', preening, dipping, dabbling, grazing and roosting. Thus the fledging period more closely resembles other periods of the annual cycle than the incubation period, in terms of the variability of behaviours.

Individual identification of parents was not practical, and identification of the parents during the fledging period was largely limited to describing their proximity to their brood. Based on their stages of moult, however, I strongly suspected that the birds most often close to their cygnets were females.

The parents of any given family differed in their behaviour depending on their distance from the brood. Within each pair, the parent closest to the brood always spent more time 'head-up' (45–55%), and the parent furthest away less (41–48%). Both parents spent very similar amounts of time preening and roosting, but in each case that furthest from the brood spent more time feeding than did the nearer parent. Although these differences were not significant when each pair was analysed separately, when the three pairs were analysed together they were highly significant (Brazil 1981c).

Different families spent differing amounts of time feeding and fed in different ways. The northern family grazed on grasses and sedges along the lakeshore. The southern pair fed by grazing and dipping, while the western pair fed mainly by dipping and also spent more time feeding than either of the other two.

Although cygnets could not be identified individually, the behaviour of each individual was recorded separately in order of their proximity to the nearest parent (nearest = 1). The total frequency of any particular behaviour was therefore divided by the number of cygnets in the brood under observation, giving the proportion of time in a given behaviour by an 'average' cygnet. Cygnets spent much less time 'head-up' than adults, and those in broods of four spent less time 'head-up' than cygnets in the brood of three. There was considerable inter-brood variation, but cygnets from all three spent significantly more time feeding (40–65%) than their parents (17–39% nearest parent; 23–45% furthest parent). Preening occupied only 4–9% of their time. Roosting was extremely variable between broods, 9% for the western pair 14% for the northern pair and 25% in the southern pair.

Cygnets engaged in eight different behaviours during the fledging stage: 'head-up', preening, dipping, feeding on emergents, picking, grazing, drinking and roosting. Four of these, 'head-up', dipping, grazing and roosting, were associated with their parents and each occurred synchronously with their parents, while the others occurred independently of their parents. There were, however, differences between the three pairs as to which behaviours were significantly synchronised. The 'head-up' behaviour of three of the northern families' cygnets was synchronised with both parents, all four of the southern family's cygnets were synchronised with both parents, and two of the western family's cygnets were synchronised with the nearest parent, while that of the third cygnet was synchronised with the furthest parent.

In addition to the links between parent and cygnet behaviour, three aspects of spacing between family members were also measured. These were 'inter-parent', 'cygnet-to-nearest-parent', and 'cygnet-to-cygnet' distances. When calculating cygnet to parent and cygnet to cygnet relationships, individual distances, i.e. distances from each cygnet to its nearest sibling or parent, and total inter-cygnet and parent to cygnet distances, were recorded. All intra-family distances proved to be rather variable, thus standard deviations were large and median results are discussed here. Considering that the area each pair occupied facilitated movement of up to several hundred metres in several directions, inter-parent distances were small. Median distances varied at 8.5–10.0 m for the three pairs. Inter-parent distances were, however, very much larger than parent to cygnet distances, which were also very similar for all three. Even the furthest cygnet from a parent was no more than about three times as far away as the nearest cygnet. Inter-cygnet distances were also, as one might expect, very small. The medians ranged from only 0.2 to 0.9 m (Brazil 1981c). Cygnets clearly maintain closer proximity to each other than to their parents, and tend to associate with a single parent, whereas parents differ markedly in their proximity to the broods.

As cygnets mature, inevitably they become bolder and explore more. Parental care declines over the same period and as a result of the combination of these factors measurable changes in the distances between parents and broods ('parent-to-cygnet' distance), and between the young ('inter-cygnet' distance) were expected. The total 'cygnet-to-nearest-parent' distance increased significantly in each of the three families during the course of fledging. Likewise, total 'inter-cygnet' distance, and mean 'inter-cygnet' distances showed increases during the fledging period, indicating the steadily growing independence of the cygnets not only from their parents, but also from each other (Fig. 8.10).

Of particular interest during this period was that the parent closest to the brood (thought to be female) fed less, and was head-up more than the further parent; this contrasts strongly with the Trumpeter in which the male fed less and was more vigilant, allowing the female and cygnets more time to feed (Grant *et al.* 1997).

THE PROXIMITY OF PARENTS AND CYGNETS

During the fledging period, parental care is again divided, but seemingly unevenly. In July 1980 I was able to collect a small amount of supplementary data on the relationships between parents and cygnets. Three families were observed on a single day. During these observations parents were sexed by eye, using the length of tail and neck, and the degree of staining on the feathers. It has been assumed that females moult first (Scott *et al.* 1953; Kinlen 1963; Brazil 1981c). Males, therefore, are identifiable not only by their longer, thinner necks and longer wings relative to the tail, but also by their

Figure 8.10. A 'mobile' family, the parents still in wing moult, with their surviving cygnet.

greater degree of staining on the head and neck. Females are not noticeably stained until later in the season, because during incubation they do very little feeding.

In each of 30 spot samples the parent to which the cygnets were closest was iden-tified and the distance between parents was recorded as less than, or greater than, five swan lengths (SL). It was impossible to determine which of the parents the cygnets were with if the parents were less than five SL apart. The parents of pair one were closer together than five SL in five of the 30 spot samples. During the rest of the period, all (four) or most (three) of the cygnets were closest to the male in 8% and to the female in 92% of the samples. The parents of pair two were closer together than five SL in 21 of the 30 samples. During the rest of the period, all (three) or most (two) of the cygnets were closest to the male in 89% and to the female in 11% of samples. The parents of family three remained more than five SL apart for the entire observation period and in 100% of samples all of the cygnets were closer to the female.

Cygnets were more often with the female in two of three families, and with the male in one. During observation of the third family, both parents were close together for most of the time, so the male bias is a result of only nine minutes observations. These results are from a very brief study, but they confirmed that cygnets tend to stay close to a single parent, and indicated that this may be the female. De Vos (1964) and Scott (1977) found the same strong association with the female during the fledging period in Trumpeter and Whistling swans, indicating that this may be a general rule among 'northern swans'. If so, this means, however, that the female's concentrated investment in her cygnets is extremely prolonged.

Hansen *et al.* (1997), also found that families in Finland keep very close together, with distances between parents and cygnets averaging eight SL, and between cygnets five. Of more interest was the discovery that families in different habitats show very different family cohesion, presumably relating to their vulnerability. Birds on water averaged ten SL between parents and cygnets, and six between cygnets, whereas on land these distances were reduced to four and three.

PARENTAL VIGILANCE AND FEEDING DURING THE FLEDGING PERIOD

Watching the brood and maintaining vigilance on their surroundings are important aspects of parental care, but inevitably impinge on other behaviours and limit the time available for feeding, which for the female may be very significant. An examination of the relationship between vigilance and feeding, therefore, was made in northern Iceland where the length of time parents remained 'head-up' and the length of the periods between 'head-up', were measured for three families. The amount of time spent vigilant may be varied by lengthening or shortening the period spent 'head-up' or by varying the period of time *between* 'head-up'.

It became clear that parents of a given family did not differ significantly in the length of 'head-up' in July–August. This similarity has also been observed among birds of differing classes in winter. However, although the lengths of vigilance bouts did not differ, the times between periods did. It transpired that the parent closer to the brood (<5 SL away) spent significantly shorter periods between 'head-up' than did the parent further away (>10 SL), thus overall the closer parent spent more time vigilant (Brazil 1981c).

Although parental care was assumed to decline as cygnets mature, it seemed questionable as to whether vigilance as an aspect of this would decline measurably in the early stages of development, although for the parents such a reduction would leave them longer periods for feeding. Time-budget data suggested that both parents of all three broods spent slightly less time 'head-up' and slightly more time feeding in August than July, offering some support for this notion. Although there was a tendency for the amount of time both parents spent 'head-up' to decline over the period, neither time spent 'head-up' nor inter 'head-up' significantly correlated with date. Clearly, much remains to be learned concerning changes in parental care among Whoopers.

Another way of examining changes in vigilance is to look at the amount of time spent feeding during this period. An increase in feeding would imply a decline in vigilance, because 'head-up' and feeding are mutually exclusive. Increasing the amount of time spent feeding necessarily restricts other behaviours, although not all of the increase in feeding may be accounted for by a reduction in 'head-up'. The lengths of feeding bouts (dips) for all three families combined were as follows: in July the parent closest to the brood had a mean dip duration (mdd) of 5.4 seconds (N=211), which was significantly shorter than that of the parent furthest from the brood (9.8 seconds, N=128. This difference was not significant in August, when the nearer parent had a mdd of 8.6 seconds (N=68) and the further parent had a mdd of 9.5 seconds (N=44). During this period then, the mean dip duration for the parent closest to the brood increased significantly (Brazil 1981c). It is apparent that parents spending most time closest to the brood increased the length of their feeding bouts between July and August. This would necessarily restrict the interval at which 'head-up' could occur. Both parents also tended to increase the total time spent feeding from July to August, which inevitably also meant a reduction in all other behaviours. It seems very likely that, coincident with this increase, there was a reduction in parental care in the form of vigilance. This is an interesting area for further study.

For one pair, individual identification of the male and female was possible, based on differences in size and the relative amounts of orange staining. This made it possible to compare aspects of behaviour of the same parent when it was at different distances from its brood. During the fledging period, the female spent more time close to the brood, and was also 'head-up' more frequently. As she moved away from her brood, however, her frequency of 'head-up' decreased significantly, although the length of time she spent vigilant at each bout did not change. In contrast, when the male moved relative to its brood, neither the length of 'head-up' nor the length of inter-'head-up' changed significantly. If this pair was at all representative, it seems that the time spent vigilant by males might be constrained by factors other than proximity to the brood, whereas for females, proximity is the important factor. The inter-'head-up' period of the female when close to the brood was significantly shorter than the male's, whereas there was no significant difference between the length of their 'head-up' bouts, as a consequence the female was more vigilant than the male when close to the brood. Even when she moved away, the female was still 'head-up' significantly more often than the male at the same distance, and for significantly longer (Brazil 1981c), clearly showing that the female is the more vigilant parent.

Marked differences in parental roles, during the incubation and fledging periods, are reflected especially in levels of vigilance. Males spend more time vigilant than females during incubation, but females are more vigilant than males during fledging.

SPACE UTILISATION AT ARNARVATN

It is difficult to distinguish between home ranges and territories among Whoopers because they generally occupy isolated sites, and interactions between territory/range-holders are thus infrequent. Despite past assumptions that pairs always nested individually at isolated patches of aquatic habitat, observations at certain locations in Iceland (Brazil 1981c) and China (Ma 1996) suggest that Whooper Swans may also defend spatio-temporal 'type A' territories (Wilson 1975) that is they defend stable territories against neighbouring conspecifics. During summer 1978 I studied how three families and one pair used their 'home ranges' (Figs. 8.11. & 12). Territorial interactions were observed between the southern and western families, and between the western family and both the pair

Figure 8.11.   Meetings betweeen families occur at some breeding sites (the females are in wing moult).

and northern family. The southern family was most consistent in its use of a limited area. The western pair used the most extensive area; the area around their nest was most unsuitable for grazing, thus the family regularly crossed the lake to meadows on the east shore. This pair moved its range to include the north shore after the disappearance of the northern pair, which had, until 31 July, vigorously defended that area. The western pair was last seen at the north end on 15 August, after which the northern family extended their range to include almost the entire area previously used by the northern pair. Generally, interactions occurred only when birds occupying adjacent ranges were close to each other, although on several occasions an individual was seen to patter or fly more than 100 m to engage another individual (Brazil 1981c).

The main function of a 'territory' would seem to be to ensure that there is a suitable area for the family to feed between hatching and fledging. Both Whooper and Trumpeter Swans, however, also nest in oligotrophic areas where limited food supplies force the female to visit suitable feeding areas during incubation breaks and then the cygnets to be led from the natal area to another site soon after hatching. In such areas, pairs are widely spaced (e.g. Bulstrode *et al.* 1973), and only rarely is there the functional requirement to defend a territory. In a rich area, however, as at Arnarvatn or the Ob Valley to both of which several pairs have been attracted, or to an extensive area such as Bayinbuluke, where many pairs occur, there is the constant provocative stimulus provided by the signal function of the white plumage (Wynne-Edwards 1962; Armstrong 1965; Johnsgard 1978) and the functional requirement for territory defence. The well-structured nature of the displays involved in such interactions suggests that successful maintenance of a territory is

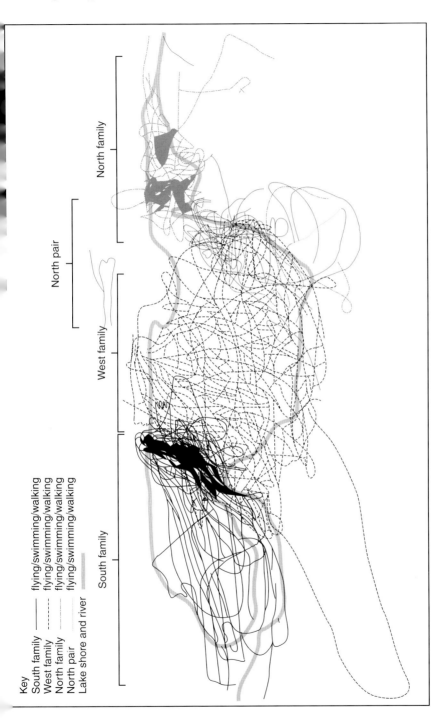

Key
South family ———— flying/swimming/walking
West family ------- flying/swimming/walking
North family ········ flying/swimming/walking
North pair ········ flying/swimming/walking
Lake shore and river ————

Figure 8.12. Home ranges of three families and one pair at Arnarvatn in 1978.

important, the value of which appears to be highest during the hatching to fledging period.

FACTORS AFFECTING BREEDING SUCCESS

Clutch size among waterfowl is extremely variable and dependent on weather, food supply and other variables. Similarly, nesting success and brood size vary between years. Nesting success is probably more dependent on the timing of spring and the conditions of the parents, whereas brood size is probably more dependent on weather and food availability in summer/autumn. The sizes of Whooper broods in winter correspond well with brood sizes on the breeding grounds during the preceding summer, indicating that reductions in family size occur through factors operating during the breeding season and autumn, rather than winter mortality. The same seems to be true of Bewick's Swan (Evans 1979).

Breeding success (in terms of cygnets reared) can be extremely variable. For example, in one region of north Iceland, in 1978, five pairs nested at three localities producing 16 cygnets initially, although one died, leaving four successful pairs with 15 young (mean = 3.75). In 1979 the coldest year of the century (Gardarsson *in litt.*), eight pairs at six localities reared just nine cygnets (mean = 1.1), and in 1980 seven pairs reared 24 young (mean = 3.4; Brazil 1981c). Among pairs nesting at these six locations, only one was a mobile pair, and given that they bred in all three years their total output was among the lowest recorded. Their nesting area was inaccessible, so the number of eggs or cygnets initially was unknown. In 1978, the family walked to Neslandavik, a journey over extremely rough terrain, but their only surviving cygnet disappeared within three days; other cygnets may have been lost en route. In 1979 and 1980 one and two cygnets were reared, but how many were lost en route is unknown. Sedentary pairs produced more young (3–4), but still experienced enormous inter-annual variation, and individual pairs (or nest sites) differed markedly in their output (see Table 8.8).

Breeding success depends not only on climate but also on the quality of breeding habitat. Gardarsson & Skarphedinsson (1984) found that the proportion of young in the Icelandic total was 18.2%, but that regionally it varied to as high as 25% in the north and 23% in the south (both regions with extensive breeding areas and few moulting birds), to as low as 8% in the southeast, a region dominated by a large moulting concentration, and with only *c*.30 breeding pairs.

Habitat quality clearly has a significant impact on productivity. In Iceland Rees *et al.* (1991a) found that birds nesting in lowlands laid more eggs (mean 4.7) than those nesting in uplands (3.8). In Finland, Ohtonen (1992) found that broods produced in fens averaged 4.4 cygnets (N=40), whereas at ponds they were larger, averaging 5.1 (N=72).

The variability in weather patterns between winters, and summers, is widely considered a crucial factor influencing the breeding success of northern waterfowl

Table 8.8. *Nesting success in an Icelandic study area (Brazil 1981c).*

No.	Locality		1978 eggs/ cygnets	1979 eggs/ cygnets	1980 eggs/ cygnets	Output
#1	Arnarvatn A (North)	S	?/4	0/0	?/4	8
#1	Arnarvatn B (West)	S	?/4	4/0	?/3	7
#1	Arnarvatn C (South)	S	?/3	3/3	?/3	9
#2	Helluvadstjorn	S	?/4	0/0	?/2	6
#3	Marshland northwest of Neslandavik	M	?/1†	?/1	?/2	3
#4	Selhagi	S		5/0		0
#5	Budlungafloi	S		?/3		3
#6	Graenilaekur	S		4/2	?/2	4
	Number of pairs	5–8	5	8	7	
	Cygnets reared	9–24	15	9	24	
	Mean Brood		3.7	1.1	3.4	

† = died
S = sedentary M = mobile

including swans. Myrberget (1981) also implicated weather as the primary factor affecting breeding success, not only in Whoopers but in other bird species too. Late thawing in spring is especially significant, as low temperatures affect primary productivity. Winter may be a period of risk, when severe weather may influence mortality, although this is not so widely documented amongst northern breeding swans. Nonetheless, severe winter weather may limit access to food, forcing Whoopers to move elsewhere, sometimes even triggering international movements. Reduced feeding opportunities and forced movement, combined with lower temperatures, may all increase energy consumption and hence lead to a reduction in body condition that can lead to winter mortality (Brazil 1991; Earnst 1991).

Nilsson (1979) was the first to demonstrate that reproduction in Whoopers is related to conditions experienced the previous winter, with nesting being more successful after mild winters. Earnst (1991) and Rees & Bowler (1991) have suggested female body condition on arrival at the nesting area as the likely mechanism by which the previous winter's weather affects reproduction in swans, for as the body condition of laying females strongly influences reproductive output, then severe winter weather may cause swans to lose body condition. Meanwhile, Elkins (1983) showed that late spring thaw depresses water temperatures, increases energetic needs and may not only delay breeding, but also render it unsuccessful if the period to autumn freeze is reduced below 130 days.

The gross impact of the northern summer climate is considered to operate on the actual length of the breeding cycle, because the period to fledging, relative to the spring thaw and autumn freeze, are crucial in determining the limits of a species' breeding distribution. In the northern part of the Whooper's range, that beyond the

Arctic Circle, Elkins (1983) found that 50% of summers were not ice-free for the necessary 130 days to complete breeding, and unsurprisingly breeding at high latitudes is often unsuccessful. This impact on breeding is further supported by the fact that northerly populations of Whoopers are smaller and produce fewer fledged young (Haapanen 1991), although, as Earnst (1991) has suggested, such differences may also be the result of habitat differences or the length of the migration.

Direct observations indicate that breeding success varies considerably between years, and that it is much lower in late springs in Iceland and at cooler higher altitudes (Brazil 1981c; Rees *et al.* 1991a). Low temperatures may operate directly, making access to nesting grounds difficult or impossible thus delaying nesting, or more indirectly by delaying and reducing the growth of potential food plants. Differences in habitat quality may also have an effect, as Earnst (1991) has suggested, though these too (particularly plant productivity) are also likely to be temperature dependent, with the growing season at higher and colder altitudes beginning later. It is this variation in plant productivity that presumably governs the different reproductive strategies adopted by the smaller bodied Whistling and Bewick's Swans, and by the larger bodied Whoopers and Trumpeters. The former are thought to carry the necessary fat reserves to their breeding grounds, furthermore both sexes cooperate during incubation, enabling both to restore energy losses during incubation breaks and thus shorten the incubation period. The larger bodied swans, breeding at somewhat lower latitudes, access important nutrition at staging areas close to their breeding grounds after arrival, thus enabling the females to undertake all incubation, with little opportunity to feed until the eggs have hatched.

In 1988, Rees *et al.* (1991a) also observed the impact of late springs in Iceland, when their upland study area was still ice- and snowbound throughout May, forcing those swans nesting there to delay laying until the end of the month. They observed that, as late as the last third of May, some pairs were still prospecting, their movements about the marshes apparently causing frequent territorial disputes, and non-breeders apparently concentrated wherever there was open water, even in farmland drainage ditches. Their discovery of particularly small mean broods at their upland study site in 1989, following an even later spring, provides further evidence that breeding success is directly linked to annual fluctuations in spring/early summer weather conditions, in addition to factors operating over the previous winter (Nilsson 1979).

Einarsson (1996) considered that climatic factors might have a substantial effect on breeding success at upland and lowland breeding sites in Iceland, particularly because spring thaw is more variable in its timing in the uplands, because spring thaw affects the timing of egg laying, and because annual variation in weather conditions are thought to have a greater effect in the highlands. He also showed that habitat quality, particularly plant biomass, was linked with the state of breeding birds and their success, and may partially explain why clutch/brood sizes were larger at Skagafjordur in 1988–91, where habitat quality is significantly better, than in the uplands. Given Ohtonen's (1992) findings from Finland concerning habitat quality, this is entirely to be expected.

Among many bird species, reproductive success is also related to the length of time pairs have been together. Rees *et al.* (1996) found that once the effects of other variables had been considered, reproductive success improved with the duration of the pair-bond in three species of swan including Whooper. Thus selecting a mate of compatible size, and remaining with that mate for a prolonged period, enables those pairs to achieve higher reproductive success.

CHAPTER 9

Non-breeding Whooper Swans

INTRODUCTION

A very large proportion of the Whooper population does not breed in any year. In fact it seems that within the species, the norm in numerical terms, is not to breed! This may come as a surprise, but as with other large waterfowl, in addition to the annual variation in young produced as a result of weather variables, a considerable proportion of the population chooses not to, or is unable to, breed in any year for a variety of reasons. Bart *et al.* (1991) went as far as to say of swans in general: 'We know of few other avian groups in which so many adult-aged birds holding territories fail to breed.'

The proportion of non-breeding Whoopers ranges from 35% on the Kola, Northern Karelia, and Kanin peninsulas, to 70–80% in northern Finland, various parts of Russia, and in Kamchatka; apparently varying in relation to latitude, vegetation zone and summer climate (Haapanen *et al.* 1973a; Haapanen 1991; Kondratiev 1991; Mathiasson 1991).

In Finland the proportion of non-breeders was 71% of the total population in 1968–70 (Haapanen *et al.* 1973a) and 73% in 1973–76. Interestingly, there was considerable variation depending on habitat type, such that in the north of the range, typified by subarctic vegetation, non-breeders may represent, at 95%, virtually the entire local population, though such areas apparently provide good summer feeding for non-breeders (Haapanen 1991). In Iceland Rees *et al.* (1991a) found that, at both lowland and upland sites, the majority (60%) did not even attempt to breed. Doubtless more attempted and failed, as found in other studies (Brazil 1981c), and they estimated that only 30–40% of the population was contributing to annual recruitment. As Haapanen (1991) has noted, all of the various populations of Whooper and Trumpeter that have been studied have very similar ratios of breeders to non-breeders, when considering all birds over one year old in spring, with *c.*30% of Whoopers attempting to breed in Eurasia, and 26–27% of Trumpeters in Alaska. In such northern areas, *c.*50% of summers are believed to be too short for Whoopers to complete their lengthy breeding cycle or provide too little food within an accessible, defendable territory to support a family (Haapanen 1991).

Since the 1970s several studies have addressed the ecology and behaviour of breeding Whoopers (Haapanen *et al.* 1973a,b, 1977; Brazil 1981b,c; Rees *et al.* 1991a; Einarsson 1996). Because non-breeders represent such a very large proportion of the population they are of particular interest, though as yet little is known of their ecology or behaviour.

The term non-breeder is in fact covers various classes, including young and sexually immature birds of pre-breeding age, as well as sexually mature individuals that attempt but fail to breed or abandon the attempt at some point between territory occupation and fledging, and even includes other sexually mature birds that, for some reason (perhaps illness or injury), do not attempt breeding. Haapanen *et al.* (1973b) considered that of the 70% of non-breeders in their study, 70% of those were 'engaged', i.e. paired. Non-breeders presumably represent a vital reservoir of potential and future breeders, ready to fill territories as they fall vacant or to replace paired individuals on death. Recruitment from the non-breeding group into the breeding population has not been studied, but it would make a very worthy subject.

In contrast to north Iceland, where I observed flocks of non-breeders in the hundreds at specific locations (Brazil 1981c), in Finland Haapanen (1991) noted that non-breeding summering birds were widely scattered among breeding areas in small groups, none comprising more than ten, as is the case in Trumpeter Swans in Alaska.

As summer gives way to autumn, the population structure changes, as two new categories appear—failed breeders and, of course, cygnets. Haapanen (1991) calculated that a theoretical autumn population would have the following structure: 19.8% successful breeders, 2.3% failed breeders, 51.8% non-breeders and 25.9% fledged cygnets. In Iceland, Rees *et al.*'s (1991a) study coincidentally tested these assumptions. They censused their Skagafjordur lowland study area in May (1988) and again in August, enabling some very interesting comparisons. In May, 378 Whoopers were present, whereas in August there were just 238. In May 126 (33.3%) were paired and occupying discrete territories, but in August only 74

(31.1%) had cygnets. Early in the season, 238 (63%) were in non-breeding flocks, and 14 (3.7%) were lone birds, whereas later 144 (60.5%) were in moulting flocks. By August, Rees *et al.* (1991a) found that whereas 39.5% of those in the Skagafjordur lowlands were paired, and 87.2% had laid eggs and 78.7% had young, 60.9% of highland birds were either singles or in non-breeding flocks.

Rees *et al.*'s (1991a) results support Gardarsson & Skarphedinsson's (1984) aerial observations of August 1982. They found that in north, south and southwest Iceland the proportion of Whoopers in discrete pairs amounted to 30% (61% with young), while 70% of the total population were in moulting flocks. For northern geese, among which the proportion of adult pairs nesting is also low, Owen & Black (1989) found that only 40% of paired Spitsbergen Barnacle Geese bred in 1986, and all studies of Whoopers to date have proved that a large proportion (perhaps related to overall population size and saturation of suitable nesting territories) does not attempt to breed. The constraints operating have yet to be elucidated, but a combination of habitat variables and the proportion of young birds are likely to be involved.

Rees *et al.* (1991a) noting that Whooper numbers declined in their lowland study area by 37% in May–August, considered that the area might also be a staging area for migrants to adjacent highland regions. Other factors may be worthy of consideration: perhaps birds are aware of the qualitative differences between different habitats and prefer to attempt to occupy territories in better habitat, until being forced out. Local moulting birds may have comprised non-breeders and early failed breeders, and some of decline in numbers might result from birds leaving the area to join moulting flocks elsewhere.

During their intensive study, Rees *et al.* (1991a) were able to demonstrate that, as one would expect, when taking habitat and food quality into account, lowland swans were significantly heavier than highland swans. More interestingly, breeders tended to be heavier than non-breeders, with breeding males in the highlands significantly heavier than their non-breeding counterparts. In the lowlands, breeding females were significantly lighter than females without young. An explanation may come from more detailed studies of behaviour during the breeding period. Brazil (1981c) found that females had little time to feed during incubation, and even after the cygnets hatched tended to spend more time than their mate in parental duties. At lowland food-rich sites, non-breeding females would have considerable extra time for feeding compared to their breeding counterparts and might, therefore be heavier. In the highlands, where habitat quality is generally presumed to be poorer, different factors may operate. There, non-breeders may be forced into even more marginal, food-poor areas.

Because of the extremely significant size of the non-breeding population, I have devoted the following chapter to them and to moult, a crucial event, which is, in effect, the major summer pre-occupation of non-breeders.

The arrival of non-breeders at their summering grounds has inevitably attracted less attention than that of breeding pairs, and the pattern may vary regionally or at least between localities. In China, *c.*1,000 Whoopers arrive in the Bayinbuluke marshes from mid March to early April, according to Ma & Cai (2000), then in April–June flocks of non-breeding and subadult swans arrive increasing the population to *c.*2,500; hatching leads to a further increase to 3,000–4,000 in July–September. As few as 400 winter at Bayinbuluke. During wing moult (mostly July in China and Russia), large flocks of 200–400 occur at Bayinbuluke, where they choose remote, sheltered areas with ample open-water plants and aquatic vegetation, and little disturbance. The extent to which an area is being used by moulting groups can be judged by the number of moulted primaries and secondaries along the shore of lakes, marshes and rivers. Groups may come ashore to rest or preen but quickly flee to water at the first sign of disturbance. Once wing feathers are being re-grown body moult proceeds. Moulting non-breeders spend *c.*34% of the day feeding, and breeders and non-breeders moult in different areas (Ma & Cai 2000).

In Fennoscandia and western Siberia, non-breeders tend to occur in widely scattered small flocks comprising no more than ten (Vengerov 1990a; Haapanen 1991). Non-breeding birds use pools and marshes until the moulting period, when they move to lakes, river channels and coastal bays (Rees *et al.* 1997a) although at some sites non-breeders arrive directly at lakes where they summer *and* moult (Fig. 9.1). In contrast, at some Icelandic sites, non-breeders occur in large flocks numbering several hundred, or even in excess of 1,000, especially on coastal bays and lagoons, and large inland lakes (Gardarsson 1979; Brazil 1981c; Gardarsson & Skarphedinsson 1984; Gardarsson *et al.* 2002). The situation in most other parts of Eurasia is unknown although in Mongolia Whoopers gather to moult on large, comparatively shallow lakes (Bold & Fomin 1996). The situation in the Myvatn area of Iceland made a detailed examination of the ecology and behaviour of non-breeders practicable. In parallel with studies of the behaviour of breeding pairs,

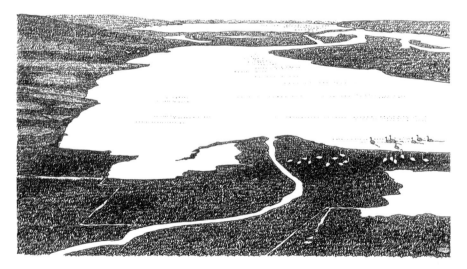

Figure 9.1. Aerial view of non-breeders at Lake Myklavatn, Iceland.

I was able to study flocks of non-breeders, focusing on their arrival, activity patterns within flocks, for comparison with other times of year, and the effect of continuous daylight on diurnal activity (Brazil 1981c). As far as I am aware, this remains the only detailed study specifically of non-breeding Whoopers. It is hoped this may shed light on the situation elsewhere, particularly given discoveries in Scandinavia concerning migration timing (see Chapter Ten).

The gathering of large flocks of non-breeders at Lake Myvatn, and at other lakes in north Iceland, is traditional. This area has been surveyed annually since 1974 as part of long-term assessment of the moulting swans there (Gardarsson *et al.* 2002). In 1978–80 I was able to observe large numbers in two distinct areas of the lake, although in other years these may have amalgamated as a single flock according to Gardarsson (*in litt.*). One of these areas, Neslandavik, a bay on the northwest side of Myvatn, was readily accessible and the swans in reasonable range, making it ideal to observe their behaviour and movements there.

FLUCTUATION IN NUMBERS
AT AN ICELANDIC MOULTING AREA

The north basin, Ytrifloi, of Lake Myvatn, and the large bay known as Neslandavik between them hold the largest inland freshwater moulting flock of Whooper Swans in Iceland (Figs. 9.2 & 9.3), though even larger numbers are now known to moult at sea bays on the west coast and brackish lagoons on the southeast coast (Gardarsson *et al.* 2002).

Figure 9.2 . Part of a moulting flock at Lake Myratn.

Gardarsson & Skarphedinsson (1984), generalising for the estimated total of *c*.6,200 Whoopers in summer moulting flocks in Iceland, assumed that non-breeding flocks usually began to form in June and built up rapidly during the first three weeks of July, with numbers stable between about 25 July to 20 August after which they decreased. It has been during this stable period of late July–August that annual moult counts have been made since 1974. These studies have shown that numbers alternate between the two main areas, which have different food supplies, possibly as a nutritional response. Between 1974 and 2000, moulting numbers have fallen by 50%, from *c*.600 to 300, a mean 2.5% per year. Strip-mining of the lakebed (commercial diatomite extraction) has reduced available shallow water areas less than 2 m deep by *c*.10 ha per year, from 7 km² in 1967 to 3 km² in 2000. These cavities increase the depth of the lake, placing macrophytes out of reach of non-diving waterfowl such as Whooper Swan (Gardarsson *et al.* 2002). As a generalisation, this describes the situation that I observed at Lake Myvatn; however, repeated daily observations revealed a rather

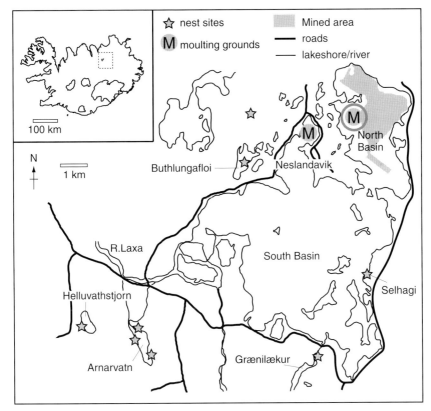

Figure 9.3. Map of the Myvatn area showing breeding and moulting grounds (after Brazil 1981c), and the mined area in 2000 (shaded) (after Gardarsson et al. 2002).

more complex pattern. I found, for example, that numbers of non-breeders did not begin to aggregate in June, but as early as late April–early May; it was also clear that various classes were involved. Peak numbers apparently occur in October in the Myvatn area (Gardarsson *in litt.*), after which numbers decline until only a small wintering population remains. Gardarsson (1979) found as many as 500–600 non-breeding summering birds moulting in two discrete areas at Myvatn, although he made no distinction between the two flocks. It was generally accepted that birds initially used Neslandavik before moving to Ytrifloi. Closer, prolonged observation, however, showed that this was not necessarily true, and that non-breeders were present in both areas simultaneously, with one group relatively stable and the other highly variable (Brazil 1981c), though there is also annual variation (Gardarsson *et al.* 2002).

The summering non-breeding Whooper Swans of Myvatn have averaged 473 birds over the period 1974–2000 (Gardarsson *et al.* 2002). A low 1 km-wide peninsula separates the two bays, Neslandavik and Ytrifloi, and there is some movement between them while birds remain capable of flight, and perhaps of some swimming between the two. Close observations of numbers using these areas over the summer (Fig. 9.4) revealed some interesting patterns. In Neslandavik, numbers that had begun to build up markedly during May reached almost 200 during June 1978 then remained relatively stable throughout late June–August and did not drop below 130. In general they fluctuated around 180. There was no significant change in numbers with date at Neslandavik. At Ytrifloi, numbers were also below 200 in June, but were more variable, fluctuating until 16 July, when they increased very rapidly to a maximum of 465 on 8 August, bringing the total number at Myvatn

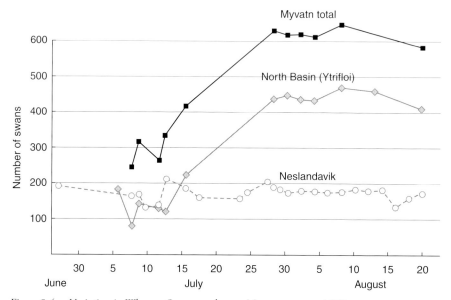

Figure 9.4.   Variation in Whooper Swan numbers at Myvatn; summer 1978.

that summer to 640 (the additional birds arrived at Ytrifloi while numbers at Neslandavik remained stable). Samples of non-breeding numbers at Myvatn in July 1980 revealed that *c*.150 were using Neslandavik and that 300 Ytrifloi, fitting well with the previous pattern. The maximum number recorded was 484 on 29 July, confirming that Lake Myvatn is indeed a focal point for a large number of non-breeders (Brazil 1981c; Gardarsson *et al*. 2002). Of particular interest is the prolonged period over which non-breeders appear. Some are present from early summer, i.e. immediately after the arrival of migratory birds, but many others do not arrive until rather late summer. A similar pattern may occur at other large moulting sites, and would be well worth looking for.

While Gardarsson *et al*. (2002) noted a strong decline in moulting numbers over the 27-year period from 1974 to 2000, their counts were all made during the period of relative flock stability in late summer (see Fig. 9.5). In addition to long-term trends in overall numbers, and inter-annual variation at particular locations, probably in relation to food, I believe that other interesting factors may affect the aggregation of moulting flocks. In the long, cold spring of 1979, following a severe winter further south, spring migration and breeding were very much delayed. Furthermore, the arrival and build-up of the Neslandavik and Ytrifloi flocks at Myvatn was delayed, with numbers climbing very slowly in Neslandavik, barely reaching 200 by 15 July, by which time birds were just beginning to arrive in Ytrifloi. That year there appeared to have been no Whoopers wintering at Myvatn or they had all dispersed by the time of my arrival. The delayed spring migration made it possible for me to observe the first arrivals into the area, and to follow their movements. In late May most of Lake Myvatn, including all of Neslandavik, was still frozen, and the first swans to arrive appeared on grass fields partially free of

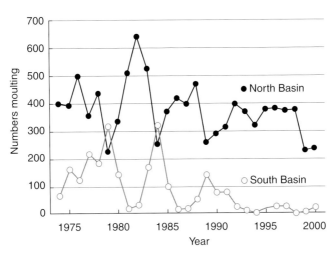

Figure 9.5. Numbers of Whooper Swans moulting in the North Basin and the South Basin of Lake Myvatn in 1974–2000 (after Gardarsson et al. 2002).

snow, near the road between Geirastathir and the Laxa River. These I have referred
to as the southern fields to distinguish them from an area 1.5–2.0 km further north
utilised later. Both are within 2–3 km of Lake Myvatn.

Numbers at the southern fields were not constant and because of the inaccessi-
bility of the region at that season, it was impossible to discover where the birds went
when they were absent. Birds may have been passing through en route to other areas
or perhaps left the fields temporarily to scout breeding territories. By the second
week of June, the flock had dwindled away, except for three isolated occurrences in
mid June. When most of the flock had already disappeared, a second flock began to
use the northern fields. This area had not previously been used and movement of the
few remaining birds from the southern fields could not account for its sudden
increase in popularity. This flock peaked at 55 on 12 June and declined rapidly as
birds began to use the partially thawed Neslandavik Bay (Brazil 1981c).

Given that the foraging fields were within a matter of a few km of potential
breeding sites, one of the most interesting aspects was that some of those in the earlier, southern field flock still had traces of juvenile plumage, were clearly consorting with adults and were apparently still in family groups. Kear (1972) had stated
that, 'Whoopers leave the wintering grounds in families but arrive at their nesting
territories in pairs. Perhaps the young are left behind at the last large lake while the
parents fly on the remaining few miles alone'. It certainly seems from my own few
observations that Kear (1972) was correct in her assumption; that is, that families
may remain together until within just a few km of potential breeding territories,
and separate at the last possible moment before pairs commence breeding. Had
Lake Myvatn thawed earlier, no doubt they would have paused there instead of on
the fields, and I might easily have missed the presence of juveniles in the flocks at
the much greater range.

Birds retaining juvenile plumage and still apparently consorting with their parents were seen only in the small, earlier, southern field flock, which therefore certainly contained experienced, successful breeders, and presumably these adults were
ready to nest. If so, this flock corresponded with what Haapanen & Hautala (1991)
termed the 'first wave' of migration. When the second flock began to form at the
northern fields, it contained no family parties and, furthermore, the dispersal of
that flock coincided with birds beginning to use the regular non-breeding area at
Neslandavik. This later-forming northern flock presumably consisted of non-
breeders preparing to moult, rather than breed, and may thus have corresponded to
Haapanen & Hautala's (1991) 'second-wave' of migrants. Spring migration occurring in two identifiable waves is a topic to which I will return.

At known breeding territories in the Myvatn area, Whooper pairs commenced
nest building as soon as the ice began to thaw, and this is the typical pattern across
the range, from Iceland to Anadyr in extreme eastern Russia (Krechmar 1982a).
Evidently those breeding pairs were already in the vicinity of, but not at, their territories, waiting for conditions to become suitable. In normal springs, nesting lakes
are probably ice-free before the swans arrive, though in Scandinavia and much of
Russia this is probably not the case. The spring of 1979 was exceptionally late and

local field feeding may just have provided early migrants with sufficient food to last them until the thaw. It certainly provided me with a very special opportunity to observe the movements of Whoopers during the very early stages of the breeding season.

It is of interest that non-breeding Whoopers moulting at Lake Myvatn use two distinct areas, Neslandavik and Ytrifloi, differing in their benthic macrophytes vegetation, with swans in Ytrifloi over pondweed and *Cladophora glomerata* beds, while those in Neslandavik feed on a loose carpet of *C. aegagropila* under Spiked Water-milfoil *Myriophyllum* (Gardarsson *et al.* 2002). Of even greater interest was the difference in the timing of the build-up of these flocks indicated by observations in 1978–80. Even though they are only 2–3 km apart, separated by the Neslandatongi peninsula (Fig. 9.3), I observed no interchange of flightless birds between the two flocks during wing moult until, that is, a drive to trap birds for ringing in July 1980 significantly disturbed the Neslandavik flock and the majority of them swam to join the flock in Ytrifloi. Except for the very earliest few arrivals at Neslandavik, when there may be no swans at all in Ytrifloi, all of those arriving in the area would be able to see both areas and both flocks simultaneously from the air. New arrivals must therefore make a choice between the two. The factors governing this remain unknown, but may relate to food quality, which may differ annually, or to disturbance.

With the exception of the movement caused by the ringing operation, interchange between the two flocks was observed only while the swans were still capable of flight and, even then, it was at a low level. During eight of 25 observation periods, no movements were observed between them, and during only four observation periods in 1978–79 did any observed movements exceed 10% of the total number present on that date in Neslandavik. Furthermore, some movements from Neslandavik to Ytrifloi occurred immediately following disturbance there and were presumably caused by it (Brazil 1981c).

During July and August the majority of non-breeders undergo full wing moult and therefore become flightless for *c.*4–6 weeks. The proportion actually capable of movement, therefore, is limited, hence the stability of flocks during this period (Gardarsson & Skarphedinsson 1984; Gardarsson *et al.* 2002). None of the flightless birds was ever observed to walk from Neslandavik, which would have involved crossing a dirt road and *c.*2 km of very rough scrubby bog, and none was observed swimming between the two, except on the occasion mentioned above.

The timing of arrival into the summering flocks suggests that the composition of the flying proportion changes during the season. In early summer, late-arriving birds no doubt included some that have failed in their breeding attempt during late incubation or had lost young cygnets, but were nevertheless still capable of flying from their territory to join a moulting flock. Large movements of flying birds recorded on 10 13 and 16 July 1978 were probably recent arrivals still settling into a flock consisting mostly of birds already moulting. By contrast, in late summer, the flying proportion was presumed to represent those swans that had begun moulting early and had recently regained their capacity to fly (Brazil 1981c).

THE STATUS OF SUMMERING WHOOPER SWANS

Whooper Swans summering at Myvatn have been observed almost exclusively in two discrete areas (see Fig. 9.3) and can rarely be seen anywhere else for very good reasons. According to bathymetric data these two are similar in that they are the only extensive areas of Lake Myvatn with a depth of 1–2 m (Adalsteinson 1979; Einarsson 1985), thus within the reach of a Whooper upending for food. Water deeper than 2 m is less suitable as the bottom flora is inaccessible, as has also been found in the Volga Delta (Rusanov 1987). There is a strong relationship between the artificial deepening of the north basin of Myvatn and the decline in moulting swans (Gardarsson *et al.* 2002) (see Fig. 9.6). Both areas provide large secure foraging areas for swans, with plentiful and accessible food, where being flightless will not compromise their ability to access food. Both are large enough to enable flocks to move well away from shore in the event of most types of disturbance, while still remaining in suitable feeding areas. Thus they provide perfect conditions for moult.

As Whoopers are flightless for several weeks, birds joining a moulting flock effectively commit themselves to remaining in one locality for that time. Moulting sites, and decisions relating to their use, are extremely significant as once made there is little or no chance of change without considerable energy expenditure and risk. Swans can be expected, therefore, to select areas where movement by swimming is adequate to reach both food and safety, so large shallow lakes, coastal lagoons, estuaries or sea bays, alternatively inland river systems, are ideal, whereas isolated

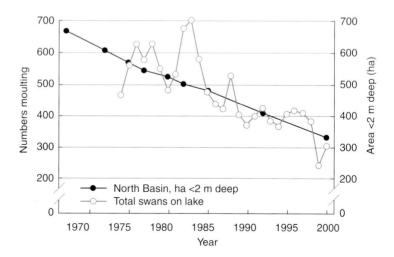

Figure 9.6.   Total numbers of Whooper Swans moulting in the North Basin plus the South Basin of Lake Myvatn in 1974–2000 compared with the area (ha) of shallow water (<2 m) in the North Basin (after Gardarsson et al. *2002).*

ponds, tarns or marshes are more risky, either in security or lack of continuous food. Lake Myvatn and other large lakes and coastal lagoons in Iceland clearly offer ideal conditions for large numbers of Whoopers to moult.

At Myvatn, there are considerable differences in the food at the two moulting localities. Neslandavik has a considerable amount of the filamentous algae *Cladophora aegagropila*, forming a carpet beneath Spiked Water-milfoil, while Ytrifloi contains mainly beds of pondweed and *C. glomerata*. Direct observation of feeding behaviour and examination of swan faeces showed that, despite its abundance, water-milfoil is normally avoided. At Neslandavik, only *Cladophora* was eaten, and only *Potomogeton filiformes* was taken at Ytrifloi. Despite its abundance, water-milfoil was not found in faeces from either site and, on several occasions, swans in Neslandavik were seen deliberately discarding water-milfoil strands they had pulled up with *Cladophora*.

It is not known why moulting Whoopers at Myvatn separate into two flocks in some years. It is possible and interesting, however, to speculate on the inter-flock differences relating to their build-up. That of a first flock, in Neslandavik, in early summer is fully consistent with it being comprised of non-breeders that have moved, more or less directly, from their wintering quarters to a summering or moulting ground. Several birds in this flock had clear traces of juvenile plumage, including a first for Iceland, a Bewick's Swan which also joined the flock (Brazil 1980). None with juvenile plumage was seen at Ytrifloi, although this could have been due to the long-distance observations. Identifying juveniles, even in Neslandavik, was difficult and dependent on good light (Brazil 1981c). Neslandavik, and areas like it might, however, be the type used by one-year-olds very recently separated from their parents, which arrive in the first wave of migrants, and thus the early gathering at Neslandavik or other such sites would be of birds with no intention of breeding. Because one-year-olds remain with their parents on spring migration, and because breeding pairs are under pressure to arrive early, of necessity one-year-olds are far more likely to be among first-wave migrants than second-wave migrants. One-year-olds are likely, therefore, to represent a higher proportion of the early non-breeding flocks.

The later, and slower, build-up of a second flock, in Ytrifloi, is consistent with it being comprised of several categories of non-breeders or of unsuccessful breeders, including swans that had spent some time at a nesting area, but had given up part way through the season, and others that moved during early summer, before settling on a moulting area. Such augmentation of moulting flocks by failed breeders also occurs among Mute Swans in Europe (Perrins & Reynolds 1967; Minton 1971), and is expected among Whoopers. This expectation and my speculation appears to be supported by Rees *et al*.'s (1991a) observations that numbers in their breeding study area declined over the summer and that some pairs that failed to breed moved on to join larger moulting flocks elsewhere.

After the late spring of 1979, the Ytrifloi flock began to build up only during mid July. Breeders, and pairs attempting to occupy territories, had been forced to wait for suitable breeding areas to thaw, and non-breeders and failed breeders were also later arriving on the moulting grounds. Counts in mid July 1980 agreed well with those made in 1978 and the pattern shown by Gardarsson (1979).

Immigrants regularly flew overhead, flying north towards Lake Myvatn (see Fig. 9.3) during observations of breeding pairs at nearby Arnarvatn. If they attempted to land at Arnarvatn, one or other pair breeding there invariably chased them off. Such behaviour not only removes intruders from active territories, but may also serve to concentrate unsuccessful breeders at suitable moulting areas, where there are no breeders, and perhaps to provide failed breeders with additional clues as to the selection of suitable nesting habitat (Kiester 1979). Both partners of successful breeding pairs moult on their breeding territory or, if they are a 'mobile' family, at a suitable feeding area nearby, while their cygnets are fledging (Haapanen *et al.* 1977; Brazil 1981c).

Gardarsson & Skarphedinsson (1984), in their model of the Icelandic population, considered that having separated from their parents when just under one year old, young may flock with other non-breeders for at least their first two summers. Thereafter, although they may attempt to breed, 3–6 year-olds are typically unsuccessful, either stopping or failing at the nest, clutch or early brood stages. Gardarsson & Skarphedinsson (1984) concluded that moulting flocks consisted largely of pre-breeders 1–6 years old. I would add that such birds not attempting to breed in a given year, should be ready to join a summering flock immediately on arrival in the breeding range. Slightly older first- and second-time breeders, and older pairs that have failed at some stage, are also likely to join them, but do so later, as the summer progresses.

On the basis of the Gardarsson & Skarphedinsson (1984) model, birds gathering in moulting flocks represent two groups. Those arriving early in the season, in May–June (such as those I observed at Neslandavik in 1978–80), are likely to be younger, perhaps 1–2 years old, and not attempting to breed. Those joining moulting flocks that congregate later, during July (such as I observed in Ytrifloi), are more likely either older birds that have attempted to breed for the first or second time and either given up or failed, or older birds that have failed. Such a flock might contain more birds 3–6 years old and older. Perhaps studies of banded birds will eventually clarify the status of birds in such moulting flocks.

As we will see, there is also evidence from migration studies that relates to this distinction between birds that make no attempt to breed, which can spend all summer in a flock, and those that may spend several or even many weeks attempting to breed before giving up or failing.

THE ACTIVITY PATTERNS OF NON-BREEDERS

The range of behaviours exhibited by non-breeders is rather more limited than that of breeders, and is much more similar to those of birds on their wintering grounds. As one would expect, the main activities of non-breeders are essential ones: feeding, roosting, preening and vigilance. In my Icelandic study, social behaviour, particularly displaying, was very infrequent, and other activities, such as wing flapping and tail wagging, so infrequent that they were included with preening, while calling was included with 'head-up' or displaying, depending on its context (Brazil 1981c).

Time budgets for July–August differed little. This was in line with prediction, as the percentage of time spent feeding was not expected to increase until September, a few weeks prior to migration, with a consequent reduction of time spent in other behaviours, and by which time birds might have moved to alternate staging areas.

One might assume that behavioural patterns of Whoopers on fresh water in Iceland during summer would closely resemble those on fresh water elsewhere in winter. There was, however, an interesting and fundamental difference between time budgets in the two situations (Owen & Cadbury 1975; Brazil 1981c). Surprisingly, summer time budgets of Icelandic non-breeders revealed morning and afternoon feeding peaks resembling those of wintering swans on agricultural land. Surely very different constraints apply in these situations and perhaps the pattern is merely coincidental for, in summer, day length is not limiting, non-breeders expend no energy on flight, and do not commute between feeding and roosting areas.

In Iceland, a decrease in feeding activity occurs during the middle of the 24-hour day; this period lasts several hours and is much longer than in winter, perhaps because the swans have access to plentiful food. In July–August, feeding peaked in early morning and was lowest around midday, and increased steadily during afternoon and evening. Despite it being light, roosting occurred most commonly at 02:00–09:00 in July, declining throughout late morning and afternoon. In August, roosting was commonest at 01:00–05:00, declined until 09:00, and peaked at 12:00–13:00, before declining to zero by 18:00. Preening occurred at a uniform low level throughout the day in July, but was more erratic in August. 'Head-up' was the most time-consuming activity, occupying 42.9% of the day or 10.3 hours each day in July and 47% or 11.3 hours per day in August.

The proportion of time spent 'head-up' was thought to be high for three reasons. Firstly, 'head-up' may be a positive activity in that it represents time spent vigilant; secondly, for swans on water, 'head-up' also represents a passive, neutral position, the resting phase between feeding and roosting, used whenever they are not engaged in any other specific activity (it does not imply that the birds are disturbed or alert); and thirdly, the day is very long, but the birds, not being very active, do not need to feed for much of the time, hence more time is spent in other activities.

Winter activity budgets, previously having been considered only during daylight, were also recalculated for a 24-hour day to facilitate comparison between winter and summer activity levels. For this, it was assumed that in winter Whoopers spend the entire night roosting. It is, however, likely that any long roosting period is interrupted by 'head-up' and preening, but nocturnal frequencies for these behaviours are unknown. When viewed over the 24-hour period, the amounts of time spent feeding in winter and summer were very similar, even though very different food plants were being eaten. There were, however, some striking differences between summer and winter.

Differences in the time spent preening, roosting and 'head-up' between winter and summer were all significant. In summer, preening and 'head-up' occupied more time, but roosting less, than in winter. Increased day length, and moderate disturbance were both expected to lead to an increase in the time spent 'head-up' compared with winter, which does appear to be the case. Preening was expected to occupy a larger proportion of time in summer, because the birds were shedding and re-growing flight feathers and also beginning to moult some of contour feathers. It seemed to me that preening also occurred briefly after each disturbance, presumably as a displacement activity by birds ready to feed. More time was spent roosting in winter, presumably because the period of darkness was longer.

Non-breeding adults in summer fed by dipping and upending. Dips were significantly shorter than upends; mean dip duration was 8.6 seconds (N=457) while mean upend duration was 10.3 seconds (N=439). As a consequence, the frequency of feeding actions was significantly faster by dipping. Mean dip rate was 4.5/minute (N=210) and the mean upend rate was 3.5/minute (N=186). Birds feeding by dipping, with a mean dip duration of 8.6 seconds and a mean dipping rate of 4.5/minute, spend a mean 39.2 seconds per minute feeding and have 20.8 seconds available for other behaviours, such as 'head-up'. Swans feeding by upending, with a mean upending duration of 10.3 seconds and a mean upending rate of 3.5/minute, spend 36 seconds per minute (60%) feeding and have 24 seconds (40%) available for other behaviours (Brazil 1981c).

SPACE UTILISATION BY MOULTING WHOOPER SWANS IN ICELAND

Moulting swans in Neslandavik did not disperse randomly throughout the entire area available. The flock tended to remain loosely together while feeding and clumped densely, far from the shore, when disturbed. Over the summer most of the bay was used at some time, but the total area can be broken readily into three zones relating to intensity of use (Fig. 9.7). The core area, including the more central region where the water is deepest, was used most, both for feeding and as a safe area when disturbed. Peripheral areas were used only occasionally, and the middle area was used to an intermediate extent. Areas of high use close to the shore were used for roosting. Some use was clearly weather related, thus on all eight occasions when the non-breeding flock was recorded on the north shore or the shore on the east side of Neslandavik (Fig. 9.7), north or northeast winds were blowing. The flock in Ytrifloi also took shelter from the wind, in the lee of islands, and spent less time feeding by upending in strong wind. Wind obviously causes Whoopers to seek shelter and shallower water in which to feed, if they are unable to upend (Brazil 1981c).

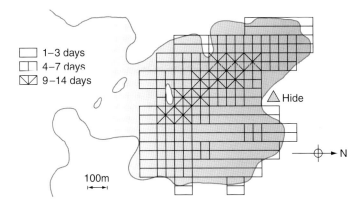

Figure 9.7. *The intensity of use of different parts of Neslandavik, Myvatn, 1978.*

MOULT

Moult of feathers and feather tracts usually occurs successively, and symmetrically, in birds. This enables them to maintain normal insulation and flight throughout the process. Among waterfowl, however, wing moult in particular has been greatly condensed. Wing feathers are moulted more or less simultaneously, the primaries usually before the secondaries. Consequently, for a short period each summer, such species entirely lose the power of flight. In many this period coincides with the brood period, that is after their ducklings, goslings or cygnets have hatched but before they have fully developed their flight feathers. During this period, male ducks adopt eclipse plumage and, both visually and behaviourally, become less conspicuous. Some, especially Common Eider, Common Shelduck, geese and swans, gather in large moulting flocks and may undertake moult migrations to reach suitable sites. Moult concentrations among swans and geese share many similarities; firstly, they are restricted to non-breeders. Mostly these are immatures because breeders remain with their young. Indeed, immatures of most swans gather in summer in flocks and undergo wing moult near the breeding grounds. They cannot be said, therefore, to undertake a moult migration, which among swans has only been described for Mute Swan (Salomonsen 1968).

A significant fact for the Whooper and the closely related Trumpeter is the annual flightless period, extending from as early as late June through August and into early September for a period of 4–6 weeks during which their wing feathers are shed and re-grown (Dement'ev & Gladkov 1967; Hansen *et al.* 1971; Kear 1972; Brazil 1981c; Vengerov 1990a) (Fig. 9.8). The equivalent period in Mute Swan may last 3–7 weeks (Kear 1972) or 5–6 weeks (Mathiasson 1973). The Whooper Swan's moult has been referred to inappropriately as 'post-breeding' (e.g. Rees *et al.* 1997a) but most of the population do not breed, they only moult, and breeding pairs moult *while* breeding. This period of restricted mobility is significant, and their

Figure 9.8.   Wing moult renders Whooper Swans flightless each summer.

behaviour must be adapted to it. During annual moult, non-breeding swans may gather in large concentrations at traditional wetlands, providing security from disturbance and predation (Hansen *et al.* 1971; Mathiasson 1973; Gardarsson 1979; Brazil 1981c).

In some areas of Iceland, Scandinavia and Russia, Whoopers moult in small flocks, but at others they gather into large flocks several hundred strong. Very little information concerning moulting in other countries has been published. Non-breeders in Finland and Sweden tend to leave their summer quarters during late June, apparently to moult elsewhere and, given the current large population and the very high proportion of non-breeders, as many as 40,000 Whoopers must moult at northern latitudes. As yet, no mass-moulting sites have been located in Fennoscandia, although it is surmised that they are likely to be in highly productive wetlands in the taiga or forest tundra of Russia (Beekman 1998). Whoopers arrive at moulting sites in the Ob River region during May and early June and on the Kola Peninsula mainly in June (Bianki 1990; Vengerov 1990a). Moulting flocks have been found on the lakes of southern Trans-Uralia during the middle third of June, but the groups are rather small, consisting of 15–36 birds. Similar-sized groups occur during July with some still flightless in the middle third of August. These moulting birds appear to favour shallow water with reedbeds (Blinova & Blinov 1997).

Mathiasson (1973) found that Mute Swans dropped their primaries before their secondaries. The primaries, although longer, have a faster growth rate, thus synchronous replacement of both primaries and secondaries is achieved, and the capacity for flight restored as swiftly as possible. This also appears to be the case in Whooper Swan and may be true of all swans. The growth rates of the primaries of several large waterfowl have been measured, Greylag Goose at 5.5 mm/day, Brent Goose at 6.0 mm/day (Mathiasson 1973) and Whooper Swan at 9.0 mm/day (Dement'ev & Gladkov 1967) and 5.3 mm/day (Brazil 1981c). These rapid growth rates, combined with the fact that waterfowl are actually capable of flying before their flight feathers have reached maximum length, serve to minimise the length of the flightless period (Owen & Ogilvie 1979). Weight loss during moult would also be an added advantage since it would facilitate earlier flight.

To understand the rate and timing of moult in Whoopers, and to establish whether males and females moulted simultaneously, I measured the growth rate of the primaries in two captive birds and then, using these as guides, calculated when moult might have started in wild-caught birds, in order to discover whether non-breeders moult at the same time as breeders (Brazil 1981c). Two injured Whoopers, obtained from South Uist, Scotland, were placed in an outdoor grass enclosure during summer 1980. The growth rate of the third primary (usually the longest) was measured as they replaced their moulted primaries. During August, the growth rate of this feather in both swans was found to be almost constant. Only in early September was growth suddenly reduced. During August there was a high correlation between date and length of the third primary (Fig. 9.9) and using data from

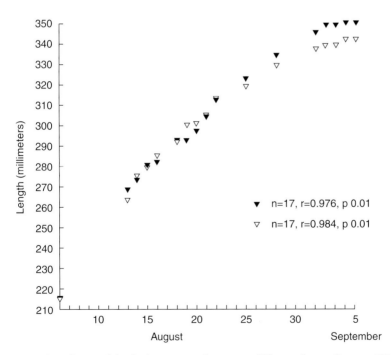

Figure 9.9.   Growth rates of the third primaries of two captive Whooper Swans; Summer 1980.

August alone, daily growth rates were 5.18 mm and 5.41 mm (mean = 5.3 mm). Actual measured daily increments during the same period averaged 6.1 mm (N=7) and 4.7 mm (N=7). All measurements were considerably less than the 9.0 mm/day reported by Dement'ev & Gladkov (1967). Perhaps the growth rates of the captive birds were, for some reason, slower than in wild birds, though they were in line with those of moulting geese.

On 25 July 1980, 46 flightless Whooper Swans were caught in a moulting flock at Neslandavik, Lake Myvatn, and the length of the third primary of 21 males and 24 females measured. No difference was found between the sexes in terms of the length of primary three, suggesting that these non-breeders were moulting concurrently (Fig. 9.10). Using the mean daily growth rate of 5.3 mm obtained from the two captives in Scotland, knowing the length of primary three on a specific date (25 July 1980), and assuming an equal and constant growth rate for males and females, it was possible to calculate when each individual had started to re-grow its primaries (Fig. 9.11; see Mathiasson 1973).

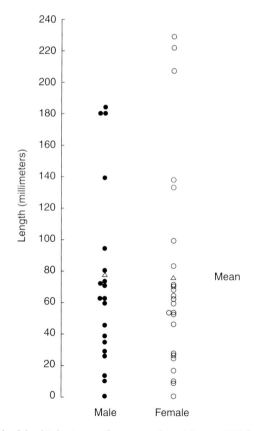

Figure 9.10.   Length of the third primary of swans caught at Myvatn, 25 July 1980.

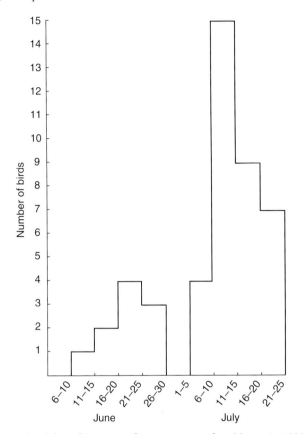

Figure 9.11.   Calculated date of initiation of primary re-growth at Myvatn in 1980.

During wing moult among waterfowl there is a hiatus, new feathers do not push out the older feathers, instead the old flight feathers are dropped, and then several days later the new feathers begin to appear. The length of this interval is unknown for the Whooper Swan, although Boyd & Maltby (1980) found that Brent Geese began re-growing their flight feathers 3–6 days after dropping the old feathers, and a similar time period may elapse in Whoopers. In the absence of information, it was not possible to calculate when feather drop actually occurred, only when re-growth started. From the increasing presence of dropped feathers along the shore of Lake Myvatn, it appeared likely that some birds in Iceland began moulting in late June.

It was expected that the date of commencement of re-growth for different individuals would be normally distributed, but the catch having been made in July was expected to have eliminated one tail of the distribution. Based on a rate of 5.3 mm/day, a few had commenced moult in mid to late June, but the majority, 78%, began primary re-growth after 5 July. The peak period was 11–15 July when 33% were estimated to have started re-growth (Brazil 1981c). Although no other data are available for comparison, Mathiasson (1973) found that most non-breeding Mutes began their wing moult slightly later, on 14 July–7 August. They may not, however, have been under such seasonal pressure to moult and migrate. On rare occasions, individuals may actually fail to moult their wing feathers in summer, retaining them for a second year (e.g. Campbell & Ogilvie 1982).

When birds from the non-breeding flock in Neslandavik were caught in late July, there was no significant difference between males and females in the length of their primaries. Assuming that feather growth rates for the sexes are the same (and there is no evidence to the contrary), then non-breeding Whooper males and females moult simultaneously, as is true for their closest relative, the Trumpeter Swan (Hansen *et al.* 1971). This simultaneous moult contrasts with that of the Mute Swan, where Mathiasson (1973) found that non-breeding females moulted later than males. It also differs from that in breeding Whoopers and Mutes, in both of which females moult first (Heinroth & Heinroth 1928; Scott *et al.* 1953; Hilprecht 1956; Kinlen 1963; Brazil 1981c; Czapulak 2002).

Non-breeders in Neslandavik had mostly begun wing feather re-growth, after moulting, in the middle third of July in 1980. The period of complete wing moult varies according to location, but can begin in mid June, in Iceland, in early July (e.g. on the Kanin Peninsula) and continues throughout July and even into mid August (e.g. on the Shantar Islands and in the lower Cis-Amur; Dement'ev & Gladkov 1967; Brazil 1981c; Babenko 2000). In Magadan, Krechmar & Krechmar (1997) considered that breeders began primary moult in the final third of July. In the Ob River region, and on the Kola Peninsula, replacement of lost primaries begins among most non-breeders during early and mid July and by 5–10 August some are able to fly and by early September all adults have finished wing moult (Bianki 1990; Vengerov 1990a). Bianki (1990) considered that breeding Whoopers moult probably slightly later than non-breeders, and females later than males. During mid-August many males with broods in the Ponoi River basin were still unable to fly. By late August, however, most adult breeders, non-breeders and young have re-grown their flight feathers and are capable of flight, though perhaps not of sustained flight (Babenko 2000). In the Magadan region of eastern Siberia, flying swans after moult were first sighted on 13–21 August (Krechmar & Krechmar 1997). By the middle third of July in 1978 and 1979 numbers had already stabilised in Neslandavik, and from their behaviour, and the extent of shed feathers on the shore, were clearly in wing moult. During the same period however, those in Ytrifloi were only just increasing, with groups of flying swans regularly arriving. Numbers there did not peak until 8 August, thus while birds in Neslandavik were already flightless, others were still arriving to join the flock at Ytrifloi. The situation, during these years at least, was indicative of there being two types of moulting birds in the Myvatn area—early and late.

Mathiasson (1973) found that not only did Mute Swans not commence feather drop immediately on joining moulting flocks, but that they usually spent 16–30 days after their arrival before commencing moult. The same may be true of Whoopers, given that birds beginning to moult in Neslandavik during mid and late

June and July were most probably the same that arrived there more than a month earlier. If there is the same initial 'waiting' period among later arriving birds, such as those reaching Ytrifloi in late July–early August, then these must have been moulting considerably later than those in Neslandavik, and perhaps explains why numbers remained high at Myvatn into October. Although the two groups moulting at Lake Myvatn were not entirely isolated, nevertheless the apparently widely differing timing of the moult among non-breeders is interesting and requires elucidation. My own belief is that those arriving later in Ytrifloi were most likely mature adults that had attempted to breed but had failed. If so, then it appears that in agreement with Kear (1972), non-breeding Whoopers moult well before breeders or failed breeders, as is the case in a range of other birds including other waterfowl such as Barnacle Goose and Mute Swan (Palmer 1972; Owen & Ogilvie 1979; Brazil 1981c; Czapulak 2002).

In Iceland, breeding Whoopers stagger their moult so that one pair member is capable of flight for almost the entire summer. For both parents to be simultaneously flightless would leave the family particularly vulnerable to disturbance from intruders or even predators. Similarly, Hansen *et al.* (1971) found that moult in breeding Trumpeters lasts over twice as long as in non-breeders. Both Brazil (1981c) and Hansen *et al.*'s (1971) observations agree with Ogilvie's (1978) suggestion that lack of synchronicity between parents exists in order to reduce the period when both parents are flightless and so presumably less able to defend their brood, rather than with Ryder's (1967) suggestion that parents delay moult to synchronise with the fledging period of their offspring. The cygnets still require parental care during the post-breeding period when the family roam the territory together. Wing moult in breeding adults cannot, therefore, be postponed until after the young have fledged because the flightless adults would be unable to follow or defend their offspring.

On the whole, geese commence moult with the underparts and finish with the neck and head (Owen 1980) and the same appears to be true of swans (Mathiasson 1973; Brazil 1981c). According to Dement'ev & Gladkov (1967) moulting occurs twice annually, a complete moult in summer and a partial one in autumn/winter. The full summer moult apparently begins with the wing feathers, which grow back *c.*1 month later. While the remiges are being shed, the feathers of the underparts and tail follow, and then the feathers of the rest of the body. I suspect, however that moult does not occur twice, but that it is instead prolonged, continuous and slow with the feathers of the head/neck not being moulted until winter. Body moult commences while the wings are still growing and continues until December (Hilprecht 1956; Brazil 1981c). The two captives mentioned above were in heavy body moult from late September/early October and continued to lose neck feathers into early winter. Body moult had apparently ceased by late December (Brazil 1981c) consistent with that observed among arriving migrants on the wintering grounds (Brazil pers. obs.) (Fig. 9.12). Rees *et al.* (1991a) were also able to confirm that male Whooper parents had shorter primaries than females, supporting Scott *et al.*'s (1953) and Brazil's (1981c)

Figure 9.12.   Winter preening releases few body feathers suggesting moult has finished.

observations that moult in breeding Whoopers is staggered, with females moulting and re-growing their flight feathers first.

Rees *et al.* (1991a) also established that annual wing moult occurs concurrently in both highland and lowland areas of Iceland. Interestingly, within their study area, they found no evidence to suggest that non-breeders moulted earlier than breeders, which differs markedly from Owen & Ogilvie's (1979) studies of geese, and my own observations of Whoopers. In looking for an explanation, perhaps we need to look at the types of birds involved. The flock at Neslandavik consisted of non-breeders, among them many perhaps just one year old, which are able to moult earlier than either breeders or failed breeders (Brazil 1981c). In Rees *et al.*'s (1991a) study, swan numbers declined considerably in May–August, as many left the area to moult. Perhaps a much greater proportion of Rees *et al.*'s (1991a) remaining moulting flock consisted of later-moulting failed breeders than young non-breeders. Clearly, moult in Whooper Swan is a subject that has been only cursorily studied to date, and is another area worthy of additional study.

Fledged cygnets commence moulting their body feathers in about October continuing until January, when they moult crown, nape and the head-side feathers. In January–March, moult proceeds with the neck, mantle, scapulars, flanks, chest and body-sides, and sometimes a few tertials, inner wing-coverts and tail feathers (Rees *et al.* 1997a). Following migration to their summering grounds they moult in sequence like adults, thus by their second winter they are usually indistinguishable from older birds.

Gardarsson & Skarphedinsson (1984) found that, whereas nearly all Icelandic Whoopers breed on fresh waters ranging from near sea level to 700 m, only 30% of moulting birds occur on fresh water (about two-thirds of which occur below 300 m, where they feed largely on pondweed, and one-third occur above 300 m). A further 28% occur on brackish lagoons, such as at Lonsfjordur in southeast Iceland (where they feed mainly on wigeongrass), but the largest proportion, *c.*41%, moult on sea bays and estuaries where they feed mainly on eelgrass and green algae.

Movements to moulting areas have been mentioned above. Gardarsson & Skarphedinsson (1984), however, noted some particularly interesting movements in Iceland during late August and September, following moult. It seems that breeders tend to move from the highlands to the lowlands, and that non-breeders move inland, from saline to fresh water. In particular, by October they found that *c.*75% of all Whoopers in Iceland were on lowland fresh water, and only *c.*20% were in saltwater habitats. They supposed that this might either be in response to the phenology of primary food plants or to a preference for more open marine habitats while flightless, for safety reasons. Thus the swans use distinctly different moulting and pre-migratory staging areas. It is possible that food availability at heavily used sites may decline forcing such a move. Both types of site provide opportunities for birds to form or re-establish social bonds, but the pre-migratory staging areas are considered likely to be the most significant as indicated by the results of ringing. Future studies of pre-migratory staging areas are likely to prove extremely interesting.

Gardarsson & Skarphedinsson (1984) discovered that the shift from moulting to staging areas was particularly noticeable in western Iceland, with none moulting in the very important staging areas of Vatnsdalsá and Borgarfjordur, and other southern wetlands. Superficially at least, it would appear adaptive to shift from summering/moulting sites to pre-migratory staging areas if, at the same time, the flocks can shift from a depleted, depleting or less nutritious food resource, to one that has not been subject to the considerable grazing pressure of a large flock of Whoopers over the prolonged summer moult. From the summer moulting grounds to the pre-migratory staging areas may involve some long-distance movements, but the longest of all, the seasonal migrations, are considered in the following chapter.

In this chapter I have been dependent on very limited information from just one small part of the breeding range, yet this already raises many challenging questions. Haapanen & Hautala (1991) have put forward the compelling idea that Whoopers arrive on their breeding grounds in two waves of spring migration. Could something similar be occurring among moulting birds? Are there really two waves of non-breeders, those not attempting to breed and those failing later in the season? Do they moult at different times? These and many other fascinating topics demand study; perhaps the information will be forthcoming from different parts of the species' immense range.

CHAPTER 10

Movements and migration

INTRODUCTION

The disappearance each year of the Whoopers from their breeding grounds may easily have led northerners to revere them as having supernatural powers. After all, mythology had it that they flew to the mountains of Valhalla or even to the moon. This delightful notion failed, inevitably, to survive the discovery of migration!

This large species must, for climatic reasons, make long annual journeys between summering and wintering areas. These follow relatively strict patterns and are discussed in detail as migrations. Swan populations, and particularly flocks within populations, are also affected by local conditions and may wander within their normal wintering range or even travel well beyond it as accidentals. These I have termed movements and are discussed only in very general terms here. The third aspect of mobility, that I have chosen to ignore, occurs on such a local scale that variation is so great as to make a special section meaningless. I am referring to the daily movements between roost and feeding areas, or even between different feeding areas within a day. This is so regionally variable, even from flock to flock, that such movements are best considered as part of feeding or roosting behaviour, and I have attempted to address them in Chapter Six.

Whooper Swans are essentially broad-front migrants, but various patterns of migration emerge (see Fig. 10.1). Mathiasson (1991) discerned four major migratory groups, approximating to Owen *et al.*'s (1986) four populations:

1. The Western group —the Icelandic population. Totalling some 20,000, mostly breeding in Iceland, and wintering in Ireland and Britain (Cranswick *et al.* 2002).
2. The Southwestern group—the West Palearctic or West Siberian population. Numbering some 60,000, which breed west of the Urals and winter in western continental Europe.
3. The Southern group—the Central Siberian population. Also numbering several tens of thousands (though Ravkin *et al.* (1990) and Ravkin (1991) suggest 300,000–400,000), breeding east of the Urals (between the Ob/Irtysh and Yenisei rivers), and wintering around the northern Black and Caspian Seas.
4. The Southeastern group—the East Siberian or East Palearctic population. Totalling more than 40,000 that breed in the Russian Far East and winter mainly in Japan, but also from Kamchatka to the Koreas. To these we should also add the Chinese birds, several thousand of which breed in the Chinese highlands of Xinjiang Province (Tibet) and winter along the Chinese coast.

Because of the relative attention that these four migratory groups or populations have received, I will focus most closely on the western, somewhat less closely on the southwestern, and address the southern and southeastern groups together.

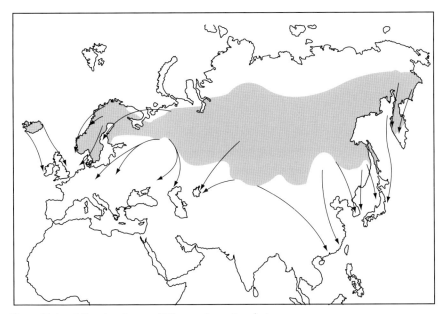

Figure 10.1.   Migration Routes of Whooper Swan Populations.

All classes undertake migrations, though there may be certain tendencies in relation to age or status that deserve fuller attention. The majority of migrants are of course non-breeders. Observations from widely separated parts of the range indicate that the first arrivals on the summering grounds are paired adults, some still accompanied by the previous year's offspring, with subadults arriving later, while in the autumn it seems that young pairs, failed breeders and non-breeders leave first, while families with cygnets leave later (Ma & Cai 2000).

Parents and their offspring migrate together, typically joining, or being joined by, other families or non-breeding adults. Autumn migration is invariably completed as a family and no doubt this is an important period for cygnets to learn both details of the route and of suitable staging areas and foraging sites (Fig. 10.2). Dement'ev & Gladkov (1967) with their characteristic brevity stated: 'Flights extremely high, by day and night; in early spring passages move as individuals, at height of traffic in groups of upwards of 10 individuals, arrayed in V-formation, from time to time emitting loud grunting honks, particularly audible at night. In spring migration rests on rivers, which thaw earlier than enclosed waters.' As they indicated, early-spring migration is by individuals, pairs and family groups, although some cygnets have become isolated from, or may have lost, parents by this stage and thus complete the migration by joining other flocks. Although family breakdown may occur at various stages along the return route, families finally separate once they reach the vicinity of their breeding area (perhaps as close as 2–3 km in some cases; see Chapter Eight), with the parents flying alone to their territory, leaving the previous year's offspring to summer and moult with others of their age or status class.

Figure 10.2.   Family members migrate together in both autumn and spring.

Until the 1970s, the migration and movements, though understood in general terms from direct observation and supposition, had been little studied. Early studies were made by Lesser (1973) and Matsui *et al.* (1981) in Japan, and Preuss (1981) and Nilsson & Nilsson (1978) in Fennoscandia. Since the 1970s, however, a number of ringing and neck-banding studies have focused on this aspect of their biology, and more recently the advantages of satellite technology have been brought to bear, thereby enormously increasing our understanding of the species' migratory habits (see Fig. 10.3). Here, I will provide an overview of our current knowledge.

The primary factor driving the Whooper Swan south in winter is the severity of the northern climate in those regions where it summers and breeds. This influence appears to operate by reducing the productivity of the plant food that they depend on. Winter severity in effect determines the ranges and distribution of the 'northern swans'. The even more northerly breeding Tundra Swans must move a long way south, while among Whoopers only relatively small numbers remain north, and those only where ice-free water remains, e.g. in the eastern Baltic or parts of Sweden, where they are restricted to small areas of open water and they depend on human provisioning. In the Far East, although the majority do migrate south, lakes that are kept ice-free by springs or hot springs serve as suitable wintering sites despite the severity of local winters, as in Kamchatka (Russia) and Hokkaido (Japan). Elsewhere, they rely on marine sites that are usually mostly ice-free. So Whoopers may change wintering locations, as do Tundra Swans, in response to winter severity and food availability (Earnst 1991).

Figure 10.3.   The Whooper Swan's migrations carry it from wild northern breeding grounds to intensively managed wintering areas.

Local weather conditions also affect their movements, and may even be the proximate trigger for migration. It has been known for a long time that swans and geese often leave the Outer Hebrides, Scotland, for Iceland in anticyclonic conditions. The attendant calm air and clear skies are presumed to assist their passage on what amounts to possibly the longest transoceanic swan flight—some 800–1,200 km, from the British Isles to Iceland, in contrast to most swan migrations which are overland (Ogilvie 1972).

Strong, high-level winds have long been presumed to assist some larger species of migratory wildfowl that are able to reach such altitudes. For example, on 9 December 1967, a party of *c.*30 migrating swans, presumed to be Whoopers, was observed, by the pilot of a civilian transport aircraft over the Inner Hebrides, Scotland, and tracked by radar during their southerly descent towards Ireland. They were flying at a ground speed of *c.*139 kph and at a record-breaking 8,200 m, the highest for any waterfowl species. They were experiencing northerly winds of 50 m/second at the edge of a jet stream, where the temperatures may have dropped to as low as −48°C (Stewart 1978). The logical assumption is that the flock departed from a coastal lagoon at sea level in Iceland at dawn, in a ridge of high pressure, and quickly reached the jet stream of the lower stratosphere (Fig. 10.4). Elkins (1983) considered that given the prevailing upper wind conditions, the flight time from Iceland to Ireland might have been only seven hours! Recent satellite telemetry confirms that even lower altitude migration may be relatively swift, lasting 12.7–101 hours between the British Isles and Iceland (Pennycuick *et al.* 1996, 1999). Given appropriate weather conditions, and this capacity of rapid long-distance migration, it is no longer surprising that arrivals at a range of sites

Figure 10.4. Though sometimes wind-assisted, migratory Whoopers are unable to glide or soar like storks or pelicans.

in Scotland and England all occur within the same timeframe, sometimes on the same or consecutive days. Whereas Bewick's Swans make a prolonged migration, stopping repeatedly at staging areas, it may be within the capability of the Whooper to travel non-stop from its pre-migratory staging area, in the vicinity of its summering/breeding grounds, to its wintering site. Obviously, not all birds make such a choice or they are unable to make such a non-stop flight when conditions are not supportive or their own body condition precludes it. Far from the norm, however, the extreme high-altitude flight noted above may well be the exception. Contrary to past assumptions, Whoopers are capable of both landing and resting on the sea, and may commonly migrate at rather low altitudes, as does the Trumpeter Swan, which generally flies low (97% below 154 m; Mitchell 1994).

Winter movements, distribution, and habitat use are all affected by severe weather (Bannerman & Lodge 1957; Brazil 1981c). Annual migrations take them south from regions of particularly severe winters, but there is no guarantee that their wintering grounds will not experience severe weather too, hence the need for flexibility in their behaviour.

The increasing severity of the weather in early winter causes further birds to migrate south from more northerly wintering areas particularly from Iceland where, in some years, up to 1,300 may winter relatively close to their summering/breeding grounds (Gardarsson & Skarphedinsson 1985; Kirby *et al.* 1992). Severe weather may also lead to more localised movements south within Scotland, or from Scotland to England or to Ireland, although the great extent of these movements, and those within Ireland in particular, suggest that movements between sites are by no means only associated with bad weather (McElwaine *et al.* 1995). Movements from Britain to Ireland are slightly commoner than in the opposite direction (55% versus 45%), which might reflect a continued southerly shift in distribution prior to direct transit back to Iceland from Ireland (Rees *et al.* 1997a,b, 2002). It has long been suspected that hard-weather movements occur between Europe and Britain, although there is little evidence for this, and such movements are presumed to account for many of the extralimital records in Mediterranean Europe and North Africa. Prolonged freezing reduces the area of both fresh and coastal brackish or even saltwater available to Whoopers in some areas, such as in northeast and eastern Asia, while snow cover prevents them from using fields or marshes and may force them further south. They may then rely on previous experience to continue to more favourable traditional feeding sites further south or, if conditions are particularly unusual, they may merely wander in search of suitable habitat (Reynolds 1982).

As one specific example, cold weather during much of early 1979 drove birds entirely away from my central Scottish study area. The weather in January 1979 was severe throughout Scotland and most inland waters were frozen for several weeks. Swans became scarce in central areas, while larger numbers were reported in the southwest. A sighting of 19 Whoopers flying southwest out to sea near Port Patrick, Wigtown, on 4 January 1979 supported what was then merely theory, that movements to Ireland sometimes occurred in hard weather (G. Sheppard in Brazil

1981c). A further example, from the opposite end of their range, Hokkaido in northern Japan, shows that, when caught by the severest of weather, birds may fail to move in time, and may even freeze to death, as happened to a flock of birds at Odaito on the east coast (Abe 1968).

Ringing, both with neck-collars and leg rings, confirmed that, as long assumed, Icelandic breeding Whoopers predominate in Ireland and Britain in winter (Brazil 1983a; Black & Rees 1984; Gardarsson 1991; Rees *et al.* 1991b). Hewson (1964) had suggested that Whoopers from Scotland moved southwest as far as Ireland during winter, and that from there they migrated directly to Iceland in spring. His hypothesis was based on the lack of spring peaks at sites in Scotland that were important in autumn. Although this might be so, Brazil (1981c) found that flock size tended to decline during late winter, and considered that the scattered small groups might use different sites from major autumn ones, and, therefore, be more easily overlooked during wildfowl counts at major wetland sites. In support of Hewson's (1964) theory, not only have birds been sighted flying towards Ireland from southwest Scotland, and birds colour-marked at WWT Caerlaverock, Scotland, have been followed to Ireland, but the majority of movements are in that direction, not in the opposite.

Interchange between Scotland and Ireland, however, is more regular than previously considered, and not entirely the result of birds moving because of severe weather in Scotland or because they pass through Scotland en route to Ireland. Very large numbers migrate directly to Ireland from Iceland. Some then move east into Scotland, as shown by an early observation of seven ringed birds that arrived in autumn at traditional points on the north coast of Ireland, before moving on to winter at WWT Caerlaverock (where they had been ringed in a previous winter; Bowler *et al.* 1993). Martin (1993) considered that there were frequent movements, usually by individuals rather than groups, between Ireland and Scotland and other parts of Britain in both directions during winter, dependent on weather and food availability, and this has largely been confirmed by subsequent ringing studies. It is likely that both Hewson's (1964) and Brazil's (1981c) theories were partially correct. Certainly, we now know that extremely large numbers migrate directly between Iceland and Ireland in both directions. Some certainly move from Scotland to Ireland during the winter, and may then migrate back to Iceland from there, while many from England and Scotland return to Iceland without visiting Ireland.

It is easy to forget that few, if any, other parts of the Whooper's range have as mild an autumn and winter, given the latitude, as Iceland, Ireland and the UK, where the climate is so distinctly ameliorated by the warming effect of the North Atlantic current with its origins in the Gulf Stream. In continental regions, however, the slide towards lower temperatures in autumn is typically continuous and may be precipitate, leading quickly to the freezing of the surface soil and most or even all bodies of standing water, and even of flowing water too (many major and most minor northern Russian, Chinese, Korean and Japanese rivers freeze) and this continues unbroken until spring thaw several months later. To be present as lakes freeze is an astonishing visual and auditory experience for warmly clad humans, but for still flightless young waterfowl it is a critically dangerous time. If polynias with sufficient food do not remain, then they perish. In the Anadyr region, for example, the autumn freeze may occur as early as 20 September, forcing families to move to larger (and thus later-freezing) lakes of the southwest Anadyr Lowland, e.g. Maiorovskoe Lake, there they join the post-moulting flocks of non-breeders at traditional autumn concentrations in late September–early October. The aggregation of up to 200 at a locality helps to prevent the lake polynias from freezing, so families benefit from being able to become stronger before finally they must depart. Occasionally, in years when the flow of water at the mouth of the Anadyr River keeps areas ice-free for longer, then families may remain there into December—but no later (Krechmar 1990).

It is easy to presume that migration must present a significant risk and cost, an assumption I had previously made myself (Brazil 1981c), until Gardarsson & Skarphedinsson (1984) showed how, relatively speaking, higher losses occurred before migration than during it. Wilmore (1974) had taken the 'migration-as-major-risk' assumption to its fanciful extreme:

'When the migrating flocks have left the mountains and the fields behind and venture upon the great flight over the expanse of ocean, many of the first-year cygnets fall exhausted into the sea, or lose their parents and consequently perish. The only resting place from Iceland to the British Isles is the Faeroes, a small group of islands, easily missed in the swans' southeast passage.'

But Matsui *et al.*'s (1981) observations of Whoopers resting on the sea off Japan, and my own of birds offshore and on sea bays around Hokkaido (Brazil 1984b; Fig. 10.5), indicated that falling 'exhausted into the sea' is quite unnecessary for a large aquatic bird that moults at brackish lagoons and also winters and feeds at tidal, saltwater habitats. Satellite telemetry has also confirmed that resting on the sea is a regular, if not common, phenomenon (Pennycuick *et al.* 1996, 1999). Furthermore, the occasional arrival of detached cygnets on the wintering grounds is an indication that to lose one's parents does not always lead to perishing!

In addition to the seasonal clues triggering migration, there may also be further local stimuli leading them to migrate under certain local weather conditions or at particular times of day. Soaring migrants such as White-naped and Hooded Cranes are often seen leaving their Asian wintering sites, particularly Arasaki in southwest Kyushu (Brazil 1987) towards the Koreas once the morning air has begun to warm, but at Welney, England, Kemp (1991) has noted that Whoopers, self-powered, rather than soaring migrants, tend to depart in the spring in the evenings on a north-westerly or northerly heading, to make a journey to Iceland of over 1,600 km. Bewick's Swans leaving from the same locality head northeast or east-northeast. Kemp's (1991) observations support Ogilvie's (1972) assertion that swan migration generally occurs at night, leaving the daylight hours for feeding, and is more frequent in clear weather as this enables them to use the moon and the stars as navigational aids.

Figure 10.5.   Far from falling 'exhausted into the sea', Whooper Swans are at home there and among sea-ice.

Furthermore, Whoopers migrate considerably later than do Bewick's. The smaller bodied Bewick's Swans breed further from their wintering grounds and are even more critically dependent on the northern ice-free period for breeding success than are Whoopers. To maximise their time in their summer range, and to be ready for breeding as soon as conditions are suitable, they commence spring migration even before the end of February, whereas Whoopers usually depart some weeks later, during March or even April (Bowler *et al.* 1992).

While long-distance, non-stop migrations are clearly possible, not all flights are such. In fact, in comparison with continental populations, Icelandic Whooper Swans migrate relatively short distances, although the 800–1,200 km of ocean that they typically cross *is* unusual. After breeding, but before setting out on the main journey south, or on initial arrival at the end of a longer flight and before continuing to other wintering or summering areas, Whoopers gather at particular staging areas. It seems that they prefer to make long, direct flights, from good feeding grounds (regardless of their location) to their wintering grounds, rather than necessarily opting for the shortest flight and beginning at the nearest coastal locality. There are, for example, significant staging areas in northwest Iceland, despite these being up to 250 km further from their early wintering areas in the northwest British Isles than are staging areas in southern coastal Iceland (Gardarsson 1991).

Severe weather may not only influence how birds move about locally during winter or trigger onward movement, but it may also affect overall timing of a migration, delaying it by days or even weeks. For example, in spring 1979, departure from central Scotland was so delayed by cold weather that birds were still present in my study area on 7 May. That year, Iceland experienced its coldest winter and spring on record, and few swans were present in southern Iceland even by the second week of May, when they would normally have been well on their way to the breeding grounds. A journey along the south coast from Reykjavik to Alftafjördur with Prof. Gardarsson (18–20 May 1979) revealed only scattered pairs and a few small flocks. At Lónsfjördur, however, in the southeast, a vast flock of *c*.3,300 Whoopers had accumulated, half of the estimated Icelandic population at the time. The area had not been watched in spring prior to 1979, but held a large concentration again in spring 1980. Suspicions that this was likely to be a regular phenomenon (Brazil 1981c; Gardarsson, pers. comm.) have been confirmed with even larger numbers, 7,000–8,000, now known to visit there (Einarsson 2000).

Lónsfjördur, a large lagoon sheltered from the sea by a ridge formed by longshore drift, is backed by a ring of mountains rising to just over 1,000 m. Valleys through the mountains may provide swans with suitable flyways to the north. It seems likely that Whoopers on spring migration choose to stage there because of its geographical and topographical situation, and the abundance of suitable food, namely wigeongrass. Two Whooper Swan stomachs, and hundreds of droppings that we examined in May 1979 confirmed that they were feeding exclusively on wigeongrass. For birds migrating from northern Scotland, the southeast corner of Iceland represents the nearest landfall. Coupled with its sheltered position and food supply, this makes it an ideal arrival site. For the same reason it is also a perfect moulting site for non-breeders, and in that capacity it has been known for longer (Boswall 1975; Brazil 1981c).

After arrival at suitable staging areas, typically still maintaining their family groups within larger flocks, Whoopers must separate so that subadults and the previous year's cygnets, are left behind, and pairs move off to their breeding territories uninhibited by the needs of their previous family. The pattern is assumed to be roughly similar throughout the range. My own observations after a particularly harsh winter, in the Myvatn area of northeast Iceland, showed that on arrival, Whoopers first move into snow-free areas, feeding there while nearby lakes thaw. From there, they then either move to breeding sites or directly to summering/ moulting areas. Haapanen *et al.* (1973a), who observed a similar pattern in Scandinavia, thought that early arrival at pre-breeding concentrations could be adaptive if it allowed breeders to occupy territories as early as the weather permitted, for at northern latitudes, where the summer ice-free period is critical, they might be more likely to achieve breeding success. It is debatable whether or not early arrival also provides more time to replenish energy reserves after migration, thus facilitating more successful breeding. If the spring is late enough to prevent birds from moving to their breeding territories, then it is also likely that the food available at pre-breeding areas will be of poorer quality than usually, because of the inhibiting effect of low temperature on spring growth. Conversely, if the weather is good, the swans are most likely to move directly onto their breeding territories.

So far I have discussed only the gross function and mechanism of migration. There is, however, the further interesting aspect of selective migration or movement, which seems worthy of further study. Brazil (1983c) considered that in Japan at least there might be a selective movement of families southwards in order to avoid the severity of winters at more northerly latitudes, as indicated by variations in the ratio of young to adults found regionally. This is by no means a new supposition. Boyd & Eltringham (1962) found that the proportion of young Whoopers in Britain was lowest in midwinter, and Reynolds (1982) noted that significantly more families than pairs or individuals emigrated from Orkney in December–January. Black & Rees (1984) considered that the lower proportion of cygnets in Britain in midwinter, combined with the short-term stays of un-ringed families, might indicate movement to and from traditional Irish wintering grounds. However, it is now known that Ireland is by far the most significant wintering area for the western population, and thus presumably there must be another explanation for this pattern of selective movement.

Reynolds (1982) surmised that selective emigration from Orkney might be related to the more demanding feeding requirements of the cygnets, which might also be more susceptible to cold weather because of their small body masses and relatively larger surface areas. It was this latter aspect, and consideration of the fact that the smaller bodied Bewick's Swan tends to winter at more southerly, and hence milder, sites than Whoopers, that led me to consider selective migration of families in Japan (Brazil 1983c). Migration itself, however, must also represent some physical risk and an energy drain on cygnets, and so where suitable ice-free areas remain relatively closer to breeding areas, then one might actually expect the reverse to be true, with proportionately more families remaining locally and not undertaking prolonged migrations. This latter supposition has now been proven in Iceland where Gardarsson & Skarphedinsson (1985) found that the proportion of young in Whooper flocks wintering there averaged *c.*23% in 1976–85; they considered that residents included a higher proportion of young families than did emigrants. It would be particularly interesting to know whether in other regions where birds winter close to their breeding areas, e.g. in Kamchatka, the same is true.

Although weather conditions clearly affect timing of seasonal migrations, landfall and subsequent movements (McElwaine *et al.* 1995), cold-weather movements should be seen in quite a different light, as operating more locally, being more transitory, and as usually occurring during the worst part of the season. Martin (1993), however, considered that it was cold weather that led small numbers of Whoopers to 'venture beyond the main wintering areas', yet birds arriving in Italy, the USA and Greenland, as he cited, are far more likely to have been overshooting migrants than cold-weather movers, and the Greenland records in particular presumably pertain to spring migrants overshooting their current normal summer range.

A further aspect has been referred to tangentially by Martin (1993), namely short-distance moult movements within Iceland. The typical pattern appears to be that many of the non-breeders settle quite quickly in early summer at sites where they will moult (although others flock in early summer at places where feeding is good, such as on new improved hayfields; Gardarsson *in litt.*), while breeders typically moult on their breeding grounds. Any moult movements, therefore, are only of failed breeders abandoning their relatively isolated breeding territories and joining the relatively safe non-breeding flocks for the rest of the summer. But whether it is specifically to moult that they move must be questionable, as they may choose to move to areas with better food supplies or which provide security in numbers.

Rees *et al.* (1997b) found that the proportion of juveniles in flocks in Britain rose from 9% in October to 14% in November and to 17% in December, consistent with early migrants in autumn being mostly non-breeders or failed breeders, followed by increasing numbers of successful breeders with their families as the season progressed. Declining numbers after December may have resulted from differential mortality of young; however, it may also result from the increasing difficulty of separating young from adults during the period when they are rapidly whitening. In midwinter, there was a decrease in the percentage of juveniles recorded at northern sites (northern Scotland and northwestern England), while the proportion further south (southwestern Scotland, southeastern Scotland, east-central England and Northern Ireland) increased, suggesting that families from more northerly sites move south during the winter. Nevertheless, in January, flocks in Scotland (except the northwest) and Northern Ireland still had a higher proportion of juveniles than those in southern England and the Irish Republic, indicating that perhaps parents with families prefer to winter at sites closest to their breeding grounds, while non-breeders migrate further south (Rees *et al.* 1997b).

Spring migration was studied in some detail during the 1950s in the Helsinki region of southern Finland. There, Jahnukainen (1963) found considerable annual variation in timing, and that when migration began early it also continued for a long period. During the early part of the migration, flocks were small, but by late April and early May the flocks were larger than average. The commonest flock size was two (presumably pairs), while only 6% contained 30+. After singles and pairs, the next commonest were flocks of 3–6, and these seem most likely to have consisted of family parties. It appears very likely, from Jahnukainen's (1963) study that in spring pairs and families tend to migrate separately, and also earlier. Jahnukainen (1963) found a significant correlation between the average air temperature, from 15 March to 15 April in Helsinki, and the median date of swan migration. He also noted that mass migration is common, meaning that on particular days large numbers pass a fixed spot, thus several hundred may be observed during a day. Jahnukainen's (1963) observations of mass migration occurred on 27 April (1958), 29 April (1960), 30 April 1960, 1 May 1958 and 1960, with totals of 144–343 during the day.

PATTERNS OF MIGRATION

Whooper Swan requires a period of *c.*3.5 months to complete its reproductive cycle from establishing a territory to leaving it with fledged young. This lengthy time requirement restricts them to a particular band of habitat across the northern temperate zone where the period between spring melt and autumn freeze at least equals or exceeds this period. Arrival on the breeding grounds before spring melt has occurred is unlikely to be advantageous, and it may even be disadvantageous in energetic terms if conditions are so severe that little food is available. While there might therefore be advantages in delaying departure from the wintering grounds, as Mathiasson (1991) has pointed out, remaining longer on the wintering grounds is likely to promote individual survival over the chances of establishing a territory and reproducing successfully. Whoopers, therefore, like other migrants, experience a conflict: they must choose whether to delay their spring migration and prolong their stay on the wintering grounds to their own benefit, or migrate so as to arrive earlier at the breeding grounds thereby improving the probability of breeding successfully. One might expect yearlings and non-breeders to opt for later migration, and breeding pairs for earlier migration. The timing and speed of migration, and the presence of suitable staging areas close to breeding sites, are all critical, all the more so for birds wintering furthest from the nesting areas.

Mathiasson (1991), looking at the various costs and benefits relating to migration, considered beneficial behaviour to include:

- migrating as short a distance as possible between summering and wintering grounds; and
- reducing mortality and increasing reproductive capacity, by breeders and non-breeders migrating at different times so as to reduce competition for food at staging areas.

While among potential costs he considered:

- breeders migrating early, and thus experiencing increased mortality from adverse weather and finding food scarce; and
- the likelihood of lowered breeding success in years with unexpected late-spring thaws or early-autumn frosts.

The potential risks of migration may be ameliorated both by individual experience and by transferred experience, by individuals repeatedly migrating via the same routes and stopping at the same staging areas that have proved reliable in previous years. Cultural transmission of this information would be facilitated particularly by cygnets accompanying their parents on their first return migration, and by flocking with geographical cohorts. Yet these habits too involve risks. Being dependent on specific routes and individual stopover localities may be catastrophic if habitats are unstable, prone to or susceptible to disturbance or destruction by man. Furthermore, in regions where either or both food and open (ice-free) water are limited, then flocking, especially in large flocks, may lead to serious intraspecific competition (Mathiasson 1991).

For each of the four populations or migratory groups that Mathiasson (1991) recognised, the essential migratory timetable is similar, although those of the central and northernmost Palearctic face the harshest summer conditions and must often wait longer for conditions to improve sufficiently for breeding. As a result, their arrival is normally somewhat later than that of western breeders. Whereas the essential seasonal timetable for migration is determined genetically, having evolved in relation to prevailing climatic conditions, it is flexible, the time of departure being dependent on an external releaser, the local weather conditions, which of course vary annually.

Timing of migration, particularly arrival and departure from breeding grounds in the far north, affects reproductive success. Late arrival both reduces the time available for birds to improve their physical condition before breeding (if such a strategy occurs), and delays the start of nesting. Furthermore, late-arriving prior breeders may find that they have been beaten to their previous year's territory, perhaps involving them in the energy consumption of reacquiring it or acquiring another. Late-arriving new breeders meanwhile may find that all of the best habitat is already occupied. Late nesting inevitably results in late hatching and, given that the fledging period is roughly constant, regardless of starting date, for eggs to hatch later than July may preclude the safe migration of the cygnets, especially given that early freezing conditions may force migration south by mid September.

Mathiasson (1991) pointed out that, as Whoopers are large, long-lived birds capable of rearing large broods in some years, within the context of lifetime reproductive success, they are able to adopt a broad migratory strategy which permits colonisation of new areas, where reproduction may not always be successful as, even within established breeding areas, the number of successful breeders fluctuates annually due to weather conditions.

As a consequence of the lengthy breeding season, it would be expected that the birds should winter as close to their breeding areas as weather conditions and winter food availability allow, so as to reduce the time and energy costs of migration, and to facilitate arrival on the breeding grounds at the most appropriate time.

If arriving late has its penalties, then so too does arriving early. Although early arrival at the breeding grounds may overcome the difficulties described above, and facilitate successful reproduction, it may be at the increased risk of encountering adverse weather conditions and poor food availability en route, both of which may negatively impact physical condition. While breeding birds face this dilemma each spring, and must depend on local weather conditions for suitable clues as to how and when to proceed, non-breeders may reduce these risks by simply departing later and moving north more slowly; they have nothing to lose by arriving later. Cygnets are faced with a different choice, however, for to delay migration and remain with non-breeders would mean that they would miss their last opportunity to learn from successful breeders, their parents, aspects of both the route and stopover sites. Migration in family units (itself dependent on individual recognition) may serve to preserve the characteristics of a particular migratory population or subpopulation, and through a combination of imprinting and cultural tradition may help maintain the geographical course of a particular route (Mathiasson 1991).

As Mathiasson (1991) has suggested those birds flying north to presumptive breeding areas beyond a certain point will not have sufficient time to reproduce before the autumn freeze, and so the genetic characteristics leading to this behaviour will not be selected for. He considered that this and other historical factors might explain the position of the main migratory dividers, the migration routes of wintering flocks originating in different geographical areas, and the breeding range.

MOVEMENTS AND MIGRATION OF ICELANDIC/IRISH/BRITISH WHOOPER SWANS

Much work has inevitably focused on the movements of the Icelandic/Irish/ British population (e.g. Brazil 1981c, 1983c; Gardarsson 1991; McElwaine *et al.* 1995), with the aim of explaining (a) why concentrations occur in certain areas and (b) the spread of ringing recoveries. Less attention has been devoted, unfortunately, to the movements of other populations. First I will address the best-known westernmost population, then those of continental Europe and finally the broad swathe of Siberia.

Some 90–95% of the Icelandic Whooper population is migratory, and almost all are assumed to winter in Ireland or the UK, although a small minority reaches continental Europe (see Fig. 10.6). Early-autumn food shortages or frosts may trigger

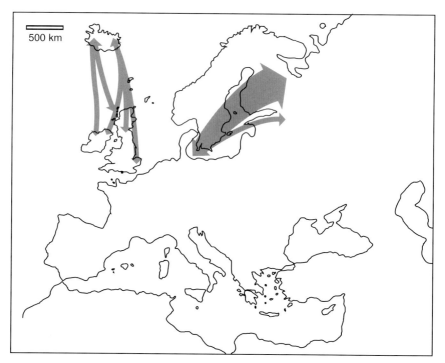

Figure 10.6.   Migration routes of Whooper Swans in western Europe (after Andersen-Harild 1984).

families to depart their breeding sites. Failed breeders and non-breeders, unconstrained by offspring, are believed to leave their summering grounds somewhat earlier than most families, as indicated by slowly rising percentages of young on the wintering grounds (Rees *et al.* 1997b). Their first stop may be at inland lakes, larger rivers or coastal lagoons in southeastern Iceland, where non-migrants may remain to winter. The migrants, however, soon continue south. The main influx into the wintering range is from late September, and lasts right through October and November. Some sites en route, however, may experience passage peaks. This appears to be the case in Shetland, where arrival begins in early October, increases rapidly during late October to an early November peak, before declining rapidly by mid November (Gardarsson & Skarphedinsson 1984). Further emigration from Iceland after the main autumn exodus may occur if winter weather intensifies, and given that the journey may take as little as seven hours, it may be worthwhile for even relatively short periods to avoid spells of serious freezing weather. The proportion remaining in Iceland varies annually, presumably in response to the weather and availability of food, but usually involves several hundreds.

The majority of Whoopers depart Iceland during October, though they will perhaps already have been on the move away from their breeding/summering areas for some weeks. Some travel directly to their chosen wintering sites in Ireland, Scotland or England, while others transit Scotland en route to sites in England, Wales or Ireland. Boyd & Eltringham (1962) noted higher numbers on the Scottish mainland in November than at other times, whereas in England, where birds tended to arrive later and leave earlier, maximum numbers occurred during December. This pattern could reflect movement south from Scotland as winter progresses or leapfrogging by later migrants to winter further south. The arrival of two Icelandic neck-banded Whoopers at Welney in early November 1980, however, proved that even some early-autumn immigrants may make the longest journeys (Brazil 1981c) and we now know that arrivals may occur during a very narrow timeframe across the British Isles. For example, first arrivals have reached Caerlaverock, Martin Mere and Welney within the space of 2–3 days in September (Bowler *et al.* 1982; Rees *et al.* 1991b), though in comparison with the total distance, that between these sites is not great.

Owen *et al.* (1986) considered that 40% arrived in Britain by mid October, that numbers peaked in November and began to decline by December to 50% of the peak by February, but increased again in March. The explanation given for this is that birds pass through Scotland on their way to and from Irish wintering grounds. In the light of the very much larger numbers now known to winter in Ireland, however, this pattern is likely to apply only to a proportion, as the majority migrate directly to Ireland, and numbers appearing in Scotland do not amount to more than a fraction of the Irish wintering total. The apparent decline in numbers during midwinter thus requires a further, more robust explanation. While none is at hand, I feel that a closer consideration of behaviour may be of value. If, as has been indicated elsewhere, feeding behaviour changes in midwinter and flock sizes decline, then the smaller groups may be far more easily missed by standard wildfowl counting techniques. Their predisposition to regroup into larger flocks at fewer locations for spring migration might explain the apparent increase in numbers later in the winter.

Departure from Britain and Ireland for the summering grounds usually occurs in March–April, and by late April the majority have arrived (a full month ahead of those breeding in northeast Russia). Many make landfall at important staging areas in the west, south and southeast, particularly at Lónsfjördur (Einarsson 2000), although stragglers may not arrive until early or mid May (Brazil 1981c; Krechmar 1982a; Owen *et al.* 1986; Martin 1993). Each year, however, some summer in the UK (e.g. Franklin 1947) and in recent decades pairs have occasionally bred in Scotland and Ireland (e.g. Gordon 1922; Thom 1986; Murphy 1992; Gibbons *et al.* 1993).

It has been shown that Whooper Swans *may* undertake extremely high-altitude migratory flights (Stewart 1978; Elkins 1983). Flight at such altitudes is only possible during the middle portion of a flight and towards the end. Inevitably it would take birds some time to climb to any great height. At altitudes of 8,000+ m, no topographical features would present a barrier to migratory flight. Those same features could, however, represent a significant barrier during the early part of a southbound flight from northern Iceland and, similarly, to birds moving north in spring from their first staging area on the Icelandic coast to inland nesting grounds, or to birds flying at far more modest altitudes, as is surely the norm. Matsui *et al.* (1981) noted that wintering Whoopers in Hokkaido use river valleys and mountain passes as flight lines, and there I too have observed them typically flying at low altitudes. Following the topographical line of least resistance is likely to be the case throughout the species' range until, that is, sufficient altitude has been attained. If we assume the same to be true in the west, then Whoopers arriving at staging posts in southeast Iceland are likely to fly around the Vatnajokull (Iceland's major glacier) rather than over it, in order to reach breeding and moulting areas in the north and east (see Fig. 10.7). Some, of course, may fly directly into their breeding areas without pausing at a coastal staging area, though this may depend on weather conditions and energy reserves. Whoopers have a maximum speed, based on recent satellite-tracking studies of 97.2 kph, although under exceptional circumstances they may travel at up to 139 kph (Stewart 1978; Whooper Facts 2001).

On the return migration, the same geographical barriers may separate the population, thus birds from the northeast third of Iceland leave from east of the glacier and those from the western two-thirds depart from west of it. If they continued on the same track, this could account for early concentrations in the Hebrides and Ireland in the west, and in Shetland, Orkney, Caithness and Aberdeenshire in the north and east of the British Isles. The lack of concentrations in the northwest Scottish mainland might solely be the result of poor coverage or a lack of suitable habitat.

Early ringing studies in Iceland suggest an interesting difference in the pattern of recoveries. It appears that swans ringed in the western highlands (Kinlen 1963) had a much more westerly distribution on the wintering grounds than those ringed at Myvatn in 1980, a pattern more strongly suggested by subsequent colour ringing, with birds from eastern Britain more likely to be recovered in

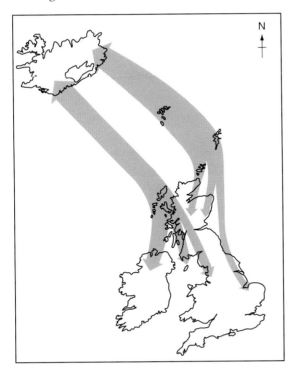

Figure 10.7. Hypothetical migration routes of Whooper Swans between Iceland and Britain (taking into account known wintering concentrations and ringing recoveries) (after Brazil 1981c).

eastern or central Iceland and vice versa, and birds from western Britain and Ireland more likely to have originated at ringing sites in northwest Iceland (McElwaine *et al.* 1995).

For more than 40 years, ringing of Whoopers from this population had taken place irregularly. Although nearly 500 were ringed in Iceland prior to 1980 the returns, in terms of information, were very poor, with only 61 (12.8%) ever recovered. It was not until the early 1980s that strides forward began to be made. Until the 1960s, almost nothing was known of the origin of Whoopers wintering in the British Isles (Kinlen 1963). Only a single ringed bird had been recovered to indicate their source; a cygnet ringed in Iceland on 24 July 1944 and found dead in South Uist in January 1948. By 1980, a total of 478 had been caught in Iceland, and marked with metal leg rings or patagial tags. This effort, mainly by Leo Kinlen's team, was sporadic, but the results were nevertheless interesting. About half were recovered in Iceland, the other half abroad, all but one in the British Isles, providing the earliest concrete evidence to link those wintering in Britain with the Icelandic breeding population. That study also provided the first evidence of movement to the European continent, when one controlled in Britain was recovered dead two years later in the Netherlands (Brazil & Petersen in press).

Of the 478 ringed prior to 1980, 229 (48%) were adults and 249 (52%) cygnets. By the end of 1982 just 61 had been reported. Thirty-seven (16%) of those ringed as adults and 28 (11%) as cygnets were later reported. Twenty seven (44%) were recovered in Iceland and 33 (54%) abroad. One controlled in Iceland was never recovered, and two further controls, one in Iceland the other in Scotland were later recovered. On the basis of these reports, it appears that mortality was lowest during the breeding season and highest in November–January, and that a mere 12% survived to breeding age (Brazil & Petersen in press).

Most of the information available from prior to 1980 comes from dead birds because sightings of the numbers on conventional tarsus rings are for all practical purposes impossible. In strong contrast, the post-1980 ringing effort has relied on visible leg rings, dyes or collars, which have provided far more information, far more easily and from many fewer birds.

In order to facilitate individual identification, studies of swans have, over the years, involved the use of various marking or recognition methods such as numbered metal or plastic tarsus rings, metal patagial tags, individual bill patterns, colour dying and plastic collars (Brazil 1983a). Metal tarsus rings have been used for more than 80 years, but only a very small proportion are ever reported because of the difficulties of reading or re-finding the rings. Metal patagial tags are even less likely to be reported, since they are inconspicuous even on birds in the hand, though they too have been used in the past. Coloured rings have been used to identify individuals in small populations or to identify age groups, subpopulations etc., and have the advantage that they are conspicuous under normal field conditions. More recently, the use of large coloured tarsus rings has been improved by engraving codes on the rings, which can then be read at ranges in excess of 100 m. Rings of this type have proven extremely useful for studies of movements or migration (Brazil 1983a).

The use of individual bill patterns as a means of recognition has proven useful, but it is limited to use at close range by skilled observers familiar with the individuals concerned or knowing the elaborate system for retrieving coded information about those individuals. The method poses problems, the biggest being that it is extremely difficult for unskilled observers to identify birds. As a result, such studies have, largely, been confined to experienced researchers of Bewick's Swan. Plastic neck-collars have the advantage that they are visible at long range and the engraved codes can be read by unskilled observers. Unlike tarsus bands they are less often out of sight under water, a major problem when using tarsus rings on largely aquatic birds such as swans. Following the extensive use of neck-collars on swans in North America, the former Soviet Union, Japan and Europe (particularly on Mute Swans), the value of neck-collars for the study of Whoopers was realised and such ringing has taken place in Finland, Denmark and Iceland. The most effective way of gathering data from the public is by using neck-collars as the codes are easily read, the likelihood of marked birds being reported is high and fewer birds need to be caught to provide information compared with other methods.

Since 1980, the style of ringing Whoopers has changed, with the Wildfowl Trust first ringing 52 birds at Caerlaverock, Scotland, in January 1980 with leg rings and picric acid (Fig. 10.8), and myself, having been inspired by the work of Preuss in Denmark, using neck-collars for the first time in Iceland in summer 1980 (Brazil 1983a). Preuss (1981) found that 116 neck-collared in early 1979 provided more data than the 266 ringed without collars in 1953–79!

Although by the late 1970s, mainly as a result of Kinlen's (1963) work, the wintering population in Britain and Ireland was considered to originate primarily from Iceland, the details of migration and the extent of movements around Britain and Ireland were still unknown. An initial small-scale project was initiated, during which birds were caught and marked at a single moulting locality in northeast Iceland, with the aim of finding answers to the following three questions:

1. Do birds moulting together migrate and winter together?
2. Do they use a single wintering site or move between sites?
3. Do birds return to the same sites in subsequent years?

The first catch of its kind was made by Gardarsson and I, and a team of Icelandic and Danish scientists at Neslandavik, on the west side of Lake Myvatn in northeast Iceland (65°35′N 17°00′W, 278 m) on 25 July 1980. Forty-six were caught and given rings and collars numbered consecutively from 1J01 to 1J46. The results were heartening, and groundbreaking at the time. Subsequent observations revealed more about their movements over just a few winters than had previously been known. By the end of May 1981, 22 (48%) had been re-sighted, several more than once. (A

Figure 10.8.   Wearing swan-jackets, some of the first birds caught at Caerlaverock await banding and dyeing.

study of 91 collared in Hokkaido in 1975–79 achieved a remarkably similar resighting rate of 47%; Yoshii 1981). Further birds may have actually wintered in Iceland, seven neck-collared birds were seen for example among a flock of 101 in the Alar area of Lake Myvatn in late October 1980 and at least two were at the same locality in April 1981 having perhaps remained there for the winter (Brazil 1983a).

The information from neck-collaring just 46 birds was much greater than could have possibly been expected had only tarsus rings been used, simply because of the visibility of the collars. Preuss (1981) had found that in the first winter after ringing, 22.4% of Danish neck-collared Whoopers were reported compared with just 9% of those ringed without neck-collars over the previous 25 years. Similarly, Brazil & Petersen (in press) found that only 13% of Icelandic Whoopers ringed without neck-collars during 1934–80 were reported. Therefore, the resighting rate of 48% during the first winter of the initial neck-collar study was extremely gratifying, especially in view of the small number ringed in relation to the population size (Brazil 1983a).

Although it had been supposed that birds moulting together might also migrate and winter together, the very wide scatter of wintering records suggested that this was not the case. Neck-collared birds were sighted as far apart as northeast Scotland, Northern Ireland, eastern England and the Western Isles. Despite sightings of several in the same areas, the only evidence of birds actually arriving together or associating with each other comes from 1J13 and 1J30, which visited three different localities together. They were sexed as male and female at ringing and are presumed to have been paired. Even at Welney, where four visited the same locality, they arrived independently. Thus, it seems that birds that moult together show no particular tendency to remain together during the winter nor even, necessarily, to migrate together. Had birds been caught at a migratory staging area the results might have been very different (but trapping would have been far more difficult). It is unknown whether non-breeders return to the same moulting ground in subsequent summers, making it uncertain whether a moulting flock represents a stable unit. Mathiasson (1973) however, showed that the rate of return of two-year-old Mutes to a moulting ground, where they had been ringed as yearlings, was 55–70%, though for older birds the rate of recurrence was lower. The ages of birds ringed at Neslandavik are unknown, but from their behaviour and time of arrival there I considered that they included first-year, pre-breeders and non-breeders, rather than failed breeders, and following Gardarsson & Skarphedinsson's (1984) model they were most likely to have been one and two-year-olds. Assuming that they behave in a similar manner to Mutes, there is a high likelihood of them returning to the same moulting ground. The breeding biology of the Whooper is still poorly known, however, and it is possible that moulting flocks also contain birds past breeding age, and those never attempting to breed, as is the case with similar Mute Swan flocks (Mathiasson 1973; Brazil 1983a). Much remains to be learned.

Of the first 46 to be neck-collared in Iceland 11 visited more than one locality, and five made prolonged stays of more than a month at a single locality. 1J20 spent 7 December 1980–12 March 1981 at Welney, having already visited three other sites. 1J25 spent 26 November 1980–13 March 1981 at Welney—the longest recorded visit of any bird at a single locality. 1J31 spent 10 November 1980–15 February 1981 also at Welney. 1J38 spent 4 February–11 March 1981 at Dungannon, Co. Tyrone, and 1J44 spent 16 November–18 December 1980 at Welney. Welney differs from most other British and Irish localities (except Martin Mere and Caerlaverock) in that food is provided and birds are closely observed. It is perhaps not surprising, therefore, that four out of five long stays were recorded there. The records show that while birds may spend long periods at a single locality, other records, of short stays or of single sightings, suggest that they may also wander freely during the winter (Brazil 1983a). The factors affecting movements within the wintering range are still poorly known, but young birds may be sampling a variety of sites so that they can select primary and backup sites for subsequent winters.

Based on the indirect evidence that there was a lack of spring peaks at those sites in Scotland which were important in autumn, it was hypothesised as early as the 1960s (Hewson 1964) that Scottish birds might move southwest to Ireland during winter, from where they returned directly to Iceland. More recent direct evidence, both from sightings of birds flying in the direction of Ireland from southwest Scotland and from observations in Ireland of some of those colour-dyed at Caerlaverock show that birds do move between the two. Ogilvie (1972) had stated that there was 'considerable interchange' between Scotland and Ireland, but cited no evidence. It was hoped that the neck-collaring study might shed some light on the subject, but none of the neck-collared birds was recorded from both Scotland (or England) *and* Ireland, although 1J39 did move west from northern England and was reported in the Outer Hebrides (Brazil 1983a). Birds migrating between Ireland and Iceland are assumed to visit or pass over the Western Isles, so it is possible that 1J39 was returning via the Hebrides having visited Ireland in December–March. Now, of course, we know of Ireland's great significance as a wintering area and we know that birds are more likely to migrate there directly from Iceland and less likely to move there from Britain.

It is not long ago that Whoopers wintering at Welney were considered to be of continental origin (Cadbury 1975; Scott 1980). Some authors (e.g. Richards 1980) took the assumption to the logical extreme and even concluded that all those in southern England must be of continental origin. That however was based on limited circumstantial evidence, some of it convincing at the time merely because of the limited information available. The most conclusive evidence was of certain individuals clearly contaminated with oil, which arrived at Welney soon after an oil spill in the Netherlands (Owen & Cadbury 1975). Furthermore, it was assumed at the time that brood sizes in Iceland were smaller than on the European continent and that large broods (Cadbury recorded one of six) were, logic dictated, not of Icelandic origin. Cadbury's (1975) 'limited evidence that Whooper Swans wintering at the Ouse Washes originate from Scandinavia or Russia, rather than Iceland', is less convincing, however, in the light of more recent information. It is no longer safe to assume that large broods cannot be from Iceland, since broods of six and eight have also been recorded in Scotland among birds almost certainly of Icelandic origin, and a brood of seven was seen in Iceland in 1978 (Brazil & Spray 1983). Similarly, the higher percentage of cygnets recorded at Welney than at most other sites may not necessarily indicate continental origin, as there is the possibility that, as in Japan, there may be a tendency for higher proportions of cygnets to occur at more southerly sites (see Brazil 1983a).

Studies of Icelandic Whoopers using neck-collars continued into the mid 1980s and have contributed to the generally accepted view (now) that the vast majority of British and Irish Whoopers originate in Iceland. In fact, sightings of four neck-banded birds at Welney (marked in Iceland in 1980) and of colour-ringed birds there in subsequent years, has provided direct evidence that at least part, if not most, of the flock wintering there is definitely of Icelandic origin (Brazil 1983a; Gardarsson 1991; Kemp 1991).

That many of the swans collared in Iceland in the 1980s were later seen in England lent considerable weight to the assumption that probably most of those wintering in England were from there. This has been further confirmed by considerable additional colour ringing of Whoopers in Britain (involving hundreds of birds) since the first at Caerlaverock in 1980 (Rees *et al.* 2002). None collared or ringed elsewhere in Europe had been sighted in Britain or Ireland by the early 1980s (Brazil 1983a) and it was presumed that there was only limited interchange with the continent, although Gardarsson & Skarphedinsson (1984), on the basis of initial ringing recoveries from the continent, felt that the possibility of some wintering regularly there, particularly in western Norway, should not be ignored (a position taken much further by Gardarsson later; see his 1991 hypothesis below).

Evidence of Icelandic Whoopers reaching the continent, was, if anything, even scarcer than of continental birds reaching Britain, although there was a single recovery of an Icelandic-ringed bird from the Netherlands, and more recently one was seen in Denmark, and another was found as far away as northwest Spain (Rees 1989; Rees *et al.* 1997a). More extensive ringing has shown, however, that birds from Iceland also occasionally reach Norway, Finland, Denmark and the Netherlands, with one Icelandic bird actually breeding in Finland in two consecutive years (Gardarsson 1991; Rees *et al.* 1997a). Of the more than 2,700 that have now been colour ringed in Britain and Iceland, only 51 (1%) have been reported from continental Europe: 38 in Denmark, seven in the Netherlands, two in Norway and one each in Finland, Germany, France and Spain (Rees *et al.* 2002). Conversely, at least 200 Finnish Whoopers are now known to winter in Britain (Laubek *et al.* 1999) and the first reached Ireland (Lough Neagh) in February 1999 (Rees *et al.* 2002), serving as a reminder that, regardless of the primary wintering range of the Icelandic breeding population, it is not isolated from that wintering in western Europe.

Coincident with the early use of neck-collars in Iceland (Brazil 1981c, 1983a; Gardarsson 1991), Whooper Swans in the Icelandic/Irish/UK population have also been marked in the UK by the WWT, first using picric acid in 1980, then more recently and regularly since, using individually coded plastic leg rings, facilitating long-term studies of individuals. One of the pitfalls for such studies, however, is ring loss, and Rees *et al.* (1990b) attempted to clarify the longevity of rings on Whoopers. They found that birds were more likely to break or lose their leg rings than were Bewick's, perhaps because of the greater internal diameter of the rings, making them easier to break, and the greater weight and size of the Whooper making it easier and more likely that they could break the rings. For males, the mean rate of ring loss was as high as 28% during the first two years after ringing, when Vynalast rings were used. For Whoopers, Vynalast thermoplastic laminate rings proved less durable than Darvic PVC bands, and even when using just one material, the quality of ring batches, and hence their longevity, varied. Similarly Gardarsson (1991) found that the number of resighted birds declined by *c.*70% annually. In retrospect, therefore, it is unsurprising that resightings of the first collared birds from Iceland fell so dramatically after just two winters, by which time collars would have also been likely to have been breaking (Brazil 1983a). Subsequent study of the Icelandic population using neck-collars has shown that nearly all such collars disappear within four years (Gardarsson 1991). Within that time, however, as many as 69% were resighted and 10% recovered. These were much higher rates than either Yoshii (1981), or Brazil (1983a) obtained, perhaps because of the increasing awareness among naturalists of the presence of banded swans, providing considerable information.

Gardarsson (1991), banded Whoopers at Vopnafjördur in northeast Iceland, where a few tens moult, at Alftafjördur in the southeast, where several hundred usually moult, and at Snæfellsnes, where over 2,000 moulted. He found that in autumn Whoopers migrate on a broad front, with a mean direction close to 140° from all three ringing localities. This study confirmed my own belief (Brazil 1981c) that birds from different areas of Iceland were likely to reach differing parts of the wintering range, by showing that west Icelandic birds tended to travel mainly to Ireland and western Scotland, and east Icelandic birds largely to Scotland, although some also reached northern parts of Ireland and southern Scandinavia. Rees (1989) too had found regional variation in the destinations of migrants, with birds ringed at Skagafjordur generally moving to Irish sites, and those from Jokuldalur in eastern Iceland occurring more frequently in England and Scotland. The mean distances travelled by birds from each of Gardarsson's (1991) study areas were similar 1,100 km from Vopnafjördur 1,122 km from Alftafjördur and 1,407 km from Snæfellsnes. Overall, east Icelandic birds moved a mean distance of *c.*1,100 km, whereas west Icelandic birds moved 1,400 km (Gardarsson 1991).

It follows that with migration occurring on a broad front, there are clear differences between movements from western and eastern Iceland. Whereas those from west Iceland move largely to Ireland, those from east Iceland tend to occur at sites further east, some even reaching the European continent, though none from west Iceland appears to have done so. For example, although within Iceland only 9% of east Icelandic and west Icelandic birds moved to the opposite half of the country, during early winter eight (19%) of those banded in east Iceland (but none from west Iceland) were found in easternmost parts of the wintering range: four in

Shetland and four in south Scandinavia (two each in Norway and Denmark). Gardarsson (1991) estimated that 600 east Icelandic Whoopers may move as far as the European continent to winter, a hypothesis that challenges any assumptions that populations might be isolated. Interestingly, Gardarsson (1991) further challenged assumptions about the birds wintering in Britain, which I will come to later in this section.

For the most part, Icelandic birds make their initial long-distance journey south during October, as do the majority right across the range. Having made a long transoceanic crossing, Icelandic Whoopers usually remain in one general area for some months, but banding has shown that many then move again in December–January. About half made further journeys of over 100 km, and typically left their early-winter sites for areas further south or southwest. These movements often took them from Scotland and north Ireland to south Ireland, though there appears to be no consistent pattern. Marked birds were in fact reported moving from south Norway to the Netherlands, from east Scotland and north Ireland to south Ireland, from Shetland to north Ireland, and from north Ireland to south Ireland (Gardarsson 1991). At a local level Brazil (1981a, 1981c) found that the size and composition of an unprovisioned wintering flock in central Scotland changed on a near-daily basis, perhaps indicating that not only do birds move again after arrival at early-winter sites, but that they may move repeatedly. The fact that, contrary to this pattern, some remain for prolonged periods at sites where food is provided, such as Caerlaverock and Welney, indicates that food availability is probably a very important factor influencing the shorter movements that occur during winter. Once the results of a study of the movements of several hundred banded in Ireland are available, we will know far more about their local movements at this season.

Although refuge-focused research has suggested that site fidelity is common (Black & Rees 1984), with Whoopers at Caerlaverock having formed a traditional attachment to the wintering site, and 78% of ringed swans returning for at least one winter, Gardarsson's (1991) considerably broader perspective has shown that winter site fidelity varies greatly. Only 45% of his banded birds were found even within 100 km of a previous winter site. To put that in perspective, a move of just 100 km would have removed all swans from my central Scottish study area (Brazil 1981c). Furthermore, 55% of his birds were found as far as 985 km from their previous winter's site, that is almost as far as the entire length of their migration, and easily far enough for them to visit a different country! This general picture is in marked contrast to the specific situations at Caerlaverock, Martin Mere and Welney, all three provisioned WWT refuges that, in catering specially for the Whoopers' daily and seasonal energy requirements, possibly even enhancing the process of reaching optimum breeding condition, provide a unique environment (Black & Rees 1984).

It should be remembered, however, that because of the practicalities involved, both Brazil's (1983a) and Gardarsson's (1991) banding studies focused on non-breeding moulting flocks. The mobility of these, and their low frequency of return in subsequent winters, may be a result of the age composition of the moulting flocks from which birds were caught. All of the non-breeding birds were at least one year old, and the majority were probably at most a few years old. Thus in both studies, relatively young birds were being sampled, and as is typical of the young of many species, these were more likely perhaps than breeding adults to make exploratory movements between wintering areas. Breeders might be expected to show more pronounced site fidelity. In support of this assumption comes Gardarsson's (1991) own discovery that although some Icelandic birds wandered to continental Europe none was reported to do so after the first winter following marking. It may be, therefore, that while there may be higher than average rates of site fidelity among successful breeders visiting provisioned refuges, other classes may move regularly between sites in response to the availability of food.

Having myself been concerned with providing evidence to show that those wintering in eastern England were also of Icelandic, not continental, origin at a time when the latter theory of origin was favoured (Brazil 1983a *contra* Richards 1980), Gardarsson (1991) has now provided an interesting counter argument, suggesting that as many as *3,000* continental Whoopers may winter in the British Isles!

The population breeding in Iceland is of course geographically isolated, and it seems reasonable to have assumed that it would form a relatively discrete unit. By the early 1980s, Gardarsson & Skarphedinsson's (1984) improved census coverage had found *c.*14,000 in Iceland in autumn, about 6,000 of these in the east and 8,000 in the west (the Icelandic breeding range is divided into two main areas by a central region very poor in swan habitat).

Gardarsson (1991) assumed, based on resightings and recoveries, that east Icelandic swans were representative of an east Icelandic population, approximately 600 of which may winter on the European continent. Given that 500–1,300 winter in Iceland, and that the British Isles are the main wintering area, only *c.*12,000 should winter in Ireland and Britain, whereas Salmon & Black (1986) had found approximately 15,000. Some having already departed from Iceland before the autumn census there may have accounted for part of the difference, but that was unlikely to be more than a fraction of the number concerned, leaving perhaps up to 3,000 wintering in the British Isles as likely to have originated from the continent. Gardarsson (1991) has now given very good reason to assume that mixed within Icelandic birds in the British Isles may be some from Fennoscandia or even Russia. It will be fascinating to see whether any banded birds from those areas are reported, and whether Gardarsson (1991) is proven correct.

Of further interest is the subject of selective migration (i.e. undertaken by one sector of the population), already suspected of occurring in Japan (Brazil 1983c) and apparently occurring to some extent in the west (Rees *et al.* 1997b). Black & Rees's (1984) study of the wintering flock at Caerlaverock revealed that, although the main arrival period is October, yearlings usually arrive first, followed by families, then by pairs and singles. The majority leave in March, but on spring migration there is no significant difference in the departure patterns of the different classes. During winter, however, there are some differences in behaviour pattern relating to class, with single birds spending a higher proportion of time away from the reserve than other classes. Up to 14% of yearlings leave very shortly after their arrival, in October–November, and by the end of January 28% of single birds had departed compared to just 17% of families and 18% of pairs. Clearly single Whoopers were less attached to Caerlaverock than other classes.

The sequence of arrivals at Caerlaverock observed by Black & Rees (1984) is interesting and deserves further examination. Their observations fit with my own belief that yearling and non-breeding Whoopers in Iceland moult earlier than either breeders or failed breeders. Being capable of flight earlier than other classes, they may then choose to migrate earlier too, not being bound to any particular site by the growth requirements of offspring. This would explain Black & Rees's (1984) observation that yearlings arrive on the wintering grounds first. The arrival of pairs and singles last into Caerlaverock may be explained if these were failed breeders. Having attempted to breed, failed breeders are very likely to have begun moulting late and are thus likely to have been delayed from migrating for longer than successful pairs and their offspring. Although highly speculative at this stage, this area would make an interesting subject for further study.

Following the initiation of conspicuous banding of Whooper Swans in Iceland with Darvic neck-collars and leg bands, the latter technique was adopted by the WWT at their reserves at Caerlaverock and Welney. Further banding in Iceland since 1988 and at Martin Mere since 1990 has led to a significant proportion of the population being ringed (more than 2,500). Resightings have contributed considerably to our understanding of movements of birds within and between years. McElwaine *et al.* (1995) were able, for example, to refer to 117 movements between Ireland and Britain, and to 187 within Ireland. Their work has shown some regional relationship between breeding and wintering areas, with swans observed in Ireland significantly more likely to have migrated there from Skagafjordur in northwest Iceland, than from two other ringing areas further east. They also found some influence of weather on behaviour, in as much as the predominant wind direction experienced during southbound migration from Iceland influenced the number and direction of subsequent movements between Ireland and Britain later in the same winter (McElwaine *et al.* 1995).

Where, as recently as 20 years ago, it was possible only to speculate on how exactly Whooper Swans migrated, whether they made continuous journeys and whether they were at risk of drowning at sea (Wilmore 1974) or whether they were able to land on the sea (Matsui *et al.* 1981) now, with the advent of satellite tracking using the ARGOS system and PTT-100 transmitters, the details are much clearer. Pennycuick *et al.* (1996, 1999) tracked a number of Whoopers migrating between Iceland and the British Isles and studied their movements in relation to weather and astronomical conditions. They showed that in general the swans migrated at very low altitudes and made frequent and sometimes prolonged stops on the sea. High-flying swans flew at 500–1,856 m; nevertheless, this was still very low compared to the maximum of 8,200 m recorded for the species (Stewart 1978). Whoopers migrated during daylight or at night, if there was a clear sky or moonlight. In poor visibility, low cloud or on dark nights with no moon or heavy cloud cover, they landed at sea. Three southbound swans flew directly from northeastern Iceland and made landfalls separately on the Outer Hebrides, the north coast of Scotland, and in Donegal Bay. Northbound migrants were affected by spring gales; however, they were able to make landfall near their staging areas on the south coast of Iceland, one 30 days after setting out and one after four days. With elapsed times for the sea-crossing taking 12.7–101 hours (Pennycuick *et al.* 1996, 1999), the capacity to land on the sea is clearly advantageous, if not essential. It seems that when crossing open sea, Whoopers will continue migrating if visibility is greater than 2 km and if either the sun or moon is higher than −4° from the horizon (otherwise they rest on the sea), indicating that they require a visible horizon in order to navigate out of sight of land.

Pennycuick *et al.*'s (1996) calculations of flight performance indicated that the amount of energy required to make the crossing between Iceland and Britain or Ireland could be supplied by an amount of fat equivalent to *c.*25% of the bird's lean mass, though this would leave little or no excess power available for climbing. They concluded that it was not credible for swans to climb, by muscle power alone, to the great height recorded previously, and that such a height could have been reached only if birds were able to soar in lee waves downwind of Iceland, in what amounts to exceptional circumstances. Pennycuick *et al.* (1996) also calculated that whereas a medium-sized female would have sufficient muscle power to fly with 25% of her lean mass as fat, this would be marginal for a large male. As a consequence of females being smaller than males, they would arrive on the breeding grounds in spring with more reserves of fat than males.

MOVEMENTS AND MIGRATION OF CONTINENTAL EUROPEAN WHOOPER SWANS

Whooper Swans summering and breeding in Fennoscandia and northwestern Russia belong to a population that is essentially (though by no means entirely) isolated from that breeding in Iceland. Spring migration begins during mid March and ends in early May, with birds travelling northeast from continental Europe and normally arriving on their breeding grounds during May. The first arrivals may be delayed by 15–20 days in cold springs; nevertheless they have been observed to arrive when temperatures were still around −20°C in western Siberia (Mathiasson 1991).

As the summer ice-free period is so crucial, the annual reproductive cycle includes a period of 4–5 weeks which early-returning breeders spend at 'waiting grounds' near their breeding sites. In all, Swedish-wintering Whoopers, for example, need *c.*2 months to reach their Russian breeding grounds.

They usually begin to leave their northwest Russian and Fennoscandian summering grounds as the temperature drops towards freezing, usually in mid or late September, occasionally as late as mid October in the west. During autumn migration, in September–October large numbers (up to 1,900 on one day) pause, for example, in the Bay of Bothnia, especially around Hailuoto, near Oulu, Finland. Migrants through Finland winter mostly in Sweden, Denmark and Germany (Ohtonen 1988; Martin 1993). Marking birds with neck-collars in Denmark has made it possible to track their movements north again, through southern Sweden into south and east Finland (Preuss 1981).

It seems that perhaps the Icelandic population discussed so far is not atypical, given that a proportion of the Fennoscandian population also remains to winter within the countries in which they breed, having moved only short distances from their breeding grounds. As with the Icelandic population, however, the majority do move considerable distances, essentially south. Many fly across the White Sea and the eastern Baltic Sea to important wintering grounds in coastal Germany, Denmark and southern Sweden, arriving usually in October–November. Some continue further south and west, as far as the Netherlands, Belgium, northern France, and perhaps also England, though the numbers doing so are believed to be weather-dependent.

The situation in Finland is particularly interesting because the population can be considered to consist of three groups. One consists of those birds that spend summer *and* winter in country. The second comprises those that summer there, but migrate south for the winter. The third consists of birds that migrate through Finland, but neither breed nor winter there, as they are en route to and from breeding grounds in western Russia. In 1973–89, peak autumn numbers pausing on migration on Hailuoto reached 4,000–4,500 (Ohtonen 1992), but the total volume of passage is unknown, though many may have originated further east than Finland. Furthermore, Whoopers breeding in different parts of Finland winter in different parts of Europe, those from the north are most likely to winter in Denmark, while those in the south are more likely to reach Britain (Laubek *et al.* 1999; Rees *et al.* 2002).

East of Finland, in Estonia, migrants formerly occurred at Matsalu Bay in considerable numbers. In spring 1957 a combined total of 40,000 Bewick's and Whoopers reportedly passed through on just one day, though this number had declined to 8,000–10,000 in 1963 and 1964, and in recent years to just 2,000 Bewick's and 4,000 Whoopers between late April and early May.

The main wintering grounds of the Fennoscandian and northwest Russian population that migrates by way of the Scandinavian peninsula or the Gulf of Bothnia, the Gulf of Finland and the Baltic, are in Denmark, Germany and southern Sweden. Cold winter weather pushes more into Germany, the Netherlands and even northern France. The regular winter range of Whooper Swans breeding in the Western Palearctic has its northeastern limit in western Sweden. Approximately 2,500 winter there, particularly along the Swedish west coast. These are eclipsed, however, by the thousands, of non-Swedish origin, that migrate through or pause in southern Sweden during autumn and spring, some of which continue south to Denmark, the Netherlands and Germany (Mathiasson 1991).

Sightings of neck-collared swans in Sweden have confirmed that: the southerly autumn migration, wintering, and the northerly spring migration, are all undertaken in family groups, though often within larger flocks. Individuals have reappeared in the same places on migration and during winter in subsequent seasons, showing both a high degree of winter site fidelity and a restricted degree of exchange between Swedish winter quarters. Young once paired, however, sometimes changed their winter quarters (Mathiasson 1991), presumably to those of their mates. Mathiasson (1991) concluded that Whoopers belonging to geographically different subpopulations follow different migratory routes and that members of these sub-populations are to be found repeatedly at certain haunts within restricted zones, that is they migrate along a narrow front, contrary to the assumptions relating to the Icelandic population. The routes of the different subpopulations, however, may converge at and diverge from certain sites (Mathiasson 1991). Until further information is available on the migratory patterns of Whooper Swan further east it will not be possible to say which is the norm.

Weather and winter severity are deemed to dictate when Whoopers leave their winter quarters, but typically they depart in late March or early April. Although flights between important staging areas are fast and direct (they rarely rest), overall they may take 3–4 weeks to reach southwestern Finland, and another month to reach northern Russia, often pausing for up to a fortnight, especially in central Sweden and western Finland, although individuals are usually present for only 2–4 days at a locality (Mathiasson 1991).

During migration they may graze on land, but usually require open water to roost. As a result, weather (especially freezing conditions and snow), determines the actual migration timetable, as safe migration can occur only as, or after, ice-covered lakes and rivers have begun to thaw. There is an apparent relationship between spring migration through Sweden, Finland and northern Kazakstan and the +3–5°C isotherm (Mathiasson 1991). I should add the caveat, however, that I have observed Whoopers roosting on the snow-covered ice of Lake Kussharo, in eastern Hokkaido, on several occasions in winter and thus the presence of ice-free areas for roosting may not be as important as others have considered. Sudden freezing even before late September may also precipitate mass autumn migration, and temperature generally is assumed to influence the timing of the autumn migration (Mathiasson 1991).

In northern Finland, above 65°N, and doubtless elsewhere too, arrival occurs so early, relative to local weather conditions, that the swans cannot move directly to their breeding sites for 2–3 weeks, sometimes even 5–6 weeks, because these are still frozen. Haapanen & Hautala (1991) have speculated, however, that the swans are capable of waiting because the local food resources at pre-breeding waiting sites are of minor value. Whoopers actually gain their most important energy resources before the spring migration even starts, by spending more time feeding just before departure, as Brazil (1981c) found for birds migrating from Scotland to Iceland; and because important food sources are available at stopover sites visited during the first part of the spring migration, such as in southern Sweden and Estonia (Haapanen & Hautala 1991).

In Sweden and Finland, and presumably elsewhere too, the Whooper's spring migration appears to occur locally in two waves, indicating that breeders and non-breeders migrate on very different schedules (Haapanen & Hautala 1991; Mathiasson 1991). Mathiasson (1991) has shown that subpopulations from different geographical areas also move on different time schedules, passing through the same areas in different waves. Under certain climatological conditions, however, these waves may contract into one.

In Finland, where the subject has been studied in detail by Haapanen & Hautala (1991), the first spring wave, presumed to be of breeders, moves north covering *c.*350 km/day. The second wave, presumably of non-breeders, moves considerably more slowly at only 20–50 km/day. The slower speed of the non-breeders results in their arriving in the northern boreal region *c.*1 month later than breeders. This pattern makes logical sense and should also be observable in other essentially continental populations. It may be masked in transoceanic migrants, such as those heading for Iceland, although there is some circumstantial evidence in support of this even from there. Separation may occur from more local staging areas. The separation of first- and second-wave migrants, essentially breeders and non-breeders, relies on non-breeders already having decided that they will not or are unable to attempt to breed before they commence migration. Presumably conditions on the wintering grounds are crucial in making this process possible. However, if birds already 'know' that they are not going to attempt to breed, it begs the question as to why they bother to migrate.

Haapanen & Hautala (1991) hypothesised that the spring migration's bimodal pattern results from breeding and non-breeding Whoopers choosing to migrate at different times, citing several factors in support of this. Birds in the first wave most commonly occur in 'flocks' of two, apparently pairs, a pattern also observed previously in Finland by Jahnukainen (1963). They move onto nesting sites as early as possible and, despite severe weather, are able to locate their old nest mound even when it is under snow. The first wave accounted for as much as 24% of the population, a figure coincidentally similar to the 30% of breeding birds in the population (Haapanen & Hautala 1991). They also considered the possibility that the bimodal spring migration might be caused by differences in the breeding or wintering areas of subpopulations. However, they found there to be a bimodal pattern throughout Finland, despite the probability of observing first-wave birds being lower because they move faster than second-wave birds. When first-wave, breeding birds reach *c.*60°N they begin to move very quickly to waiting areas in or near their summer range. In northern Finland, above 65°N, they may have to wait several weeks before they are able to move onto their territories, whereas in more southerly areas, below 64–65°N, they are able to occupy them immediately, at least in mild years.

For second-wave, non-breeders, spring migration is less critical in its timing. Through Finland they follow, roughly, the +5°C isotherm, and consequently they find sufficient open water (ice-free areas of even shallow lakes and snow-melt flooded areas) for them to be less dependent on crucial staging areas than are first-wave birds. One consequence of the bimodal migration pattern is that non-breeders, currently comprising up to 70% of the population, don't compete with breeders for the limited food resources available early in the season. Rushing ahead in the first wave also ensures that breeders are able to establish their occupancy of the best breeding areas before the non-breeders arrive (Haapanen & Hautala 1991).

Particularly interesting is that first-wave birds comprised a significantly higher proportion (57%) of the total population during the period 1946–56 than in 1956–65 (21%). The lower proportion of second-wave, non-breeders during the earlier decade coincides with a period when the productivity of the population was very poor, in fact it was then close to extinction. When the overall population level was very low, the proportion of non-breeders was particularly low (Haapanen & Hautala 1991). The same was also true of the Trumpeter Swan when its population was low (Banko 1960).

The use of neck-collars to mark Whoopers reared in Finland (south of Oulu) in 1987–94 has shown that they winter in the Jutland region of Denmark, mostly at Limfjorden (eight of 73 birds marked), Sweden (three), Norway (four) and south-west Finland (one). Birds continue moving during the winter. One moved to Norway in December 1991, subsequently continuing south to Denmark in February 1992 (Ohtonen 1996). During spring migration, neck-collared birds were resighted in Sweden and Finland with summer observations occurring in Sweden, Finland and Russia (Murmansk; Ohtonen 1996).

A 27-year study of migration through southern Estonia by Zhelnin (1981) revealed that in general Whoopers are broad-front spring and autumn migrants that concentrate at sites with existing waterbodies or where waterbodies formerly existed. In particular, Zhelnin (1981) studied the diurnal migration of Bean Geese, Whooper Swans and Common Cranes along the Kavilda Valley around Elva. Of post-glacial origin, this valley, aligned north–south and approximately 2–4 km wide provides an ideal route for migrants. Observations from the same point over the 27 autumns 1950–76, revealed that migration began in late September and continued until early December, although in most years main migration was during the second half of October and November. Numbers recorded varied considerably between years, but generally flocks were small, 5–123 birds, and total numbers 25–1,930. 1976, however, was an exceptional year with 374 flocks, one of them of 320 birds, and totalling an unprecedented 12,650, most of them on 10–11 November, indicating that migration routes and stopover sites may have great value during extreme years.

More recent work on swan migration through Estonia, e.g. that of Rootsmae (1990) in 1975–84 based on a network of observers, unfortunately did not distinguish between Bewick's and Whoopers. The very early arrival in spring in coastal areas (between mid February and mid March, averaging 3 March) may have been biased by the presence of one or other species migrating earlier. The average spring arrival for the whole country was 30 March, with birds appearing later (during April) further from the sea. Passage was most intensive in west Estonia and less visible in the southeast, with Matsalu and Haapsalu Bays being the most significant stopover sites in spring with birds numbering in the thousands. Inland, concentrations were considerably smaller, typically just a few hundred at most (Rootsmae 1990). Migrating flocks were on the whole small, the earliest travelling singly, in pairs or in small flocks averaging fewer than ten in March 18 in April and 36 in May (averaging 18.4 for the entire spring); maximum flock sizes were 40 in March 160 in April and 200 in May, but these may have been of either species, or mixed.

As one would expect, the general direction of migration in spring was north (35.9%), with some travelling northeast (23.1%) and fewer east (20.1%), although the direction varied considerably with the location. It has been assumed that the main migration takes place at night (Paakspuu 1967; Kastepold & Paakspuu 1985), during Rootsmae's (1990) study of diurnal migration, however, it was observed from before sunrise until after sunset, though daytime movements comprised 93% of all recorded movements. Intensive passage began soon after sunrise and continued into late morning, was lowest in midday and early afternoon, and became more intensive again towards late afternoon and evening, continuing into twilight and even darkness. The last migrant swans leave Estonia during May (average 16 May; 1975–84), with the later birds usually occurring along west and northwest shores (Parnu, Haapsalu and Harju districts). The average departure date for the country was 27 April.

In autumn, the first swans arrive in Estonia between 30 August and 2 October (average 12 September), but mostly during the first half of September in coastal waters. Earlier arrivals are typically in coastal waters of west Estonia and on Saaremaa Island, while in southeastern and southern Estonia the first may not appear until late October or even November. The extent and intensity of autumn passage is highly variable, but the bulk seems to occur during the last third of September and the first third of October, with numbers declining through October–November, and passage is rare in December. As in spring, migration is more noticeable in west Estonia than further east. The most important sites are Matsalu and Haapsalu Bays, and although numbers at any one time at these may not much exceed 1,000 (occasionally 2,000 and exceptionally 3,000), the overall total is considerable. While most autumn migrants are found along coasts or at lakes, they also occur inland on fields and flooded meadows (Rootsmae 1990).

In autumn, swan flocks are larger earlier in the season, declining towards winter, thus in September average flock size was 45.1, in October 26.1, in November 18. 2 and in December just 8.6; the average for the whole autumn being 28.7. The largest flocks were of 300 in September, 400 in October and 140 in November. The majority headed south (32.9%) and southwest (32.4%), and somewhat less frequently west (26.1%). A shift in directions, from south during September and early October, to southwest and even west in October–November was thought by Rootsmae (1990) to be the result of Bewick's Swans arriving earlier and from the north, while Whoopers came later and headed mainly southwest and west. According to Paakspuu (1967) in autumn, swan migration occurs more often at night than in spring, and this was confirmed by Rootsmae (1990) who recorded twice as many flights before sunrise, after sunset and after dark than in spring (14.6%). During daytime, movements were most intense between 09:00 and

14:00, with some increase again in late afternoon or evening. The last migrants usually leave Estonia in December (average 16 December; 1975–84), with birds leaving inland districts earlier. The average time of departure for the whole of Estonia over the study was 14 November. Relatively few Whoopers winter in coastal Estonia, amounting to fewer than 100, although sometimes several hundreds in mild winters (Rootsmae 1990).

In the Ukraine, in Northern Priazovie (the Cis-Azov Sea area) both Mute and Whooper Swans are common migrants and winter visitors, while Mutes also summer there (Lysenko 1987). Although Orlov (1965 in Lysenko 1987) recorded Whoopers during summer too, Lysenko (1987) found none summering in 1968–79. In spring, Whooper Swan migrates through this region between 10 and 21 March, while in autumn they migrate in late October–early December. On migration they usually rest on extensive shallow waterbodies with plentiful macrophytes (including eelgrass and Horned Pondweed), but occasionally they also rest along coasts of the Azov Sea and in the wide bends of the Bolshoi Utlyuk, and Molochnaya rivers. Whoopers sometimes winter along the north shore of the Azov Sea, if the weather is not severe.

MOVEMENTS AND MIGRATION OF EASTERN WHOOPER SWANS

The swathe of northern Russia across which Whooper Swans summer and breed is so enormous as to be difficult to imagine. The factors affecting migration within this area no doubt vary almost as enormously, but for all its size, relatively little information is available on swan migration here. Some of that which is, fails to distinguish reliably between the species involved, and much of what has been published has focused on birds of east Asia, stimulated by a banding programme there and in Japan. It has been necessary therefore to make sweeping generalisations and assumptions that I trust the reader will forgive.

In particular, information is generally lacking from the West Siberian–Caspian Sea flyway, west of the Urals, although the great decline in numbers and contraction of the wintering range in Kazakhstan (Kovshar 1996) suggests that this population may be experiencing problems. Birds are known to migrate through the Ili River delta and Lake Balkhash, Kazakhstan, at both seasons, though the size of this migration has not been quantified (Zhatkanbayev 1996). Whoopers also migrate through the Kurgaldzhino Nature Reserve, in the Tselinograd region, during late March and early April, with the peak passing on 7–12 April, however numbers are small, just a few tens. In autumn they are more numerous and migration begins as early as the last the third of August and continues until late October, with the peak on 27–29 September (Andrusenko 1987).

On the waterbodies of the Irgiz–Turgai system (northeast of the Aral Sea in Kazakhstan), Auezov *et al.* (1987) studied both spring and autumn migration during the 1970s. They found that the first Whoopers appeared in late March as mean daytime air temperature rose above 0°C. As floods form numerous shallow pools on the steppe, the swans appear with largest numbers during the first third of April. In this region most were in pairs or small flocks of 3–6 (presumably families), and singles were also often recorded. In the Turgai Depression, while Whooper Swans are passage migrants, Mutes arrive in late April and begin nesting during May, while in May–June they moult on the lakes of the Irgiz–Turgai system. Migrant Whoopers commonly pass during early morning (07:00–09:00, 38.5%, with most passing 13:00–16:00). In contrast to spring when Whoopers amount to only 4.1% of all swans (the rest being Mutes), in autumn Whooper Swans are very variable depending on weather conditions and may even predominate (e.g. 82.7% of all swans in autumn 1975). They do appear later however, with the first passing on or about 30 September, with highest numbers in the first two-thirds of October. Autumn migration (mostly to the south; 76.7%) was commonest at 08:00–10:00 and again 17:00–19:00 with none seen 12:00–17:00. Singles were rare, with most occurring in small groups of 2–8, and most often in groups of 5–6 (Auezov *et al.* 1987). Information on the central Siberian and Mongolian population is sparse.

In the northwest of the region, the first Whoopers are seen each spring in April, even before lakes are free of ice. Haapanen & Hautala (1991) considered that this sector of the north Russian population probably migrates a shorter distance than those nesting in Fennoscandia. For example, in Matsalu Bay, Estonia, much larger numbers can be seen than in Finland. Further east, large numbers of swans migrate through the Ukraine, especially along the shores of the Black, Sivash and Azov Seas. However, although large concentrations of migrating and wintering swans occur in these areas, Serebryakov *et al.* (1991) who studied swans in the region, failed to distinguish reliably between Mutes and Whoopers, rendering their work of little use in this context.

While the general routes of migrating birds may be elucidated by direct observation, by observing collared birds or tracking birds by satellite, the factors governing the timing of migration or even triggering migration through given areas remain unclear. Both in the Tiumen Oblast and over the Kandalaksha Gulf, Russia, at least it seems that timing of spring migration during late March and early April coincides with the transition of mean daily air temperatures from subzero to above zero (Antipov 1999; Bianki 1999). Most swans migrate along the large rivers (Antipov 1999). In the southern part of Trans-Uralia for example, Whooper Swan is an early-spring migrant which arrives before the end of March and in early April once daytime air temperatures have risen above freezing, despite local waterbodies still being ice-covered (Blinova & Blinov 1997).

Migration of Whoopers in northwest Siberia has been studied both along the lower reaches of the Ob River (1968–74), east of the Urals by Braude (1987), and in southern Yamal by Kalyakin & Vinogradov (1987). Along the lower Ob, Braude (1987) found that Whoopers arrive from the south, parallel to the Urals, flying along the floodplain and river valleys joining the Ob, and heading mainly north, but also northeast and even east. Spring migration there begins with the arrival of individuals, pairs and small flocks, and coincides with the appearance of the earliest signs of spring, on 9–24 April, with mean arrival date being 14 April (Braude 1987). Migration typically lasts 40–52 days, but depends very much on the weather, being delayed by up to 15–20 days in cold springs, and duration being greatly reduced. In normal years migration occurs in several waves. Migration comprises singles (up to 5%), pairs (15%) and, mostly, small flocks of 3–12 birds (up to 80%), with young flying together with adults. Clearly the number of birds is considerable, as through the Dvuob'ie region alone some 3,000–5,000 Whoopers pass on migration (Braude 1987). The migration of Whoopers along the Lower Ob River is on a broad front up to 30–50 km wide, but at no great height, with most averaging just 50–100 m above the plain, although in afternoons and evenings birds commonly flew slightly higher, and up to 200 m. Migration heights are lowered in strong headwinds, heavy precipitation or frost (Braude 1987).

Further north, Kalyakin & Vinogradov (1987), studying Whooper migration in the Shchuch'ya River basin at the base of the Yamal Peninsula during the 1970s, found that spring arrival was 6–29 May, but that only small numbers were involved (12–32 depending on year). Most spring arrivals were of pairs, less often singles, and much more seldom in threes and fours (perhaps families), and only once a flock of ten. Nevertheless, considerable numbers of non-breeders occurred in some years during early July, concentrating in the wider lower reaches of the Shchuch'ya River, apparently in their thousands (but including both Whooper and Bewick's; Kucheruk *et al*. 1975; Kalyakin & Vinogradov 1987). Small numbers of Whoopers nest in the Shchuch'ya River delta, but have disappeared from areas often visited by people, where 70–100 years ago they were common (Finsh & Brehm 1882; Kucheruk *et al*. 1975). Autumn migration is less visible, because in this region it takes place mainly at night, at 20:00–23:00, with the main migration occurring on 1–20 September (Kalyakin & Vinogradov 1987).

On the Ural-Tobol plateau, the timing of the Whooper's migration varies annually, but in autumn it seems to be driven by sharp falls in temperature and, in particular, the fall of daytime temperatures from 5°C to 0°C. In 1979–94, analysis of flock structure revealed that 41% of migrant flocks consisted of fewer than ten birds, flocks of 11–20 comprised 32.3%, 21–50 18.6%, and 50–1,000 6.1 % of flocks. Surveys made during mid October 1983–87 indicated that 67% of the swans passing through this region are Whooper, and 33% Mute Swans (Samigullin & Parasich 1995).

In the Omsk Region, in autumn, Whoopers begin to leave the northern forest-steppe on 20 September–15 October. They fly in a southerly direction at a height of 100–400 m and flocks consist of several tens or even hundreds. Small flocks stop to roost on lakes in the taiga zone and may remain for several days (Yakimenko 1997).

In Yakutia, along the Lena River valley (between the Olekmei and Munoi tributaries) and at Troitskoe Lake (67°N) in the Kolyma River valley Labutin *et al.* (1987) studied swan migration during the 1970s. They found that whereas Bewick's Swan is an extremely rare migrant, Whooper is commoner. In central Yakutia they found that the earliest arrived on 23 April, while further north first arrivals were not until late April or early May. During the same period (30 April–10 May) migrant Whoopers were found along the middle and lower reaches of the Kolyma River (Krechmar *et al.* 1990). During the period 1974–77 the passage occurred in two unequal waves. Along the Lunkhe River 40.5% passed on 4 May 1976, with a further 25.3% two weeks later, on 17 May. Along the Dyanyshke River 30.6% passed on 17 May 1976 and 50% on 21–24 May, while in 1977 44.9% passed on 16–17 May and 14% on 20 May. Along the middle reaches of the Kolyma migration starts weakly, and ceases during cold spells and snow falls, but becomes more intense in late May (20–27 May) when one (1979) or two (1978) waves of passage accounted for 50–76% of all birds recorded, before weakening again at the end of May. Labutin *et al.*'s (1987) results indicated that spring passage lasts 25–30 days in this region and is most intensive in the first two-thirds of May along the Lena River valley, and during the second half of May along the Kolyma. During this period, migrants arrive as pairs (21.2–28.5%) or in small flocks of up to ten (39.4–52.6%). During intensive passage larger flocks of up to 40 (10.5–17.6%) may also occur.

In autumn, during late September and early October, passage along the Lena River valley is not significant, however along the Kolyma River large numbers may be encountered (e.g. 762 in autumn 1978 1,854 in autumn 1979) in flocks up to 55. Pairs are far less common in autumn than spring, and the most common are groups of 5–20 birds. The main wave of autumn migration, during which 33–43% pass, appears to occur as the floodplain lakes freeze in late September. Although flocks are occasionally sighted into early October, the main migration ceases by the end of September. In this region, migrants typically follow river valleys and fly at heights no greater than 150–300 m (Labutin *et al.* 1987).

In south Primorye, during early-spring migration, they pass through in small flocks of 10–15, but as the intensity of migration increases, so flock size also increases, perhaps further indication of a double-wave spring migration as in Europe. The majority (57%) pass through in flocks of 30–70 birds, and the largest flock was of 122 on 29 March 1989. Although some were noted at night, the majority (51%) passed during the evening between 18:00 and 20:00. Although weather conditions seemed not to affect passage intensity, the height of passage (normally *c.*150–300 m) was lower in windy weather and higher (500+ m) in clear weather with light or no wind (Gorchakov 1996).

Timing of migration depends on latitude, with birds in spring seen as early as late February near Bukhara, although more typically it occurs in March, and until mid or late April. During this season they are most often seen in pairs and more seldom in small groups, whereas during winter singles and pairs are more common (Kashkarov 1987).

In China, two separate migratory routes have been identified, eastern and western routes. In the west, birds which breed in the northwest, in Xinjiang, migrate southeast over Gansu, Guizhou, Ningxia, Shaanxi and Shanxi, and winter in the lower reaches of the Chanjiang River (Yuan & Guo 1992). In the east, birds that have bred in Siberia migrate via the Wusuli River, passing over Heilongjiang, Jilin and Liaoning, and then cross the Bohai Gulf to winter on the Shandong Peninsula (Li 1996) where they are on a similar latitude to birds wintering in the Koreas and in northern Honshu, Japan.

At Bayinbuluke, despite its altitude, arrival is early, mid-March to mid-April, even though large waterbodies do not thaw until mid May and most vegetation is not green until June, and departure is late (October–November), with birds staying up to 210 days (Ma & Cai 2000). Migration takes place at no great height, usually 50–400 m above ground, and usually at night along river valleys, possibly more commonly around the full moon when waterways reflect moonlight making navigation easy. The Kaidu River forms the main migration route for the Bayinbuluke population and some may winter in Qinghai Lake, Qinghai, while others travel along the Kunes River towards Central Asia and Russia (Ma & Cai 2000).

Thankfully, the Whooper Swans of east Asia have attracted more attention. Those breeding in extreme northeast Russia, e.g. in the Anadyr-Penzhina lowlands and other regions, including the Loyma basin, appear to be isolated, with little or no interaction between birds breeding in the different regions. Furthermore, it has not been established whether they have contact during winter. Their movement away from the extreme northeast is of particular interest. Two main migration routes were documented by Kondratiev (1991). Birds that arrive to breed in the Anadyr-Penzhina lowland during May fly across Kamchatka and those marked in the Anadyr region have not been observed on the well-watched Japanese wintering grounds. It is assumed that they winter in observer-poor Kamchatka, where migrating Whoopers are numerous (Lobkov 1987; Kondratiev 1991), and where Lobkov (1987) reported small winter gatherings and more recent research has shown that a substantial number, more than 7,500, winter (Gerasimov & Alekseev 1990), and also to the Kurile Islands (though this being the case one might expect them also to reach east Hokkaido, Japan, via this route).

The remaining birds, those nesting in the Yana-Taui Depression, the Kolyma Basin, and in the Chaun lowlands, along with birds nesting in southern Magadan region and Yakutia, migrate via the western Sea of Okhotsk and Sakhalin to winter on Hokkaido and Honshu, Japan (Kondratiev 1991; Ostapenko 1991).

Kondratiev (1991) estimated, on the basis of observing up to 800 on spring migration, and taking into account the intensity of migration, that Whoopers in the extreme northeast number 2,000–2,200. These figures would appear to reflect only a small proportion of the population of northeast Asia, however, given other available figures.

Both on the polynias and at the breeding sites in the floodplain landscape of the Anadyr River, Whooper Swans usually appear first in pairs or in small groups of 3–5, which are presumably pairs with their previous year's offspring (Krechmar 1990) (Fig. 10.9). Further south, in the area of the Amur River and the Shantar Islands, both Whooper and Bewick's occur, although the latter do so only in small numbers (several tens up to 300). Migrant Whoopers pass from late April until early June, with most during the first half of May, though small numbers remain to breed at sites such as in the Bolon Lake basin (5–6 pairs in 1970; 28 pairs in 1979;

Roslyakov 1987). During spring, most arrive from China into the Lower Cis-Amur River area by way of the valleys of the Sungary and Amur, and some along the Ussuri, thence to Bolon Lake, where they rest before continuing along the Amur Valley to Komsomol'sk-na-Amure, where they diverge, some heading north to Evoron Lake and others northeast to Udyl' Lake and Amgun' River mouth, and the Shantar Islands. The numbers of Whoopers on spring passage have increased considerably since 1969. In spring 1970 there were *c.*500 on Bolon Lake, in 1973 over 1,500 and in 1975 there were *c.*5,000. During the period 1976–80 numbers fluctuated at 5,000–6,000 in spring (Roslyakov 1987).

In autumn, they depart the Lower Cis-Amur River area from mid September until the Amur River freezes and snow falls, usually by mid November. Whereas in spring, when *c.*6,000–9,000 pass, in autumn as many as 10,000–13,000 may occur. On migration they rest for prolonged periods of days or even weeks in the region of the Shantar Islands, in the Amur estuary and on large lakes of the Lower Cis-Amur River area (Roslyakov 1987).

Whoopers also migrate along both coasts of Sakhalin according to Ostapenko (1991), who suggests that these routes merge in the north of the island (where, apparently, they formerly bred; Gizenko 1955), and also with a continental route used by birds from Lake Khanka and the Amur lowlands. During the 1950s–60s the numbers migrating over Sakhalin were estimated to be about 5,000, but by 1990 this was considered to have increased to no fewer than 15,000, although according to Ostapenko (1991), Japanese ornithologists claim a much higher figure of *c.*30,000. Despite the enormous discrepancy between them, either figure indicates that the east Asian Whooper population is much larger now than it was 30 years ago.

Figure 10.9.   Polynias are important resting and feeding areas in spring and autumn.

On autumn migration, birds using the west Okhotsk flyway diverge, taking one of the coastal Sakhalin routes or continuing southwest for Lake Khanka and beyond. It is perhaps birds following this route that eventually reach the Koreas. Or is there another undescribed route drawing on birds breeding further west in Russia? Lake Khanka is an important stopover for migratory Whoopers, though it was considered to form the extreme southeast of the current Eurasian range by Glushenko & Bocharnikov (1991) who seemed to have ignored both Korean and Japanese sites. Both Whooper and Bewick's migrate through the Prichankayskaya lowland, though Bewick's represent only a small proportion (6–14%) of the total. Spring migration occurs in March–May, though it is most intensive during April. Although only small numbers of swans are seen daily (averaging 3–5 flocks/day, and fewer than 50/flock), large numbers gather in the shallows of Lake Khanka and other lowland lakes in northeast China and the southern Russian Far East (Glushenko & Bocharnikov 1991).

Aerial surveys in 1987–88 found that numbers varied markedly through April, with just 375 on 2–3 April 1,133 on 11–12 April, a low of 585 on 17–19 April and then a massive increase to 3,689 on 21–23 April. Numbers declined rapidly during May, and by mid May most had left, e.g. there were only 35 on 23–25 May 1987 (Glushenko & Bocharnikov 1991). This bimodal pattern, with a smaller peak earlier in April and another three times larger later in April, appears to represent the only evidence from Asia of the two-wave spring migration (Haapanen & Hautala 1991; Mathiasson 1991). Taking the two peaks (11–12 and 21–23 April), if, as in Europe, these represent periods when the peak numbers of breeders and non-breeders are on the move, then the non-breeding proportion of this population would appear to be *c.*69%, a figure coincidentally almost identical to Haapanen & Hautala's (1991) estimate of 70% of those migrating through Finland. Again, following up the comparison with Haapanen & Hautala's (1991) work, the large size of the non-breeding population may indicate that either they belong to a large successful population or to one that has saturated the available breeding habitat.

In autumn, two movements are recognised. First is that of part of the summering population away from the Prichankayskaya lowland by mid September, despite not all of the young having fledged by then. This is followed by the main autumn migration, in October–early November. Surveys in October 1987 recorded 192 swans on 9–12 October and 1,138 on 28–29th. Of the latter 19% were cygnets indicating that breeding success had been very good (Glushenko & Bocharnikov 1991). In both spring and autumn, Glushenko & Bocharnikov (1991) observed relatively few swans migrating during the day, perhaps circumstantial evidence that, in that region, they typically migrate at night.

Across the Okhotsk Sea, in Kamchatka, spring migration along the peninsula begins in the last third of March, and also usually occurs in two waves, with the main mass in April according to E. G. Lobkov (in Gerasimov & Alekseev 1990). In other parts of the peninsula, e.g. at Kharchinskoe Lake, in the centre, though birds are present in March they are still in wintering flocks with migration not getting underway there until early May. During this season flocks appear at polynias in the frozen rivers and at unfrozen lakes, and there may be confusion between wintering flocks and early-spring migrants. In autumn, migration begins in October, with concentrations appearing annually, particularly at larger rivers, for example the Moroshechnaya (midway along the west coast). At such places concentrations may be brief but large, thus for example V. G. Mironova, a hunter, reported as many as 1,000–1,500 on 15–20 October 1983, with all leaving during the night of 21 October (Gerasimov & Alekseev 1990). In other years, smaller numbers (400 on 2–4 October 1979, 550 on 15 October 1987) have been reported. In northeast Kamchatka small numbers (10–20) have been observed migrating over the Litke Strait as early as the start of October (V. K. Karacharov, in Gerasimov & Alekseev 1990). Migration occurs along the coast, and birds visit coastal lagoons, such as Malamvayam, river mouths, bays and lakes such as Stolbovoe and Nerpich'ye, where they may stage for some time. On the shallow waters of the Kirun Gulf, huntsman A. V. Nazarov counted up to 2,000, with another 300 on Vereshchaginskaya Spit, on 12 November 1987; two days later they had all gone. In central Kamchatka, on Kharchinskoe Lake, tens appeared in the first third of October in 1983. In the middle third of October 1979 at lakes in the Kamchatka River valley flocks of 200–300 were present, while in late October–early November 1974 there were at least 500 at Azhabach'ye Lake alone (E. G. Lobkov, in Gerasimov & Alekseev 1990), suggesting that migration is quite prolonged.

Anecdotal reports from Kamchatka, and from northernmost Hokkaido, and my own observations from western Hokkaido, indicate that both Whooper and Bewick's Swans commonly migrate at night in this region, and do so at no great altitude, being distinctly audible from the ground.

In Japan, where the wintering population had reached almost 14,000 by 1988 and more than doubled to *c.*30,000 by the late 1990s (WBSJ Research Division 1988; Miyabayashi & Mundkur 1999), the main migration route brings Whoopers into the country in autumn via Hokkaido to its main wintering areas in northern-most Honshu. They are less common further south, in central Honshu for example, uncommon in southwest Honshu, where it is replaced by Bewick's Swan, and a rarity further south, such as in Kyushu (Brazil 1983c, 1991). It is generally assumed that birds travel through the length of the Japanese archipelago, from north to south in autumn and returning in spring, rather than crossing from the Asian continent via the Sea of Japan, though small numbers reaching Kyushu may arrive there via the Korean Peninsula.

In autumn, the first reach Hokkaido during early October from their staging posts in southern Sakhalin. One concentration winters in southeast Hokkaido, the vanguard arriving in late October and as many as 5,000–6,000 gather there in mid November, remaining until coastal waters freeze in December, then moving south along the Pacific coast. In spring, as early as March, flocks gather on lakes and marshes along the Okhotsk Sea coast of northern Hokkaido before moving north during April. At Utonai-ko, in southwest Hokkaido, Whoopers first arrive in late October–early November, and numbers peak from late February to mid March, with northward movement commencing in late March of birds which began to move north in Honshu in mid March. By mid to late April the majority have departed, although individuals do occasionally summer at various localities in Hokkaido (Brazil 1991).

Whereas Hokkaido once held the bulk of the migrants reaching Japan, approximately 8,000 now move further south to sites in northern Honshu. There they arrive in early November, numbers peaking in late November–December. A few move along the Sea of Japan coast, as far south as Niigata Prefecture, but it is unusual to find Whoopers south of northern Japan except in severe winters, although stragglers (overshooting autumn migrants) have even reached the Ogasawara Islands and the Nansei Shoto.

The significance of certain sites becomes particularly relevant on migration when many may pass through the same site in just a few days. At Toufutsu-ko, in northeast Hokkaido, a small number winters (Tamada 1981; Brazil 1987) but between 25 October and 10 November 1,000–3,000 Whoopers may arrive in a single day. Numbers then fluctuate until ice on the lake in mid December forces them south. Banding has shown that many continue to Aomori Prefecture in northernmost Honshu. With spring melt the swans return, their numbers peaking around 20 April when Tamada (1981) reported up to 4,000 there. The total there is likely to be considerably more than this, making this site of crucial importance for the population wintering in Japan. Northbound migration occurs mainly early in the morning in groups of 400–1,500, sometimes 2,000 (Tamada 1981), which, contrary to Kemp (1991) and Ma & Cai (2000), suggests that migration, here at least, may be commonest during daytime rather than at night.

Matsui *et al.* (1981) paced migrating swans at 100 kph in windless conditions (with favourable winds, they presumably fly faster). They established the existence of migration routes from Aniva Point in southeast Sakhalin to eastern Hokkaido, to sites such as Toufutsu-ko, Furen-ko and Akkeshi. The route then swings west to Cape Erimo, with some crossing the Hidaka Mountains to Lake Utonai, while others cross the Tsugaru Strait to the Shimokita Peninsula of Honshu, and to Ominato, Kominato and into Miyagi Prefecture to Izunuma (Brazil 1987).

Matsui *et al.* (1981) observed several hundred resting on the sea near Cape Shiretoko in spring (though I have failed to find any there in autumn), and they also saw flocks on the floating ice in the Sea of Okhotsk, and on the sea, up to 5 km off Ishikari Bay in western Hokkaido. These sightings were of particular interest as prior to Kanai *et al.*'s (1997) and Pennycuick *et al.*'s (1999) satellite-tracking studies, they were the only evidence that Whoopers were willing to land at sea and thus need not necessarily make long oceanic crossings non-stop, although this was generally assumed to be the case between the UK and Iceland (e.g. Ogilvie 1972).

Evidence for cold-weather movements between sites in the Far East was first provided by Nakanishi (1981), who revealed a correlation between falling temperatures in eastern Russia and Hokkaido and increasing numbers of Whoopers at Hyoko in Niigata Prefecture, along with confirmation of birds from Hokkaido reaching Hyoko, from observations of neck-collars. Preuss (1981) noted that collar over-icing which is 'often seen in *C. olor* under cold and windy conditions, has never been noticed in *C. cygnus.*' Perhaps this is true in the west. However, I have observed over-icing of collars on a number of occasions and on several different birds, in the dry-cold atmosphere of Hokkaido. Feather icing also occurs under such extreme low-temperature conditions (Fig. 10.10).

Japan's wintering Whoopers have long been assumed to migrate by way of Sakhalin and the west Okhotsk flyway, to breed in areas of central-eastern Russia (Mathiasson 1991), and satellite technology has now helped to confirm this. Whooper Swans banded in Aomori Prefecture, the northernmost part of Honshu, and fitted with PTT transmitters, were tracked for 3–5 months and found to migrate via Hokkaido and Sakhalin to the mouth of the Amur River, and thence to their breeding grounds in northern Siberia (H. Higuchi *in litt.*). These satellite data confirm the general routing that was previously known from field observations, and also confirm that birds pause frequently en route. No information was available, however, on the height of flight. More than likely migration would have been at relatively low levels, between sea level and *c.*2,000 m. I have observed Whoopers on migration in early winter and seen them in flight between their first arrival on the coast of Hokkaido and sites further south, crossing the rim of Lake Mashu, a *c.*700 m-high caldera rim. Birds migrating between Siberia and through Japan have various mountain ranges and volcanic regions to avoid, to pass over or alternatively to find valley routes through (Fig. 10.11). No doubt further satellite work will help elucidate just how they make these journeys, but from direct observation they appear to follow lines of least resistance—along major valleys and over low passes.

Satellite tracking of swans began in Japan in 1991 (Higuchi *et al.* 1992), though with limited success, because of damage caused to radio collars by the swans. In

Figure 10.10.   Icing of plumage, with or without collars, may occur under extreme conditions.

Figure 10.11.   Migrating Whooper Swans typically fly at no great altitude.

1994–95 several were successfully banded and tracked (Kanai *et al.* 1997). Swans were caught at Kominato, a well-known wintering site in Aomori Prefecture, where 400–500 regularly winter. Six were caught on 21 February 1994 and a further nine on 23 February 1995; they were each fitted with backpack transmitters attached using Teflon coated ribbon.

Satellite tracking provided a considerable amount of data which conclusively revealed that, for birds from Kominato at least, they migrated via rivers and lakes in eastern Hokkaido, east of the Hidaka and Daisetsu mountains, which dominate the centre of Hokkaido, before moving north via southern and central Sakhalin. Three of the first six were tracked as far as their potential breeding areas in north-eastern Russia. Two summered on the lower Amur River, west of the northern tip of Sakhalin, while a further bird crossed the Okhotsk Sea to Chukotka and summered on the Indigirka River (Kanai *et al.* 1997) (Fig. 10.12).

In addition to the general information on the route taken by these birds, Kanai *et al.*'s (1997) study provided valuable information on the style of migration undertaken in east Asia. Birds are quite likely to travel short distances quickly and although they may make short stops of 1–2 days at several localities en route, they are as likely to make very prolonged stops. For example, one began its journey by leaving Kominato on 21 February 1994. It crossed to Hokkaido, then spent 25 days on the Tokachi River, in southern Hokkaido, remaining there until early April. It then moved north again, first via Abashiri Lake on the opposite coast of Hokkaido and thence to southern Sakhalin, before staying 16 days at central Aniva Bay, from 14 April. It then made short journeys to the lower Amur River where it remained until its transmitter quit on 13 June. A second made a similar journey with stops of six, eight, ten 13 and 20 days at various sites in Hokkaido, Sakhalin,

Figure 10.12.   Migration route of Whooper Swans satellite tracked from Kominato, northern Japan, 1994 & 1995 (after Kanai et al. 1997).

the Amur River, and the northern Sea of Okhotsk, before reaching the Indigirka River on 6 June. After moving about the Indigirka for 12 days it settled at an area of marshes where it was still present when its transmitter quit on 25 July. The third followed a similar route and also spent 25 days resting from 1 April before completing its migration, via Sakhalin, to the lower Amur River. Some birds, it seems, may take more than 70 days to reach their summering grounds, others may only travel *c*.30 days, spending the summer at much lower latitudes. I suspect that these studies may only have involved non-breeders that lacked any urgency to return north quickly. The northbound movements of breeding adults might be very different.

Five of nine swans were successfully tracked to their Russian summering grounds in 1995. They followed much the same route, from Kominato in northern Honshu, to eastern Hokkaido, then via southern and central Sakhalin to the lower Amur River. Some travelled from central Sakhalin to northern Sakhalin then across the Okhotsk Sea to the northern Okhotsk region and thence to the Indigirka River, while others travelled from central Sakhalin to the lower Amur River first, then north across the Okhotsk Sea and finally to the Kolyma River. Tracking in 1995 also confirmed rest stops of 1–4 days to as many as 8–34 days en route. Kanai *et al*.'s (1997) work showed that in addition to a number of lesser sites, highly favoured sites, where birds spent long stays, were on the Tokachi River, at Aniva Bay in central Sakhalin and the lower Amur River. The latter is also a significant resting site for migratory Bewick's Swans that travel the Sea of Japan coast of Hokkaido to Sakhalin, whereas Whoopers travel via the Pacific side of Hokkaido.

MARKING METHODS

In concluding this chapter it is of interest to note the differing degrees of information returned by different means of marking birds. The enormously costly method of attaching satellite transmitters inconveniences a very small number of birds but provides a considerable amount of data, and is of particular value in regions where ground access is difficult or impossible, and where ornithologists are few. The use of neck-collars is widespread and provides plentiful data in regions where there are sufficient ornithologists and keen birdwatchers willing to spend time watching for and reading their numbers. As with transmitters, inconveniencing a relatively small number of birds nevertheless provides considerable date. Furthermore, the visibility of collars makes them of use where birds are not approachable and they are suitable whether birds spend time on water or on land, or on both. Ringing or banding of birds only with tarsus bands is the least effective method. It requires the capture and ringing of a large number of individuals in order to provide returns, and even where the leg rings are large and clearly numbered, reading them is at worst virtually impossible where birds spend their time on water, and at best difficult and time-consuming on land. Although practical in heavily populated areas with many dedicated observers, in the majority of the species' range there are simply too few observers, and conditions are such that resightings are likely to be few or non-existent if only using tarsus bands. The cost of transmitters makes their use prohibitively expensive, while tarsus bands require most effort for least return, leaving collars as the most cost-effective form of individually marking Whoopers (Fig. 10.13). But given that, what of the impacts on the birds of attaching bands, collars or transmitters? I know of only one study that has actually examined this for Whooper Swans. Ueta *et al.* (1997) found that, within two days, the amount of time the birds spent pecking at their transmitters declined from 100–200 seconds in 15 minutes to only 3–23 seconds in 15 minutes by the second day. They concluded that transmitters did not have any significant effect on behaviour. Given that transmitters and their harnesses weigh up to 85 g, neck collars or tarsus rings are a considerably smaller burden for a swan, thus they too can be deemed insignificant in their affect on behaviour. I have heard concerns

Figure 10.13. Neck collars are visually effective, but catching birds using net and hook is exhausting work.

voiced among members of the public in the UK and Japan that collars sometimes appear tight, but in winter when neck feathers are fluffed out the collars merely appear less loose than in summer (Fig. 10.14). I have heard similar concerns voiced about icing. I have certainly observed icing of collars, but I have also observed icing on neck feathers when they did not have collars, and in both instances icing was probably only temporary.

I am unaware of any studies examining the relative merits and demerits of using neck-collars on Whoopers, although Spray & Bayes (1992) made a very relevant

Figure 10.14   Fluffed out neck feathers in winter can make a collar appear tight.

study of neck-collars on Mute Swans. They found that while birds with collars fed significantly less often by upending than those without, there was no effect on the frequency of feeding, their weights or on breeding success. Furthermore, there was no indication of damage to neck feathers or collars, and there was no indication of any consistent pattern of preening between those with and without collars, suggesting that they have little or no impact, especially given that collars usually break and fall off within a few years.

The use of neckbands or collars for marking waterfowl has been ongoing for more than 50 years, but their use on swans was pioneered by Bill Sladen in North America. Their use has been extensive in some countries (particularly North America) and far more restricted in others (e.g. the UK). The various marking techniques that are available involve either dying the plumage, attaching markers to the beak (nasal discs or saddles), attaching tags to the patagium, using leg rings (coloured or otherwise), and using neck-collars. It is the collars that have the most significant advantage of being readily visible, even when birds are on water (typical behaviour) when leg rings are rarely visible much less legible, and numbers engraved on the collars are legible at ranges up to 600 m, meaning that birds can be observed easily without disturbing them (Fig. 10.15). In open habitats, and when studying nervous species that will not tolerate close approach by observers, neck-collars provide the only practical means of study. Although there has been opposition to the use of collars on swans, there is little evidence that they have long-term negative affects on any aspects of the biology of the species concerned (Spray & Bayes 1992). High rates of return, in terms of observations for a given number of birds marked in this way, make neck-collars invaluable in the study of

Figure 10.15. Using neck collars renders swans far more visible than using leg rings.

Whooper Swan migration, with up to 65% resighted within the first year, and considerably higher rates compared with other methods (e.g. Brazil 1983a; Gardarsson 1991). For example in Denmark, Preuss (1981) found that of 226 swans marked with leg rings in the 25-years up to 1979 only 9% were recovered abroad, whereas in the first year of collaring just 116 Whooper Swans, 22.4% were resighted abroad in just one year.

The factor generally weighed against the use of neck-collars for research is an entirely aesthetic one—the risk of an adverse reaction from the viewing public. This risk has not, however, been examined or quantified, and in Japan, where collared Whoopers are regularly found among flocks hand-fed by the public, I have rarely heard any such negative reactions despite many hours of observations in such situations, and their presence seems to generate considerable extra positive interest in the lives of the birds.

CHAPTER 11

In sickness and in health

INTRODUCTION

The Whooper is a wild swan, choosing, on the whole, remote summering grounds and breeding localities away from human disturbance. The same could have been said, in the past, of its wintering grounds too, but in areas where food has been provided it has become increasingly accustomed to man. In some countries, such as the UK, food is provided officially at a handful of wildfowl refuges so the birds become familiar with a few people appearing at certain times, but are brought by the food close to observation hides. In other countries, such as Japan, where wintering sites are now relatively few because of habitat loss, food is also given, but by members of the public at the majority of wintering sites. There are no hides, and large numbers of people directly approach the water's edge, as a result the birds have become virtually fearless and are even willing to feed from the hand (Fig. 11.1). Yet even where they are not provisioned, they commonly winter at sites on agricultural land or at man-made or man-adapted waters and waterways. So, despite their natural wildness, Whoopers are now accustomed to man throughout much of the winter range.

Figure 11.1.   Hand-feeding is a common way to interact with Whooper Swans in Japan.

Their greater approachability is not now confined only to their wintering range. For example, Haftorn in his *Norges Fugle* (1971), still the standard work on north Norwegian birds, described the Whooper as very shy and avoiding people in the nesting season. The most conspicuous development of the last 25 years is that in many places, including northern Norway, Whoopers appear to have lost their fear of man, which earlier made them a symbol of unspoiled nature. At several places, including around Tromsø, they now nest close to well-travelled roads and villages (Wim Vader *in litt.*).

Human activities ranging from hunting and egg collecting to land reclamation for agriculture increasingly affect the summering and breeding range of the Whooper Swan, but ultimately the changes wrought by global warming will have the greatest impact, reshaping the northern limit of the range in both summer and winter.

Whoopers are wild birds, but though they have no history of domestication they may become tame. The earliest record of one in human company dates from the 12th century, which was tame towards St Hugh, Bishop of Lincoln, yet aggressive and unmanageable towards everyone else. Apparently it followed St Hugh, even entering the manor and climbing the stairs, and it learned to search his pockets for bread (Kear 1990). Nowadays, birds become relatively approachable wherever they are provisioned during winter, coming close to observatories or even to people and their vehicles (Fig. 11.2). At some localities this has become a focus for sightseeing, at

Figure 11.2.   Whoopers are approachable where they are regularly fed.

others, such as WWT centres, floodlighting is even provided for evening viewing, though such displays have their critics (Kear 1990). I frequent several *rotenburo* (outdoor hot-spring pools) in Hokkaido where I can soak in pleasurable warmth stargazing while surrounded by knee-high snow and lake ice, accompanied by the soft contact calls of nearby Whoopers. It is not so much that they are tame; it is just that they have not learnt fear.

Whereas little is known of the natural lifespan in the wild, other than that it may be for decades, the accidents it meets with, and the factors affecting its mortality in general, are better known. Given these factors, and because of its wide range, large population, popular appeal and unthreatened status, I will attempt to weave together information on longevity, habitat change and destruction, and conservation, into a single concluding chapter.

In parts of its range, the mythologies surrounding the Whooper Swan have been taken seriously. Thus, at Kominato, in Japan, where it was revered as a messenger of Raiden (god of thunder and lightning), harming them was believed to bring divine punishment. As a result, sick swans were nursed to health at the local shrine, and a hunting ban was introduced in the vicinity as early as 1896. In 1920, the village and harbour were declared a sanctuary, and in 1931 were made a natural monument and officially protected (Austin 1949; Austin & Kuroda 1953; Brazil 1991). To this day, the area not only holds wintering swans, but also has been the focus for studies including the satellite tracking of birds to their Russian summering grounds (Kanai *et al.* 1997).

On the one hand, man's influence may be positive, in Japan, for example; on the other, in Russia, human influences may be less positive. The economic development of the far eastern region is affecting both Whooper and Bewick's Swans negatively, according to Kondratiev (1991), because of the transformation of their breeding habitats. Furthermore, leaking oil pipelines may have serious local or even regional effects if oil reaches major rivers, as was feared in an October 1994 spill which reached the Kolva River, a tributary of the Pechora, the delta of which is an important breeding area for Whoopers and a migratory staging area for large numbers of both swans (Bowler *et al.* 1994). Disturbance, resulting from the developing oil and gas industries on the Yugor Peninsula has led to a significant decline in Whooper numbers in the area (Anon 2001).

Even in Japan, where Whoopers have achieved popular appeal, feeding at popular sightseeing spots has meant that it is now extremely difficult to find wintering flocks foraging in an entirely natural way (Fig. 11.3). Furthermore, the food, consisting mainly of cereals, rice, waste tea and bread, provides bulk and carbohydrates but is low in nutritional value and considered inappropriate (Hatakayama 1981). It should be noted, however, that wild Whoopers in Scotland selected carbohydrate-rich cereal grains during autumn and midwinter, even where freshwater lakes with aquatic vegetation were available (Brazil 1981c). It may be that the food given in Japan, while not the most suitable, is sufficient and certainly it serves to attract birds and leads to ice-free areas being maintained for them. The downside of winter provisioning in regions where winter weather can be extremely severe is that birds may be tempted to remain at more northerly sites where food is provided rather than moving further south as local weather conditions might dictate in the absence of such food. By saving them the risks of onward migration, artificial feeding may contribute to their survival, but, in areas prone to sudden freezing, it may have the adverse effect of retaining birds that might otherwise have escaped extremely severe conditions. Large numbers are known to have frozen to death at

Figure 11.3.   Most wintering flocks in east Asia receive at least some food from people.

Odaito, east Hokkaido, on at least two occasions in the 1960s and 1970s (Abe 1968), most recently in February 1976, when *c*.500 died, though Ohmori (1981b) denies that winter feeding contributed to the mortality, because it was begun in earnest only as a result of the deaths, not prior. There have certainly been no major mortality events at Odaito since 1976, and perhaps even if there were they should not cause undue concern. Many of those affected were likely to have been non- or failed breeders if, as is suspected, there is some selective migration by families away from the northernmost sites, thus such mortality is unlikely seriously to impact the population. Hypothermia has also been reported from Khyrgas Lake, Mongolia, where several dozen died during the cold winter of 1967–68, at a site where provisioning was certainly not implicated (Batbayar 2000), and in the especially cold winter of 1983–84, of approximately 200 wintering on the non-freezing channels connecting Khar-nur and Khara-us-nur, only 53 survived (Bold & Fomin 1996). In Chukotka, research into the timing of fledging and autumn freezing of lakes, and timing of migration, led Krechmar (1990) to conclude that some young inevitably die as a result of early-autumn freezes.

Whereas in Japan there has been some concern over the nutritional value of some of the foods provided, in the British Isles Whoopers predominantly feed on natural vegetation or on agricultural crops, though they and Eurasian Wigeon, among others, have been implicated in the spreading of alien aquatic plants including the escaped aquarium plants from America, Canadian pondweed and Esthwaite waterweed (Kear 1990).

LONGEVITY AND MORTALITY

The Whooper, like Tundra and Mute Swans, is generally long-lived, with a potential lifespan in excess of 20 years, and low levels of annual mortality. Among 478 Whoopers ringed in Iceland prior to 1980, the oldest known-aged bird recovered was just under 12 years at death, while of those ringed when full-grown, the oldest was recovered 14 years 8 months after ringing, making it at least 16 years old (Brazil & Petersen in press). At the time of recovery these were both new records for the species in the wild. The longest previously published was 7.5 years (Cramp & Simmons 1977). Andersen-Harild (pers. comm.) has noted that a bird ringed in 1963 in Denmark, when already 1.5 years old, was re-sighted in January 1982, making it at least 20.5 years old, and an Icelandic-ringed bird of 22+, the current record holder, has also been reported (Rees *et al.* 1997a). In captivity they are known to live for 25+ years (Scott & The Wildfowl Trust 1972; Spray 1980), but Honda (1981) suggested that, like Bewick's Swan, the normal lifespan was likely to be in the region of ten years. Thus far, Trumpeters have the edge over Whoopers, reaching more than 24 years in the wild and 32.5 years in captivity (Mitchell 1994). However, it is unlikely that either species commonly reaches such ages, and normal life expectancy may be 10–15 years. Though Boyd & Eltringham (1962) estimated the Whooper's gross annual mortality to be 20%, more recent long-term studies have provided estimates of minimum survival rates for all three species, with Whoopers having the highest at 85% (Brown *et al.* 1992b) (see Table 11.1).

Einarsson (1996) calculated minimum annual survival rates among Icelandic birds to be 82.6% for birds from their first to second winter, 80.1% for birds 2–6 years old, and 78.3% of those of unknown age in the eight years after ringing; nevertheless few Whoopers reach old age.

It would be of particular interest to know more clearly during which seasons and which stages of life Whoopers are most prone to mortality in the wild. In captivity, mortality occurs during winter and spring, but not in summer, however the sample size was very small (N=10; Brown *et al.* 1992a). One might surmise that stresses experienced during breeding, migration and severe winter weather or food shortages might all significantly contribute to mortality. Studies in Whoopers specifically are unfortunately few, but those that have addressed several species provide some interesting comparisons. Macdonald *et al.* (1987 1990), for example, found that among 70 Mute and 14 Whoopers in Scotland, flying accidents were the single most significant cause of death of both adults and juveniles, causing 22% of adult deaths and 23% of juveniles. Lead poisoning was the next most significant factor, causing the loss of a further 21% of adults and 10% of juveniles. When various factors are grouped into general categories, 30% were caused by adverse environmental factors, 25% by injury, 20% by infections and 10% by parasitism. More recently, the causes of death reported for 143 ringed birds included 68% as a result of human circumstances (including collisions), a further 13% were shot or taken by people, while just 6% were considered to have died naturally, seven birds were killed by predators and five by pollutants (Rees *et al.* 2002).

As among young Trumpeters, low temperatures combined with rain during the first two weeks of life may kill cygnets. Haapanen *et al.* (1973b) were first to show that mortality among cygnets of mobile pairs is higher than those of pairs nesting at localities capable of sustaining the family until fledging. Haapanen (1991) also demonstrated that although the mean monthly mortality of cygnets in Finland was 10%, mortality varied regionally with lower mortality in the south (3–6%), and significantly higher mortality (10–14%) in the north. In contrast, adult mortality during spring and summer was just 2%. Taking that further, Haapanen (1991) showed that, given the assumptions that first breeding is at six years, and that therefore 92% of non-breeders are immature, mortality of first-years is *c*.30%,

Table 11.1. Swan survival rates (after Brown et al. *1992b).*

Species	Minimum annual survival rate
Bewick's Swan	>80%
Mute Swan	81.8%
Whooper Swan	85.1%

second-years is 25% and for older birds it is 12%. Putting that in terms of breeding potential, a pair producing an average brood from 4.4 eggs, will produce only 0.8 cygnets that survive long enough to breed.

In Iceland, Rees *et al.* (1991a) found that by midsummer cygnets in their highland study area were less well developed than those in the lowlands, and consequently they were likely to be less well prepared for autumn migration and subsequent survival. If, as seems likely, highland swans typically lay later than in the lowlands, then autumn weather is likely to significantly affect cygnet survival there. Certainly, several studies in Russia have shown that the timing and severity of autumn freeze is significant in its impact on cygnet survival.

It had been assumed that autumn migration was a period of serious risk, and that juvenile mortality during it was likely to be high, especially where long trans-oceanic flights were involved (Wilmore 1974; Brazil 1981c; Owen & Black 1989, 1991). Though migration must be considerably energy consuming, contrary to past assumptions, there is little evidence that it actually contributes significantly to mortality. Gardarsson & Skarphedinsson (1984) considered that migration to the British Isles from Iceland was not particularly hazardous. Their examination of brood sizes showed that prior to emigration from Iceland, in September–October, broods were already significantly smaller than in early August, suggesting that the most significant mortality of young (after the first few days) also occurs prior to fledging, rather than during migration or early winter. This is supported by data from Finland. Bart *et al.* (1991) found that the percentage of eggs surviving from laying to hatching was 74%, of cygnets surviving from hatching to fledging 90%, and of eggs/cygnets surviving from laying to fledging 59%. In Finland, cygnet production per pair in late August was very stable at 3.2 cygnets in 1964–70, with regional differences. Whereas in southern Finland in early September it remained high at 3.2, in Lapland and Kuusamo it was 2.5 (Haapanen 1991), suggesting that mortality may be regionally variable in response to local weather conditions during critical periods.

Rees *et al.* (1991a) were able to provide evidence in support of Gardarsson & Skarphedinsson's (1984) belief, when they showed that in May–August, lowland breeders lost 34% of their potential offspring where the mean clutch size was 4.7 eggs and mean brood size 3.1 cygnets, yet between August and when the families were re-sighted, in their winter range, they lost only a further 14.3% of the original clutch potential, i.e. 0.67 cygnets. They also concluded that, given that other studies of migratory swans have revealed annual mortality rates less than 35% for second-year birds and less than 20% thereafter (Evans 1979), the most hazardous period, for lowland cygnets at least, is the time prior to fledging, not autumn migration. In northwest Russia, Mineev (1995) was of the opinion that considerable numbers of cygnets perish in the first weeks of life as a result of predation and unfavourable conditions, while in northeast Russia unfavourable weather has been implicated in cygnet mortality (Kondratiev 1990, 1991).

An earlier study tracing the recoveries of 478 Whoopers ringed in Iceland up to 1980 (Brazil & Petersen in press) found that of the 61 (12.8%) recovered, approximately half were from Iceland and half from the British Isles. The monthly distribution of recoveries indicated that mortality was highest in winter (Fig 11.4). The high proportion of deaths on the wintering grounds could of course be the result of observer bias. Nevertheless, the results reveal interesting patterns: firstly that mortality in Britain was largely due to collisions with overhead wires, whereas such mortality was much rarer in Iceland, where the proportion shot was much larger. Concerning the mortality among known-aged birds, 64% died in their first year. Of those recovered abroad, 39.4% were known (or assumed) to have been killed by collision with overhead wires (also the commonest cause of mortality among Mutes and Whoopers studied by Macdonald *et al.* 1987), whereas in Iceland this accounted for only 14.8%, reflecting no doubt the differing extent of overhead wires in the two countries. Conversely, far greater numbers had been shot in Iceland (37%; despite full protection) than in the British Isles (12.1%). The figure for birds shot in Iceland may be higher as it was suspected that some reported merely as 'found dead' had been shot by the 'finder'. Only two (7.4%) were reported as having died of disease and/or malnutrition, both in Iceland. The year and season of death were known for 58 birds. Of these, 25 were of known age, having been ringed as cygnets. Mortality was particularly heavy in the first winter when 16 (64%) of the 25 were recovered. A further three were recovered in the second year, representing a third of the survivors.

Winters in the British Isles in particular, and even in continental western Europe, are considerably milder than in much of the species' range, nevertheless winter mortality is still the most significant. In colder areas (Russia, China, Mongolia, Korea and Japan), extreme cold, loss of open water and hence access to food have at times caused catastrophic mortality. Similar mortality has been noted among Trumpeters in North America.

The age of first breeding for the Whooper Swan still seems debatable, with Cramp & Simmons (1977) assuming it to be four years, although Delacour (1954)

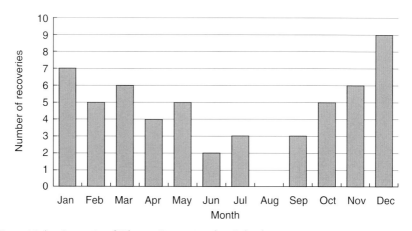

Figure 11.4.   Recoveries of Whooper Swans ringed in Iceland prior to 1980.

and Boyd & Eltringham (1962) believed it to be five, and more recently Haapanen (1991) has considered six the norm. Interestingly, Andersen-Harild (1981) found that fewer than 10% of Mutes bred in their fourth year and he suspected that some never join the breeding population. I suspect that the same may well prove true for Whooper Swan. Four years should perhaps be regarded as the probable minimum age of first breeding in the wild. This being the case, only a very small proportion (three or 12%) of known-age birds from Brazil & Petersen's (in press) study survived long enough to breed.

FLYING ACCIDENTS

While examination of recoveries of Whooper Swans ringed in Iceland prior to 1980 indicated that first-winter mortality was particularly high (64%) among known-aged birds, overall winter mortality on the Ouse Washes, England, was found by Owen & Cadbury (1975) to average just 3% (the same as Mutes, but just 1.4% for Bewick's). The single most significant mortality factor in previous studies was collision with powerlines (30% on the Ouse Washes) (Fig. 11.5), though that was very closely followed (29%) by lead poisoning from ingested pellets and fishing weights (Owen & Cadbury 1975). In contrast, shooting, though of course illegal, accounted for just 8% of deaths.

Both Whooper and Bewick's are far more likely than Mutes to die in October–March as a result of flying accidents (Table 11.2). This is presumably

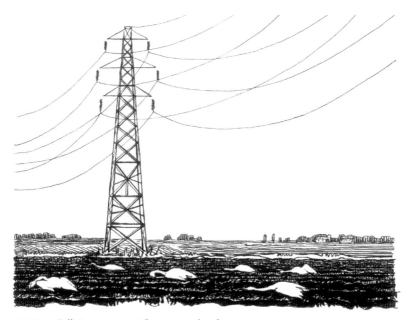

Figure 11.5.   Collisions are a significant mortality factor.

Table 11.2.   Swan mortality resulting from collision
(after Brown et al. *1992b).*

	Mortality resulting from collision
Mute Swan	18%
Whooper Swan	25%
Bewick's Swan	27%

because of their field-feeding habits, which mean that they spend part of each day flying between their roosting and foraging areas (often in low-light levels and poor visibility), whereas Mute Swan tends to be more sedentary, remaining at the same sites both day and night. Power and other overhead cables are a major hazard for flying swans, especially when moving between roost and foraging grounds at dawn and dusk. In an incident I observed in central Scotland, the actual cause of death appeared to have been electrocution rather than impact. A massive blue flash was observed on impact and the young Whooper was found below the wires with deep burn marks across its neck, presumably from having connected with two wires (Fig. 11.6).

Long-necked swans are, it seems, particularly susceptible to collisions with wires. Although they have a wide field of monocular vision to each side, suitable for detecting disturbance in any direction, they have only a narrow zone of binocular vision to the front and rear, and it appears they are rather poor at detecting thin horizontal objects ahead of them.

Figure 11.6.   A first winter Whooper Swan beneath power lines, following electrocution.

Myrberget (1981) documented a pair of Whoopers colliding with and being killed by powerlines in July in north Norway, and although Brazil & Petersen (in press) found that aerial collisions were the commonest cause of mortality on their wintering grounds in Britain (39.4%), nevertheless 14.8% of mortality was also due to this cause in Iceland, indicating that collisions are neither confined to winter nor to densely populated regions.

Kemp (1991) confirmed the general assumption of the dangers of overhead wires, applied specifically in the English Fens. Like Owen & Cadbury (1975), he also found that collisions were the main cause of mortality on the Ouse Washes. Furthermore, such incidents were more likely to occur during foggy weather or in gale-force winds (e.g. over 50 Bewick's and Whoopers died in this way in November/ December 1989). The placing of visible markers, 'bird diverters' to the earth wire above the national grid wires to alert swans to their presence does reduce mortality (e.g. Bowler *et al.* 1995) but have only limited value as swans returning from field feeding to roost sites often do so as it becomes dark or in winter fog, making it unlikely that they will be able to see them. Similarly, poor weather obscures the markers. The only guaranteed way of reducing such mortality would be to bury underground those cables on major flight paths, and to recognise the need for buried cables in environmental impact assessments of new developments in areas frequented by swans.

In Japan, where both power and telephone lines are strung above ground, the risks of injury to cranes, swans, geese and other waterfowl are considerable. At the world-famous crane site at Arasaki in southwest Kyushu, I found a very unfortunate Hooded Crane that had amputated both legs on overhead wires near the roost that now attracts over 10,000 Hooded and White-naped Cranes. Accidents occur there in many winters. In Hokkaido, a number of Whoopers with seriously broken wings has accumulated in recent years at Utonai-ko. As at Arasaki, powerlines ring the entire area and the swans are very likely to have sustained their injuries in collisions. As a result, the survivors are now resident there. Similar injuries may have led to records of Whooper Swans summering at other sites in other parts of their range.

The proliferation of wind farms using tall turbines in many parts of the Whooper's west European wintering range poses a new threat to those moving between feeding and roosting areas. Larsen & Clausen (2002) found that on the basis of recorded heights of flocks in flight, wind parks with medium-sized turbines posed a greater risk than those with large rotors.

POISONING

Lead poisoning is also a serious cause of mortality, being only slightly less significant than flying accidents (Owen & Cadbury 1975). Poisoning results from ingesting discarded lead fishing weights or shotgun pellets, both of which accumulate at wetlands that are fished or shot over. Lead weights and pellets are retained with grit in the gizzard, probably having been deliberately ingested by mistake instead of grit, to aid digestion (Fig. 11.7). They become ground down, over time. Whereas the harmless grit eventually passes right through the digestive system and is soon replaced, the lead is ground down in the gizzard forming soluble lead salts that are absorbed into the bloodstream. Because of the time involved in gathering a lethal accumulation of lead in the system, adults are more likely to be found suffering from, or to be killed by, lead poisoning than are cygnets. Overall, lead poisoning accounted for 21% of deaths among adults and 10% of cygnets, although in specific areas or incidents the proportions can vary considerably (Table 11.3).

Lead may also make itself felt less noticeably via sub-lethal effects, including impacts on the nervous system. Such effects may be responsible for reduced coordination and thus cause, for example, birds suffering from sub-lethal lead poisoning to be more susceptible to collisions with powerlines (O'Halloran *et al.* 1991). Although since 1982 UK anglers have been encouraged to adopt a voluntary code of practice promoting careful use and disposal of lead weights, and since 1 January 1987 the sale or import of lead weights has been banned by law, the amount of lead

Figure 11.7.   While ingesting grit is common, this young Whooper presumably had no intention of ingesting such a large pebble.

Table 11.3.   Main causes of death among Whooper Swans (after Brown et al. 1992b).

Age	N	Flying accident	Lead poisoning	Aspergillosis	No diagnosis	Predated	Others
Adult	16	25%	50%	0%	0%	6.2%	18.8%
Juvenile	7	14%	0%	14%	14%	14%	44%

remaining in British waterways ensures that the risks of lead poisoning will not disappear quickly. Lead poisoning will remain a significant potential cause of mortality for years after countries impose voluntary or compulsory bans on the use of lead by sportsmen. Furthermore, it is not only lead from fishing weights that swans are likely to ingest. In areas where shooting has a strong tradition, spent gunshot is also a major source of potentially lethal lead, and may be the main cause of lead poisoning among Whoopers in Scotland.

A substantial number of Whooper Swans died in Scotland in 1980–86. Post mortems implicated lead poisoning, mostly from shotgun pellets, in a massive 47%. Whoopers were clearly more at risk than Mutes, which over the same period experienced only 13% of deaths linked to lead. Comparisons of median blood lead levels in Iceland (31 ug 100 ml-1) and Scotland (48 ug 100 ml-1) indicated that lead shot was being ingested on the Scottish wintering grounds (Milne & Ramsay 1986; Spray & Milne 1988). The greatly differing mortality rates resulting from lead poisoning, high for Whooper and low for Mute, was considered by Spray & Milne (1988) to result from their differing exposure to lead as a result of their differences in feeding location, diet and grit requirements.

Whoopers in northwest Europe, including the British Isles, have, in recent decades, altered their feeding habits and now spend considerable time foraging on agricultural land. This change has not been without risk, as Thom (1986) noted. In January–March 1968, nearly 40, c.33% of the local Blairgowrie population was found dead or dying (MacMillan 1968). They had been feeding on young winter wheat seedlings and had been poisoned by the toxic mercury in the seed dressing still carried on the grain. Similar cases of large-scale mortality also occurred in Wigtown, Scotland, in 1969 and Roxburgh, Scotland, in 1979 (Badenoch 1980), although the latter may have been the result of lead poisoning from the ingestion of lead gunshot, as has occurred at Possil Marsh and the Ythan Estuary. Poisoning incidents such as these may have lasting impact on local Whooper distribution; the Blairgowrie herds have never fully recovered, perhaps because, as Thom (1986) thought, a regular wintering flock might represent the bulk of a local breeding population, or perhaps because the use of a specific wintering area may be passed on culturally from generation to generation. With a significant loss of birds from a particular site it may take some time or chance occurrence for birds to rediscover its value.

According to Spray & Milne (1988), two northeast Scottish sites certainly have sufficiently high densities of lead pellets, resulting from high-intensity shooting, to be potentially dangerous to waterfowl. Both are frequently used by swans for gritting, leaving them fatally exposed. In comparison with England, there is a relatively low incidence of split shot used in game fishing found in poisoned swans in Scotland. Spray & Milne (1988) suggested that this was because it is coarse fishing that is more popular in Scotland. Thus, regional differences in human activities can have implications for the mortality of wild swans—even if they are legally protected from all forms of hunting.

Lead poisoning is by no means confined to the UK. O'Halloran *et al.* (1988 1991) reported the first case in Ireland, which resulted from the ingestion of gunshot used almost two decades earlier. Lead poisoning has also been documented among Norwegian and Swedish Whoopers (Herredsvela 1984; Mathiasson 1993), and at the other extreme of the species' range, has also been extensively reported from Japan, by Hattori (1981), Jin *et al.* (1989), Honda *et al.* (1990), Ochiai *et al.* (1992) and Asakawa *et al.* (1997), in each case as a result of ingesting lead shotgun pellets or fishing weights. Mortality has been particularly significant at Miyajima-numa, Hokkaido where, during spring 1989, 33 Whoopers died. They were diagnosed as having sub-acute lead poisoning, with ingested shotgun pellets implicated as the source. Lead concentrations in their livers were 5.5–44.3 mg/kg wet weight. It was concluded that exposure occurred after they had arrived to winter in Japan, with mortality probably occurring within 30 days of exposure to, and retention of, lead shot in the gizzard. This conclusion is of particular interest given that O'Halloran *et al.* (1988) implicated 20-year-old lead shot in mortality, and that others have assumed that adults are more likely to be susceptible than young. If poisoning can occur so rapidly, and if poisonous lead shot and weights are still prevalent after 20 years then we can anticipate that such incidents will continue for many years and will affect all sectors of the population. Miyajima-numa is a migratory stopover in both spring and autumn for as many as 60,000 White-fronted Geese, as well as for hundreds of Bewick's and Whooper Swans. Migrants visiting there may be presumed to be energetically stressed, and thus perhaps particularly susceptible to poisoning. Lethal or sub-lethal lead poisoning is likely to occur wherever Whoopers spend time at sites that are regularly shot over or fished (Fig. 11.8).

Figure 11.8.   A weak-necked Whooper Swan suspected of suffering from lead poisoning (see Birkhead & Perrins 1986).

Nor is lead poisoning confined to Whooper Swan. Its impact is widely known among Mutes (e.g. Birkhead 1983), but recently the Whooper's close New World relative, the Trumpeter Swan has also been found suffering in Canada and the USA. In Idaho, Montana and Wyoming, ingestion of lead shotgun pellets and fishing weights accounted for *c.*20% of known mortality, 45.7% in Minnesota and in western Washington for a staggering nearly 50%. Similar mortality has been reported from British Columbia (Canada), where despite a ban on the use of lead shot for waterfowl hunting since 1990 at the main wintering areas, individuals continue to die from lead poisoning. Other causes of mortality included gunshot wounds, disease, parasitic infection and traumatic injury, particularly impact (Blus *et al.* 1989; Langelier *et al.* 1990; Degernes & Frank 1991; Wilson *et al.* 1998). Lead, in the form of shot and weights, persists in wetlands for very long periods, and as a result Whoopers and other waterfowl may continue to accumulate poisons long after its use has been banned.

Lead is by no means the only potential poison in environments used by swans. In Japan, Mute Swans have suffered mass mortality from copper poisoning in Honshu (Kobayashi *et al.* 1992), at a site which could easily be visited by Whoopers. In Europe, indeterminate chemical poisoning has caused mortality among Whoopers in the Czech Republic (Norek & Sveda 1990), while heavy metals and organochlorines have caused deaths among those wintering in western Sweden (Mathiasson 1993). A sample of 20 Whoopers found dead or dying in western Sweden analysed for the presence of various heavy metals and organochlorines provided some startling evidence of poisoning. Five were suffering from acute lead poisoning, and another five had medium lead loads, while several others had well above average levels of aluminium (two), manganese (one), nickel (one) and tin (two) in their livers and kidneys. In terms of lead poisoning, the issue appeared to be local, with birds wintering on fresh water at greater risk, where they were ingesting lead in the form of fishing lead sinkers. Other chemicals appeared, in general, in proportion to background levels of contamination with the exceptions already noted, which were presumed to have been specifically exposed at an unknown contamination source (Mathiasson 1993).

HUNTING

Hunting Whoopers and collecting their eggs dates back at least to the Iron Age, and probably well beyond. Nets, dogs and in recent centuries guns, have been the methods used from Iceland to Iran, England to Russia. In Iceland Whooper hunts (álftavei(ar) and whooper drives (álftaslagur; of flightless swans) were traditional, as was eating eggs (álftaregg). The collecting of eggs for food apparently lasted until relatively recently, ceasing as a result of agricultural improvements (Harrison 1973), though a wide range of other waterfowl eggs are still harvested.

In north Iceland swans were hunted from horseback using long-handled whips (Gardarsson *in litt.*), while another fascinating hunting method apparently used in Iceland relied on flying as a learned reflex action. Bub (1991; quoting Macpherson's *A History of Fowling* 1897 and Timmerman's *Die Vogel Islands* 1938), describes it in some detail, 'Several of the swan hunting farmers took up their places in spots where they knew from experience that post-moult swans and their young would fly close by on their way to the sea. Once the birds were sighted and had come within reasonable distance, the farmers started, upon a pre-arranged signal, an ear-splitting noise by yelling and screaming, barking of dogs and banging on all sorts of instruments. The young swans, scared to death, would fall to earth where they could then be grabbed while the adults seemed to pay no attention whatsoever to the incident and continued on the journey.' The explanation given for this strange behaviour was that, just like riding a bicycle or dancing, flying at first seems to be learned consciously and can only become reflex after sufficient practice. Thus, the young swans are unable to fly and pay attention to other things, so that when their attention is overwhelmingly attracted through some unusually strong stimulus, they lose control and fall to the ground. A somewhat similar technique has been practiced, I believe, for taking migrating cranes in Pakistan.

Swan hunting has long been practiced in Russia and adjacent regions. Around the Caspian Sea, Cossacks and Kazaks hunted them during moult, as was also done apparently on the Black Sea. Teams of people in light boats drove the swans, until they were exhausted and could be killed with oars, dragged into the boats and skinned with as many as 300 killed per boat in a day. The skins were powdered with salt or dampened with brine and turned inside out to preserve them, while the primaries were used for quill pens and the meat was salted and used in soup. In Siberia, hunters attracted swans close enough to kill by means of rough decoys. In the White Sea regions, Karels used to catch swans during spring floods by means of traps set to the bottom in shallow water and using bunches of grass as bait. When the swans dipped or upended for food they triggered the spring trap, which drowned them. Elsewhere, of course, they are shot along with many other waterfowl. Although it is claimed that swan hunting was not on a commercial scale, market records belie this, with, for example 320 skins registered at Obdorskaya fair in trans-Uralia in 1864, 200 in 1870, 3,000 in 1877–78 and 5,175 in 1881. These figures, clearly increasing through the years, would have represented only a small fraction of the swans taken throughout the country, and must have had a serious impact on populations in the late-19th century (Kaverznev 1999; Sotnikov 1999; Brokhaus & Efron 2001).

In the past, hunting, egg collecting and disturbance have directly extirpated some local populations, e.g. in the Orkneys, off mainland Scotland, where the species was wiped out in 1760–80 (Reynolds 1982), and south Greenland (Bent 1962). Bewick (1797–1804) noted that in Iceland the flightless, moulting birds were 'taken advantage of by the natives, who kill them with clubs, shoot, and hunt them down with dogs, by which they are easily caught. The flesh is highly esteemed by them as a delicious food, as are also the eggs.' Though described by Wilmore (1974) as 'leathery and rather oily' the taste and texture depend as much on the cooking method as on the individual, and it is far less fatty, and also darker than goose meat. I can certainly attest to the deliciousness of the meat having sampled a number of casualties, and having done so, I am not at all surprised that they have been widely hunted, given that one bird can easily feed a whole family for 2–3 meals. 'The Icelanders, Kamtchatdales, and other natives of the northern world', according to Bewick (1797–1804), 'dress their skins with the down on, sew them together, and make them into garments of various kinds: the North American Indians do the same, and sometimes weave the down as barbers weave the cawls for wigs, and then manufacture it into ornamental dresses for the women of rank, while the larger feathers are formed into caps and plumes to decorate the heads of the chiefs and their warriors. They also gather the feathers and down in large quantities, and barter or sell them to the inhabitants of more civilised nations.'

Later, Seebohm (according to Hudson 1926) considered that: 'The extermination of the Whooper in so many of its breeding places has arisen from the unfortunate habit, which it evidently acquired years ago, before men came upon the scene—a habit which it shares with the goose. . . .Swans and geese. . .drop nearly all their flight feathers at once, and for a week or two, before the new feathers have grown, are quite unable to fly. In some localities the Whoopers have had the misfortune to breed where the natives have been clever enough to surround them at the critical period of their lives, and stupid enough to avail themselves of the opportunity this afforded of killing the geese that laid the golden egg.'

In Japan too, the availability of firearms after 1868 led to the decline of the Whoopers that were formerly abundant as far south as Tokyo in winter. They were killed so extensively for their down and for their quills that numbers were soon severely reduced and the species became scarce. Protection in 1925 led to a halt in the decline (Austin 1949), although hunting did not entirely cease and enjoyed something of a resurgence during the American occupation of Japan after 1945. By the time hunting was eventually halted, entirely different factors were also beginning to seriously impact the population. Although numbers have subsequently increased as a result of the cessation of hunting, considerable habitat loss, affecting the majority of Japan's wetlands since 1945, has meant that their once extensive wintering range in Japan can never be recolonised because so few potential sites remain (Brazil 1991). Target of the increasingly well-armed hunter, the range contracted.

Now, hunting for food and even sport do not rate as significant threats, but local poaching may be detrimental as evidenced by that for swanskin caps (Semenov-Tyan-Shantskiy & Gilyazov 1990). Legal protection prevents such direct slaughter, but the spread of the human population with its attendant industry, pollution and powerlines, and particularly the conversion of wetlands, are no less significant a threat to the Whooper Swan.

Whooper, Bewick's and Mute Swans all receive legal protection to varying degrees under national laws. In Britain they are protected under the Wildlife and

Countryside Act 1981 Schedule 1, making hunting illegal. Furthermore, both Whooper and Bewick's are theoretically protected throughout their migratory range under the 1980 Bonn Convention, and by legislation in the various range countries concerned. Nevertheless, despite this 10% of Whoopers and as many as 40% of Bewick's X-rayed while caught for ringing in Britain in 1988–89 had one or more lead pellets in their body tissues (not their gizzards) (Rees *et al.* 1990a). These had been shot at deliberately either in the UK or along their migration routes. Despite EU legislation, shooting and killing of Whoopers occurs in various countries, and 'massacres' of swans (Mutes and Whoopers) have been reported in Greece (Anon 1985; Handrinos 1996). The problem is by no means new, as Haapanen *et al.* (1969) documented 34 Whoopers that had been illegally shot in Finland, mostly in autumn, in 1951–68. However, the media response to the 1993 swan massacre in Greece was such that it led to reduced illegal hunting and an increased awareness of the need for wetland conservation (Handrinos 1996), thus some long-term good resulted.

Despite hunting controls in North America since 1918, illegal, unintentional and malicious shooting of swans remains a problem, confirming the need for environmental education long after legal protection of a species is implemented. In some range countries hunting remains a persistent threat, thus in the Cis-Amur, Khabarovsk Krai hunters shoot both Whooper and Bewick's Swans, on migration through the Tartar Strait, from shore or boats equipped with outboard engines that enable hunters to approach flocks resting on the water (Roslyakov 1990). In the Yakutsk region shooting of both Whooper and Bewick's has recently been legalised under regional law (Rees & Bowler 1997). Unfortunately in China, not only is hunting continued, it is encouraged. As recently as April 1997, a China Centre Television news report showed tourists being taught how to hunt swans near Beijing. Swans have also been shot in Fuzhou, and hundreds continue to be captured for zoos (Ma 1997). Furthermore, in Xinjiang province, home to China's largest breeding and wintering populations of Whoopers, they are at risk from hunting, egg theft, nest destruction and disturbance from fishing with explosives and overstocking with domestic animals (Ma & Cai 2002).

Even where Whoopers are protected by law, shooting of other species may result in significant disturbance, despite the swans not being the direct object of hunting. At Lake Constance (Bodensee), on the border between Germany, Switzerland and Austria, the southernmost regular wintering ground of the Whooper in central Europe, numbers have increased considerably since they began to winter there in the 1950s. When a ban on shooting was placed on one particular bay (Ermatinger Becken) in 1985, that area rapidly became the most important site at the lake, despite only being infrequently used previously. Whooper Swan is now a protected species in all three countries. Hunting disturbance had precluded the exploitation of a considerable food supply in the area by the swans, whereas following the shooting ban the carrying capacity of the lake has increased (Schneider-Jacoby *et al.* 1991).

DISTURBANCE

Disturbance at the nest by man or predators has affected breeding success and even population distribution, and is deemed to have been the cause for the serious depletion or loss of some local populations. In the Russian Far East, in what many Europeans might consider an extremely remote region, the establishment of the Magadan reserve in 1982 led Whooper numbers to increase in the Yana-Taui depression, suggesting that hunting had previously been restricting population growth. In the same region, it has also been noted that Whoopers rarely breed at sites regularly visited by people, with abandoned nests indicative of desertion through disturbance (Kondratiev 1991). At high latitudes between the Kola Peninsula and Pechora, Mineev (1995) considered that the prolonged presence of people in breeding and moulting regions was one reason why the species had declined there, and in regions surrounding the Pechora, Nikolaev (1998) found Whoopers to be susceptible not only to poaching but also disturbance during incubation, with some marshes and lakes being abandoned due to disturbance. In the Yamal-Nenets Autonomous District, northeast of the Urals, rapid population growth combined with intensive development of swan habitats has negatively affected numbers of summering and moulting Whoopers as a result of habitat loss, disturbance from increased water transport to previously inaccessible waterbodies, persecution and nest robbing (Braude 1987).

In northern Karelia, Bianki (1988) found that forestry logging also negatively affected Whooper breeding, with pairs abandoning lakes and bogs around which forest had been cleared. After the construction of the Murmansk–St Petersburg road made the eastern part of the Lapland Nature Reserve accessible Whoopers have not bred there because of disturbance, while it has not bred since 1932 along the lower Chuna River (also in the reserve) through disturbance caused by researchers visiting permanent study plots (Semenov-Tyan-Shantskiy & Gilyazov 1990). In Mongolia too, intensive development and associated disturbance, including the loss of nests from fires, and nest predation by domestic/feral dogs has been implicated in the species' steady decline, especially in areas close to anthropogenic influences (Bold & Fomin 1996). If in such 'remote' regions disturbance is a significant factor, then it is unsurprising that it is a problem in more populated regions of the West. Just outside Iceland's capital, Reykjavik, I observed two pairs of Whoopers nesting at Ellidavatn. Both nests were active in May but unfortunately were deserted during June, most likely as a result of regular incidental disturbance by members of the public visiting the lake for recreation.

Where small populations are attempting to re-establish themselves, as in Scotland, or where they are attempting to occupy new regions, e.g. in Poland, disturbance is a major threat particularly as tourists, travellers and naturalists travel ever further afield in search of wildlife or photographic adventures. Martin (1993) considered that in such sensitive regions it is vital that breeding areas are kept secret, but that is only one possible response. Another, and perhaps more effective approach in the long term is to use such sites as educational opportunities, allowing visitors access to safe watchpoints where disturbance is avoided, but where they can experience this magnificent bird at its breeding haunts. There seems to be a steady trend towards increasing numbers of Whoopers in western Europe, which should permit more people to come into contact with this species, one hopes with a measure of respect and awe.

At staging areas or at wintering grounds, disturbance may be equally significant. Hunting disturbance has been shown to have a very significant impact in Germany. A further ramification of that disturbance is that shooting increases waterfowl flight distances and may lead Whoopers into other conflict areas, particularly with recreational activities, intensifying the problem. At Lake Constance, although Whooper Swan is a protected species, the increased popularity of watersports, and extension of the sport season to include winter, makes such activities a serious threat to all wintering waterfowl there (Schneider-Jacoby *et al.* 1991).

Most significant, however, is habitat destruction. Many areas once occupied by Whoopers are no longer available to them. This is especially true in the wintering grounds, for while their breeding sites are largely removed from concentrations of human population, the wintering range coincides with enormous human populations in some of the most densely populated regions of Earth. In China, Korea and Japan, for example, where most freshwater and coastal wetlands have already been destroyed, the Whooper's overall range has contracted considerably. Some remaining sites may well be able to hold larger numbers than formerly because of eutrophication leading to greater vegetative growth (e.g. at Lake Constance in Germany; Schneider-Jacoby *et al.* 1991) or because of the popularity of feeding swans, as in Japan. Overall, however, available wintering habitat declined considerably during the second half of the 20th century.

In the Far East, in contrast to northwest Europe where Whoopers usually occur in small flocks, they are typically encountered in large flocks, numbering hundreds if not thousands. This is doubtless a result of habitat loss forcing birds to rely on a limited network of surviving sites, each one increasingly significant as others are destroyed.

PREDATION

Predation in such a large species appears not to be a significant cause of mortality or at least little conclusive material is available other than largely anecdotal reports. None of the 61 recoveries of 478 pre-1980 ringed birds from Iceland appeared to have been predated (Brazil & Petersen in press), and although Brown *et al.* (1992b) found that 6.2% of adults and 14% of juveniles had been predated, the identity of the predators was unclear and if they had been killed by predators or consumed after death was also uncertain. I have, for example, seen a number of Whooper and Bewick's Swans in the English Fens apparently partially eaten by Red Foxes, but in each case the cause of death was impact with overhead powerlines, and the foxes had been attracted to their carcasses as scavengers.

Wilmore (1974) ambiguously described cygnets as surviving predation by eagles, Gyrfalcons, Arctic Foxes and pike *Esox lucius*, though whether there is documented evidence of these attempts or whether they simply occur in their range is uncertain, although Degtyarev (1987) speculated that pike predated downy chicks in Yakutia. Throughout most of the species' range, both Red and Arctic Foxes occur, while in some parts wolves are also present, as are feral dogs. While swans on water are presumably always safe from potential mammalian predators, adults at the nest may be susceptible to foxes, as are the eggs and young cygnets. Parents would, however, in all probability be able to defend their young from such attacks by retreating quickly to water. Those parents moving their very young families between nest sites and feeding areas, as reported in Iceland and Finland and presumably elsewhere, can be supposed to be more susceptible to mammalian predation, as the young cygnets often have to cross dry ground en route to suitable feeding areas.

In northern Norway, both Golden Eagles and Red Foxes have shown interest in incubating Whooper Swans, though predation was not proven (Myrberget 1981). In North America, both Black and Brown Bears have been reported to prey on Trumpeter Swan nests (Banko 1960; Hansen *et al.* 1971; Henson & Grant 1992), and as the ranges of Brown Bears and Whooper Swans overlap across much of Eurasia, so bears must be listed as possible predators of Whoopers, as may perhaps even various of the larger mustelids, including Eurasian Wolverine. Coyotes, Raccoons, American Wolverine and American Mink have killed Trumpeter Swan cygnets, and Canadian Otters may also predate them (Siferd 1982; Mitchell 1994). Although Coyote does not occur in Eurasia, foxes and feral dogs are of an equivalent size and pose a similar threat; in Mongolia, for example, dogs are known to destroy clutches of eggs and nestlings (Bold & Fomin 1996). Furthermore, American Mink has been widely introduced in Eurasia and must surely pose a threat to Whoopers in some parts of its range, as may the wide-ranging European Mink and River Otter. Additional predators of Trumpeter cygnets (less than four months old) have included: snapping turtles *Chelhydra serpentina*, California Gull and even Great Horned Owl, while larger cygnets and adults have been taken by Golden Eagle, Bobcat, Red Fox and Coyote (Mitchell 1994).

In northern Russia, from the Kola to the Pechora, the Red Fox, White-tailed Eagle and Golden Eagle are considered the main threats, and large predators such as Wolverines and Brown Bear may prey on moulting birds, and Northern Raven and Hooded Crow have been observed attacking clutches (though after the parent had been flushed from the nest by people). Loss of whole clutches to foxes and Badgers has been reported from the Tyumen region of western Siberia, their strategy being to steal one egg at a time during a series of visits (presumably during incubation breaks; Semenov-Tyan-Shanskiy & Gilyazov 1990; Vengerov 1990b; Mineev 1995).

Few other species could even remotely be charged with threatening the Whooper Swan, but Aristotle, in his *Historia Animalium*, made an early observation that suggests raptors may certainly attempt to take them: 'If the eagle attacks them they will repel the attack and get the better of their assailant, but they are never the first to attack.' In Hokkaido, Japan, where both White-tailed and Steller's Sea Eagles are common in winter, I have only once observed attempted predation. That incident involved a pair of White-tailed Eagles attacking a family of swans in a coordinated manner, seemingly attempting to separate the cygnets from their parents, though ultimately, as Aristotle had also observed, unsuccessfully because of the efforts of the parents. In Finland, Haapanen (1973b) noted that while eagles nest in areas of high swan density, there are no records of predation. That eagles are capable of killing Whooper Swans has, however, been confirmed in Norway where a young White-tailed Eagle was seen to attack and eat a Whooper on Hirta (Overskaug 1986). In Montana, five Trumpeter cygnets were killed by Golden Eagles in 1944–1947 (Sharp 1951) suggesting that where its range overlaps that of the swan, it too must be considered a potential predator. If White-tailed Eagles are capable of preying on Whoopers, and Golden Eagles are able to kill Trumpeters, then the very much larger Steller's Sea Eagle, the range of which overlaps that of the Whooper in the Russian Far East in summer and winter, and in Hokkaido in winter, must also be capable of killing swans, although such predation has not been documented.

Ma & Cai (2000) recorded Red Fox, Wolf, Wild Boar, Northern Raven, other corvids and raptors as preying on Mute Swan cygnets. Such predation is likely to be entirely opportunistic, and could as equally well apply to Whooper Swans, which must guard their nests against these, and potential gull *Larus* spp. predators (Ma 1996), although no doubt human activities are far more damaging.

Actual predation on Whoopers is rare, though perhaps commonest among pairs that move their cygnets while very young, at which times birds of prey, Red Fox and Northern Raven are the most serious threats to eggs and young cygnets. Around Bayinbuluke in China, local herdsmen have witnessed attacks on Whooper Swans by various raptors including Upland Buzzard, Imperial Eagle, Pallas's Sea Eagle and even, apparently, Bearded Vulture (Ma & Cai 2000). In addition to larger raptors, gulls, and skuas *Stercorarius* spp. are known egg and cygnet predators (Whooper Facts 2001).

Human activities are far more threatening, however, than predation. In the 1970s soldiers stationed in the Bayinbuluke area supplemented their rations by collecting baskets of swan eggs. Later, zoo collectors made considerable impact, now disturbance by muskrat hunters, fishers, herb and mushroom collectors is affecting the population, and in the 1980s large numbers of cygnets died after the area was sprayed against locusts (Ma & Cai 2000).

ILLNESS

Whooper Swans have been reported as suffering from a wide range of ailments, not only heavy metal poisoning and sometimes incur non-life threatening injuries (Fig. 11.9). They sometimes suffer from food matter catching and accumulating beneath the tongue, causing visible swelling of the skin there. Oesophageal

Figure 11.9.   Minor injuries are difficult to document, but a torn web is conspicuous.

impaction also occurs and can be relieved by surgery (Cooke 1982), but they have been known to die as a result of impaction of the gizzard (Jennings 1959). Macdonald (1963) reported the results of his examinations of four dead adults found in poor condition. The single dead male weighed 5.45 kg, while the weights of three dead females ranged from 3.6 to 5.7 kg (mean 4.7 kg). One dead first-winter in poor condition weighed 5.2 kg. He considered them all grossly underweight and discounted acute poisoning; in one case an infestation of parasitic *Capillaria* spp. was considered the probable cause of death.

A wide range of parasites have been reported from the species, and according to Beer & Ogilvie (1972), Wilmore (1974), and Hattori (1981), Whoopers have been found with nematodes in the heart and stomach, flukes (Trematodes) in their orbits, intestines and caecum, and mites in their nasal cavities. Whoopers clearly suffer from a multitude of parasites, including gizzard worms *Amidostomum* spp., nematodes of the genus *Epomidiostomum*, horny-headed worms *Acanthocephala* spp., particularly *Polymorphus boschadis* and *Fillicollis anatis* (Petrova 1989; Nakamura & Asakawa 2001; Asakawa *et al.* 2002). In 1997, the first record of a *Hyptiasmus* sp. trematode was obtained from an immature female, which was also suffering from lead poisoning (Asakawa *et al.* 1997). Ten of the trematodes were found in the ventral air sacs. This seems to be only the second record of *Hyptiasmus* sp. from the genus *Cygnus* (the first of *H. magnus* was obtained from a Black Swan in New Zealand; Charleston 1979). The snails, *Lymnaea* spp. and *Helicella* spp., that normally host the trematode, are widely found among aquatic vegetation in Japan, and may easily be ingested when Whoopers are feeding on such vegetation.

Trumpeter Swans are sometimes parasitised by leeches (up to 12% of birds caught in winter) (Banko 1960; Bangs *et al.* 1981; Mitchell 1994) as are other waterfowl, and Whooper Swan is also likely to experience this form of parasitism, while Trumpeters have been reported dying from multiple parasitism (Cowan 1946).

Tapeworms (*Cestoda*) have been described from the Whooper Swan, the commonest being *Hymenolepis* spp. (Beer & Ogilvie 1972). However, an entirely new cestode, *Dicranotaenia microcephala*, was isolated from a Whooper Swan in Japan by Sawada (1987). Another new species, *Gastrotaenia paracygni*, was described from Whooper and Mute Swans (Czaplinski 1966). Whooper Swan is not only prone to infestations of internal parasites, as are other swans, but also attracts mallophaga (Beer & Ogilvie 1972), which, like internal parasites, may proliferate in already weakened individuals, causing irritation and a further decline in condition.

Other illnesses are more difficult to define and identify, though most regular swan watchers will have seen Whoopers with tumour-like growths on the head (see Fig. 11.10), typically above one of the eyes, or on the feet. Their cause remains unknown. Swans in general are also susceptible to a range of diseases that may

Figure 11.10.   A rare head tumour.

prove fatal, including avian cholera, avian tuberculosis, avian pox, and probably duck viral enteritis (e.g. Mullié *et al.* 1980; Mitchell 1994; Asakawa *et al.* 2002).

Though not illness, predation, parasite or injury, one incident reveals that strange accidents can happen to swans. Along the coast of east Hokkaido, Japan, I once came across a Whooper with what appeared to be an enormous bill deformity. On closer observation it proved to be a very large clam, which was very firmly attached to the bird's lower mandible (see Fig. 11.11). I can only assume that while seeking food in shallow water the swan had accidentally slipped its beak inside the clam, which must have been partly open. In this fluke incident, the surprised clam would no doubt immediately have closed its valves, catching a very much larger meal than ever before! The harder the swan tried to dislodge it, the harder the clam would have clamped shut, as anyone who has attempted to open clams can attest. Presumably the clam would have remained clamped to the swan's beak until it died, but whether the swan survived that long is anyone's guess.

Figure 11.11. Vigorous shaking failed to free this bird of the accidentally caught clam.

LEGAL PROTECTION AND CONSERVATION

Local and national laws protect wildlife, sometimes including specifically the Whooper Swan in many range states, from Iceland and the British Isles to China and Japan. Internationally, it is protected under the EC Birds Directive, the Bonn Convention and the Berne Convention. However, such legal status is insufficient to guarantee its safety as continued disturbance of its breeding grounds, and agricultural and industrial development of its wintering grounds continue to reduce its range. Furthermore, illegal hunting including shooting continue in both Russian and European parts of its range, with as many as 10% of birds X-rayed in Scotland having lead shotgun pellets in their body tissues (Rees *et al.* 1990a). Strict enforcement of regulations, along with a network of protected breeding, wintering and migration sites, is essential. Lampio (1980) proposed year-round protection for various waterfowl, including Whooper Swan in the whole of Europe. He also proposed sanctuaries in the immediate vicinity of shooting areas to provide refuge to disturbed birds, along with active management measures, including feeding, predator control, and control of lead-poisoning pollution and disturbance. Although some countries have effective legislation, further protection measures are necessary, e.g. in the British Isles where the species is fully protected, Batten *et al.* (1990) still considered further conservation measures necessary, especially control of disturbance both at the few breeding sites and at the wintering grounds, and the reduction of lead shot on certain wintering sites. In northern Norway Myrberget (1981) mentions what seems to be the only documented attempt to encourage Whooper Swans to nest—a unique attempt to hasten spring snow melt by artificially building a small nesting islet.

International law, public sympathy and the enormous extent of the breeding and wintering ranges of the Whooper Swan all serve to protect it (Fig. 11.12). Although regional conflicts over conservation occur, and though locally it may be in decline and habitat destruction may have a significant negative impact, the overall population is believed to be stable or even increasing. I trust, therefore, that for a long time to come the beautiful Whooper Swan will grace man's world, bringing beauty and wonder to the wilderness, awe and inspiration to the swan watcher and swan photographer, and providing an endless stream of new questions to the swan biologist.

Figure 11.12. From Iceland to Japan people have taken Whooper Swans to their hearts, protecting them with international laws and public sympathy.

Scientific names of bird and mammal species appearing in the text

Scientific names of mammals follow Corbet & Hill (1991) while those of birds follow the latest published world checklist by Clements (2000).

MAMMALS

Coyote	*Canis latrans*
Wolf	*Canis lupus*
Feral Dog	*Canis familiaris*
Arctic Fox	*Alopex lagopus*
Red Fox	*Vulpes vulpes*
Black Bear	*Ursus americanus*
Brown Bear	*Ursus arctos middendorffii*
Raccoon	*Procyon lotor*
American Mink	*Mustela vison*
European Mink	*Mustela lutreola*
Eurasian Wolverine	*Gulo gulo*
American Wolverine	*Gulo luscus*
Badger	*Meles meles*
River Otter	*Lutra lutra*
Canadian Otter	*Lutra canadensis*
Bobcat	*Lynx rufus*
Caspian Seal	*Phoca caspica*
Wild Boar	*Sus scrofa*
Giraffe	*Giraffa camelopardalis*
American Beaver	*Castor canadensis*
Muskrat	*Ondatra zibethicus*

BIRDS

Cattle Egret	*Bubulcus ibis*
Mute Swan	*Cygnus olor*
Whistling Swan	*Cygnus columbianus columbianus*
Bewick's Swan	*Cygnus columbianus bewickii*
Whooper Swan	*Cygnus cygnus*
Trumpeter Swan	*Cygnus buccinator*
Black Swan	*Cygnus atratus*
Black-necked Swan	*Cygnus melanocoryphus*

Coscoroba Swan	*Coscoroba coscoroba*
Cape Barren Goose	*Cereopsis novaehollandiae*
Bean Goose	*Anser fabalis*
Pink-footed Goose	*Anser brachyrhynchus*
White-fronted Goose	*Anser albifrons*
Greylag Goose	*Anser anser*
Snow Goose	*Chen caerulescens*
Ross's Goose	*Chen rossii*
Canada Goose	*Branta canadensis*
Barnacle Goose	*Branta leucopsis*
Brent Goose	*Branta bernicla*
Common Shelduck	*Tadorna tadorna*
Wood Duck	*Aix sponsa*
Mandarin Duck	*Aix galericulata*
Eurasian Wigeon	*Anas penelope*
American Wigeon	*Anas americana*
Common Teal	*Anas crecca*
Mallard	*Anas platyrhynchos*
Northern Pintail	*Anas acuta*
Garganey	*Anas querquedula*
Northern Shoveler	*Anas clypeata*
Red-crested Pochard	*Netta rufina*
Canvasback	*Aythya valisineria*
Greater Scaup	*Aythya marila*
Tufted Duck	*Aythya fuligula*
Common Eider	*Somateria mollissima*
Long-tailed Duck	*Clangula hyemalis*
Bufflehead	*Bucephala albeola*
Common Goldeneye	*Bucephala clangula*
Barrow's Goldeneye	*Bucephala islandica*
Goosander	*Mergus merganser*
Smew	*Mergus albellus*
Bearded Vulture	*Gypaetus barbatus*
Pallas's Sea Eagle	*Haliaeetus leucoryphus*
White-tailed Eagle	*Haliaeetus albicilla*
Steller's Sea Eagle	*Haliaeetus pelagicus*
Upland Buzzard	*Buteo hemilasius*
Imperial Eagle	*Aquila heliaca*
Golden Eagle	*Aquila chrysaetos*
Gyrfalcon	*Falco rusticolus*
Eurasian Coot	*Fulica atra*
American Coot	*Fulica americana.*
Common Crane	*Grus grus*
Hooded Crane	*Grus monacha*
Sandhill Crane	*Grus canadensis*
White-naped Crane	*Grus vipio*
Japanese Crane	*Grus japonensis*
Demoiselle Crane	*Anthropoides virgo*
California Gull	*Larus californicus*
Great Horned Owl	*Bubo virginianus*
Blakiston's Fish Owl	*Ketupa blakistoni*
European Wood Pigeon	*Columba palumbus*

Common Cuckoo	*Cuculus canorus*
Eurasian Treecreeper	*Certhia familiaris*
Winter Wren	*Troglodytes troglodytes*
Common Dipper	*Cinclus cinclus*
European Starling	*Sturnus vulgaris*
Rook	*Corvus frugilegus*
Hooded Crow	*Corvus corone cornix*
Northern Raven	*Corvus corax*

Swan Biometrics

Species	Length (mm)	Wing (mm)	Tarsus (mm)	Bill (mm)	Sex	N	Mean	Std dev	Range
Mute Swan	1,250–1,550	533–623	99–118	69–88	M	59	11,800.0	890.0	9,200.0–14,300.0
Cygnus olor					F	35	9,670.0	640.0	7,600.0–10,600.0
Black Swan	1,150–1,400	416–543	90–105	56–79	M	270	6,200.0		4,600.0–8,700.0
C. atratus					F	243	5,100.0		3,700.0–7,200.0
Black-necked Swan	1,020–1,240	400–450	78–88	71–86	M	8	5,400.0		4,500.0–6,700.0
C. melanocoryphus					F	7	4,000.0		3,500.0–4,400.0
Whooper Swan	1,400–1,650	562–635	104–130	92–116	B	12	9,350.0		7,400.0–14,000.0
C. cygnus									
Trumpeter Swan	1,500–1,800	604–680	110–140	105–120	M	27	11,400.0	727.0	
C. buccinator					F	47	10,300.0	1,230.0	
Tundra Swan	1,200–1,500	501–569	95–115	92–107	M	76	7,100.0		4,700.0–9,600.0
C. columbianus					F	86	6,200.0		4,300.0–8,200.0
Bewick's Swan	1,150–1,400	469–548	92–116	82–102	M	96	6,400.0		4,900.0–7,800.0
C. c. bewickii					F	95	5,700.0		3,400.0–7,200.0
Coscoroba Swan	900–1,150	427–480	88–98	63–70	M	2	4,600.0		3,800.0–5,400.0
Coscoroba coscoroba					F	3	3,800.0		3,200.0–4,500.0

(after Madge & Burn 1988; Dunning 1993)

Whooper Swan Biometrics

	Male			Female		
	Wing (UK; Japan in brackets)			**Wing (UK; Japan in brackets)**		
	Mean (mm)	*Range*	*N*	*Mean (mm)*	*Range*	*N*
Adult	612 (604)	553–674	534 (31)	596 (583)	521–674	589 (41)
Yearling	605	550–636	49	597	548–623	53
Juv	591 (573)	504–647	156 (8)	557 (553)	452–628	197 (18)
	Skull			**Skull**		
Adult	175	138–192	630	170	140–189	678
Yearling	176	155–190	64	174	161–184	75
Juv	176	144–191	177	172	140–188	217
	Tarsus (UK; Japan in brackets)			**Tarsus (UK; Japan in brackets)**		
Adult	124 (120)	107–142	411 (8)	119 (124)	102–133	453 (12)
Yearling	124	113–131	38	121	112–130	58
Juv	123 (136)	112–129	95 (2)	119	113–130	136
	Weight Winter (UK)			**Weight Winter (UK)**		
	Mean (kg)	*Range*	*N*	*Mean (kg)*	*Range*	*N*
Adult	10.2	7.2–13.5	655	9.2	5.6–13.1	718
Yearling	9.8	7.4–12.4	69	9.2	7.3–12.0	85
Juv	9.0	5.2–12.1	177	8.4	5.6–11.3	220
	Weight Winter (Denmark)			**Weight Winter (Denmark)**		
Adult	11.0	8.6–15.5	48	9.0	7.8–10.9	39
Juv	9.5	7.4–11.1	20	8.5	7.0–10.3	20
	Weight Winter (Japan)			**Weight Winter (Japan)**		
Adult	9.8		31	9.0		40
Juv	8.0		8	7.1		18
	Weight Summer (Iceland)			**Weight Summer (Iceland)**		
Adult	9.8	8.0–12.3	109	8.2	6.4–11.0	99
	Weight Summer (Finland)			**Weight Summer (Finland)**		
Adult	10.7	9.0–12.9	16	8.9	7.0–10.8	35

(after Rees *et al.* 1997a; Albertsen *et al.* 2002)

APPENDIX 3

Foods Eaten by Whooper Swans

English Name

Scientific Name

Aquatic/Marshland Vegetation
"Algae"
Water Shield	*Brasenia shreberi*
Water Starwort	*Callitriche sp.*
Filamentous Green Algae	*Cladophora aegagropila*
Filamentous Green Algae	*Cladophora glomerata*
Common Spike Rush	*Eleocharis palustris*
Canadian Pondweed	*Elodea canadensis*
Gut Weed	*Enteromorpha intestinalis*
Common Horsetail	*Equisetum arvense*
Water Horsetail	*Equisetum fluviatile*
Scouring Rush	*Equisetum hyemale*
Marsh Horsetail	*Equisetum palustre*
Common Cottn Grass	*Eriophorum angustifolium*
Prickly (Gorgon) Waterlily	*Euryale ferox*
Red Seaweed	*Gloiopeltis furcata*
Rush sp.	*Juncus sp.*
Ivy-leaved (Star) Duckweed	*Lemna trisulca*
Water Forget-me-not	*Myosotis scorpioides*
Sacred Lotus	*Nelumbo nucifera*
Japanese Pond Lily	*Nuphar japonicum*
Dwarf Water Lily	*Nymphaea tetragona*
Common Reed	*Phragmites communis*
Pondweed	*Potamogeton sp.*
Slender-leaved Pondweed	*Potamogeton filiformis*
Pondweed	*Potamogeton franchetti*
Fennel (Sago) Pondweed	*Potamogeton pectinatus*
Broad-leaved Pondweed	*Potamogeton natans*
Common Water Crowfoot	*Ranunculus aquatilis*
Celery-leaved Buttercup	*Ranunculus sceleratus*
Thread-leaved Water Crowfoot	*Ranunculus trichophyllus*
Wigeon Grass	*Ruppia sp.*
Beaked Tasselwood	*Ruppia maritima*
Beaked Wigeon Grass	*Ruppia rostellata*
Arrowhead	*Sagittaria sagittifolia*
Bur Reed	*Sparganium emersum*
Water Chestnut	*Trapa natans*
Bulrush (Cattail)	*Typha latifolia*
Green Algae	*Ulva pertusa*
Green Algae	*Ulva recticulata*
Horned Pondweed	*Zannichelia palustris*
Asiatic Wild Rice	*Zizania latifolia*
Eelgrass	*Zostera sp.*

| Eelgrass | *Zostera marina* |
| Eelgrass | *Zostera asiatica* |

Terrestrial Vegetation

Creeping Bent Grass	*Agrostis palustris*
Creeping Bent	*Agrostis stolonifera*
Marsh Foxtail	*Alopecurus geniculatus*
Rock Cress	*Arabis spp.*
Lyngby's Sedge	*Carex lyngbyei*
Mountain Bog Sedge	*Carex rariflora*
Bottle Sedge	*Carex rostrata*
Stonewort	*Chara sp.*
Thistles (seeds)	*Cirsium sp.*
Orchard Grass	*Dactylis glomerata*
Crowberry (berries)	*Empetrum nigrum*
Scheuchzer's Cottongrass	*Eriophorum scheuzeri*
Slender Cottongrass	*Eriophorum gracile*
Fescue sp.	*Festuca sp.*
Sweet Grass	*Glyceria aquatica*
Sweet Grass	*Glyceria fluitans*
Reed Sweet Grass	*Glyceria maxima*
Rye Grass	*Lolium perenne*
Timothy Grass	*Phleum pratense*
Bluegrass	*Poa pratensis*
Alpine Bistort	*Polygonum viviparum*
Marsh Yellow Cress	*Rorippa palustris*
Dwarf Willow	*Salix herbacea*
Woolly Willow	*Salix lanata*
Dwarf Bamboo/Bamboo Grass	*Sasa sp.*
Seaside Bulrush	*Scirpus maritimus*
Bittersweet	*Solanum lyratum*
Clover sp.	*Trifolium sp.*
White Clover	*Triolium repens*
Blueberry (berries)	*Vaccinium myrtillus*

Crop Plants

Grasses (various)	
Sugar Beet	*Beta vulgaris*
Swede/Turnip	*Brassica napobrassica*
Barley	*Hordeum vulgare*
Wheat	*Triticum aestivum*
Rye	*Secale cereale*
Potato	*Solanum tuberosum*
Rape	*Brassica napus*

Others

"Aquatic insects"	
"Berries"	
"Fruit"	
"Refuse"	
Midges spp.	*Chironomid spp.*
Caddis Flies spp.	*Trichoptera spp.*

"Worms"
"Shellfish"
Marine Mussel   *Mytilus edulis*
Freshwater Mussel   *Unio pictorum*
"Frogs"
"Fish"
Sockeye Salmon (eggs)   *Oncorhynchus nerka*
Rat   *Rattus sp.*
Mallard   *Anas platyrhynchos*

(after Scott & The Wildfowl Trust 1972; Lesser 1973; Cramp & Simmons 1977; Brazil 1981c, 2002; Hatakayama 1981; Rees *et al.* 1997a)

Notes

CHAPTER 1

1 And one, the Coscoroba Swan has recently been relegated to the sidelines as 'not really a swan'.
2 Although a collaborative research programme between UK scientists and Russian scientists has now been ongoing for ten years (the WWT/Russian Institute for Nature Conservation study on the Russki Zavorot Peninsula).
3 The Ornithobius complex of the mallophagan family Philopteridae.
4 Long extinct.
5 Both the lowland boreal zone and the higher arctic-alpine tundra of Iceland differ from their continental equivalents of continental or Siberian taiga and tundra in being oceanic with cool summers and cool rather than cold winters.
6 Incubation constancy refers to the percentage of total time available that the eggs are actually incubated.

CHAPTER 2

1 My modern Icelandic colleagues question the authenticity of such a tale.
2 An element strongly echoed in the Irish myth "The Wooing of Étaíne" dating probably to the 8th century in which Midir escapes with Étaíne from the hall at Tara via the smoke-hole and they fly off as swans.
3 In another example of the transference of a common belief but to a different species, in Japan marital fidelity is embodied by the Mandarin Duck and presentation of a pair of these in symbolic form to young married couples is not uncommon.
4 Álft, also spelled álpt, plural alftir or elftur, is both the colloquial and now the official name of *Cygnus cygnus*. It appears even in folk poetry such as the old lullaby:

> Bí-bí og blaka, álftirnar kvaka.
> Ég laet sem ég sofi, en samt mun ég vaka".
> [Bee-bee and flap the wings, the whoopers quack.
> I pretend to sleep, although I shall stay awake".]

5   Curiously, in Japan, the very same theme of metamorphosis from human to avian form has been transferred across the species barrier to the Japanese Crane.

6   The swan's neck is a record-breaker, containing 24–25 cervical vertebrae, compared with only 18–19 for geese. The longest-necked mammal, the giraffe, derived from an entirely different lineage, has a mere seven vertebrae despite its incredible length.

7   The Children of Lir, one of the 'Three Sorrows of Storytelling', is a tale based on a theme that reached Ireland from the British kingdom of Strathclyde in the 8th century (Armstrong 1958).

8   Eve was Daughter of King Bov the Red.

CHAPTER 3

1   Original birch forest was removed by human activitiy.

2   In this book the term is used in a purely geographical sense, to refer easily to the archipelago including the two countries of Ireland and the United Kingdom.

CHAPTER 4

1   Attu is one of the outer Aleutian Islands, closest to Kamchatka.

2   The provisioning of food at sites such as WWT Caerlaverock, WWT Martin Mere and WWT Welney in the UK, and at many sites in Japan, may in fact be creating quite an artificial situation.

3   Some sites not covered during the census were presumed to have held birds during the count period, thus numbers there were estimated.

4   A record 4,641 Bewick's Swans also arrived the same winter (13 January 1992).

5   The number of birds missed, and estimated, in the 1982 census is thought to have been underestimated. Recalculation, based on two later counts indicates at more likely total of 13,507 (10,478 adults, 2,079 young, with no increase in the population during the early 1980s (Gardarsson, *in litt.* 2001).

CHAPTER 5

1 Ravkin *et al.*'s (1988) higher estimates are virtually impossible to reconcile with others made from outside the region.

2 In publications prior to 1924 the name St Petersburg was used, however publications dating from the period 1924–1990 used the name Leningrad, while since 1990 it has been renamed St Petersburg.

3 Syktyvkar, capital of the Komi Republic, is situated at approximately 62°N 51°E.

4 Polynia, though often used in a strictly oceanographic sense, means an area of open water otherwise surrounded by ice. It may remain ice-free for various reasons including the presence of upwellings or currents, but its availability is important to wildlife.

5 The central portion of the huge region situated east of Moscow and stretching north from the Caspian towards the Barents Sea.

6 Previously known as Cheremisskiy Territory

7 The capital of Kurgan Oblast, Kurgan, is situated at c.56°N 66°E.

8 East of Chelyabinsk.

9 In the region of Tyumen.

10 Assessing the significance and validity of the figures given in some papers has been extremely difficult. I quote them here so that the reader may consider the various possibilities themselves. For those with the inclination, time and funding, clearly Russia is the place to focus on Whooper Swan research, so much remains to be clarified.

11 What is not clear from Ravkin *et al.*'s (1988) work is whether the high 'first half of summer numbers' included very large numbers of spring migrants, although this seems the only reasonable explanation for the considerable drop to 'second half of summer' numbers.

12 Sors are shallow, often saline swampy lakes in Kazakhstan and southern west Siberia, and in floodplains of large rivers; also shallow bays of lakes.

13 The capital is Barnaul at 53°N 83°E.

14 The Barzugin River flows into the eastern side of Lake Baikal.

15 A region spanning the Arctic Circle and stretching east from the Kolyma River and lying north of Kamchatka.

16 Within the borders of the Penzhinsk administrative district of the Kamchatka Region.

17 Quite possibly the same family as that reported at Summer Lake, Oregon, in November 2000.

CHAPTER 8

1   Arnarvatn southwest of Myvatn.
2   Time allocated to different behaviours has been recorded as the percentage of time each female was in view, unless the behaviour could occur only in the area in view, in which case it was recorded as a percentage of the observation period (Brazil 1981c).
3   As there is no known locality Baraka in the ex-USSR or modern Russia, it is possible that it is a misspelling of Baraba, the Baraba forest-steppe around Novosibirsk.

Bibliography

ABE, M. 1968. Some notes on the swans and on the main factors that caused their extensive death at Odaito Bay, Nemuro, Hokkaido. *Tori* 18, 379–91.

ABSOLOM, A. F. & PERRINS, C. M. 1999. Double-brooded Mute Swans. *Br. Birds* 92, 365–6.

ADALSTEINSSON, H. 1979. Zooplankton and its relation to available food in Lake Myvatn. *Oikos* 32, 162–94.

ADAM, A. M., BAYANDIN, O. V., GUREEV, S. P., DUBOVIK, A. D., KILIN, S. V., LELIN, V. G. & MILOVIDOV, S. P. 1990. The Whooper Swan in Tomsk Cis-Ob River area. In: Syroechkovski, Y. E. (ed.) [*Ecology and Conservation of Swans in the USSR.*] Science Press, Moscow.

ADAMIAN, M. S. & KLEM, D. 1999. *Handbook of the Birds of Armenia.* American University of Armenia, Oakland.

AHMAD, A. 1985. Sighting of Whooper Swans (*Cygnus cygnus*) in Baluchistan. *J. Bombay Nat. Hist. Soc.* 82, 192–3.

AIREY, A. F. 1955. Whooper Swans in southern Lakeland. *Bird Study* 2, 143–50.

ALI, S. & RIPLEY, S. D. 1968. *Handbook of the Birds of India and Pakistan,* vol. 1. Oxford University Press, London.

ALI, S. & RIPLEY, S. D. 1983. *A Pictorial Guide to the Birds of the Indian Subcontinent.* Oxford University Press, Delhi.

ALBERTSEN, J. O. 1997. Whooper Swans in northern Japan. *Swan Spec. Group Newsl.* 6, 13–4.

ALBERTSEN, J. O. & KANAZAWA, Y. 2002. Numbers and ecology of swans wintering in Japan. *Proc. Fourth Intern. Swan Symp., 2001.* Waterbirds 25, Special Publication 1, 74–85.

ALBERTSEN, J. O., ABE, Y., KASHIKAWA, S., OOKAWARA, A. & TAMADA, K. 2002. Age and sex differences in biometrics data recorded for Whooper Swans wintering in Japan. *Proc. Fourth Intern. Swan Symp., 2001.* Waterbirds 25, Special Publication 1, 334–339.

AL-ROBAAE, K. H. & SALEM, Y. A. 1996. Status of migratory wildfowl (Anatidae) in Iraq. *Gibier Faune Sauvage* 13, 275–83.

ALSTRÖM, P. & OLSSON, U. 1989. The pattern on the sides of the lower mandible of Whooper Swan and Bewick's Swan. *Vår Fågelvärld* 48, 24.

ANDERSEN-HARILD, P. 1981. Population dynamics of *Cygnus olor* ringed in Denmark. *Proc. 2nd Int. Swan Symp.* Sapporo, Japan, 1980, 176–191.

ANDERSEN-HARILD, P. 1984. Whooper and Bewick's Swans on migration. *Fugle* 4 (4): 16. [In Danish.]

ANDERSON, J. A. 1944. Whooper Swans grazing. *Br. Birds* 38, 37.

ANDREWS, J, & CARTER, S. (eds.) 1993. *Britain's Birds in 1990–91: The Conservation and Monitoring Review.* BTO & JNCC, Thetford & Peterborough.

ANDRUSENKO, N. N. 1987. [On the ecology and numbers of swans in the Tselinograd region.] In: [*Ecology and Migrations in the USSR.*] Nauka, Moscow. [In Russian.]

ANDRUSENKO, N. N. 1990. New data on swans of Kurgaldzhino nature reserve. In: Syroechkovski, Y. E. (ed.) [*Ecology and Conservation of Swans in the USSR.*] Science Press, Moscow.

Anon 1955. The Whooper Swan at Lake Hyoko, Suibara, Niigata Prefecture. The story of its taming. *Tori* 1955, 14–23.

Anon 1985. The swan massacre. *Elleniki Ornithologiki Etairia Euntaktes* 2, 17, 22.

Anon 1994. European news. *Br. Birds* 87, 1–15.

Anon 1995a. European news. *Br. Birds* 88, 26–45.

Anon 1995b. European news. *Br. Birds* 88, 263–80.

Anon 1996a. European news. *Br. Birds* 89, 23–45.

Anon 1996b. European news. *Br. Birds* 89, 247–266.

Anon 1997a. European news. *Br. Birds* 90, 79–93.

Anon 1997b. European news. *Br. Birds* 90, 238–250.

Anon 1999. www.ping.be/cr-birding/cr-swans.htm#bookmark_WhBe9899UK.

Anon 2000. A list of Latvian bird species. http://www.putni.lv/lvp_list_of_lvbsp.htm.

Anon 2001. http://www.hmao.wsnet.ru/obsved/priroda/redvid.htm.

ANTIPOV, A. M. 1999. Terms of migration of the Whooper Swans in the Tiumen Oblast. *Casarca* 5, 210–13.

ANTIPOV, A. M. 2001a. [Population structure and dynamics of the Whooper Swan in the 'Elizarovskiy' Federal Nature Refuge (Khanty-Mansi Autonomous Area) in 1977–1985.] In: Popovkina, A. B. (ed.) [*Problems of Study and Conservation of Anseriformes of Eastern Europe and Northern Asia.*] Moscow. [In Russian.]

ANTIPOV, A. M. 2001b. [Density dynamics and population structure of the Whooper Swan in the Middle Zaural'ie in 1977–1985.] In: Popovkina, A. B. (ed.) [*Problems of Study and Conservation of Anseriformes of Eastern Europe and Northern Asia.*] Moscow. [In Russian.]

ARDAMATSKAYA, T. B. & KORZYUKOV, A. I. 1991. Numbers and distribution of Mute Swans *Cygnus olor*, Whooper Swans *C. cygnus* and Bewick's Swans *C. bewickii* in the Black Sea area of the Ukraine, USSR. *Wildfowl Suppl.* 1, 53–5.

AOU 1998. *Check-list of North American Birds*, Seventh edn. American Ornithologists' Union, Washington DC.

ARMSTRONG, E. A. 1947. *Bird Display and Behaviour: An Introduction to the Study of Bird Psychology.* Lindsay Drummond, London.

ARMSTRONG, E. A. 1955. *The Wren.* Collins, London.

ARMSTRONG, E. A. 1958. *The New Naturalist. The Folklore of Birds.* Collins, London.

ARMSTRONG, E. A. 1963. *A Study of Bird Song.* Oxford University Press, Oxford.

ARMSTRONG, E. A. 1965. *Bird Display and Behavior: An Introduction to the Study of Bird Psychology.* Dover, New York.

ARTEM'YEV, Y. T. & POPOV, V. A. 1977. [*Birds of Volga-Kama Territory.*] Nauka, Moscow. [In Russian.]

ARVIDSSON, B. L. 1987. [Distribution and population size of Whooper Swan *Cygnus cygnus* in Sweden.] *Vår Fågelvärld* 46, 248–55. [In Swedish.]

ASAKAWA, M., TANIYAMA, H., NAKADE, T. & KAMEGAI, S. 1997. First record of the Cyclocoelid trematode, *Hyptiasmus* sp., from Whooper Swan in Japan. *Jap. J. Orn.* 46, 133–5.

ASAKAWA, M., NAKAMURA, S. & BRAZIL, M. A. 2002. An Overview of Infectious and Parasitic Diseases in Relation to the Conservation Biology of the Japanese Avifauna. *J. Yamashina Inst. Orn.* 34, 200–221.

ATKINSON-WILLES, G. L. 1964. The wildfowl situation in England, Scotland, and Wales. *Proc. Eur. Meeting Wildl. Conserv.* 1, 25–9.

ATKINSON-WILLES, G. L. 1981. The numerical distribution and the conservation requirements of swans in north-west Europe. *Proc. Second Intern. Swan Symp., Sapporo, Japan, 1980*, 40–8.

AUBRY, Y. & BANNON, P. 1995. Quebec region. *Field Notes* 49, 902.

AUSTIN, O. L. 1949. Waterfowl in Japan. Natural Resources Section, GHQ of the Supreme Commander for the Allied Forces (in occupied Japan), vol. 18.

AUSTIN, O. L. & KURODA, N. 1953. The birds of Japan: their status and distribution. *Bull. Misc. Comp. Zool. Harvard* 109, 279–613.

AUEZOV, E. M. & GRACHEV, V. A. 1987. [Swans on waterbodies of the Alakol' hollow.] In: [*Ecology and Migrations in the USSR.*] Nauka, Moscow. [In Russian.]

AUEZOV, E. M., KHROKOV, V. V. & BEREZOVSKIY, V. G. 1987. [The passage of swans in lower Turgai.] In: [*Ecology and Migrations in the USSR.*] Nauka, Moscow. [In Russian.]

BABENKO, V. G. 2000. [*Birds of Lower Cis-Amur River Area.*] Prometei, Moscow. [In Russian.]

BADENOCH, C. O. 1980. A report on the wintering and one mortality incident among Whooper Swans (*Cygnus cygnus*) on the River Teviot, Roxburgh. *Orn. Hist. Berwickshire Nat. Club* 41, 221–6.

BAILEY, S. F. & SINGER, D. 1996. Middle Pacific Coast region. *Natl. Audubon Soc. Field Notes* 50, 218–220.

BAILEY, T., BANGS, E. & BERNS, V. 1980. Back carrying of young by trumpeter swans. *Wilson Bull.* 92, 413.

BALDAEV, K. F. 1990. The Whooper Swan in Mari Territory (Krai). In: Syroechkovski, Y. E. (ed.) [*Ecology and Conservation of Swans in the USSR.*] Science Press, Moscow. [In Russian.]

BANGS, E. E., BERNS, V. D. & BAILEY, T. N. 1981. Leech parasitism of trumpeter swans in Alaska. *Murrelet* 62, 24–6.

BANKO, W. E. 1960. *The Trumpeter Swan.* North Amer. Fauna no. 63. US Fish & Wildl. Service, Washington DC.

BANNERMAN, D. A. & BANNERMAN. W. M. 1983. *The Birds of the Balearics.* Croom Helm, Beckenham.

BANNERMAN, D. A. & LODGE, G. E. 1957. *The Birds of the British Isles*, vol. 6. Oliver & Boyd, London.

BANNON, P. & DAVID, N. 1999. Quebec region. *North Amer. Birds* 53, 256.

BARABASHIN, T. O., BORODIN, O. V., & KIRYASHIN, V. V. 2001. [Rare Anseriformes of Ul'yanovsk region.] In: Popovkina, A. B. (ed.) [*Problems of Study and Conservation of Anseriformes of Eastern Europe and Northern Asia.*] Moscow. [In Russian.]

BARANOV, A. A. 1990. Data on distribution of the Whooper Swan in Tuva. In: Syroechkovski, Y. E. (ed.) [*Ecology and Conservation of Swans in the USSR.*] Science Press, Moscow.

BART, J., EARNST, S. & BACON, P. J. 1991. Comparative demography of the swans: a review. *Wildfowl Suppl.* 1, 15–21.

BARTOSZEWICZ, M. 1999. Importance of the Slonsk reserve, western Poland for wintering and migrating Whooper Swans *Cygnus cygnus* in 1991–1998. *Swan Spec. Group Newsl.* 8, 9–10.

BATBAYAR, N. 2000. A report on a waterbird Survey in the Khar Us Nuur National Park, Mongolia, 19–21 January 2000. WWF Mongolia.

BATTEN, L. A., BIBBY, C. J., CLEMENT, P., ELLIOTT, G. D. & PORTER, R. F. 1990. *Red Data Birds in Britain.* Poyser, London.

BAUER, W. & MÜLLER, G. 1969. Zur Avifauna des Ewros-Delta. *Beitr. naturkd. Forsch. SW-Deutschl.* 28, 33–52.

BAUMANIS, J. 1975. Nesting of the Whooper Swan in Latvia. *Zool. Muz. Rak. P. Stuckas Latvij. Valsts Univ.* 12, 79–81. [In Latvian.]

BAXTER, E. V. & RINTOUL, L. J. 1953. *The Birds of Scotland*, vol. 2. Oliver & Boyd, Edinburgh.

BEEKMAN, J. H. 1997. International censuses of the north-west European Bewick's Swan population, January 1990 and 1995. *Swan Spec. Group Newsl.* 6, 7–9.

BEEKMAN, J. H. 1998. Research, conservation and monitoring of Arctic migratory birds in the Baltic and White Sea ('RECMAB') workshop, Helsinki, Finland, 2–4 April 1998: Bewick's and Whooper Swan workshop. *Swan Spec. Group Newsl.* 7, 20–1.

BEEKMAN, J. H., LAUBEK, B. & CRANSWICK, P. 1999. International Bewick's and Whooper Swan census 15–16 January 2000. *Swan Spec. Group Newsl.* 8, 4–6.

BEER, J. V. & OGILVIE, M. A. 1972. Mortality. In: Scott, P. & The Wildfowl Trust. *The Swans.* Michael Joseph, London.

BELIK, V. P. 1990. Swans in Rostov region. In: Syroechkovski, Y. E. (ed.) [*Ecology and Conservation of Swans in the USSR.*] Science Press, Moscow.

BELL, M. V. 1979. Whooper Swan breeding success in 1978. *N. E. Scot. Bird Rep.* 1978, 40–1.

BELLROSE, F. C. 1976. *Ducks, Geese and Swans of North America.* Stackpole, Harrisburg.

BENGTSON, S. A. 1971. Food and feeding of diving ducks at Lake Myvatn, Iceland. *Ornis Fennica* 48, 77–92.

BEN'KOVSKIY, L. M., & BEN'KOVSKAYA, I. L. 1990. Wintering grounds of swans on Sakhalin and Iturup islands. In: Syroechkovski, Y. E. (ed.) [*Ecology and Conservation of Swans in the USSR.*] Science Press, Moscow.

BENT, A. C. 1962. *Life Histories of North American Wildfowl,* vol. 2. Dover, New York.

BEREZOVIKOV, N. N. & BELIALOV, O. V. 1999. Nesting of Whooper Swan in the central Tien Shan. *Casarca* 5, 214–5.

BERGMAN, D. L. & HOMAN, H. J. 1995. Sighting of a Whooper Swan in North Dakota. *South Dakota Bird Notes* 47(2), 34–5.

BERRY, J. 1997. Notes on the Essex County whooper swans, 1993–1997. *Bird Observer (Belmont)* 25, 240–5.

BESZTERDA, P., MAJEWSKI, P. & PANEK, M. 1983. Wintering of the Mute Swan *Cygnus olor* and the Whooper Swan *Cygnus cygnus* in flooded areas of the Warta River mouth. *Acta Ornithologica (Warsaw)* 19, 217–225.

BEWICK, T. 1797–1804. *History of British Birds.* Beilby and Bewick, Newcastle.

BIANCHI, E., MARTIRE, L. & BIANCHI, A. 1969. Gli uccelli della provincia di Varese (Lombardia). *Rev. Ital. di Orn.* 1969, 125–7.

BIANKI, V. V. 1990. On the Whooper Swan's ecology in Kola Peninsula. In: Syroechkovski, Y. E. (ed.) [*Ecology and Conservation of Swans in the USSR.*] Science Press, Moscow.

BIANKI, V. V. 1999. Migration dynamics of the Whooper Swan over the Kandalaksha Gulf. *Casarca* 4, 207–13.

BIANKI, V. V. & SHUTOVA, E. V. 1987. [Distribution and number of swans in the north of European part of the USSR.] In: [*Ecology and Migrations in the USSR.*] Nauka, Moscow. [In Russian.]

BIRKHEAD, M. 1983. Lead levels in the blood of Mute Swans *Cygnus olor* on the River Thames. *J. Zool. Lond.* 199, 59–73.

BIRKHEAD, M. & PERRINS, C. 1986. *The Mute Swan.* Croom Helm, London.

BIVAR, A. D. H. 1947. Occurrence of the Whooper Swan [*Cygnus cygnus* (Linn.)] and Greater European Bustard (*Otis tarda* Linn.) in the Punjab. *J. Bombay Nat. Hist. Soc.* 46, 731–2.

BLACK, J. M. 1988. Pre-flight signalling in swans: a mechanism for group cohesion and flock formation. *Ethology* 79, 143–57.

BLACK, J. M. & OWEN, M. 1999a. Agonistic behaviour in barnacle goose flocks: assessment, investment and reproductive success. *Anim. Behav.* 37, 199–209.

BLACK, J. M. & OWEN, M. 1999b. Parent–offspring relationships in wintering barnacle goose. *Anim. Behav.* 37, 187–98.

BLACK, J. M. & REES, E. C. 1984. The structure and age of the Whooper Swan population wintering at Caerlaverock, Dumfries and Galloway, Scotland: an introductory study. *Wildfowl* 35, 21–36.

BLINOVA, T. K. & BLINOV, V. N. 1997. [*Birds of Southern Trans-Uralia. Forest-steppe and Steppe.*] Nauka, Novosibirsk. [In Russian.]

BLOMGREN, A. 1974. *Sångsvan.* Bonniers, Stockholm.

BLUS, L. J., STROUD, R. K. REISWIG, B. & MCENEANY, T. 1989. Lead poisoning and other mortality factors in Trumpeter Swans. *Environ. Toxicol. Chem.* 8, 263–71.

BOBROWICZ, G., GRABINSKI, W. & RANOSZEK, E. 1986. New breeding record of Whooper Swans *Cygnus cygnus* in Poland. *Ptaki Slaska* 4, 80–4. [In Polish.]

BOETTISCHER, H. VON. 1943. Was ist *Cygnus davidii* Swinh.? *Zool. Anzieger* 138, 71–85.

BOEV, Z. 2000. *Cygnus verae* sp. n. (Anseriformes: Anatidae) from the Early Pliocene of Sofia (Bulgaria). *Acta Zool. Cracoviensia* 43, 185–92.

BOLD, A. & FOMIN, V. E. 1996. [Birds.] In: [*Rare Animals of Mongolia (Vertebrates).*] IPEE RAN, Moscow. [In Russian.]

BONISCH, R. 1992. Teil uber sommerung eines Singschwanes *Cygnus cygnus* in der nordlichen Oberpfalz. *Orn. Anzeiger* 31, 72–3.

BOREIKO, V. E., & GRISHCHENKO, V. N. 1999. [*Ecological Traditions, Popular Beliefs, Religious Views of Slavic and Other Nations*], vol. 2. Second edn. Nature-Protection Propaganda, 12. [In Russian.]

BOSWALL, J. 1975. Icelandic swan saga. *Wildl.* 17, 508–11.

BOWLER, J. M. 1992. The growth and development of Whooper Swan cygnets. *Wildfowl* 43, 27–39.

BOWLER, J. M., REES, E. C. & BUTLER, L. 1992. Bewick's and Whooper Swans *Cygnus columbianus bewickii* and *Cygnus cygnus*: the 1991–92 season. *Wildfowl* 43, 225–31.

BOWLER, J. M., BUTLER, L. & REES, E. C. 1993. Bewick's and Whooper Swans *Cygnus columbianus bewickii* and *Cygnus cygnus*: the 1992–93 season. *Wildfowl* 44, 191–9.

BOWLER, J. M., BUTLER, L., LIGGETT, C. & REES, E. C. 1994. Bewick's and Whooper Swans *Cygnus columbianus bewickii* and *Cygnus cygnus*: the 1993–94 season. *Wildfowl* 45, 269–75.

BOWLER, J. M., BUTLER, L., HESKETH, C., HESKETH, R., LIGGETT, C. & REES, E. C. 1995. Bewick's and Whooper Swans *Cygnus columbianus bewickii* and *Cygnus cygnus*: the 1994–95 season. *Wildfowl* 46, 176–87.

BOYD, H. 1972. Classification. In: Scott, P. & The Wildfowl Trust. *The Swans*. Michael Joseph, London.

BOYD, H. & ELTRINGHAM, S. K. 1962. The Whooper Swan in Great Britain. *Bird Study* 9, 217–41.

BOYD, H. & MALTBY, L. S. 1980. Weights and primary growth of Brent Geese *Branta bernicla* moulting in Queen Elizabeth Islands, N. W. T., Canada 1973–1975. *Ornis Scand.* 11, 135–41.

BRADSHAW, D. 1986. Swans. *Kansai Time Out, November*, 44–5.

BRAGIN, A. B. 1987. [Conservation and breeding of the Whooper Swan in the Lapland Nature Reserve.] In: [*Ecology and Migrations in the USSR.*] Nauka, Moscow. [In Russian.]

BRAUDE, M. I. 1987. [The migrations of the Whooper Swan and Bewick's Swan in lower Ob River.] In: [*Ecology and Migrations in the USSR.*] Nauka, Moscow. [In Russian.]

BRAUDE, M. I. 1998. [The swan increase in Kurgan Region.] In: [*Materials on Bird Distribution in the Urals, in Cis-Uralia and West Siberia.*] Ekaterinburg. [In Russian.]

BRAZIL, M. A. 1980. A record of Bewick's Swan (*Cygnus columbianus bewickii*) from Iceland. *Natturufraedingurinn* 50, 57–60.

BRAZIL, M. A. 1981a. The behavioural ecology of *Cygnus cygnus cygnus* in central Scotland. *Proc. Second Intern. Swan Symp., Sapporo, Japan, 1980*, 273–91.

BRAZIL, M. A. 1981b. Summer behaviour of *Cygnus cygnus cygnus* in Iceland. *Proc. Second Intern. Swan Symp., Sapporo, Japan, 1980*, 272–3.

BRAZIL, M. A. 1981c. The behavioural ecology of the Whooper Swan (*Cygnus cygnus cygnus*). Ph.D. thesis, Stirling University.

BRAZIL, M. A. 1981d. Geographical variation in the bill patterns of Whooper Swans. *Wildfowl* 32, 129–31.

BRAZIL, M. A. 1983a. Preliminary results from a study of Whooper Swan movements using neck collars. *J. Coll. Dairying Ebetsu (Japan)* 10, 79–90.

BRAZIL, M. A. 1983b. A case of unusual aggression by a Whooper Swan. *Tori* 32, 155.

BRAZIL, M. A. 1983c. The breeding success and distribution of Whooper Swans *Cygnus cygnus* wintering in Japan. *Strix* 2, 95–103.

BRAZIL, M. A. 1984a. Winter feeding methods of the Whooper swan (*Cygnus cygnus*). *J. Yamashina Inst. Orn.* 16, 83–6.

BRAZIL, M. A. 1984b. The behaviour of Whooper Swans (*Cygnus cygnus*) wintering in a tidal environment. *Strix* 3, 40–9.

BRAZIL, M. A. 1984c. The year of the Whooper Swan. *Birds* 10(4), 42–5.

BRAZIL, M. A. 1987. *A Birdwatcher's Guide to Japan.* Kodansha International & Wild Bird Society of Japan, Tokyo.

BRAZIL, M. A. 1990. Pure enough to grace the sacred waters. *Okura Lantern* 28, 35–9.

BRAZIL, M. A. 1991. *The Birds of Japan.* Christopher Helm, London.

BRAZIL, M. A. 2002a. Brood amalgamation in Bewick's Swan *Cygnus columbianus bewickii*: a record from Japan. *J. Yamashina Inst. Orn.* 33, 204–9.

BRAZIL, M. A. 2002b. An addition to the diet of the Whooper Swan *Cygnus cygnus* from Japan. *J. Yamashina Inst. Orn.* 33, 210–212.

BRAZIL, M. A. & KIRK, J. 1979. The current status of Whooper Swans in Great Britain and Ireland. Unpublished report to The Wildfowl Trust, Slimbridge.

BRAZIL, M. A. & PETERSEN, A. in press. Mortality of Whooper Swans *Cygnus cygnus* ringed in Iceland. Casarca x,xx–xx.

BRAZIL, M. A. & SHERGALIN, J. 2002. The status and distribution of the Whooper Swan *Cygnus cygnus* in Russia I. *J. Yamashina Inst. Orn.* 34, 162–199.

BRAZIL, M. A. & SHERGALIN, J. in press. The status and distribution of the Whooper Swan *Cygnus cygnus* in Russia II. *J. Yamashina Inst. Orn.* 34, xx–xx.

BRAZIL, M. A. & SPRAY, C. J. 1983. Large clutch and brood sizes of Whooper Swans. *Scott. Birds* 12, 226–7.

BRAZIL, M. A. & YAMAMOTO, S. 1989. Blakiston's Fish Owl in Hokkaido, with particular reference to vocalisations. In: Meyburg, B.-U. & Chancellor, R. D. (eds.) *Raptors in the Modern World.* World Working Group on Birds of Prey, Berlin, London & Paris.

BROKHAUS, F. A. & EFRON, I. A. 2001. http://library.boom/ru/encyclo/brokgauz/b00000150.htm.

BRODIE, J. 1980. Estimation of winter roosting time of the Starling. *Bird Study* 27, 117–9.

BROWN, L. H., URBAN, E. K. & NEWMAN, K. 1982. *The Birds of Africa,* vol. 1. Academic Press, London.

BROWN, M. J., LINTON, E. & REES, E. C. 1992a. Diseases of swans in captivity. *Wildfowl* 43, 58–69.

BROWN, M. J., LINTON, E. & REES, E. C. 1992b. Causes of mortality among wild swans in Britain. *Wildfowl* 43, 70–9.

BUB, H. 1991. *Bird Trapping and Bird Banding.* Cornell University Press, Ithaca.

BUCKLAND, S. T., BELL, M. V. & PICOZZI, N. 1990. *The Birds of North-east Scotland.* North-East Scotland Bird Club, Aberdeen.

BULSTRODE, C. J. K., CORBETT, E. S. & PUTMAN, R. J. 1973. Breeding of Whooper Swans in Iceland. *Bird Study* 20, 37–40.

BYRD, G. V., GIBSON, D. D. & JOHNSON, D. L. 1974. Birds of Adak Island, Alaska. *Condor* 76, 288–300.

CADBURY, C. J. 1975. Populations of swans at the Ouse Washes, England. *Wildfowl* 26, 148–59.

CAITHAMER, D. F. 1995. *Survey of Trumpeter Swans in North America.* US Fish & Wildl. Service, Laurel, Maryland.

CAITHAMER, D. F. 2001. *Trumpeter Swan Population Status, 2000.* US Fish & Wildl. Service, Laurel, Maryland.

CAMPBELL, B. 1960. The Mute Swan census in England and Wales 1955–56. Bird Study 7, 208–223.

CAMPBELL, C. R. G. & OGILVIE, M. A. 1982. Failure of Whooper Swan to moult wing feathers. *Br. Birds* 75, 578.

CASTELLANI, R., PARODI, R. & PERCO, F. 1985. Primo caso accertato di svernamento di Cigni selvatici *Cygnus cygnus* in Italia. *Atti Convegno Ital. Orn.* 1985, 249–50.

CELMINS, A., BAUMANIS, J. & MEDNIS, A. 1993. *List of Latvian Bird Species.* Riga.

CHARLESTON, W. A. G. 1979. Preliminary observation on the parasites of the black swan (*Cygnus atratus*). *New Zealand J. Zool.* 6, 653.

CHARMAN, K. 1977. The grazing of *Zostera* by wildfowl in Britain. *Aquaculture* 12, 229–33.

CHISOLM, H. & SPRAY, C. 2002. Habitat Usage and Field Choice by Mute and Whooper Swans in the Tweed Valley, Scotland. *Proc. Fourth Intern. Swan Symp., 2001.* Waterbirds 25, Special Publication 1, 177–182.

CLARK, S. L. & JARVIS, R. L. 1978. Effects of winter grazing on yield of ryegrass seed. *Wildl. Soc. Bull.* 6, 84–7.

CLELAND, J. B. 1906. Mechanical advantages of the shape of swans for feeding. *Emu* 5, 206.

CLEMENTS, J. F. 1991. *Birds of the World: A Checklist.* Fourth edn. Ibis Publishing, Vista, California.

CLEMENTS, J. F. 2000. *Birds of the World: A Checklist.* Fifth edn. Ibis Publishing, Vista, California.

COLHOUN, K. 1997. The wintering ecology of Icelandic Whooper Swans *Cygnus cygnus* (L.) in north-west Ireland. Ph.D. thesis, University of Ulster.

COLHOUN, K., DAY, K. R. & REES, E. C. 1996. Habitat utilisation and the effects of grazing by Whooper Swans in Ireland. *Swan Spec. Group Newsl.* 5, 11.

COLHOUN, K. & DAY, K. R. 2002. Effects of grazing on grasslands by wintering whooper swans. *Proc. Fourth Intern. Swan Symp., 2001.* Waterbirds 25, Special Publication 1, 168–176.

CONANT, B., HODGES, J. I., GROVES, D. J. & KING, J. G. 2002. Census of Trumpeter Swans on Alaskan Nesting Habitats, 1968–2000. *Proc. Fourth Intern. Swan symp., 2001.* Waterbirds 25, Special Publication 1, 3–7.

COOKE, S. W. 1982. Surgical relief of oesophageal impaction in a whooper swan. *Vet. Rec.* 111, 166.

COOPER, J. A. 1979. Trumpeter Swan nesting behaviour. *Wildfowl* 30, 55–71.

CORBET, G. B. & HILL, J. E. 1991. *A World List of Mammalian Species. Third Edition.* Oxford University Press, Oxford.

COWAN, I. M. 1946. Death of a Trumpeter Swan from multiple parasitism. *Auk* 63, 248–9.

COWARD, T. A. 1920. *The Birds of the British Isles and their Eggs.* Warne, London.

CRAIK, J. C. A. 1991. Results of bird-ringing in Argyll in 1990. *Argyll Bird Rep.* 7, 56–66.

CRAMP, S. & SIMMONS, K. E. L. (eds.) 1977. *The Birds of the Western Palearctic*, vol. 1. Oxford University Press, Oxford.

CRANSWICK, P. A. 1996. International census of the Icelandic Whooper Swan population: January 1995. *Swan Spec. Group Newsl.* 5, 8–9.

CRANSWICK, P. A. & POLLITT, M. 1994. Wildfowl counts in the UK, 1992–93. *Wildfowl* 45, 282–92.

CRANSWICK, P. A., BOWLER, J. M., DELANY, S., EINARSSON, O., GARDARSSON, A., McELWAINE, J. G., MERNE, O. J., REES, E. C. & WELLS, J. H. 1996. Numbers of Whooper Swans *Cygnus cygnus* in Iceland, Ireland and Britain in January 1995: results of the international Whooper Swan census. *Wildfowl* 47, 17–30.

CRANSWICK, P. A., COLHOUN, K., EINARSSON, O., McELWAINE, G., GARDARSSON, A., POLLITT, M., & REES, E. C. 2002. The status and distribution of the Icelandic Whooper Swan *Cygnus cygnus* population: results of the International Whooper Swan Census 2000. *Proc. Fourth Intern. Swan Symp., 2001.* Waterbirds 25, Special Publication 1, 37–48.

CURRY-LINDAHL, K. 1964. The situation of ducks, geese and swans in Norway, Sweden, and Finland. *Proc. Eur. Meeting Wildl. Conserv.* 1, 3–13.

CZAPLINSKI, B. 1966. *Gastrotaenia paracygni* sp. n. (Hymenolepididae). *Acta Parisitol. Pol.* 13, 35–9.

CZAPULAK, A. 1991. Status of the Mute Swan *Cygnus olor* and the Whooper Swan *Cygnus cygnus* in Silesia (SW Poland). *Wildfowl Suppl.* 1, 39–43.

CZAPULAK, A. 2002. Timing of Primary Moult in Breeding Mute Swans. *Proc. Fourth Intern. Swan Symp., 2001.* Waterbirds 25, Special Publication 1, 258–267.

CZAPULAK, A. & PLATA, W. 1998. Bill pattern variation in the Whooper Swan *Cygnus cygnus* wintering in Poland. *Notatki Ornitologiczne* 39, 27–32.

CZAPULAK, A. & WITKOWSKI, J. 1996. Breeding of the Whooper Swan in the Barycz Valley. *Ptaki Slaska* 11, 153–5.

DANE, C. W. 1966. Some aspects of breeding biology of the Blue-winged Teal. *Auk* 83, 389–402.

DATHE, H. 1984. Singshwäne, *Cygnus cygnus*, im Sommer auf dem Saaler Bodden. *Beitr. zur Vogelkunde* 30, 74.

DAVID, N. 1996. *Liste commentée des oiseaux du Québec.* Association québécoise des groupes d'ornithologues, Québec.

DAVID, N., AUBRY, Y. & BANNON, P. 1999. Quebec region. *North Amer. Birds* 53, 361.

DAVIS, D. E. 1956. Population changes and roosting times of Starlings. *Ecol.* 36, 423–30.

DEGEN, A., FLORE, B.-O., LUDWIG, J. & SUDBECK, P. 1996. Numbers of Mute, Bewick's and Whooper Swans *Cygnus olor, C. columbianus bewickii* and *C. c. cygnus* in Niedersachsen: results of countrywide counts in January and March 1995. *Vogelkdl. Ber. Niedsachs.* 28, 3–18.

DEGERNES, L. A. & FRANK, R. K. 1991. Causes of mortality in Trumpeter Swans *Cygnus buccinator. Wildfowl Suppl.* 1, 352–5.

DEGTYAREV, A. G. 1987. [Whooper Swan.] In: [*Red Data Book of Yakutia.*] Nauka, Novosibirsk. [In Russian.]

DEGTYAREV, A. G. 1990. [The Whooper Swan in northeast Yakutia.] In: Syroechkovski, Y. E. (ed.) [*Ecology and Conservation of Swans in the USSR.*] Science Press, Moscow. [In Russian.]

DELACOUR, J. 1954. *The Waterfowl of the World.* Country Life, London.

DELACOUR, J. & MAYR, E. 1945. The family Anatidae. *Wilson Bull.* 57, 3–55.

DELANY, S., REYES, C., HUBERT, E., PIHL, S., REES, E., HAANSTRA, L. & VAN STRIEN, A. 1999. *Results from the International Waterbird Census in the Western Palearctic and Southwest Asia, 1995 and 1996.* Wetlands Intern. Publ. no. 54, Wageningen.

DEMENT'EV, F. P. & GLADKOV, N. A. 1967. *Birds of the Soviet Union,* vol. 4. Israel Program for Scientific Translations, Jerusalem.

DEUTSCHMANN, H. 1997. Der Singschwan *Cygnus cygnus* als neuer deutscher Brutvogel. *Limicola* 11, 76–81.

DEUTSCHMANN, H. & HAUPT, H. 1992. Sommerbeobachtungen des Singschwans (*Cygnus cygnus*) in Ost-Brandenburg. *Orn. Mitt.* 44. 109.

DEUTSCHMANN, H. & HAUPT, H. 1994. Übersommernde Singschwäne (*Cygnus cygnus*) in Ost-Brandenburg. *Orn. Mitt.* 46, 132–3.

DIMOND, S. & LAZARUS, J. 1974. The problem of vigilance in animal life. *Brain Behav. Evol.* 9, 60–79.

DIRKSEN, S. & BEEKMAN, J. H. 1991. Population size, breeding success and distribution of Bewick's Swans *Cygnus bewickii* wintering in Europe in 1986–87. *Wildfowl Suppl.* 1, 120–4.

DIRKSEN, S., BEEKMAN, J. H. & SLAGBOOM, T. H. 1991. Bewick's Swans *Cygnus columbianus bewickii* in the Netherlands: numbers, distribution and food choice during the winter season. *Wildfowl Suppl.* 1, 228–37.

DORJIEV, T. Z. & ELAEV, E. N. 2001. [Distribution and number of swans in Baikal Siberia.] In: Popovkina, A. B. (ed.) [*Problems of Study and Conservation of Anseriformes of Eastern Europe and Northern Asia.*] Nauka, Moscow. [In Russian.]

DORST, J. 1971. *The Life of Birds.* Weidenfeld & Nicolson, London.

DROBOVTSEV, V. I. & ZABORSKAYA, V. N. 19807. [On migrations, number and distribution of the Whooper Swan and Mute Swan in the forest-steppe of northern Kazakhstan.] In: [*Ecology and Migrations in the USSR.*] Nauka, Moscow. [In Russian.]

DRENT, R. & SWIERSTRA, P. 1977. Goose flocks and food finding: field experiments with Barnacle Geese in winter. *Wildfowl* 28, 15–20.

DREWIEN, R. C. & BENNING, D. S. 1997. Status of Tundra Swans and Trumpeter Swans in Mexico. *Wilson Bull.* 109, 693–701.

DUNNING, J. B. (ed.) 1993. *CRC Handbook of Avian Body Masses.* CRC, Boca Raton.

DUQUET, M. (ed.) 1992. *Inventaire de la Faune de France. Vertébrés et principaux Invertébrés.* Éd. Nathan et Muséum national d'histoire naturelle, Paris.

DYMOND, N. 1981. Where Whoopers winter. *Birds* 7, 15.

DYRCZ, A., GRABINSKI, W., STAWARCZYK, T. & WITKOWSKI, J. 1991. *Ptaki Slaska.* Wroclaw.

EAJ 1999. *Report on the 30th Annual Census of Waterfowl (Anatidae) in January 1999.* Environment Agency of Japan, Tokyo.

EARNST, S. L. 1991. The Third International Swan Symposium. *Wildfowl Suppl.* 1, 7–14.

EBBINGE, B., CANTERS, K. & DRENT, R. 1975. Foraging routines and estimated daily food intake in Barnacle Geese wintering in the northern Netherlands. *Wildfowl* 26, 1–19.

EGOROV, V. A., SAMUSEV, I. F., & BEREZOVIKOV, N. N. 2000. [Waterfowl of the Kalba upland (eastern Kazakhstan).] *Selevinia* 1–4, 117–24. [In Russian.]

EINARSSON, A. 1985. The bottom of Lake Myvatn: past, present and future. *Natturfraedingurinn* 55, 153–73.

EINARSSON, Ó. 1996. Breeding biology of the Whooper Swan and factors affecting its breeding success, with notes on its social dynamics and life cycle in the wintering range. Ph.D. thesis, University of Bristol.

EINARSSON, Ó. 2000. Iceland. In: Heath, M. F. & Evans, M. I. (eds.) *Important Bird Areas in Europe: Priority Sites for Conservation.* BirdLife International, Cambridge.

EINARSSON, Ó. & REES, E. C. 2002. Occupancy and Turnover of Whooper Swans on Territories in Northern Iceland: Results of a Long-term Study. *Proc. Fourth Intern. Swan Symp., 2001.* Waterbirds 25, Special Publication 1, 202–210.

EISENHAUER, D. I. & KIRKPATRICK, C. M. 1977. *Ecology of the Emperor Goose in Alaska.* Wildl. Monogr. no. 57. US Fish & Wildl. Service, Washington DC.

ELAEV, E. N., ESHEEV, V. E., MITUPOV, C. T., WEIGL, S., WEGLEITNER, S. & JAMSAEV, B. G. 2000. [On the avifauna of Toreiskaya hollow (depression) (southeast trans-Baikalia).] *Ornitologicheskie issledovaniya v Rossii* 2, 54–73. [In Russian.]

ELKINS, N. 1979. High altitude flight by swans. *Br. Birds* 72, 238–39.

ELKINS, N. 1983. *Weather and Bird Behaviour.* Poyser, Waterhouses.

ELLISON, W. G. & MARTIN, N. L. 1999. New England region. *North Amer. Birds* 53, 30.

EMEL'YANOV, V. I. & SAVCHENKO, A. P. 1990. Distribution and number of the Whooper Swan in the period of seasonal migrations in the south of middle Siberia. In: Syroechkovski, Y. E. (ed.) [*Ecology and Conservation of Swans in the USSR.*] Science Press, Moscow.

EMEL'YANOV, V. I. & SAVCHENKO, A. P. 2001. [Modern status of the subfamily Anserinae in the south of Cis-Enisei Siberia.] In: Popovkina, A. B. (ed.) [*Problems of Study and Conservation of Anseriformes of Eastern Europe and Northern Asia.*] Nauka, Moscow. [In Russian.]

ERICKSON, R. A. & HAMILTON, R. A. 2001. Report of the California Bird Records Committee: 1998 records. *Western Birds* 32(1), 13–49.

EVANS, M. E. 1975. Breeding behaviour of captive Bewick's Swans. *Wildfowl* 26, 117–30.

EVANS, M. E. 1977. Recognising individual Bewick's Swans by bill pattern. *Wildfowl* 28, 153–8.

EVANS, M. E. 1979. Aspects of the life cycle of the Bewick's Swan, based on recognition of individuals at a wintering site. *Bird Study* 26, 149–62.

EVANS, M. E. & DAWNAY, A. 1972. The swan in mythology and art. In: Scott, P. & The Wildfowl Trust. *The Swans.* Michael Joseph, London.

EVANS, M. E. & KEAR, J. 1978 Weights and measurements of Bewick's Swans during winter. *Wildfowl* 29, 118–22.

EVANS, M. E. & LEBRET, T. 1973. Leucistic Bewick's Swans. *Wildfowl* 24, 61–2.

EVANS, M. E. & SLADEN, W. J. L. 1980. A comparative analysis of the bill markings of Whistling and Bewick's Swans and out-of-range occurrences of the two taxa. *Auk* 97, 697–703.

FAGAN, B. 2000. *The Little Ice Age. How Climate made History. 1300–1850.* Basic Books, New York.

FEA, J. 1775. *The Present State of the Orkney Islands Considered.* Brown, Edinburgh.

FEARE, C., DUNNET, G. M. & PATTERSON, I. J. 1974. Ecological studies of the Rook (*Corvus frugilegus* L.) in north-east Scotland: food intake and feeding behaviour. *J. Appl. Ecol.* 11, 867–96.

FEDOROV, E. G. & KHODKOV, G. I. 1987. [The Whooper Swan on waterbodies of central Baraba (west Siberia).] In: [*Ecology and Migrations in the USSR.*] Moscow. [In Russian.]

FEFELOV, I. V., TUPITSYN, I. I., PODKOVYROV, V. A. & ZHURAVLEV, V. E. 2001. *Ptitsky del`ty Selengi.* [The Birds of the Selenga River Delta]. Eastern-Siberian Publishing House, Irkutsk.

FIEBIG, J. 1993. Drie jährige ornithologische Studien in Nordkorea. 1. Allgemeiner Teil und Non-Passeriformes. *Mitt. Zool. Muz. Berlin* 69 Suppl.: Ann. Orn. 19, 43–99.

Finland.org. 2000. Finland's national nature symbols. http://www.finland.org/natseng.html

FINSH, O. & BREHM, A. 1882. [*Journey in West Siberia.*] Moscow. [In Russian.]

FLINT, P. R. & STEWART, P. F. 1992. *The Birds of Cyprus. B.O.U. Check-list No. 6.* Second edn. British Ornithologists' Union, Tring.

FLOYD, J. F. M. 1946. The wild swans of Erin. *Avicult. Mag.* 52, 189–93.

FORSMANN, D. 1990. Kurzer Sommer in der Taiga. *Ornis* 5, 29–31.

FRANCOIS, J. 1981. Sur la presence hivernale en Lorraine du cygne sauvage (*Cygnus cygnus*) et du cygne de Bewick (*Cygnus bewickii*) au cours de la derniere decennie. *Ciconia* 5, 83–95.

FRANKLIN, M. H. 1947. Late Whooper Swans in Co. Sligo. *Irish Nat. J.* 9, 98.

FRAZER, J. G. 1922. *The Golden Bough. A Study in Magic and Religion.* Macmillan, London.

FRAZIER, A. & NOLAN, V. 1959. Communal roosting of the Eastern Bluebird in winter. *Bird Banding* 30, 219–26.

FRENCH, T. 1997. The Essex County whooper swans: the MDFW position. *Bird Observer (Belmont)* 25, 246–7.

FRERE, H. T. 1846. On the rusty tinge of the plumage of wild swans (*Cygnus bewickii*). *Zoologist* 4, 130.

FRIDRIKSSON, S., MAGNUSSON, B. & GUNNARSON, T. 1977. The grazing of swans and geese in agricultural land at the farm Nedri Halsu i Kjos. *Fjölrit Rala* 13, 32–4.

GARDARSSON, A. 1975. Votlendi. *Rit Landverndar* 4, 100–34.

GARDARSSON, A. 1979. Waterfowl populations of Lake Myvatn and recent changes in numbers and food habits. *Oikos* 32, 250–70.

GARDARSSON, A. 1991. Movements of Whooper Swans *Cygnus cygnus* neck-banded in Iceland. *Wildfowl Suppl.* 1, 189–94.

GARDARSSON, A. & SIGURDSSON, J. B. 1972. [*Report on Pink-footed Goose Studies in Thjorsarver 1971*]. Orkstofnun Raforkudeild, Reykjavik. [In Icelandic.]

GARDARSSON, A. & SKARPHEDINSSON, K. H. 1984. A census of the Icelandic Whooper Swan population. *Wildfowl* 35, 37–47.

GARDARSSON, A. & SKARPHEDINSSON, K. H. 1985. [The wintering of Whooper Swans *Cygnus cygnus* in Iceland.] *Bliki* 4, 45–56. [In Icelandic.]

GARDARSSON, A., EINARSSON, A. & THORSTEINSEN, S. 2002. Whooper Swans moulting at Lake Myvatn, Iceland 1974–2000. *Proc. Fourth Intern. Swan Symp., 2001.* Waterbirds Special Publication 1, 49–52.

GERASIMOV, N. N. & ALEKSEEV, S. A. 1990. [The Whooper Swan in Kamchatka.] In: Syroechkovski, Y. E. (ed.) [*Ecology and Conservation of Swans in the USSR.*] Science Press, Moscow. [In Russian.]

GERASIMOV, Y. N. 2000. [The 'Kharchinskoe Ozero' Game Reserve as a halting site for waders in the spring migration period.] In: [*Waders of Eastern Europe and Northern Asia at the Turn of the Century.*] Moscow. [In Russian.]

GERASIMOV, Y. N., SAL'NIKOV, G. M. & BUSLAEV, S. V. 2000. [*Birds of Ivanovo Region.*] Moscow. [In Russian.]

GERMINY, G. DE. 1937. Esiste o ne il Cigne di David? (*Cygnus davidii*). *Rassegna faunistica* 4, 32.

GIBBONS, D. W., REID, J. B. & CHAPMAN, R. A. 1993. *The New Atlas of Breeding Birds in Britain and Ireland: 1988–1991.* Poyser, London.

GIBBONS, D. W., AVERY, M. I. & BROWN, A. F. 1996. Population trends of breeding birds in the United Kingdom since 1800. *Br. Birds* 89, 291–305.

GILL, F. B. 1994. *Ornithology.* Second edn. Freeman & Co., New York.

GILLIGAN, J., SMITH, M. ROGERS, D. & CONTRERAS, A. 1994. *Birds of Oregon.* Cinclus Publ., Oregon.

GIZENKO, A. I. 1955. [*Birds of the Sakhalin Region.*] Izd-vo AN SSSR, Moscow. [In Russian.]

GJERSHAUG, J. O., THINGSTAD, P. G., ELDØY, S. & BYRKJELAND, S. 1994. *Norsk Fugleatlas. Hekkefuglenes utbredelse og bestandsstatus i Norge.* Norsk Ornitologisk Forening, Oslo.

GLOVER, S. A., TERRILL, S. B., SINGER, D. S. & ROBERSON, D. 2001. Regional reports: middle Pacific Coast. *North Amer. Birds* 55, 223–6.

GLUSHENKO, Y. N. 1990. Status of the Whooper Swan population in Khankai-Sungachinskaya lowland in 1987. In: Syroechkovski, Y. E. (ed.) [*Ecology and Conservation of Swans in the USSR.*] Science Press, Moscow.

GLUSHENKO, Y. N. & BOCHARNIKOV, V. N. 1991. The current status of Whooper Swan *Cygnus cygnus* and Bewick's Swan *Cygnus bewickii* populations in the Prichankayskaya lowland, USSR. *Wildfowl Suppl.* 1, 77.

GODFREY, W. E. 1986. *The Birds of Canada.* Revised edn. National Mus. Nat. Sci., Ottawa.

GODO, G. 1986. Whooper Swan breeding in south Norway. *Var Fuglefauna* 9, 34. [In Norwegian.]

GOLOVAN, V. I. & KONDRATIEV, A. V. 1999. [The Whooper Swan *Cygnus cygnus* breeding in Leningrad region.] *Russian J. Orn.* 86, 11–12.

GOODMAN, S. M. & MEININGER, P. L. (eds.) 1989. *The Birds of Egypt.* Oxford University Press, Oxford.

GORCHAKOV, G. A. 1996. [Spring migration of Anseriformes in the mouth of Razdolnaya River (south Primorye.)] In: Litvinenko, N. M. (ed.) [*Birds of the Wetlands of the Southern Russian Far East and Their Protection.*] Dal'nauka, Vladivostok. [In Russian.]

GORDIENKO, N. S. 2001. [Modern status of swans and geese in southern trans-Uralia.] In: Popovkina, A. B. (ed.) [*Problems of Study and Conservation of Anseriformes of Eastern Europe and Northern Asia.*] Moscow. [In Russian.]

GORDON, A. 1922. Nesting of the Whooper Swan in Scotland. *Br. Birds* 15, 170–1.

GORE, M. E. J. & WON, P.-O. 1971. *The Birds of Korea.* Roy. Asiatic Soc., Seoul.

GORELOV, M. S., GORSHKOV, Y. A., EMELIN, G. A. & PROKHOROV, E. V. 1987. [On swan nesting in Kuibyshev region.] In: [*Ecology and Migrations in the USSR.*] Nauka, Moscow. [In Russian.]

GORMAN, G. 1996. *The Birds of Hungary.* Christopher Helm, London.

GRANT, T. A., HENSON, P. & COOPER, J. A 1994. Feeding ecology of Trumpeter Swans breeding in south central Alaska. *J. Wildl. Management* 58, 774–80.

GRANT, T. A., HENSON, P. & COOPER, J. A. 1997. Feeding behaviour of Trumpeter Swans *Cygnus buccinator. Wildfowl* 48, 6–15.

GRAVES, R. 1961. *The White Goddess.* Faber & Faber, London.

GRIBBIN, J. & GRIBBIN, M. 2001. *How a change of climate made us human.* Penguin Books, London.

GRIMMETT, R., INSKIPP, C. & INSKIPP, T. 1999. *Birds of the Indian Subcontinent.* Christopher Helm, London.

GRINCHENKO, A. B., ZHMUD, M. E., KOSHELEV, A. I., KORZYUKOV, A. I., LYSENKO, V. I. & PERESADKO, L. V. 1988. Wintering of swans in Ukrainian Cis-Black Sea area during 1984–1988. In: Syroechkovski, Y. E. (ed.) [*Ecology and Conservation of Swans in the USSR.*] Science Press, Moscow.

GROTE, H. 1943. Was ist *Cygnus davidii* Swinh? *Orn. Mber.* 51, 90–1.

GRUSON, E. S. 1976. *Checklist of the Birds of the World.* Collins, London.

GUNN, W. W. H. 1973. Environmental stress on the Whistling Swan. *Wildfowl* 24, 5–7.

GUSAKOV, E. S. 1987. [Number and population of the Whooper Swan in Penzhino-Parapol region.] In: [*Ecology and Migrations in the USSR.*] Nauka, Moscow. [In Russian.]

HAAPANEN, A. 1982. The life history of a female Whooper Swan *Cygnus cygnus. Ornis Fennica* 59, 153–4.

HAAPANEN, A. 1987. [The Whooper Swan population in Finland.] *Lintumies* 22, 146–50. [In Finnish.]

HAAPANEN, A. 1991. Whooper Swan *Cygnus cygnus* population dynamics in Finland. *Wildfowl Suppl.* 1, 137–41.

HAAPANEN, A. & HAUTALA, H. 1991. Bimodality of spring migration of the Whooper Swan *Cygnus cygnus* in Finland. *Wildfowl Suppl.* 1, 195–200.

HAAPANEN, A. & NILSSON, L. 1979. Breeding waterfowl populations in northern Fennoscandia. *Ornis Scand.* 10, 145–219.

HAAPANEN, A., HELMINEN, M. & SUOMALAINEN, H. K. 1969. Joutsenen ampumista paukset 1951–68. *Suomen Riista* 21, 76–81.

HAAPANEN, A., HELMINEN, M. & SUOMALAINEN, H. K. 1973a. The spring arrival and breeding phenology of the Whooper Swan *Cygnus cygnus* in Finland. *Finnish Game Res.* 33, 31–8.

HAAPANEN, A., HELMINEN, M. & SUOMALAINEN, H. K. 1973b. Population growth and breeding biology of the Whooper Swan *Cygnus cygnus* in Finland in 1950–1970. *Finnish Game Res.* 33, 39–60.

HAAPANEN, A., HELMINEN, M. & SUOMALAINEN, H. K. 1977. The summer behaviour and habitat use of the Whooper Swan *Cygnus cygnus. Finnish Game Res.* 36, 50–81.

HAFTORN, S. 1971. *Norges Fugle.* Universitetsforlaget, Oslo-Bergen-Tromsø.

HAGEMEIJER, E. J. M. & BLAIR, M. J. 1997. *The EBCC Atlas of European Breeding Birds: Their Distribution and Abundance.* Poyser, London.

HAGERUP, A. T. 1891. *The Birds of Greenland.* Boston.

HALDAAS, S. 1985. A breeding record of the Whooper Swan *Cygnus cygnus* in Trondelag [Norway]. *Fauna (Oslo)* 38, 60–2.

HALL-CRAGGS, J. 1974. Controlled antiphonal calling by Whooper Swans. *Ibis* 116, 218–32.

HAMMOND, M. C. & MANN, G. E. 1956. Waterfowl nesting islands. *J. Wildl. Manage.* 25, 242–248.

HAMPTON, P. D. 1981. The wintering and nesting behaviour of the Trumpeter Swan. MSc. thesis, University of Montana.

HANDRINOS, G. 1996. The numbers and distribution of swans (*Cygnus* sp.) wintering in Greece. *Gibier Faune Sauvage, Game & Wildl.* 13, 463–76.

HANDRINOS, G. & AKRIOTIS, T. 1997. *The Birds of Greece.* Christopher Helm, London.

HANSEN, B., MOSGAARD, B. S., & LAUBEK, B. 1997. Breeding behaviour of Whooper Swans in Finland, late summer 1996. *Swan Spec. Group Newsl.* 6, 15.

HANSEN, H. A., SHEPHERD, P. E. K., KING, J. G. & TROYER, W. A. 1971. *The Trumpeter Swan in Alaska.* J. Wildl. Management. Monogr. Ser. no. 26.

HARRISON, J. 1973. *A Wealth of Wildfowl.* Corgi, London.

HART, A., HART, L. & DENNIS, W. 1984. Vigilance and scanning patterns in birds. *Anim. Behav.* 32, 1216–25.

HARVEY, N. G. 1998. Evolutionary and population genetics of swans. *Swan Spec. Group Newsl.* 7, 15–16.

HARVEY, N. G. 1999. A hierarchical genetic analysis of swan relationships. Ph.D. thesis, University of Nottingham.

HATAKAYAMA, M. 1981. The feeding for the swans and its result. *Proc. Second Intern. Swan Symp., Sapporo, Japan, 1980,* 247–8.

HATTORI, K. 1981. Post-mortem examination of *Cygnus cygnus cygnus* in Japan. *Proc. Second Intern. Swan Symp., Sapporo, Japan, 1980,* 312.

HAUGE, K. O. 1990. [Wintering swans (*Cygnus cygnus, C. olor, C. columbianus*) in Norway 1987/88.] *Cinclus* 13, 65–74. [In Norwegian.]

HEIM DE BALSAC, H. & MAYAUD, N. 1962. *Les Oiseaux du nord-ouest de l'Afrique.* Lechevalier, Paris.

HEIN, D. 1961. Wood Duck roosting flights at Paint Creek, Iowa. *Proc. Iowa Acad. Sci.* 68, 264–70.

HEIN, D. & HAUGEN, A. O. 1966. Illumination and Wood Duck roosting flights. *Wilson Bull.* 78, 301–8.

HEINROTH, O. & HEINROTH, M. 1928. *Die Vögel Mitteleuropas,* Bd. 3. Berlin.

HEMSLEY, D. 1998. Whooper Swan with white bill-base. *Br. Birds* 91, 233.

HENSON, P. & COOPER, J. A. 1992. Division of labour in breeding trumpeter swans *Cygnus buccinator. Wildfowl* 43, 40–8.

HENSON, P. & GRANT, T. A. 1992. Brown Bear, *Ursus arctos middendorffii,* predation on a Trumpeter Swan, *Cygnus buccinator,* nest. *Canadian Field Naturalist* 106, 128–30.

HENTY, C. J. 1977. The roost flights of Whooper Swans in the Devon Valley (central Scotland). *Forth Naturalist & Historian* 2, 31–5.

HERREDSVELA, H. 1984. Lead poisoning of mute and whooper swans at Jaeren, southwest Norway, 1984. *Var Fuglefauna* 7, 149–53.

HEWSON, R. 1964. Herd composition and dispersion in the Whooper Swan. *Br. Birds* 57, 26–31.

HEWSON, R. 1973. Changes in a winter herd of Whooper Swans at a Banff loch. *Bird Study* 20, 41–9.

HIGUCHI, H., OZAKI, K., FUJITA, G., SOMA, M., KANMURI, N. & UETA, M. 1992. Satellite tracking of the migration routes of cranes from southern Japan. *Strix* 11, 1–20.

HILDEN, O. 1964. Ecology of duck populations in the island group of Valassaaret, Gulf of Bothnia. *Ann. Zool. Fennica* 1, 153–279.

HILPRECHT, A. 1956. *Höckerschwan, Singschwan, Zwergschwan.* Ziemsen, Wittenberg.

HINDE, R. A. 1952. The behaviour of the Great Tit (*Parus major*) and some other related species. *Behav. Suppl.* 2, 1–201.

HOAR, W. S. 1956. Photoperiodism and thermal resistance of goldfish. *Nature Lond.* 1978, 364–5.

HÖHN, E. O. 1948. Courtship display and species recognition in Whooper Swan. *Br. Birds* 41, 54.

HÖHN, E. O. 1955. Evidence of iron staining as the cause of rusty discoloration of normally white feathers in Anserine birds. *Auk* 72, 414.

HØRRING, R. & SALOMONSEN, F. 1941. Further records of rare or new Greenland birds. *Medd. Orn. Grøn.* 131(5), 11–12.

HOKLOVA, T. Y. & ARTEMYEV, A. V. 2002. Modern distribution of the Whooper Swan *Cygnus cygnus* (L.) in the north-west of Russia. *Proc. Fourth Intern. Swan Symp., 2001.*

HOLLOWAY, S. 1996. *The Historical Atlas of Breeding Birds in Britain and Ireland 1875–1900.* Poyser, London.

HOLMGREN, V. & KARLSSON, J. 1982. [Further increases of the Whooper Swan *Cygnus cygnus* population in Skåne, south Sweden. Results from a census in 1978–79.] *Anser* 21, 163–8. [In Swedish.]

HONDA, K. 1979. *Hakucho no Iru Fukei.* Nippon Hoso Shuppan Kyokai, Tokyo.

HONDA, K. 1981. The lifespan of swans. *Proc. Second Intern. Swan Symp. Sapporo, Japan, 1980,* 80–1.

HONDA, K., LEE, D. P. & TATSUKAWA, R. 1990. Lead poisoning in swans in Japan. *Environ. Poll.* 65, 209–18.

HORIUCHI, M. 1981. Ten years of swan counts in Japan. *Proc. Second Intern. Swan Symp., Sapporo, Japan, 1980,* 14–15.

HOWARD, R. & MOORE, A. 1991. *A Complete Checklist of the Birds of the World.* Second edn. Academic Press, London.

HOWELL, S. N. G. & PYLE, P. 1997. Twentieth report of the California Birds Records Committee: 1994 records. *Western Birds* 28, 117–41.

HOWELL, S. N. G. & WEBB, S. 1995. *A Guide to the Birds of Mexico and Northern Central America.* Oxford University Press, Oxford.

DEL HOYO, J., ELLIOTT, A. & SARGATAL, J. (eds.) 1994. *Handbook of the Birds of the World,* vol. 1. Lynx Edicions, Barcelona.

HUDSON, W. H. 1926. *British Birds.* Longman, London.

HUTCHINSON, C. 1979. *Ireland's Wetlands and their Birds.* Irish Wildbird Conservancy, Dublin.

HUTCHINSON, C. D. 1989. *Birds in Ireland.* Poyser, Calton.

HUTCHINSON 1996. *Hutchinson's Multimedia Encyclopaedia.* Hutchinson, London.

IAPICHINO, C. & MASSA, B. 1989. *The Birds of Sicily. B.O.U. Check-list No. 11.* British Ornithologists' Union, Tring.

IIJIMA, Y. 1984. Strange feeding behaviour of the Whooper Swan *Cygnus cygnus. Tori* 33, 80.

ILLISON, R. 1990. The whooper swan breeds in Estonia. *Eesti Loodus* 5, 305.

INGLIS, I. R. & ISAACSON, A. J. 1978. The responses of Dark-bellied Brent Geese to models of geese in various postures. *Anim. Behav.* 26, 953–9.

INGREMEAU, D. & INGREMEAU, M. 1992. Cas d'hybridation entre *Cygnus cygnus* et *Cygnus olor.* Une premiere? *Geai* 15–16, 94–5.

INSKIPP, C. & INSKIPP, T. 1985. *A Guide to the Birds of Nepal.* Croom Helm, London.

JABOUILLE, P. 1936. Le cygne de David (*Cygnus davidii*) en Siberie. *Oiseaux* 6, 152–5.

JACKSON, R. D. 1952. Whooper Swan in Co. Mayo in June 1951. *Irish Nat. J.* 10, 216.

JAHNUKAINEN, M. 1963. On the spring migration of the Whooper Swan (*Cygnus cygnus*) in the Helsinki region in the years 1950–61. *Ornis Fennica* 40, 1–12.

JAMES, R. D. 1991. *Annotated Checklist of the Birds of Ontario.* Second edn. Roy. Ontario Mus., Toronto.

JEACOCK, A. A. 1945. Rare visitor. *Field* 185, 481.

JENNINGS, A. R. 1959. Diseases of wild birds, 5th report. *Bird Study* 6, 19–22.

JIN, K., OHYAMA, T., KATOH, Y., CHIBA, Y. & TSUZUKI, T. 1989. Lead poisoning in Whooper Swans at Miyajima swamp in Hokkaido. *Rep. Hokkaido Inst. Public Health,* 107–9.

JOENSEN, A. H. 1974. Waterfowl populations in Denmark 1965–73. *Danish Review Game Biol.* 9(1).

JOHANSEN, H. 1961. Revised list of the birds of the Commander Islands. *Auk* 78, 44–56.

JOHNSGARD, P. A. 1960. Hybridisation in the Anatidae and its taxonomic implications. *Condor* 62, 25–33.

JOHNSGARD, P. A. 1965. *Handbook of Waterfowl Behaviour.* Cornell University Press, Ithaca.

JOHNSGARD, P. A. 1973. Proximate and ultimate determinants of clutch size in Anatidae. *Wildfowl* 24, 144–9.

JOHNSGARD, P. A. 1974. The taxonomy and relationships of the northern swans. *Wildfowl* 25, 155–63.

JOHNSGARD, P. A. 1978. *Ducks, Geese and Swans of the World.* University of Nebraska Press, Lincoln & London.

KAKIZAWA, R. 1977. A female 'crested' Whooper Swan *Cygnus cygnus* wintering at Hyoko in Niigata, 1976–77 winter. *Misc. Rep. Yamashina Inst. Orn.* 9, 284–5.

KAKIZAWA, R. 1981. Hierarchy in the family group and social behaviour in wintering *Cygnus cygnus. Proc. Second Intern. Swan Symp., Sapporo, Japan, 1980,* 210–11.

KALMINS, M. 1997. Short communications about swans widespread in Latvia. *Swan Spec. Group Newsl.* 6, 9–11.

KALYAKIN, V. N. 1999. [Birds of Novaya Zemlya Region and Franz Josef Land.] In: [*Materials on Bird Distribution in Ural, Cis-Ural Area and West Siberia*.] Ekaterinburg. [In Russian.]

KALYAKIN, V. N. & VINOGRADOV, V. G. 1987. [The swans in southern Yamal.] In: [*Ecology and Migrations in the USSR*.] Nauka, Moscow. [In Russian.]

KANAI, Y., SATO, F., UETA, M., MINTON, J., HIGUCHI, H., SOMA, M., MITA, N. & MATSUI, S. 1997. The migration routes and important rest sites of Whooper Swans satellite-tracked from northern Japan. *Strix* 15, 1–13.

KARELIN, G. S. 1883. [Journey of G. S. Karelin via the Caspian Sea.] *Zapiski Russkogo Geogr. O-va* 10.

KARYAKIN, I. V. & KOZLOV, A. A. 1999. [*Preliminary Cadastre of Birds of Chelyabinsk Region*.] Manuscript Publishing House, Novosibirsk. [In Russian.]

KASHKAROV, D. Y. 1987. [Whooper Swan.] In: [*The Birds of Uzbekistan*, vol. 1.] Fan Publishers, Tashkent. [In Russian.]

KASTEPOLD, T. & PAAKSPUU, V. 1985. *Lindude ranne Matsalu alal. Matsalu-rahvusvahelise tahtusega margala*. Tallinn.

KASYBEKOV, E. S. 1993. *The Birds of Eastern Part of Issyk-Kul Region (Kyrghyzstan, Central Asia)*. Kyrghyz Ornithological Society, Bishkek.

KAVERZNEV, V. N. 1999. [Hunting of feathered game.] Zavigar Press, Minsk. [In Russian.]

KEANE, E. M. & O'HALLORAN, J. 1992. The behaviour of a wintering flock of mute swans *Cygnus olor* in southern Ireland. *Wildfowl* 43, 12–19.

KEAR, J. 1963. The history of potato-eating by wildfowl in Britain. *Wildfowl Trust Ann. Rep.* 14, 54–65.

KEAR, J. 1964. The changing status of the Greylag Goose and the Whooper Swan on agricultural land in Iceland. Unpublished report to Ministry of Agriculture, Reykjavik.

KEAR, J. 1970a. The experimental assessment of goose damage to agricultural crops. *Biol. Conserv.* 2, 206–12.

KEAR, J. 1970b. The adaptive radiation of parental care in waterfowl. In: Crook, J. (ed.) *Social Behaviour in Birds and Mammals*. Academic Press, London.

KEAR, J. 1972. Reproduction and family life. In: Scott, P. & The Wildfowl Trust. *The Swans*. Michael Joseph, London.

KEAR, J. 1990. *Man and Wildfowl*. Poyser, London.

KEMP, J. B. 1991. Status and habits of Whooper Swans in Norfolk. *Norfolk & Norwich Nat. Soc. Trans* 29, 99–100.

KEMP, J. B. & REVETT, J. 1992. 'Water-boiling' display by Whooper and Bewick's Swans. *Br. Birds* 85, 463–4.

KENDEIGH, S. C. 1961. Energy of birds conserved by roosting in cavities. *Wilson Bull.* 73, 140–7.

KENNEDY, P. G., RUTTLEDGE, R. F. & SCROOPE, C. 1954. *The Birds of Ireland*. Oliver & Boyd, Edinburgh.

KENWARD, R. E. & SIBLY, R. M. 1977. A Woodpigeon (*Columba palumbus*) feeding preference explained by a digestive bottleneck. *J. Appl. Ecol.* 14, 815–26.

KENYON, K. W. 1963. Further observations of Whooper Swans in the Aleutian Islands, Alaska. *Auk* 80, 540–2.

KESSEL, B. & GIBSON, D. D. 1978. *Status and Distribution of Alaska Birds*. Cooper Orn. Soc., Camarillo, California.

KHOKHLOV, A. N. 1995. [*Ornithological Observations in Western Turkmenia*.] Stavropol State Ped. University, Stavropol. [In Russian.]

KHOKHLOVA, T. Y. & ARTEM'YEV, A. V. 2000. [The Whooper Swan *Cygnus cygnus* nesting in Kenozero National Park (Arkhangelsk region).] *Russian J. Orn.* 102, 22–3.

KIESTER, A. R. 1979. Conspecifics as cues: a mechanism for habitat selection in the Panamanian Grass Anole (*Anolis auratus*). *Behav. Ecol. Sociobiol.* 5, 323–30.

KINLEN, L. 1963. Ringing Whooper Swans in Iceland. *Wildfowl Trust Ann. Rep.* 14, 107–14.

KIRBY, J. S., REES, E. C., MERNE, O. J. & GARDARSSON, A. 1992. International census of Whooper Swans *Cygnus cygnus* in Britain, Ireland and Iceland: January 1991. *Wildfowl* 43, 20–6.

KIRWAN, G. M. & MARTINS, R. P. 1994. Turkey bird report 1987–91. *Sandgrouse* 16, 76–117.

KIRWAN, G. M. & MARTINS, R. P. 2000. Turkey bird report 1992–1996. *Sandgrouse* 22, 13–35.

KISHCHINSKIY, A. A. 1988. [*Avifauna of Northeast Asia. History and Modern State.*] Nauka, Moscow [In Russian.]

KLIMOV, S. M., ZEMLYANUKHIN, A. I., SARYCHEV, V. S. & MEL'NIKOV, M. V. 2001. [The birds of the upper Don basin: Anseriformes.] *Russ. J. Orn.* 149, 523–43. [In Russian.]

KLOMP, H. 1970. The determination of clutch size in birds. A review. *Ardea* 58, 1–112.

KNUDSEN, H. L., LAUBEK, B. & OHTONEN, A. 2002. Growth and survival of Whooper Swan cygnets reared in different habitats in Finland. *Proc. Fourth Intern. Swan Symp., 2001.* Waterbirds 25, Special Publication 1, 211–220.

KOBAYASHI, Y., SHIMADA, A., UMEMURA, T. & NAGAI, T. 1992. An outbreak of copper poisoning in mute swans (*Cygnus olor*). *J. Vet. Med. Sci.* 54, 229–33.

KOFFIJBERG, K. 1998. Review of staging swans and geese in the Netherlands. *Swan Spec. Group Newsl.* 7, 4–5.

KONDRATIEV, A. Y. 1990. Swans in the extreme north-east of the USSR (status of populations, results and prospects of study). In: Syroechkovski, Y. E. (ed.) [*Ecology and Conservation of Swans in the USSR.*] Science Press, Moscow.

KONDRATIEV, A. Y. 1991. The distribution and status of Bewick's Swans *Cygnus bewickii*, Tundra Swans *C. columbianus* and Whooper Swans *C. cygnus* in the 'extreme northeast' of the USSR. *Wildfowl Suppl.* 1, 56–61.

KORNEV, S. V. & KORSHIKOV, L. V. 1995. [On the Whooper Swan in Orenburg steppe, trans-Uralia.] In: [*Wildlife of the Southern Ural and Northern Caspian Sea Area.*] Orenburg State Ped. Institute Press, Orenburg. [In Russian.]

KORPI, R. 1998. Field notes: eastern Oregon, winter 1997–98. *Oregon Birds* 24, 93–4.

KOSKIMIES, P. 1989. *Distribution and Numbers of Finnish Breeding Birds. Appendix to Suomen lintuatlas.* Lintutieto Oy, Helsinki.

KOTYUKOV, Y. V. 1990. Swans in Ryazan region. In: Syroechkovski, Y. E. (ed.) [*Ecology and Conservation of Swans in the USSR.*] Science Press, Moscow.

KOVSHAR, A. F. (ed.) 1996. *The Red Data Book of Kazakstan*, vol. 1. Third edn. Ministry of Ecology & Bioresources, Kazakhstan.

KOVSHAR, A. F. & LEVIN, A. S. 1982. [*Catalogue of the Oological Collection of the Institute of Zoology of Academy of Sciences of Kazakh SSR.*] Nauka, Alma-Ata. [In Russian.]

KREBS, C. J. 1972. *Ecology. The Experimental Analysis of Distribution and Abundance.* Harper Row, New York.

KREBS, C. J. & DAWKINS, R. 1984. Animal signs: Mind-reading and Manipulation. In: Krebs, J. R. & Davies, N. B. (eds) *Behavioural Ecology. An Evolutionary Approach.* Blackwell Scientific Publications, Oxford.

KRECHMAR, A. V. 1982a. [Ecology of *Cygnus cygnus* in the Anadyr River basin (USSR).] *Zool. J.* 61, 402–10. [In Russian.]

KRECHMAR, A. V. 1982b. Ecology of incubation in *Cygnus cygnus* at the extreme north-east of its range. *Zool. J.* 61, 1385–95.

KRECHMAR, A. V. 1990. [Whooper Swan (*Cygnus cygnus*) ecology in extreme northeast Asia.] In: Syroyochkovski, Y. E. (ed.) [*Ecology and Conservation of Swans in the USSR.*] Science Press, Moscow. [In Russian.]

KRECHMAR, A. V. & KRECHMAR, E. A. 1997. [Waterfowl of the Kava River drainage.] In: [*Species Diversity and Population Status of Waterside Birds in Northeast Asia.*] Magadan. [In Russian.]

KRECHMAR, A. V. & KONDRATIEV, A. Y. 1986. Comparative ecological analysis of Bewick's and Whooper Swans nesting biology. In: Andreev, A. V. & Krechmar, A. V. (eds.) [*Experimental Methods and Results of their Application in Northern Birds Studies.*] Acad. Sci. USSR, Vladivostock.

KREN, J. 2000. *Birds of the Czech Republic.* Christopher Helm, London.

KRIVENKO, V. G., MOLOCHAEV, A. V., BORSHCHEVSKIY, V. G., AZAROV, V. I. & MARTYSHIN, T. N. 1984. [Results of the Whooper Swan census in Yamal-Nenets Autonomous District.] In: [*Modern Status of Waterfowl Resources.*] Moscow. [In Russian.]

KRIVOSHEEV, V. G. 1963. [Bird migrations and hunting of them during passage in the Kolyma River valley.] In: [*Problems of Nature Conservation in Yakutia.*] Yakutsk. [In Russian.]

KRIVTSOV, S. K. & MINEEV, Y. N. 1990. Daily budgets of time and energy of the Whooper Swan and Bewick's Swan in nesting period. In: Syroechkovski, Y. E. (ed.) [*Ecology and Conservation of Swans in the USSR.*] Science Press, Moscow.

KRIVTSOV, S. K. & MINEYEV, Y. N. 1991. Daily time and energy budgets of Whooper Swans *Cygnus cygnus* and Bewick's Swans *Cygnus bewickii* in the breeding season. *Wildfowl Suppl.* 1, 319–21.

KUCHERUK, V. V., KOVALEVSKIY, Y. V. & SURBANOS, A. G. 1975. [Changes of bird population and bird fauna of the southern Yamal over the last 100 years.] *Bull. Moscow Nat. Soc., Dept. of Biol.* 80, 112–8.

KUCHIN, A. P. 1988. [Overwintering of the Whooper Swan in the Altai.] In: Shvetsov, Y. G. (ed.) *Redkie nazemnye pozvonochnye Sibiri.* Nauka, Novosibirsk. [In Russian.]

KUCHIN, A. P. 2001a. [Century-long and seasonal dynamics of climate of Altai and its influence on migrations and wintering of birds.] In: [*Actual Problems of Study and Conservation of Birds of Eastern Europe and Northern Asia.*] Matbugat Jorty Press, Kazan. [In Russian.]

KUCHIN, A. P. 2001b. Red Data Book of Altai. http://www.gasu.gorny.ru/virt/rb/7/d710. html.

KUCHIN, A. P. & KUCHINA, N. A. 1990. The Whooper Swan in Altai Krai (Territory). In: Syroechkovski, Y. E. (ed.) [*Ecology and Conservation of Swans in the USSR.*] Science Press, Moscow.

KURANARI, E. 1961. A report on swans from Kyushu. *Tori* 1961: 377–80.

KURESOO, A. O. 1988. On wintering of swans in Estonia on the basis of many winter censuses in 1967–1987. In: Syroechkovski, Y. E. (ed.) [*Ecology and Conservation of Swans in the USSR.*] Science Press, Moscow.

KURESOO, A. 1991. Present status of Mute Swans *Cygnus olor* Whooper Swans *C. cygnus* and Bewick's Swans *C. bewickii* wintering in the eastern Baltic region. *Wildfowl Suppl.* 1, 214–7.

KURODA, N. 1931a. On the migration of certain birds in Tokyo and the vicinity (I). *Tori* 1931, 15–40.

KURODA, N. 1931b. On the migration of certain birds in Tokyo and the vicinity (II). *Tori* 1931, 112–66.

KURODA, N. 1968. Summary distribution of the Anserine birds found in the Japanese islands. *Tori* 18, 392–405.

KYDYRALIEV, A. A. 1990. [*Birds of Lakes and Mountain Rivers of Kirghizia.*] Ilim, Frunze. [In Russian.]

LABUTIN, Y. V., DEGTYAREV, A. G., POZDNYAKOV, V. I. & GERMOGENOV, N. I. 1987. [Migration peculiarities and numbers of swans in Yakutia.] In: [*Ecology and Migrations in the USSR.*] Nauka, Moscow. [In Russian.]

LABUTIN, Y. V., GERMOGENOV, N. I. & POZDNYAKOV, V. I. 1988. [*Birds of Near-water Landscapes of Lower Lena River Valley.*] Nauka, Novosibirsk. [In Russian.]

LACK, D. 1967. The significance of clutch size in waterfowl. *Wildfowl Trust Ann. Rep.* 18: 125–8.

LACK, D. 1968. *Ecological Adaptations for Breeding in Birds.* Methuen, London.

LACK, D. 1974. *Evolution Illustrated by Waterfowl.* Blackwell Scientific Publications, Oxford.

LACK, P. (ed.) 1986. *The Atlas of Wintering Birds in Britain & Ireland.* Poyser, Calton.

LADYGIN, A. V. 1991. [Wintering birds in Lake Kuriiskoe basin (south Kamchatka) and their relationships with salmon spawning.] *Bull. Moscow O-Va ispyt. Prir. Otd. Biol.* 96(5): 17–22. [In Russian.]

LAMPIO, T. 1980. Management of waterfowl populations. *Acta Ornithologica* 17, 127–45.

LANGELIER, K., TOLKSDORF, M., HERUNTER, S., CAMPBELL, R. & LEWIS, R. 1990. Poisoning by lead shot in trumpeter swans. *Canadian Vet. J.* 31, 221–222.

LARSEN, J. K. & CLAUSEN, P. 2002. Potential wind park impacts on Whooper Swans in winter: the risk of collison. *Proc. Fourth Intern. Swan Symp., 2001.* Waterbirds 25, Special Publication 1, 327–330.

LAUBEK, B. 1995a. Udbredelse og faenologi hos rastende og overvintrende Sang- og Pibesvaner i Danmark 1991–93. *Dansk Orn. Foren. Tidsskr.* 89, 67–82.

LAUBEK, B. 1995b. European continental Whooper Swan *Cygnus cygnus* population estimate. *Swan Spec. Group Newsl.* 4, 5–6.

LAUBEK, B. 1995c. Habitat use by Whooper Swans *Cygnus cygnus* and Bewick's Swans *Cygnus columbianus bewickii* wintering in Denmark: increasing agricultural conflicts. *Wildfowl* 46, 8–15.

LAUBEK, B. 1996. Status of Whooper Swan (*Cygnus cygnus*) and Bewick's Swans (*Cygnus columbianus bewickii*) staging and wintering in Denmark. *Gibier Faune Sauvage, Game & Wildl.* 13, 1333–5.

LAUBEK, B. 1998. The northwest European Whooper Swan (*Cygnus cygnus*) population: ecological and management aspects of an increasing waterfowl population. Ph.D. thesis, University of Åarhus.

LAUBEK, B., NILSSON, L. WIELOCH, M., KOFFIJBERG, K., SUDFELT, C. & FOLKESTAD, A. 1999. Distribution, numbers and habitat choice of the NW European Whooper Swan (*Cygnus cygnus*) population: results of an international census in January 1995. *Vogelwelt* 120, 141–154.

LAZARUS, J. 1972. Natural selection and the function of flocking in birds: a reply to Murton. *Ibis* 114, 556–8.

LAZARUS, J. 1978. Vigilance, flock size and domain of danger size in the White-fronted Goose. *Wildfowl* 29, 135–45.

LAZARUS, J. & INGLIS, I. R. 1977. The breeding behaviour of the Pink-footed Goose: parental care and vigilant behaviour during the fledging period. *Behav.* 69, 62–87.

LEE, W. KOO, T. & PARK, J. 2000. *A Field Guide to the Birds of Korea.* LG Evergreen Foundation, Seoul.

LEIBAK, E., LILLELEHT, V. & VEROMANN, H. (eds.) 1994. *Birds of Estonia. Status, Distribution and Numbers.* Estonian Academy Publishers, Tallinn.

LEMIEUX, L. 1959. The breeding biology of the Greater Snow Goose on Bylot Island, Northwest Territories. *Canadian Field Naturalist* 73, 117–28.

LESSER, R. 1973. Hyoko: winter habitat of wild swans at Suibara. *Wildfowl* 24, 33–41.

LEVALLEY, R. & ROSENBERG, K. V. 1983. The winter season. Middle Pacific coast region. *Amer. Birds* 38, 352–6.

LEWARTOWSKI, Z. 1992. The locality of *Cygnus cygnus* near Lawsko in the North Podlasie Lowland. *Chronmy Przyrode Ojczysta* 48, 101–2.

LI, X. 1996. Numerical distribution and conservation of Whooper Swans (*Cygnus cygnus*) in China. *Gibier Faune Sauvage, Game & Wildl.* 13, 477–86.

LIANG, G. 1982. The swan natural reserve in Bayinbluk. *Chinese Wildl.* 3, 25.

LIAO, Y.-F. 1985. [Whooper Swans wintering in Ching Hai Lake.] *Chinese Wildl.* 3, 24–6. [In Chinese.]

LIEGL, C. 1998. Waterfowl in Har Us Nuur National Park and at Ayrag Nuur (western Mongolia), Report on two expeditions in June and September 1998. WWF Project Office, Mongolia.

LIMPERT, R. J., ALLEN, H. A. & SLADEN, W. J. L 1987. Weights and measurements of wintering Tundra Swans. *Wildfowl* 38, 108–13.

LIMPERT, R. J. & EARNST, S. L. 1994. Tundra Swan (*Cygnus columbianus*). In: Poole, A. & Gill, F. (eds) *The Birds of North America*, no. 89. The Academy of Natural Sciences Philadephia & American Ornithologists' Union, Washington DC.

LINNÉ, C. VON 1758. *Systema Naturae*, Ed. 10, 122.

LIPSBERG, Y. K. 1983. [*Birds of Latvia. Territorial Distribution and Numbers.*] Zinatne Press, Riga. [In Russian.]

LIPSBERG, Y. K. 1988. Distribution and dynamics of the Mute Swan and Whooper Swan breeding in Latvia. In: Syroyochkovski, Y. E. (ed.) [*Ecology and Conservation of Swans in the USSR.*] Science Press, Moscow.

LIU, T.-Y. 1987. [Observations on the overwintering habits of the Whooper Swan in the Bohai Bay, Shandong Province.] *Chinese Wildl.* 6: 24–5. [In Chinese.]

LOBKOV, E. G. 1987. [Nesting, migration and overwintering of Whooper Swans in Kamchatka.] In: Syroechkovksi, E. V. (ed.) [*Ecology and Migration of Swans in the USSR.*] Nauka, Moscow.

LOCKMAN, D. C., MITCHELL, C., REISWEG, B. & GALE, R. 1990. Identifying potential winter habitat for Trumpeter Swans. *Proc. Eleventh Trumpeter Swan Soc. Conf., 3–6 February 1988, Everett, Washington*, 20–2.

LÖHRL, H. 1955. Schlalgewohnheitan der Baumlaufer (*Certhia familiaris* and *C. brachydactyla*) und anderer kleinvogel in kaltern winter nachten. *Vogelwarte* 18, 71–6.

LOVEGROVE, R., WILLIAMS, G. & WILLIAMS, I. 1994. *Birds in Wales.* Poyser, London.

LUDLOW, F. 1945. The Whooper Swan. *J. Bombay Nat. Hist. Soc.* 45, 421.

LUIGUJÕE, L., KURESOO, A. & LEIVITS, A. 2002. Numbers and distribution of Whooper Swans breeding, wintering and on migration in Estonia 1990–2000. *Proc. Fourth Intern. Swan Symp., 2001.* Waterbirds 25, Special Publication 1, 61–6.

LUMSDEN, H. G. 1988. Productivity of Trumpeter Swans in relation to condition. *Proc. Tenth Trumpeter Swan Soc. Conf., Maple Plain*, 150–4.

LUMSDEN, H. G. 2002. Laying and Incubation Behavior of Captive Trumpeter Swans. *Proc. Fourth Intern. Swan Symp., 2001.* Waterbirds 25, Special Publication 1, 293–95.

LUND, H. M.-K. 1963. Recent records of wild swans in Norway. *Fauna* 16: 10–16 [In Norwegian.]

LYSENKO, V. I. 1987. [Dynamics of numbers and some peculiarities of swan ecology in the northern cis-Azov Sea area.] In: [*Ecology and Migrations in the USSR.*] Nauka, Moscow. [In Russian.]

LYSENKO, E. V., LAPSHIN, A. S., SIMONOV, D. V. & KOLYGANOVA, M. V. 1997. [Rare birds of Mordovia.] In: [*Fauna, Ecology and Conservation of Rare Birds of the Middle Volga River Area.*] Saransk. [In Russian.]

MA, M. 1996. The status and breeding ecology of Whooper Swans *Cygnus cygnus* in Bayinbuluke, Xinjiang, China. *Swan Spec. Group Newsl.* 5, 15–19.

MA, M. 1997. Status of Whooper Swans in China: 1996–97. *Swan Spec. Group Newsl.* 6, 9.

MA, M. & CAI, D. 2000. *Swans in China.* Trumpeter Swan Soc., Maple Plain.

MA, M. & CAI, D. 2002. Threats to Whooper Swans in Xinjiang, China. *Proc. Fourth Intern. Swan Symp., 2001.* Waterbirds 25, Special Publication 1, 331–33.

MACDONALD, J. W. 1963. Mortality in wild birds. *Bird Study* 10, 91–108.

MACDONALD, J. W., CLUNAS, A. J., EATOUGH, C. J., ISHERWOOD, P. & RUTHVEN, A. D. 1987. Causes of death in Scottish swans (*Cygnus spp.*). *State Vet. J.* 41, 165–175.

MacDonald, J. W., Goater, R., Atkinson, N. K. & Small, J. 1990. Further causes of death in Scottish swans (*Cygnus* spp.). *State Vet. J.* 44, 81–93.

MacKinnon, J. & Phillipps, K. 2000. *A Field Guide to the Birds of China.* Oxford University Press, Oxford.

MacMillan, A. T. 1968. Whooper deaths. *Scott. Birds* 5, 111–12.

Madge, S. & Burn, H. 1988. *Wildfowl.* Christopher Helm, London.

Markgren, G. 1963. Studies on wild geese in southernmost Sweden. Part I. *Acta Vertebratica* 2, 299–418.

Marriott, R. W. 1973. The manurial effect of Cape Barren Goose droppings. *Wildfowl* 24, 131–3.

Martin, B. P. 1993. *Wildfowl of the British Isles and North-west Europe.* David & Charles, Newton Abbot.

Marven, D. 1999. Rare birds in Canada in 1999: British Columbia. *Birders J.* 9, 262–3.

Mathiasson, S. 1973. A moulting population of non-breeding Mute Swans with special reference to flight-feather moult, feeding ecology and habitat selection. *Wildfowl* 24, 43–53.

Mathiasson, S. 1980. Weight and growth of morphological characters of *Cygnus olor. Proc. Second Intern. Swan Symp., Sapporo, Japan, 1980*, 379–389.

Mathiasson, S. 1991. Eurasian Whooper Swan *Cygnus cygnus* migration, with particular reference to birds wintering in southern Sweden. *Wildfowl Suppl.* 1: 201–8.

Mathiasson, S. 1992. [Hybridization between Mute and Whooper Swans.] *Göteborg Natur. Mus. Arstryck* 1992, 43–59. [In Swedish.]

Mathiasson, S. 1993. Heavy metals and organochlorines in Whooper swans (*Cygnus cygnus*) wintering in western Sweden. *Ring* 15, 147–53.

Matsui, S., Yamanouchi, N. & Suzuki, T. 1981. On the migration route of swans in Hokkaido, Japan. *Proc. Second Intern. Swan Symp., Sapporo, Japan, 1980*, 60–70.

Maybank, B. 1995. Atlantic provinces region. *Field Notes* 49, 222.

Mayfield, H. 1952. Captive Whooper Swans kill other waterfowl. *Auk* 69, 461–2.

McCaskie, G. & San Miguel, M. 1999. Report of the California Bird Records Committee: 1996 records. *Western Birds* 30, 57–85.

McKelvey, R. W. 1979. Swans wintering on Vancouver Island 1977–1978. *Canadian Field Naturalist* 93, 433–6.

McKelvey, R. W. 1981. Some aspects of the winter foraging ecology of Trumpeter Swans at Port Alberni and Comox Harbour, British Columbia. MSc. thesis, Simon Fraser University.

McElwaine, J. G., Wells, J. H. & Bowler, J. M. 1995. Winter movements of Whooper Swans visiting Ireland: preliminary results. *Ir. Birds* 5, 265–78.

Meissner, W. 1993. Wintering of the Mute Swan (*Cygnus olor*) and the Whooper Swan (*Cygnus cygnus*) of the Gulf of Gdansk during the seasons of 1984/85 to 1986/87. *Notatki Ornitologiczne* 34, 39–54.

Mel'nikov, Y. I. 2000. [On the avifauna of the lower Tunguska Basin within the boundaries of the Irkutsk region.] *Russian J. Orn.* 89, 10–16.

Mel'nikov, Y. I., Tanichev, A. I. & Zharov, V. A. 1990a. [Number and spatial structure of the Whooper Swan population in Verkhneangarskaya (Upper Angara) depression]. In: Syroechkovski, Y. E. (ed.) [*Ecology and Conservation of Swans in the USSR.*] Science Press, Moscow. [In Russian.]

Mel'nikov, Y. I., Vodop'yanov, B. G., Naumov, P. P. & Mel'nikova, N. I. 1990b. The Whooper Swan number change in Irkutsk Region in connection with intensive development of region. In: Syroechkovski, Y. E. (ed.) [*Ecology and Conservation of Swans in the USSR.*] Science Press, Moscow. [In Russian.]

Merila, E. & Ohtonen, A. 1987. [Mixed feeding of Whooper Swans and ducks during migration.] *Suomen Riista* 34, 52–8. [In Finnish.]

MERNE, O. J. 1986. Whooper Swan. In: Lack, P. (ed.) *The Atlas of Wintering Birds in Britain and Ireland*. Poyser, Calton.

MERNE, O. J. & MURPHY, C. W. 1986. Whooper Swans in Ireland, January 1986. *Ir. Birds* 3, 199–206.

MIKAMI, S. 1989. First Japanese records of crosses between Whistling *Cygnus columbianus columbianus*, and Bewick's swans *Cygnus columbianus bewickii*. *Wildfowl* 40, 131–4.

MIKHALEVA, E. V. 1997. [The Whooper Swan *Cygnus cygnus* attempted to nest on the Valaam archipelago (Ladoga Lake) in June 1997.] *Russian J. Orn.* 25, 19–20. [In Russian.]

MILLER, I. D. & SKALON, O. V. 1990. New data on the Mute Swan breeding in Tula region. In: Syroyochkovski, Y. E. (ed.) [*Ecology and Conservation of Swans in the USSR.*] Science Press, Moscow.

MILNE, H. 1976. Some factors affecting egg production in waterfowl populations. *Wildfowl* 27, 141–2.

MILNE, H. & RAMSAY, N. F. 1986. Poisoning of Whooper Swans (*Cygnus cygnus*) by ingested lead in NE Scotland. Nature Conservancy Council CSD Report no. 673.

MINEEV, Y. N. 1988. [Whooper Swan (*Cygnus cygnus*) in the European northeast USSR.] *Zool. J.* 67, 1430–4. [In Russian.]

MINEEV, Y. N. 1995. [Whooper Swan.] In: [*Fauna of European Northeast Russia*, vol. 1.] Nauka, St Petersburg. [In Russian.]

MINEEV, O. Y. 2000. [Data on the waterfowl fauna of the Indiga River basin.] *Casarca* 6, 299–301. [In Russian.]

MINTON, C. D. T. 1971. Mute Swan flocks. *Wildfowl* 22: 71–88.

MINTON, C. D. T. 1974. Wader weights. *Wash Wader Ringing Group Rep.* 1973–1974.

MITCHELL, C. D. 1994. Trumpeter Swan (*Cygnus buccinator*). In: Poole, A. & Gill, F. (eds.) *The Birds of North America*, no. 105. Philadelphia: The Academy of Natural Sciences, Philadelphia & American Ornithologists' Union, Washington DC.

MITCHELL, C. D. 1998. Whooper Swans (*Cygnus cygnus*) in North America. *Swan Spec. Group Newsl.* 7, 7–8.

MITCHELL, C. D. & ROTELLA, J. J. 1995. Post-hatch brood amalgamation in Trumpeter Swans. *Swan Spec. Group Newsl.* 4, 8.

MITCHELL, C. D. & ROTELLA, J. J. 1997. Brood amalgamation in Trumpeter Swans. *Wildfowl* 48, 1–5.

MIYABAYASHI, Y. & MUNDKUR, T. 1999. *Atlas of Key Sites for Anatidae in the East Asian Flyway*. Wetlands International, Japan, Tokyo & Kuala Lumpur.

MLIKOVSKY, J. & SVEC, P. 1989. Review of the Tertiary waterfowl. (Aves: Anseridae) of Czechoslovakia. *Casopis Pro Mineralogii A Geologii* 34, 199–203.

MLODINOW, S. & TWEIT, B. 2001. Regional reports: Oregon–Washington region. *North Amer. Birds* 55, 93–7.

MOLODOVSKIY, A. V. 1965. [The Whooper Swan in southern Mangyshlak.] *Orn.* 7, 481. [In Russian.]

MONVAL, J.-Y. & PIROT, J.-Y. 1989. *Results of the IWRB International Waterfowl Census, 1967–1986*. Intern. Waterfowl Research Bureau Spec. Publ. no. 8, Slimbridge.

MONROE, B. L. & SIBLEY, C. G. 1993. *A World Checklist of Birds*. Yale University Press, New Haven.

MORSE, D. H. 1977. Feeding behaviour and predator avoidance in hetero-specific groups. *Biosci.* 27, 332–9.

MOSS, S. 1998. Predictions of the effects of global climate change on Britain's birds. *Br. Birds* 91, 307–25.

MULLIÉ, W. C., SMIT, T. & MORAAL, L. 1980. Swan mortality caused by avian cholera in the Dutch delta area in 1979. *Watervogels* 5, 142–7.

MUNHTOGTOH, O. 1995. A censuring of the Whooper Swan (*Cygnus cygnus*). *Intern. Conf. 'Asian Ecosystems and their Protection', Ulaanbaatar, Mongolia, 21–25 August 1995*, 123.

MURASE, M. 1990. [Whistling Swan and Bewick's Swan pair and family overwintering.] *Strix* 9, 213–217 [In Japanese].

MURASE, M. 1991. [Whistling Swan and Bewick's Swan pair and family overwintering – second record.] *Strix* 10, 274–279. [In Japanese].

MURPHY, C. 1992. First recorded breeding record of Whooper Swan in Ireland. *Ir. Birding News* 3, 29–31.

MURPHY, C. 1993. The luck of the Irish. *Birdwatch* 2(1), 30–1.

MURTON, R. K. 1965. *The Woodpigeon.* Collins, London.

MYRBERGET, S. 1981. Breeding of *Cygnus cygnus* in a coastal area of northern Norway. *Proc. Second Intern. Swan Symp., Sapporo, Japan, 1980,* 171–5.

NAGEL, J. 1965. Field feeding of Whistling Swans in northern Utah. *Condor* 67, 446–7.

NAKAMURA, S. & ASAKAWA, M. 2001. New records of parasitic nematodes from five species of the order Anseriformes in Hokkaido, Japan. *Japanese J. of Zoo and Wildlife Medicine* 6, 27–33.

NAKANISHI, A. 1981. The cold air current in the Khabarovsk area, and the correlation with swan numbers at Lake Hyoko, Japan. *Proc. Second Intern. Swan Symp., Sapporo, Japan, 19080,* 70–73.

National Geographic 1999. *Field Guide to the Birds of North America.* Third edn. National Geographic Soc., Washington, DC.

NAUMANN, J. A. 1897–1905. *Naturgeshichte der Vögel Mitteleuropas.* Gera, Köhler.

NECHAEV, V. A. 1991. [*Birds of Sakhalin Island.*] Inst. of Biol. & Pedology, Vladivostok. [In Russian.]

NEHLS, H. in press. Whooper Swan. In: Marshall, D. B., Hunter, M. H. & Contreras, A. (eds.) *Birds of Oregon: A General Reference.* Oregon State University Press, Corvallis.

NEILL, P. 1806. *A Tour of Orkney.* London.

NICE, M. M. 1935. Some observations on the behaviour of starlings and grackles in relation to light. *Auk* 52, 91–2.

NIGHTINGALE, B. & ALLSOPP, K. 1997. The ornithological year 1996. *Br. Birds* 90, 538–48.

NIKIFOROV, M. E., KOZULIN, A. V., GRICHIK, V. V. & TISHECHKIN, A. K. 1997. [*Birds of Belarus at the Start of the 21st Century.*] Korolev, Minsk. [In Russian.]

NIKOLAEV, V. I. 1998. [*Birds of Marsh Landscapes of National Park 'Zavidovo' and the Upper Volga River Area.*] Tver'. [In Russian.]

NILSSON, L. 1975. Midwinter distribution and numbers of Swedish Anatidae. *Ornis Scand.* 6, 83–108.

NILSSON, L. 1979. Variation in the production of young of swans wintering in Sweden. *Wildfowl* 30, 129–34.

NILSSON, L. 1991. Utbredning, beståndsstorlek samt låntidföräandringar i beståndens storlek hos övervintrande sjöfåglar i Sverige. *Ornis Svecica* 1, 11–28.

NILSSON, L. 1994. [Thirty years of midwinter counts of waterfowl along the coasts of Scania, 1964–1993.] *Anser* 33, 245–56. [In Swedish.]

NILSSON, L. 1997. Changes in numbers and habitat utilization of wintering Whooper Swans *Cygnus cygnus* in Sweden 1964–1997. *Ornis Svecica* 7, 133–42.

NILSSON, L. 2002. Numbers of Mute Swan and Whooper Swans in Sweden, 1967–2000. *Proc. Fourth Intern. Swan Symp., 2001.* Waterbirds 25, Special Publication 1, 53–60.

NILSSON, L., ANDERSSON, O., GUSTAFSSON, R. & SVENSSON, M. 1998. Increase and changes in distribution of Breeding Whooper Swans *Cygnus cygnus* in northern Sweden from 1972–75 to 1997. *Wildfowl* 49, 6–17.

NILSSON, S. G. & NILSSON, I. N. 1978. Resting of the Whooper Swan *Cygnus cygnus* in south Sweden during spring migration in relation to recent habitat changes. *Anser Suppl.* 3, 195–7.

NOREK, K. & SVEDA, M. 1990. Useless death of eight swans. *Veterinarstvi* 40, 115.

NORTHCOTE, E. M. 1982. Size, form and habit of the extinct Maltese Swan *Cygnus falconeri.* *Ibis* 124, 148–58.

NORTHCOTE, E. M. 1988. An extinct 'swan-goose' from the Pleistocene of Malta. *Palaeontology (Lond.)* 31, 725–40.

NYGÅRD, T., LARSEN, B. H., FOLLESTAD, A. & STRANN, K.-B. 1988. Numbers and distribution of wintering waterfowl in Norway. *Wildfowl* 39, 164–76.

OCHIAI, K., JIN, K., ITAKURA, C., GORYO, M., YAMASHITA, K., MIZUNO, N., FUJINAGA, T. & SUZUKI, T. 1992. Pathological study of lead poisoning in Whooper Swans (*Cygnus cygnus*) in Japan. *Avian Disease* 36, 313–23.

O'DONOGHUE, P. D. & O'HALLORAN, J. 1994. The behaviour of a winter flock of Whooper Swans *Cygnus cygnus* at Rostellan Lake, Cork. *Proc. Roy. Ir. Acad.* 94, 109–118.

OED 1971. *The Compact Edition of the Oxford English Dictionary*. Clarendon Press, Oxford.

OGILVIE, M. A. 1967. Mute Swan population and mortality. *Wildfowl Trust Ann. Report* 18, 64–73.

OGILVIE, M. A. 1972. Distribution, numbers and migration. In: Scott, P. & The Wildfowl Trust. *The Swans*. Michael Joseph, London.

OGILVIE, M. A. 1978. *Wild Geese*. Poyser, Berkhamsted.

OGILVIE, M. A. 1979. *The Bird-watcher's Guide to the Wetlands of Britain*. Batsford, London.

OGILVIE, M. A. & PEARSON, B. 1994. *Wildfowl Behaviour Guide*. Hamlyn, London.

OGILVIE, M. A. & the Rare Breeding Birds Panel. 1994. Rare breeding birds in the United Kingdom in 1991. *Br. Birds* 87, 366–93.

OGILVIE, M. & the Rare breeding Birds Panel. 1995. Rare breeding birds in the United Kingdom in 1992. *Br. Birds* 88, 67–93.

OGILVIE, M. & the Rare Breeding Birds Panel. 1996a. Rare breeding birds in the United Kingdom in 1993. *Br. Birds* 89, 61–91.

OGILVIE, M. & the Rare Breeding Birds Panel. 1996b. Rare breeding birds in the United Kingdom in 1994. *Br. Birds* 89, 387–417.

OGILVIE, M. & the Rare Breeding Birds Panel. 1998a. Rare breeding birds in the United Kingdom in 1995. *Br. Birds* 91, 417–47.

OGILVIE, M. & the Rare Breeding Birds Panel. 1998b. Rare breeding birds in the United Kingdom in 1996. *Br. Birds* 92, 120–54.

OGILVIE, M. & the Rare Breeding Birds Panel. 1999a. Non-native birds breeding in the United Kingdom in 1997. *Br. Birds* 92, 472–6.

OGILVIE, M. & the Rare Breeding Birds Panel. 1999b. Rare breeding birds in the United Kingdom in 1997. *Br. Birds* 92: 389–428.

OGILVIE, M. & the Rare Breeding Birds Panel. 2000. Non-native birds breeding in the United Kingdom in 1998. *Br. Birds* 93, 428–33.

O'HALLORAN, J., MYERS, A. A. & DUGGAN, P. F. 1988. Lead poisoning in swans and sources of contamination in Ireland. *J. Zool. (Lond.)* 216, 211–24.

O'HALLORAN, J., MYERS, A. A. & DUGGAN, P. F. 1991. Lead poisoning in Mute Swans *Cygnus olor* in Ireland: a review. *Wildfowl Suppl.* 1, 389–95.

O'HALLORAN, J., RIDGWAY, M. & HUTCHISON, C. D. 1993. A Whooper Swan *Cygnus cygnus* population wintering at Kilcolman Wildfowl Refuge, Co. Cork, Ireland: trends over 20 years. *Wildfowl* 44, 1–6.

Ohio Bird News 2000 www.aves.net/obr.

OHLSEN, B. 1972. Beobachtungen an Zwerg- und Singschwanen. *Falke* 19, 60–2.

OHMORI, T. 1981a. The wild swans of Lake Inawashiro, Japan: some aspects of their migration. *Proc. Second Intern. Swan Symp., Sapporo, Japan, 1980*, 74–80.

OHMORI, T. 1981b. Artificial feeding of swans in Japan. *Proc. Second Intern. Swan Symp., Sapporo, Japan, 1980*, 244–6.

OHTONEN, A. 1988. Bill patterns of the Whooper Swan in Finland during autumn migration. *Wildfowl* 39, 153–4.

OHTONEN, A. 1992. [Increasing trend in the Whooper Swan population.] *Suomen Riista* 38, 34–44. [In Finnish.]

OHTONEN, A. 1996. Resightings of the Whooper Swan (*Cygnus cygnus*) marked with neck-collars on their breeding grounds in Finland. *Gibier Faune Sauvage, Game & Wildl.* 13, 1347–8.

OHTONEN, A. & HUHTALA, K. 1991. Whooper Swan *Cygnus cygnus* egg production in different nesting habitats in Finland. *Wildfowl Suppl.* 1, 256–9.

OKILL, D. 1987a. 1986 Whooper Swan survey. *Shetland Bird Club Newsl.* 66, 6.

OKILL, D. 1987b. Whooper Swan count 1987. *Shetland Bird Club Newsl.* 70, 5.

OKILL, D. 1988. Where were all the Whoopers? *Shetland Bird Club Newsl.* 75, 2–3.

OKILL, D. 1990. 1989 Whooper Swan count. *Shetland Bird Club Newsl.* 80, 3–4.

OLIVER, W. R. B. 1955. *New Zealand Birds.* Reed, Auckland.

ONAMI, B. 1961. The Whooper Swan in Fukushima. *Tori* 1961, 375–7.

OSJ (Ornithological Society of Japan). 1974. *Checklist of Japanese Birds.* Fifth edn. Gakken, Tokyo.

OSJ (Ornithological Society of Japan). 2000. *Checklist of Japanese Birds.* Sixth revised edn. Ornithological Society of Japan, Obihiro.

OSTAPENKO, V. A. 1991. Migration of Bewick's Swans *Cygnus bewickii* and Whooper Swan *Cygnus cygnus* wintering in Japan through Sakhalin Island and adjacent territories, USSR. *Wildfowl Suppl.* 1, 224–6.

OSTAPENKO, V. A., BOGDANOVICH, G. G., VISHNEVSKAYA, L. M. & SPITZYN, V. V. 1990. Reproduction of five swan species at Moscow Zoo. In: Syroechkovski, Y. E. (ed.) [*Ecology and Conservation of Swans in the USSR.*] Science Press, Moscow.

OVERSKAUG, K. 1986. [Eagle attacking a Whooper Swan.] *Var Fuglefauna* 9, 98–9. [In Norwegian.]

OWEN, M. 1971. Selection of feeding sites by White-fronted Geese. *J. Appl. Ecol.* 8, 893–905.

OWEN, M. 1972. Some factors affecting food intake and selection in White-fronted Geese. *J. Anim. Ecol.* 41, 79–92.

OWEN, M. 1977. *Wildfowl of Europe.* Macmillan, London.

OWEN, M. 1980. *Wild Geese of the World.* Batsford, London.

OWEN, M. & BLACK, J. M. 1989. Factors affecting the survival of barnacle geese on migration from the breeding grounds. *J. Anim. Ecol.* 58, 603–18.

OWEN, M. & BLACK, J. M. 1990. *Waterfowl Ecology.* Blackie, Glasgow.

OWEN, M. & BLACK, J. M. 1991. Geese and their future fortune. *Ibis* 113, 28–35.

OWEN, M. & CADBURY, C. J. 1975. The ecology and mortality of swans at the Ouse Washes, England. *Wildfowl* 26, 31–42.

OWEN, M. & KEAR, J. 1972. Food and feeding habits. In: Scott, P. and The Wildfowl Trust. *The Swans.* Michael Joseph, London.

OWEN, M. & OGILVIE, M. A. 1979. Wing molt and weights of Barnacle Geese in Spitsbergen. *Condor* 81, 42–52.

OWEN, M., ATKINSON-WILLES, G. L. & SALMON, D. G. 1986. *Wildfowl in Great Britain.* Second edn. Cambridge University Press, Cambridge.

OWENS, S. 1995. Mute Swan/Whooper Swan hybrid. *BBC Wildlife* 13(6), 45.

PAAKSPUU, V. 1967. [Observations of swan passage along the western coast of Estonia.] *Communications Baltic Comm. Study Bird Migr.* 4, 46–53. [In Russian.]

PALMER, R. S. 1972. Patterns of moulting. In: Farner, D. S. & King, J. R. (eds.) *Avian Biology.* Academic Press, New York.

PALMER, R. S. 1976. *Handbook of North American Birds,* vol. 2. Yale University Press, New Haven.

PANOV, Y. N. 1973. [*Birds of Southern Ussuriland. Fauna, Biology, Behaviour.*] Nauka, Novosibirsk. [In Russian.]

PATTEN, M. A. 2000. The winter season, 1999–2000: warm weather and cross-continental wonders. *North Amer. Birds* 54, 146–9.

PATTON, D. L. H. & FRAME, J. 1981. The effects of grazing in winter by wild geese on improved grassland in west Scotland. *J. Appl. Ecol.* 18, 311–25.

PEARSON, H. J. 1904. *Three Summers Among the Birds of Russian Lapland.* Porter, London.

PEJA, N. & BINO, T. 1996. Historic and current data on Anatidae in Albania. *Gibier Faune Sauvage, Game & Wildl.* 13, 1383–4.

PENNYCUICK, C. J., EINARSSON, O., BRADBURY, T. A. M. & OWEN, M. 1996. Migrating Whooper Swans *Cygnus cygnus*: satellite tracks and flight performance calculations. *J. Avian Biol.* 27, 118–34.

PENNYCUICK, C. J., BRADBURY, T. A. M., EINARSSON, O. & OWEN, M. 1999. Response to weather and light conditions of migrating Whooper Swans *Cygnus cygnus* and flying height profiles, observed with the Argos satellite system. *Ibis* 141, 434–43.

PERENNOU, C., MUNDKUR, T. & SCOTT, D. A. 1994. *The Asian Waterfowl Census 1987–91: Distribution and Status of Asian Waterfowl.* Intern. Waterfowl Research Bureau Publ. no. 24, Slimbridge.

PERFIL'YEV, V. I. 1972. [Resources of huntable waterfowl of the northeast and their wise management.] In: [*Nature of Yakutia and its Conservation.*] Yakutsk. [In Russian.]

PERFIL'YEV, V. I. 1976. [*New Data on Bird Distribution in Lower Lena.*] Yakutsk. [In Russian.]

PERFIL'YEV, V. I. 1979. [Influence of anthropogenous factors on number of waterfowl of the north of Yakutia.] In: [*Conservation and Wise Management of Wildlife in the Natural Environment of Yakutia.*] Yakutsk. [In Russian.]

PERFIL'YEV, V. I. 1987. [The Whooper Swan and Bewick's Swan in the north of Yakutia.] In: [*Ecology and Migrations in the USSR.*] Nauka, Moscow. [In Russian.]

PERKINS, S. 1994. New England region. *Field Notes* 48, 275.

PERKINS, S. 1999. New England region. *North Amer. Birds* 53, 258.

PERRINS, C. M. & REYNOLDS, C. M. 1967. A preliminary study of the Mute Swan, *Cygnus olor. Wildfowl Trust Ann. Rep.* 18, 74–84.

PETERSEN, W. R. 1997. New England region. *Field Notes* 51, 972.

PETROVA, K. 1989. [Representatives of the genus *Epomidiostomum* Skrjabin, 1915 (Nematoda, Amidostomatidae) encountered in Bulgaria: species composition and morphology.] *Khelmintologiya* 27, 20–32. [In Bulgarian.]

PILCHER, R. E. M. & KEAR, J. 1966. The spread of potato-eating in Whooper Swans. *Br. Birds* 59, 160–1.

POPOV, V. V., MURASHOV, Y. P., OLOVYANNIKOVA, N. M., STEPANENKO, V. N. & USTINOV, S. K. 1998. [Rare bird species of Baikal-Lena Nature Reserve.] *Proc. Baikal-Lena State Nature Res.* 1, 95–98.

PORTER, R. F., CHRISTENSEN, S. & SCHIERMACKER-HANSEN, P. 1996. *Field Guide to the Birds of the Middle East.* Poyser, London.

PORTENKO, L. A. 1981. *Birds of the Chukchi Peninsula and Wrangel Island.* Vol. 1. Smithsonian Institution and National Science Foundation, Washington, D.C.

POSLAVSKIY, A. N. 1972. [On summer migrations of Anseriformes in northern cis-Caspian Sea area.] *Orn.* 10, 288–96.

POTTER, S. & SARGENT, L. 1973. *The New Naturalist. Pedigree: Words from Nature.* Collins, London.

POYARKOV, N. D. 2001. [Trends of Anseriformes number changes in lower Cis-Amur River area.] In: [*Actual Problems of Study and Conservation of Birds of Eastern Europe and Northern Asia.*] Matbugat Jorty Press, Kazan.

POYARKOV, N. D. & JOHNSON, S. R. 1996. The unique productivity of the Whooper Swan in the middle reaches of the Ob River (Khanty-mansi Autonomous District). *Casarca* 2, 230–5.

POYSA, H. & SORJONEN, J. 2000. Recolonisation of breeding waterfowl communities by the Whooper Swan: vacant niches available. *Ecography* 23, 342–8.

PREUSS, N. O. 1981. Preliminary results of neck-collared *Cygnus cygnus cygnus.* In: *Proc. Second Intern. Swan Symp., Sapporo, Japan, 1980,* 141–4.

PRICE, A. L. 1994. *Swans of the World. In Nature, History, Myth and Art.* Council Oak Books, Tulsa.

PRIDATKO, V. I. 2001. http://vkgu.ukg.kz/vk3_1.dbp.

PROFUS, P. 1999. Numbers and territorial expansion of the breeding population of the Whooper Swan *Cygnus cygnus* in central Europe. *Swan Spec. Group Newsl.* 8, 11–12.

RAFFAELE, H., WILEY, J., GARRIDO, O., KEITH, A. & RAFFAELE, J. 1998. *A Guide to the Birds of the West Indies.* Princeton University Press, Princeton.

RANDLER, C. 2001. Field identification of hybrid waterfowl. *Alula* 7, 42–8.

RAVELING, D. G. 1969. Preflight and flight behaviour of Canada Geese. *Auk* 86, 671–81.

RAVELING, D. G., CREWS, W. E. & KLIMSTRA, W. D. 1972. Activity patterns of Canada Geese in winter. *Wilson Bull.* 84, 278–95.

RAVKIN, Y. S. 1991. Number and distribution of Mute Swans *Cygnus olor*, Bewick's Swans *C. bewickii*, and Whooper Swans *C. cygnus* in the west Siberian plain. *Wildfowl Suppl.* 1, 68–72.

RAVKIN, Y. S., VARTAPETOV, L. G., MILOVIDOV, S. P., ADAM, A. M. & FOMIN, B. N. 1990. Number and distribution of swans in west-Siberian plain (forest-tundra, forest zone, forest-steppe and steppe). In: Syroechkovski, Y. E. (ed.) [*Ecology and Conservation of Swans in the USSR.*] Science Press, Moscow.

REES, E. C. 1981. The recording and retrieval of bill pattern variations in *Cygnus columbianus bewickii. Proc. Second Intern. Swan Symp. Sapporo, Japan, 1980,* 105–19.

REES, E. C. 1989. Whooper Swan *Cygnus cygnus* research programme. *Wildfowl* 40, 161–2.

REES, E. C. & BOWLER, J. M. 1991. Feeding activities of Bewick's Swans *Cygnus columbianus bewickii* at a migratory site in the Estonian SSR. *Wildfowl Suppl.* 1, 249–55.

REES, E. C. & BOWLER, J. M. 1997. Fifty years of swan research and conservation by the Wildfowl and Wetlands Trust. *Wildfowl* 47, 248–63.

REES, E. C., BOWLER, J. M. & BUTLER, L. 1990a. Bewick's and Whooper Swans; the 1989–90 season. *Wildfowl* 41, 176–81.

REES, E. C., OWEN, M., GITAY, H. & WARREN, S. 1990b. The fate of plastic leg rings used on geese and swans. *Wildfowl* 41, 43–52.

REES, E. C., BLACK, J. M., SPRAY, C. J. & THORISSON, S. 1991a. Comparative study of the breeding success of Whooper Swans *Cygnus cygnus* nesting in upland and lowland regions of Iceland. *Ibis* 133, 365–73.

REES, E. C., BOWLER, J. M. & BUTLER, L. 1991b. Bewick's and Whooper Swans *Cygnus columbianus bewickii* and *C. cygnus*: the 1990–91 season. *Wildfowl* 42, 169–75.

REES, E. C., EINARSSON, O. & LAUBEK, B. 1997a. *Cygnus cygnus* Whooper Swan. *BWP Update* 1, 27–35.

REES, E. C., KIRBY, J. S. & GILBURN, A. 1997b. Site selection by swans wintering in Britain and Ireland; the importance of habitat and geographic location. *Ibis* 139, 337–52.

REES, E. C., BOWLER, J. M. & BEEKMAN, J. H. 1997c. *Cygnus columbianus* Bewick's Swan and Whistling Swan. *BWP Update* 1, 63–74.

REES, E. C., LIEVESLEY, P., PETTIFOR, R. A. & PERRINS, C. M. 1996. Mate fidelity in swans: an interspecific comparison. In: Black, J. M. (ed.) *Partnerships in Birds: The Study of Monogamy.* Oxford University Press, Oxford.

REES, E. C., COLHOUN, K., EINARSSON, O., McELWAINE, G., PETERSEN, A. & THORSTENSEN, S. 2002. Whooper Swan. In: Wernham, C. V., Toms, M. P., Marchant, J. H., Clark, J. A., Siriwardena, G. M. & Baillie, S. R. (eds.) *The Migration Atlas: Movements of the Birds of Britain and Ireland.* Poyser, London.

REYNOLDS, C. M. 1965. The survival of Mute Swan cygnets. *Bird Study* 12, 128–9.

REYNOLDS, P. 1982. Wintering Whooper Swans. *Orkney Bird Rep.* 1981, 49–54.

REZANOV, A. G. 1990. Quantitative indices of feeding behaviour of swans. In: Syroechkovski, Y. E. (ed.) [*Ecology and Conservation of Swans in the USSR.*] Science Press, Moscow.

RICHARDS, A. J. 1980. *The Birdwatcher's A–Z.* David & Charles, London.

RIDGWAY, M. & HUTCHINSON, C. D. 1990. *The Natural History of Kilcolman.* Kilcolman Wildfowl Refuge, Cork.

ROBERSON, D. 1986. Ninth report of the California Bird Records Committee. *Western Birds* 17, 49–77.

ROBERTS, B. B. 1934. Notes on the birds of central and south-east Iceland with special reference to food habits. *Ibis* 76, 239–64.

ROBERTS, J. L. 1964. Metabolic responses of freshwater sunfish to seasonal photoperiods and temperatures. *Helgol. Wiss. Meeresunter* 9, 459–73.

ROBERTS, T. J. 1991. *The Birds of Pakistan*, vol. 1. Oxford University Press, Oxford.

ROGACHEVA, E. V. 1988. [*Birds of Middle Siberia.*] Nauka, Moscow. [In Russian.]

ROGACHEVA, E. V. 1992. *The Birds of Central Siberia.* Husum.

ROOTSMAE, L. 1990. [On swan passage in Estonia.] *Communications Baltic Comm. Study Bird Migr.* 23, 86–104. [In Russian.]

ROSE, P. M. 1995. *Western Palearctic and South-West Asia Waterfowl Census 1994.* Intern. Waterfowl Research Bureau Spec. Publ. no. 35, Slimbridge.

ROSE, P. M. & SCOTT, D. A. 1997. *Waterfowl Population Estimates.* Second edn. Wetlands Intern. Publ. no. 44, Wageningen.

ROSLYAKOV, G. E. 1987. [Swans of the lower cis-Amur River area and the Shantar Islands.] In: [*Ecology and Migrations in the USSR.*] Nauka, Moscow. [In Russian.]

ROSLYAKOV, G. E. 1990. Swans in Khabarovsk territory. In: Syroechkovski, Y. E. (ed.) [*Ecology and Conservation of Swans in the USSR.*] Science Press, Moscow.

ROSLYAKOV, A. G. & VORONOV, B. A. 2001. [The breeding of geese and swans in Khabarovsk territory.] In: Popovkina, A. B. (ed.) [*Problems of Study and Conservation of Anseriformes of Eastern Europe and Northern Asia.*] Moscow.

RUDENKO, A. G., YAREMCHENKO, O. A. & RYBACHUK, K. I. 2000. [Wintering of waterfowl in the Black Sea Biosphere Reserve.] *Casarca* 6, 302–14. [In Russian.]

RÜGER, A., PRENTICE, C. & OWEN, M. 1986. *Results of the IWRB International Waterfowl Census 1967–1983.* Intern. Waterfowl Research Bureau Spec. Publ. No. 6, Slimbridge.

RUSANOV, G. M. 1987. [The status and role of wintering habitats of the Whooper Swan in the northern Caspian Sea.] In: [*Ecology and Migrations in the USSR.*] Nauka, Moscow.

RUSANOV, G. M. 1990. The success of wintering of Whooper Swan in the Volga River delta. In: Syroechkovski, Y. E. (ed.) [*Ecology and Conservation of Swans in the USSR.*] Science Press, Moscow.

RUSANOV, G. M., REUTSKIY, N. D., KRIVONOSOV, G. A., GAVRILOV, N. N., LITVINOVA, N. A. & BONDAREV, D. V. 1999. [*Vertebrates of Astrakhan Nature Reserve.*] Series 'Flora and Fauna of Nature Reserves'. Moscow.

RUSTAMOV, E. A. 1994. The wintering waterfowl of Turkmenistan. *Wildfowl* 45, 242–7.

RUTTLEDGE, R. F. 1963. Whooper Swans feeding on refuse dump. *Br. Birds* 56, 340.

RUTTLEDGE, R. F. 1966. *Ireland's Birds.* Witherby, London.

RUTTLEDGE, R. F. 1974. Winter distribution of Whooper and Bewick's Swans in Ireland. *Bird Study* 21, 141–5.

RYDER, J. P. 1967. The breeding biology of Ross' Geese in the Perry River region, Northwest Territories. *Can. Wildl. Serv. Report* 3, 1–56.

SADKOV, V. S. & SAFRONOV, N. N. 1990. Migrations and dynamics of the Whooper Swan number in the northern Baikal and in upper Angara hollow (depression). In: Syroechkovski, Y. E. (ed.) [*Ecology and Conservation of Swans in the USSR.*] Science Press, Moscow.

SALMON, D. G. & BLACK, J. M. 1986. The January 1986 Whooper Swan census in Britain, Ireland and Iceland. *Wildfowl* 37, 172–4.

SALOMONSEN, F. 1968. The moult migration. *Wildfowl* 19, 5–24.

SAMIGULLIN, G. M. 1990. Distribution and number of swans in Orenburg Region. In: Syroechkovski, Y. E. (ed.) [*Ecology and Conservation of Swans in the USSR.*] Science Press, Moscow.

SAMIGULLIN, G. M. & PARASICH, O. M. 1995. [The Whooper Swan in the steppes of southern Ural.] In: [*Problems of Ornithology.*] Barnaul. [In Russian.]

SANDBERG, R. 1992. *European Bird names in Fifteen Languages.* Sandberg, Lund.

SARYCHEV, V. S., VOROB'YOV, G. P., KLIMOV, S. M. & NEDOSEKIN, Y. V. 1990. Modern number of swans in upper Cis-Don River area. In: Syroechkovski, Y. E. (ed.) [*Ecology and Conservation of Swans in the USSR.*] Science Press, Moscow.

SAWADA, I. 1987. A new hymenolepidid cestode, *Dicranotaenia microcephala*, new species, from the Whooper Swan *Cygnus cygnus. Japanese J. Parasitol.* 36, 268–70.

SCARLETT, R. J. 1972. *Bones for the New Zealand Archaeologist.* Canterbury Mus. Bull. 4.

SCHIÖLER, E. L. 1925. [*Birds of Denmark*, vol. 1.] Glydendal, Copenhagen. [In Danish.]

SCHLATTER, R., SALAZAR, J., VILLA, A. & MEZA, J. 1991. Demography of Black-necked Swans *Cygnus melanocoryphus* in three Chilean wetland areas. *Wildfowl Suppl.* 1, 88–94.

SCHNEIDER-JACOBY, M., FRENZEL, P., JACOBY, H., KNÖTZCH, G. & KOLB, K.-H. 1991. The impact of hunting disturbance on a protected species, the Whooper Swan *Cygnus cygnus* at Lake Constance. *Wildfowl Suppl.* 1, 378–82.

SCOTT, D. A. & ROSE, P. M. 1996. *Atlas of Anatidae in Africa and Western Eurasia.* Wetlands International Publ. no. 41, Wageningen.

SCOTT, D. K. 1977. Breeding behaviour of wild Whistling Swans. *Wildfowl* 28, 101–6.

SCOTT, D. K. 1978. Identification of individual Bewick's Swans by bill pattern. In: Stonehouse, B. (ed.). *Animal Marking.* Macmillan, London.

SCOTT, D. K. 1980a. The behaviour of Bewick's Swans (*Cygnus cygnus bewickii*) at the Welney Wildfowl Refuge, Norfolk, England, UK, and on the surrounding fens: a comparison. *Wildfowl* 31, 5–18.

SCOTT, D. K. 1980b. Winter behaviour of wild Whistling Swans: a comparison with Bewick's Swans. *Wildfowl* 31, 119–21.

SCOTT, D. K. 1980c. Functional aspects of prolonged parental care in Bewick's Swans. *Anim. Behav.* 28, 938–52.

SCOTT, D. K. 1980d. Functional aspects of the pair bond in winter in Bewick's Swans (*Cygnus columbianus bewickii*). *Behav. Ecol. & Sociobiol.* 7, 323–7.

SCOTT, D. K. 1981. Geographical variation in the bill patterns of Bewick's Swans. *Wildfowl* 32, 123–8.

SCOTT, P. 1950. The swans, geese and ducks of the British Isles. *Severn Wildfowl Trust Rep.* 1949–1950, 121–9.

SCOTT, P. 1980. *Observations of Wildlife.* Phaidon, Oxford.

SCOTT, P., FISHER, J. & GUDMUNDSSON, F. 1953. The Severn Wildfowl Trust expedition to central Iceland. *Wildfowl Trust Ann. Rep.* 5, 79–115.

SCOTT, P. & The Wildfowl Trust. 1972. *The Swans.* Michael Joseph, London.

SEIBERT, H. C. 1951. Light intensity and the roosting flights of herons in New Jersey. *Auk* 68, 63–74.

SEMENOV-TYAN-SHANSKIY, O. I., & GILYAZOV, A. S. 1990. The Whooper Swan ecology in Lapland Nature Reserve. In: Syroechkovski, Y. E. (ed.) [*Ecology and Conservation of Swans in the USSR.*] Science Press, Moscow.

SEREBRYAKOV, V. V., GRISHCHENKO, V. N. & POLUDA, A. M. 1991. The migration of swans, *Cygnus* spp., in the Ukraine, USSR. *Wildfowl Suppl.* 1, 218–23.

SERIE, J. R., LUSZCZ, D. & RAFTOVICH, R. V. 2002. Population Trends, Productivity, and Harvest of Eastern Population Tundra Swans. *Proc. Fourth Intern. Swan Symp., 2001.* Waterbirds 25, Special Publication 1, 32–36.

SHARP, W. M. 1951. Observations on predator prey relations between wild ducks, trumpeter swans and golden eagles. *J. Wildl. Management* 15, 224–6.

SHARROCK, J. T. R. 1980. Rare breeding birds in the United Kingdom in 1978. *Br. Birds* 73: 5–26.

SHAW, G. 1979. Functions of dipper roosts. *Bird Study* 26, 171–8.

SHAW, T. H. 1938. On the alimentary canal of the Whooper Swan. *Bull. Fan. Mem. Inst. Biol.* 8, 387–402.

SHCHADILOV, Y. M. & BELOUSOVA, A. V. 2001. [Distribution and number dynamics of three swan species in Russia: comparative analysis.] In: Popovkina, A. B. (ed.) [*Problems of Study and Conservation of Anseriformes of Eastern Europe and Northern Asia.*] Moscow. [In Russian.]

SHCHADILOV, Y. M., REES, E. C., BELOUSOVA, A. V. & BOWLER, J. M. 2002. Annual variation in the proportion of Whooper Swans and Bewick's breeding in Northern European Russia. *Proc. Fourth Intern. Swan Symp., 2001.* Waterbirds 25, Special Publication 1, 86–94.

SHEPPARD, J. R. 1981. Whooper and Bewick's Swans in northwest Ireland. *Ir. Birds* 2, 48–59.

SHEPPARD, J. R. 1993. *Ireland's Wetland Wealth.* Irish Wildbird Conservancy, Dublin.

SHEVCHENKO, V. L., DEBELO, P. V., GAVRILOV, E. I., NAGLOV, V. A. & FEDOSENKO, A. K. 1993. [On ornithofauna of the Volga-Ural interstream area] In: [*Fauna and Biology of Birds of Kazakhstan.*] Almaty. [In Russian.]

SHIIREVDAMBA, T. (ed.) 1997. *Mongolian Red Book.* Ministry for Nature & Environment of Mongolia, Ulan Bator.

SHINKARENKO, A. V., MEL'NIKOV, YU. I., PODKOVYROV, V. A., ZHURAVLEV, V. E. & PYZH'YANOV, S. V. 1990. Character of the presence and number of swans in southern and middle Baikal. In: In: Syroechkovski, Y. E. (ed.) [*Ecology and Conservation of Swans in the USSR.*] Science Press, Moscow.

SHIRIHAI, H. 1996. *The Birds of Israel.* Academic Press, London.

SIEGFRIED, W. R. 1971. Communal roosts of the Cattle Egret. *Trans. Roy. Soc. S. Afr.* 39, 419–43.

SIEGFRIED, W. R., FROST, P. G. H., BALL, I. L. & MCKINNEY, D. F. 1977. Evening gatherings and night roosting of African Black Ducks. *Ostrich* 48, 5–16.

SIFERD, T. D. 1982. Mink, *Mustella vison,* attacks Trumpeter Swan *Cygnus buccinator* cygnet. *Canadian Field Naturalist* 96, 357–8.

SIKORA, A. 1994. The whooper swan (*Cygnus cygnus*) – a new breeding species of Pomerania. *Notatki Ornitologiczne* 35, 179–80.

SIKORA, A. 1995. [Mixed brood of the whooper swan (*Cygnus cygnus*) and mute swan (*Cygnus olor*) in the province of Suwalki.] *Notatki Ornitologiczne* 36, 368–70. [In Polish.]

SLADEN, W. J. L. 1991. Swans should not be hunted. Third IWRB International Swan Symposium. *Wildfowl Suppl.* 1, 368–75.

SMIDDY, P. & O'HALLORAN, J. 1991. The January 1991 swan census in Co. Cork. *Cork Bird Rep.* 1991, 85–9.

SMIDDY, P. & O'SULLIVAN, O. 1993. Whooper Swan *Cygnus cygnus.* Fortieth Irish bird report, 1992. *Ir. Birds* 5, 83.

SOFRONOV, Y. N. 2000. [On the avifauna of the Lena River delta.] [*Ornithlogical Studies in Russia,* 2.] Ulan-Ude. [In Russian.]

SOODLA, I. 1990. A summer encounter with the whooper swan. *Eesti Loodus* 4, 251.

SOODLA, I. 1992. A whooper swan nested on Lake Uru. *Eesti Loodus* 7–8, 400.

SOOTHILL, E. & WHITEHEAD, P. 1978. *Wildfowl of the World.* Blandford Press, London.

SOTNIKOV, V. N. 1999. [*Birds of Kirov Region and Adjoining Territories,* vol. 1.] Triada-S Press, Kirov. [In Russian.]

SOULTANOV, E. MOSLEY, P., PAYNTER, D. & AARVAK, T. 1998. Results of geese and swans count during winter in Azerbaijan in 1993 and 1996. *Berkut* 7: 30–1.

SOWLS, L. K. 1955. *Prairie Ducks.* Stackpole Co. & Wildl. Management Inst., Harrisburg & Washington DC.

SPARCK, R. 1958. An investigation of the food of swans and ducks in Denmark. *Danish Review Game Biol.* 3, 45–7.

SPENCER, R. & the Rare Breeding Birds Panel. 1993. Rare breeding birds in the United Kingdom in 1990. *Br. Birds* 86, 62–90.

SPILLING, E., BLOH, M., BÖTTCHER, M., DEGEN, A. & FRICKENHELM, D. 1998. Quantifying food consumption by swans foraging on fields of oil-seed rape. *Swan Spec. Group Newsl.* 7, 17–19.

SPRAY, C. J. 1980. The wild Whooper. *Scots Magazine* 114, 261–8.

SPRAY, C. J. & BAYES, K. 1992. The effect of neck collars on the behaviour, weight and breeding success of Mute Swans *Cygnus olor. Wildfowl* 43, 49–57.

SPRAY, C. J. & MILNE, H. 1988. The incidence of lead poisoning among Whooper and Mute swans *Cygnus cygnus* and *C. olor* in Scotland. *Biol. Conserv.* 44, 265–82.

SQUIRES, J. R. 1991. Trumpeter Swan food habits, forage processing, activities and habitat use. Ph.D. thesis, University of Wyoming.

STANFORD, F. 2000. The ornithology of Anglo-Saxon England. http://www.kami.demon. co.uk/gesithas/birdlore/fugsrc.html.

STEIMAR, F. 1979. Observations de cygnes de Bewick (*Cygnus bewickii*) et des sauvages (*Cygnus cygnus*) dans les inundations du Ried Centre de Alsace. *Ciconia* 3, 121–2.

STEPANYAN, L. S. 1990. [*Conspectus of Ornithological Fauna of the USSR.*] Nauka, Moscow.

STEPANYAN, L. S. 1998. [Materials on ornithological fauna of the Korean Peninsula.] Orn. 28,114–19. [In Russian.]

STEVENSON, H. M. & ANDERSON, B. H. 1994. *The Birdlife of Florida.* University Press of Florida, Gainesville.

STEWART, A. G. 1978. Swans at 8,000 metres. *Br. Birds* 71, 459–60.

ST LOUIS, M. J. 1995. Whooper Swan at Summer Lake Wildlife Area, Oregon, and California wintering areas. *Oregon Birds* 21, 35–6.

SVAZAS, S., STANEVICIUS, V. & CEPULIS, M. 1997. The status numbers and distribution of swans and geese in Lithuania. *Acta Zoologica Lituanica Orn.* 6, 66–78.

SWANN, R. 1975. Seasonal variation in suburban Blackbird roosts in Aberdeen. *Ring. & Migr.* 1, 37–42.

SWINHOE, R. 1870. Zoological notes of a journey from Canton to Peking and Kalgan. *Proc. Zool. Soc. Lond.*, 427–51.

SYKES, P. & SONNEBORN, D. W. 1998. First breeding records of Whooper Swan and Brambling in North America at Attu Island, Alaska. *Condor* 100, 162–4.

SYROECHKOVSKI, E. 2002. Distribution and population estimates for swans in the Siberian Arctic in the 1990s. *Proc. Fourth Intern. Swan Symp., 2001.* Waterbirds 25, Special Publication 1, 100–13.

TAMADA, M. 1981. The wild swans at Lake Tofutsu-ko, Hokkaido, Japan. *Proc. Second Intern. Swan Symp., Sapporo, Japan, 1980*, 322–3.

TATE, J. & TATE, D. J. 1966. Additional records of Whistling Swans feeding in dry fields. *Condor* 68, 398–9.

TEMMINCK, C. T. & SCHLEGEL, H. 1845–50. Aves. In: Siebold's *Fauna Japonica.* Siebold, Leiden.

Texas Ornithological Society 1995. *Checklist of the Birds of Texas.* Third edn. Capital Printing, Austin.

THÉVENOT, M., BEAUBRUN, P., BAOUAB, R. & BERGIER, P. 1982. Compte-rendu d'or-nithologique marocaine. Année 1981. *Doc. Inst. Sci. Maroc* 7, 1–120.

THÉVENOT, M., VERNON, J. D. R. & BERGIER, P. 2002. *The Birds of Morocco. B.O.U. Check-list No. 20.* British Ornithologists' Union, Tring.

THOM, V. M. 1986. *Birds in Scotland.* Poyser, Calton.

THOMPSON, D. Q. & LYONS, M. D. 1964. Flock size in a spring concentration of Whistling Swans. *Wilson Bull.* 76, 282–5.

THOMPSON, W. A., VERTINSKY, I. & KREBS, J. R. 1974. The survival value of flocking in birds: a simulation model. *J. Anim. Ecol.* 43, 785–820.

THOMSEN, P. & JACOBSEN, P. 1979. *The Birds of Tunisia.* Nature-Travels, Copenhagen.

THORPE, W. H. 1972. Duetting and antiphonal song in birds: its extent and significance. *Behav. Monogr. Suppl.* 18.

TICE, B. 1998. Field notes: western Oregon, winter 1997–98. *Oregon Birds* 24, 95–9.

TILLERY, J. R. 1969. Notes on nesting and hatching behaviour of Trumpeter Swans *Cygnus cygnus buccinator* at Great Bend Zoo. *Intern. Zoo Yearbook* 9, 122–4.

TIMMERMAN, G. 1963. Fragen der Anatidensystematik in parasitologischer Sicht. In: *Proc. Thirteenth Intern. Orn. Congr.*, 189–97.

TKACHENKO, E. E. 1990. Number and distribution of swans on wintering in Azerbaijan. In: Syroechkovski, Y. E. (ed.) [*Ecology and Conservation of Swans in the USSR.*] Science Press, Moscow.

TOBISH, T. G. 1996. Spring season. March 1–May 31, 1996. Alaska region. *Natl. Audubon Soc. Field Notes* 50, 318–21.

TODD, F. S. 1979. *Waterfowl. Ducks, Geese & Swans of the World.* Sea World Press, San Diego.

TOMEK, T. 1999. *The birds of North Korea. Non-passeriformes. Acta Zoologica Cracoviensia* 42, 1–217.

TOMIALOJC, L. 1976. *Birds of Poland. A List of Species and Their Distribution.* Natl. Centre for Scientific, Techn. & Eco. Inf., Warsaw.

TOMIALOJC L. 1990. [*The Birds of Poland. Their Distribution and Abundance.*] Second edn. Warsaw. [In Polish.]

TOMIALOJC, L. & STAWARCZYK, T. in press. *The Birds of Poland. Faunistic Monogr.*

TROCHLELL, D. 2001. Regional reports: Idaho–western Montana region. *North Amer. Birds* 55, 325–7.

TWEIT, B. & JOHNSON, J. 1992. Oregon/Washington region. Winter 1991–92. *Amer. Birds* 46, 472–5.

TWEIT, B. & GILLIGAN, J. 1995. Whooper Swan. Oregon–Washington region. *Field Notes* 49, 92.

TWEIT, B. & GILLIGAN, J. 1998. Whooper Swan. Oregon–Washington region. *Field Notes* 52, 115.

UETA, M., NIPPASHI, K. & HIGUCHI, H. 1997. Effect of transmitters on the behavior of wild and captive Whooper Swans. *Strix* 15, 133–7.

US Fish & Wildlife Service & Canadian Wildlife Service 1996. *Waterfowl Population Status 1996.* US Fish & Wildl. Service, Washington DC.

Various 2001. Whooper Swan discussions. http://www.birdingfaqs.com/Whooper.htm.

VARTAPETOV, L. G. 1984. [*Birds of the Taiga Interstream Areas of West Siberia.*] Nauka, Novosibirsk. [In Russian.]

VARTAPETOV, L. G. 1998. [*Birds of Northern Taiga of the West-Siberian Plain.*] Nauka, Novosibirsk. [In Russian.]

VAURIE, C. 1965. *The Birds of the Palearctic Fauna: Non-passeriformes.* Witherby, London.

VAURIE, C. 1972. *Tibet and its Birds.* Witherby, London.

VENABLES, L. S. V. & VENABLES, U. M. 1950. The Whooper Swans of Loch Spiggie Shetland. *Scott. Nat.* 63, 142–52.

VENGEROV, M. P. 1990a. Moult of the Whooper Swan in the flood-land of the lower Ob River (Khanty-Mansi district of Tyumen region). In: Syroechkovski, Y. E. (ed.) [*Ecology and Conservation of Swans in the USSR.*] Science Press, Moscow.

VENGEROV, M. P. 1990b. The Whooper Swan breeding in Khanty-Mansijsk district of Tyumen region. In: Syroechkovski, Y. E. (ed.) [*Ecology and Conservation of Swans in the USSR.*] Science Press, Moscow.

VINOGRADOV, V. G. & AUEZOV, E. M. 1991. Numbers, distribution and breeding success of Whooper Swans *Cygnus cygnus*, and Mute Swans *C. olor*, in central Kazakhstan, USSR, in 1985 and 1987. *Wildfowl Suppl.* 1, 73–6.

VOROB'YOV, K. A. 1954. [*Birds of the Ussurian Taiga.*] Acad. Sci., USSR, Moscow. [In Russian.]

DE VOS, A. 1964. Observations on the behaviour of captive Trumpeter Swans during the breeding season. *Ardea* 52, 166–89.

VYSOTSKIY, V. G. 1998. [The Whooper Swan *Cygnus cygnus* nesting on the southern bank of the Ladoga Lake.] *Russian J. Orn.* 33, 10–11. [In Russian.]

WADA, K. 1961. Swan's natural history. *Tori* 1961, 348–54.

WALTERS, M. 1980. *The Complete Birds of the World.* Reed, Sydney.

WARD, P. & ZAHAVI, A. 1974. The importance of certain assemblages of birds as information centres for food finding. *Ibis* 115, 517–34.

WANG, Y. 1987. The habits and construction of nests of the Whooper Swan. *Study of Nature Resources* 3, 61–2.

WAYRE, P. 1965. Somersaulting during bathing by geese and swans. *Br. Birds* 63, 383–4.

WBSJ Research Division 1982. The numerical distribution of geese, swans and ducks in Japan. The first waterfowl count (1982) of the Wild Bird Society of Japan. *Strix* 1: 43–55.

WBSJ Research Division 1984. Results of the nationwide counts of waders and waterfowl conducted by the Wild Bird Society of Japan. *Strix* 3, 101–12.

WBSJ Research Division 1988. Results of the nationwide counts of waterfowl. *Strix* 7, 301–4.

WBSJ (Wild Bird Society of Japan) 1990. Results of the 9th nationwide waterfowl census 1990. *Strix* 9, 255–63.

WBSJ (Wild Bird Society of Japan) 1991. Results of the 10th nationwide waterfowl census 1991. *Strix* 10, 301–14.

WBSJ (Wild Bird Society of Japan) 1992a. Results of the annual nationwide waterfowl census 1982–1992. *Strix* 11, 361–6.

WBSJ (Wild Bird Society of Japan) 1992b. Results of the annual nationwide waterfowl count 1992. *Strix* 11, 366–74.

WBSY (Wild Bird Society of Yaeyama) 1982. [Checklist of the birds of Yaeyama.] In: [*10-years anniversary issue of the Wild Bird Society of Yaeyama.*] Wild Bird Society of Yaeyama, Yaeyama.

WELLS, J. H., MCELWAINE, J. G. & COLHOUN, K. 1996. Irish Whooper Swan study. *Swan Study Newsl.* 6, 10–11.

Wharfedale Naturalists 1999. http://www.wharfedale-nats.org.uk/ornith99.htm.

WHITLOCK, R. 1979. The roosting strategies of waders. Unpublished report to Edward Grey Inst. Orn. Conf., Oxford.

Whooper Facts 2001. http://www.bbc.co.uk/education/archive/heading_south/whooperfaq. htm.

WIELOCH, M. 1990. Report on winter Whooper and Bewick's Swan count. *Notatki Orn.* 31, 138–40. [In Polish.]

WILKE, F. 1944. Three new bird records for St. Paul Island, Alaska. *Auk* 61, 655–6.

WILLIAMS, G. R. 1964. Extinction and the Anatidae of New Zealand. *Wildl. Trust Ann. Rep.* 15, 140–6.

WILLIAMS, M. 1981. The demography of New Zealand's *Cygnus atratus* population. *Proc. Second Intern. Swan Symp., Sapporo, Japan, 1980*, 147–61.

WILMORE, S. B. 1974. *Swans of the World.* David & Charles, London.

WILSON, E. O. 1975. *Sociobiology, the New Synthesis.* Belknap Press, Cambridge.

WILSON, L. K., ELLIOT, J. E., LANGELIER, K. M., SCHEUHAMMER, A. M. & BOWES, V. 1998. Lead poisoning of Trumpeter Swans *Cygnus buccinator*, in British Columbia. *Canadian Field Naturalist* 112, 204–11.

WITHERBY, H. F., JOURDAIN, F. C. R., TICEHURST, N. F. & TUCKER, B. W. 1940. *Witherby's Handbook of British Birds*, vol. 3. Witherby, London.

WON, P.-O. 1981. Present status of swans wintering in Korea and their conservation. *Proc. Second Intern. Swan Symp., Sapporo, Japan, 1980*, 15–19.

WRIGHT, E. N. & ISAACSON, A. J. 1978. Goose damage to agricultural crops in England. *Ann. Appl. Biol.* 88: 334–8.

WYNNE-EDWARDS, V. C. 1962. *Animal Dispersion in Relation to Social Behaviour.* Oliver & Boyd, Edinburgh.

YAKIMENKO, V. V. 1997. [Distribution and number of the Whooper Swan (*Cygnus cygnus*) in Omsk Region.] *Bull. Goose and Swan Study Group of E. Europe & N. Asia* 3, 272–9.

YAMAMOTO, H. 1961. Records of swans from Iwate. *Tori* 1961, 372–5.

YAMASHINA, Y. 1982. *Birds in Japan, a Field Guide.* Shubun International, Tokyo.

YARRELL, W. 1830. On a new species of wild swan. Trans. *Linn. Soc. Lond.* 16, 445–54.

YEATS, W. B. 1994. *The Collected Poems of W. B. Yeats.* The Wordsworth Poetry Library, London.

YEE, D. G., BAILEY, S. F. & DEUEL, B, E. 1992. The autumn migration. August 1–November 30, 1991. Middle Pacific coast region. *Amer. Birds* 46, 142–7.

YOSHII, M. 1981. Neck-banding of swans in Japan. *Proc. Second Intern. Swan Symp., Sapporo, Japan, 1980*, 80–81.

YUAN, G. & GUO, L. 1992. The distribution and protection of swans in Xinjiang, China. *Arid Area Study* 9, 60–3.

YUMOV, B. O. 2001. [On nesting ecology of some waterfowl in upper parts of Barguzin River (northeastern cis-Baikalia).] In: Popovkina, A. B. (ed.) [*Problems of Study and Conservation of Anseriformes of Eastern Europe and Northern Asia.*] Moscow. [In Russian.]

YURKO, V. V. 2001. [Wintering of swans in Minsk city.] In: Popovkina, A. B. (ed.) [*Problems of Study and Conservation of Anseriformes of Eastern Europe and Northern Asia.*] Moscow. [In Russian.]

ZABELIN, M. M. 1996. [On the Taiga Bean Goose and Whooper Swan in the Turukhan District, Krasnoyarsk territory.] *Casarca* 2, 308–12.

ZHALAKEVICHIUS, M. (comp.) 1995. [Birds of Lithuania. Status, number, distribution (breeding, migration, wintering.] *Acta Orn. Lituanica Spec. Issue* 11. [In Lithuanian and English.]

ZHATKANBAYEV, A. Z. 1996. Swans in the Ili River Delta and Lake Balkhach. *Gibier Faune Sauvage, Game & Wildl.* 13, 1383–4.

ZHELNIN, V. 1981. [Autumn migration of the Bean Goose and Whooper Swan in south Estonia in the course of 27 years.] *Communications Baltic Comm. Study of Bird Migr.* 12.

ZHURAVLEV, D. V. & PAREIKO, O. A. 1999. [Observations of Whooper Swans (*Cygnus cygnus*) in the Palessia State Radio-ecological Reserve.] *Subbuteo* 2, 43–4. [In Russian.]

ZIMMER, R., ERDTMANN, B., THOMAS, W. K. & QUINN, T. W. 1994. Phylogenetic analysis of *Coscoroba coscoroba* using mitochondrial srRNA gene sequences. *Mol. Phyl. & Evol.* 3, 85–91.

ZUBKO, V. N. 1988. Ecology of swans in Askania-Nova. In: Syroechkovski, Y. E. (ed.) [*Ecology and Conservation of Swans in the USSR.*] Science Press, Moscow.

ZYKOV, V. B. & REVYAKINA, Z. V. 1996. [Migration of swans on Sakhalin Island.] In: [*Birds of the Wetlands of the Southern Russian Far East and Their Protection.*] Dalnauka, Vladivostok. [In Russian.]

Index

Note: page numbers in *italics* refer to figures and tables